Climate Resilience and Environmental Sustainability Approaches

Anubha Kaushik • C. P. Kaushik • S. D. Attri
Editors

Climate Resilience and Environmental Sustainability Approaches

Global Lessons and Local Challenges

Springer

Editors
Anubha Kaushik
University School of Environment Management
Guru Gobind Singh Indraprastha University
Delhi, India

C. P. Kaushik
Environment Committee
PHD Chamber of Commerce and Industry
Delhi, India

S. D. Attri
India Meteorological Department, Ministry of Earth Sciences
Government of India
Delhi, India

ISBN 978-981-16-0901-5 ISBN 978-981-16-0902-2 (eBook)
https://doi.org/10.1007/978-981-16-0902-2

This Springer imprint is published by the registered company Springer Nature Singapore Pte Ltd.
The registered company address is: 152 Beach Road, #21-01/04 Gateway East, Singapore 189721, Singapore

Foreword

Human development has brought in tremendous changes on the earth, and one of the most important impacts has been global climate change caused by increasing emissions of greenhouse gases due to anthropogenic activities. Global warming/ climate change is no longer a myth. It is a major challenge that has far-reaching impacts on agriculture, food security, ecosystems, economy, safety, and human health. As our economy is mainly carbon based, there is constant increment in carbon emissions through various developmental activities. The climate change projections made by various climate models over the globe suggest that in the years to come, large-scale impacts of climate change will be seen on temperature, sea level rise, rainfall pattern, drought, cyclones, and other extreme weather events along with appearance of new pests and diseases that may affect our crops and human health. Climate change is a global phenomenon, but its impacts are going to vary across different nations depending upon several factors including ecological, geographic, and socioeconomic conditions. As the global population is growing exponentially, the demand for energy and other natural resources is also growing at a rate that makes our development unsustainable. The 17 Sustainable Development Goals (SDGs) adopted by all UN member nations call for an action to put an end to poverty and hunger, remove inequality, achieve effective climate action, provide safe water and alternate energy, prevent environmental degradation, and lead to balanced socioeconomic and environmental sustainability. Despite predicted global warming consequences, there is a need for learning from the climate feedbacks and adapting and evolving to make life as sustainable as possible.

This book uniquely discusses these key issues in the light of knowledge at the global level and puts forth the challenges faced at the local level based on case studies, mainly from Asian countries. This book will be of great interest and use to researchers, policy makers, academicians, and students. I congratulate the editors for their endeavor to publish this book at a time when global attention is focused on various aspects mentioned above, especially from the Asian countries which are bearing the brunt of environmental degradation and climate change.

I am sure this book will be useful to all those interested in current environmental issues for their academic pursuits and implementation of environment policies.

Ministry of Earth Sciences, Government of India
Delhi, India
M. Rajeevan

Preface

Climate change is one of the most alarming issues the world is facing in the present times. It has already started showing its impact across the world in the form of a rise in global temperature, erratic rainfall patterns, and extreme weather events, which pose serious challenges to agriculture, food security, safety, and human health including outbreak of new diseases. Various development activities by humans have changed the face of the earth, depleted various natural resources, polluted the earth's environment, and hampered the life support systems on the earth putting sustainability at stake. While all the member countries of the United Nations have committed to work for achieving the Sustainable Development Goals (SDGs) for providing food security, clean water, and affordable and clean energy, evolving strategies for mitigation and adaptations to climate change, and developing sustainable communities, the actions to be taken are challenging. While some of the issues are global in nature, the challenges differ depending upon the ecological, geographical, demographic, and socioeconomic conditions of different countries. With an aim to share and exchange experiences on these aspects, an international conference "Knowledge and Policy for Sustainable Development: Global Lessons and Local Challenges" was organized under the aegis of the Directorate of International affairs, Guru Gobind Singh Indraprastha University, New Delhi, India, during September 27–29, 2019, in collaboration with various government bodies including the Ministry of Earth Sciences, All India Council for Technical Education, Indian Council of Social Science Research, Defence Research and Development Organization, and Council of Scientific and Industrial Research. This book includes selected papers presented in the conference and some invited papers from different countries encompassing various dimensions of climate resilience and environmental sustainability.

The novelty of the book is that it discusses the key issues considering knowledge at the global level and puts forth the challenges faced at the local level based on case studies, mainly from various Asian countries. The book discusses impacts of climate change, vulnerability, and adaptations, especially in the agriculture sector, and specific sustainable development issues and practices related to the urban sector. Original case studies in the book provide insights into ancient wisdom and new dimensions of sustainable environment management. There is a focus on policy implications and sustainable approaches with current scenarios and future projections. A holistic approach has been taken to address the subject with new research and critical appraisal of knowledge and policy.

The book is divided into two sub-themes: (A) Climate Change, Mitigation, and Sustainable Agriculture and (B) Environmental Sustainability: The theme of the book has been presented as Introduction in Chap. 1.

The sub-theme "Climate Change, Mitigation, and Sustainable Agriculture" has nine chapters dealing with various aspects of the subject. Chapter 2 discusses progress in climate change simulation in Southeast Asia based on downscaling simulations (RCDS) and critically reviews the RCDS activities in countries such as Malaysia, Thailand, the Philippines, Indonesia, Vietnam, and China relating them to vulnerability, impact, and climate adaptations in the region. Chapter 3 presents an interesting account of climate change signatures over Schirmacher Oasis, Antarctica, wherein the author discusses as to how the oasis offers a suitable site to understand the climate processes and impact of global warming in Antarctica. REDD+ and its implementation in the Indian context have been discussed in Chap. 4, with a major emphasis on reducing emissions from deforestation through improved conservation efforts. In Chap. 5, the authors have analyzed the climate change impacts, vulnerability, and mitigation in the Indian Ocean Region and put forth policy suggestions. Chapters 6–10 are focused on sustainability in agriculture, which is one of the most important sectors affected by climate change. Chapter 6 emphasizes on agricultural adaptations to climate change for achieving sustainability, while the significance of enhanced climate service delivery mechanisms has been discussed in Chap. 7. A comprehensive appraisal of agrometeorological services for climate-resilient agriculture is presented in Chap. 8. Chapter 9 discusses how microbial diversity and multifunctional microbial bio-stimulants help in agricultural sustainability, and Chap. 10 suggests vertical farming technique for sustainable agriculture.

The second sub-theme "Environmental Sustainability" discusses various aspects of alternate energy, eco-friendly waste management, and various sustainable approaches in the next 11 chapters. Chapters 11–13 deal with some renewable energy alternatives and their feasibility in view of their important role in reducing carbon emissions and achieving sustainable development goals. Chapter 11 discusses the economic viability and sustainability of biofuels, while Chap. 12 critically analyzes the prospects and challenges of microbial fuel cell technology for energy production from wastewater treatment. Challenges and solution for renewable energy development are discussed as a case study of Uttarakhand, India, in Chap. 13. A critical analysis of carbon footprinting of an industry is presented as a case study in Chap. 14 to understand the impacts of industrial activities on climate change.

Sound and eco-friendly management of wastewaters as well as solid wastes is of immense importance for environmental sustainability. Challenges and prospects of the microbe-mediated bioremediation approach for removal of toxic pollutants like reactive dyes from wastewaters have been discussed in Chap. 15, while the degradation and biotransformation of recalcitrant chlorinated phenols by microorganisms is discussed in Chap. 16. Solid waste management issues and practices in India have been critically evaluated in Chap. 17 in the light of policy provisions, while the role of social enterprises as an emerging platform in waste management in India has been discussed in Chap 18.

The linkages between sustainability and traditional ethnoecological knowledge (TEK) have been analyzed in Chap. 19 based on a case study of Odisha, India. Various challenges and preparation for achieving Sustainable Development Goals (SDGs) in Bangladesh have been highlighted in Chap. 20. Sustainable health care under Ayushman Bharat Initiative in India has been examined in Chap. 21 with a view to assess the role of institutional CSR. A summary at the end of each section provides insight into the lessons, challenges, and prospects in these areas of climate resilience and environmental sustainability. The book is primarily meant for faculty, students, and researchers and is also useful for the policy makers and government agencies.

Kind patronage and encouragement from Padmashree Prof. (Dr.) Mahesh Verma, Vice-Chancellor, Guru Gobind Singh Indraprastha University, New Delhi, India, for the preparation of the book is gratefully acknowledged. The editors are grateful to the invited authors and coauthors from various countries: Dr. Russ Shnell, National Oceanic and Atmospheric Administration (NOAA), USA; Prof. Fredolin Tangang, University of Malaysia, Malaysia; Dr. Siva Kumar, Geneva, Switzerland; and Dr. G. Srinivasan, Regional Integrated Multi Hazard Early Warning Systems, Thailand. The editors also put on record the valuable inputs from our panel of esteemed reviewers. Finally, the timely submission of scholarly articles in final form by our valued authors, which has been the key to successful completion of the book keeping with the timeline, is thankfully acknowledged. The editors are thankful to the team of Springer Nature, Singapore, for handling the process of the publication of the book in an efficient and amicable way.

Delhi, India

Anubha Kaushik
C. P. Kaushik
S. D. Attri

Contents

About the Editors

Anubha Kaushik is Professor and Former Dean, University School of Environment Management, and Director, International Affairs, GGS Indraprastha University, New Delhi, India. She obtained MSc and PhD from Kurukshetra University, Kurukshetra. She was earlier Chairperson and Dean in GJ University of Science and Technology, Hisar, and Kurukshetra University, and Visiting Fellow in the Department of Biosystems Engineering, University of Manitoba, Canada. Her research areas include ecotechnology, bioremediation, bioenergy, sustainable development, and environmental impact assessment. She won 9 gold medals and several awards in her educational and professional career, has more than 180 publications in international and national journals and 5 books and several conference proceedings, is editor/member editorial board of several journals, reviewer of international journals, expert member in various national, international, and state level bodies related to promotion of research, environment protection, river rejuvenation, biodiversity conservation, pollution prevention, and environmental jurisprudence.

C. P. Kaushik is Co-Chairman, Environment Committee, PHD Chamber of Commerce and Industry, New Delhi, and Former Dean and Professor, GJ University of Science and Technology, Hisar, India. He received his masters from KU Kurukshetra and MPhil and PhD degrees from the University of Delhi, India. His areas of interest include environmental pollution and remediation, EIA, climate change, and sustainable development. He has been officiating Vice-Chancellor, Dean Academic Affairs, Dean and Chairman, and coordinator of the World Bank's TEQIP at GJUS&T, Hisar. He was member, NABET of Quality Council of India, State EIA Authority, and various international, national, and state level committees related to environment, education, administration, and policy making. He has published over 100 research papers in national and international journals, is editor/editorial board member of international and national journals, reviewer of foreign and Indian journals, authored 5 books, and edited many proceedings of international and national conferences.

S. D. Attri is Scientist "G," and Additional Director General of Meteorology, India Meteorological Department, Ministry of Earth Sciences, Government of India, New Delhi. He obtained his masters from Haryana Agriculture University, Hisar, and PhD from GJ university of Science and Technology, Hisar. His areas of interest include climate change, environment, urban meteorology, and agrometeorological services. He has been UNDP/WMO Fellow; Member, Commission for Atmospheric Sciences Management Group, WMO; and Expert Member Task force for Global Framework for Climate Services & Commission for Agricultural Meteorology, WMO. He was member Environmental Expert Committees, Government of India, Secretary and Vice-President, Indian Meteorological Society, and presently, Executive Editor of the international journal *Mausam*. He has published 100 research papers, monographs, books, and reports, has been honored by Hon'ble Prime Minister, India, for his contribution to IPCC, is the joint winner of the Nobel Peace Prize, 2007, and has received Best IMD Officer and Pride of University, GJUS&T awards.

Climate Resilience and Environmental Sustainability Approaches: An Introduction

1

Anubha Kaushik, S. D. Attri, C. P. Kaushik, and Russ Schnell

Abstract

Human development and growth in the past one and a half century have been largely driven by fossil fuel derived energy and the carbon-based economy has been mainly responsible for increased greenhouse gas emissions leading to the global climate change problem. Unsustainable lifestyle of the modern society has already started showing visible signs of impacts on our planet in the form of global warming, natural disasters, disease outbreaks, pollution-related health issues, eroded biodiversity, and degraded natural ecosystems. While the seventeen Sustainable Development Goals (SDGs) of United Nations come as a global call to end poverty, initiate climate action, protect the planet, and ensure that all people enjoy peace and prosperity by 2030, the capacity, challenges, and preparedness to achieve the same vary in different parts of the globe. The chapter introduces the concepts of positive climate actions, climate resilient and sustainable agriculture, renewable energy alternatives, sustainability principles, practices, and challenges.

A. Kaushik (✉)
University School of Environment Management, Guru Gobind Singh Indraprastha University, Delhi, India
e-mail: aks.es.10@gmail.com

S. D. Attri
India Meteorological Department, Ministry of Earth Sciences, Government of India, Delhi, India

C. P. Kaushik
Environment Committee, PHD Chamber of Commerce and Industry, Delhi, India

R. Schnell
National Oceanic and Atmospheric Administration (NOAA), Boulder, CO, USA

A. Kaushik et al. (eds.), *Climate Resilience and Environmental Sustainability Approaches*, https://doi.org/10.1007/978-981-16-0902-2_1

Keywords

Climate action · Energy alternatives · Sustainability principles · SDGs

1.1 Human Development and Environment

A nation's energy consumption is generally taken as an index of its development. Ever since humans appeared on Earth, there has been an increasing demand for energy. Initially, humans used the Sun's energy for agriculture, biomass energy for food, energy from domesticated animals for getting work done on farms, and then they learned to harness the energy from flowing waters and winds. With the advent of the Industrial Revolution in 1850s, energy demands spiked, and fossil fuels became the primary source of energy in the form of coal and oil, followed by natural gas and some nuclear energy. Emissions from the burning of fossil fuels steadily increased carbon concentrations in the global atmosphere, raising them from 280 parts per million (ppm) in pre-industrial times (IPCC 2013) to 415.26 ppm in 2019 at Mauna Loa Observatory (Dockrill 2019). Rapid development, based mainly on fossil fuels, led to the present-day challenges of global climate change with severe influences on human communities as well the planet's flora and fauna.

Over the past two decades, scientific research and global media reports have been emphasizing the importance of countries worldwide decarbonizing their economies and moving to alternate sources of energy to make life sustainable on the planet earth in the form we presently know it. Growth and development are key to raising the quality of life of human societies. But in the quest for growth, humans have jeopardized Earth's life support systems. The impacts of modern societies' unsustainable lifestyles are becoming evident in the form of global warming, natural disasters, disease outbreaks, pollution-related health issues, depletion of natural resources, degradation of natural ecosystems, and erosion of biodiversity. As development progresses, greenhouse gas emissions will continue to increase, accelerating climate change and putting our planet on track to reach highly unsustainable levels by the end of this century. According to John Houghton, the Co-Chair of scientific assessment of the UN Inter-governmental panel on climate change "*The impacts of global warming will be such that I have no hesitation in describing it as a weapon of mass destruction.*"

1.2 Climate Action

Rapid changes in climate are resulting in increased frequency and intensity of extreme weather events including heavy rainfall, heat waves, droughts, floods, and tropical cyclones. These extreme events are exacerbating water management problems, decreasing agricultural production and food security, increasing health risks, and impacting critical infrastructure and basic services. These impacts are being observed worldwide from equator to poles, from land to oceans, and in small countries to large ones. To protect our planet, concerted efforts are needed to foster general awareness, innovations, and good practices at local levels, and to fulfill the climate commitments made under the Paris Agreement, as well as other international agreements. Responsible climate action will provide huge opportunities to modernize our infrastructures, create new jobs, and promote greater prosperity in the world. The Green Climate Fund (GCF), established within the framework of the UN Framework Convention on Climate Change, was aimed to mobilize US$100 billion annually by 2020 to address the needs of developing countries to both adapt to climate change and to invest in low-carbon development. Coordinated efforts are required to achieve the targets enumerated under the 17 Sustainable Development Goals (SDGs) agreed upon in 2015, by global leaders to end poverty, conquer inequality, and fix climate change by 2030. The following targets have been envisaged under the 13th SDG of Climate Action:

- Strengthen resilience and adaptive capacity to climate-related hazards and natural disasters (Target 13.1)
- Integrate climate change measures into national policies, strategies, and planning (Target 13.1)
- Build knowledge and capacity to meet climate change (Target 13.3)
- Implement the UN Framework Convention on Climate Change (Target 13A)
- Promote mechanisms to raise capacity for planning and management (Target 13B)

According to the recent report of the Food and Agriculture Organization of the United Nations, climate variability and extremes are negatively affecting all dimensions of food security: food availability, access, utilization, and stability. This scenario will impact directly the global hunger problem, making solutions more difficult without addressing the climate problem through adaptation and mitigation. For example, following the Climate Smart Agriculture approach will allow communities to handle the challenges of a changing climate by sustainably increasing agricultural productivity, fortifying food systems, adapting and building resilience, reducing greenhouse gas emissions, and promoting better knowledge, good practices, and innovations (FAO, IFAD, UNICEF, WFP, and WHO Report 2018).

However, the COVID-19 pandemic has introduced new challenges by putting additional stresses on economies and the environment. The pandemic has underscored the importance of better nutrition pointing to the need to boost one's immunity and maintain good health to better resist viral attacks. This crisis has created an opportunity to push for consumption of locally available, nutritious foods, consumer education, and nutrition literacy. Governments can facilitate these outcomes by placing a high priority on policies to incentivize technology start-ups, and to develop agri-logistics to strengthen the value chains. During the current global scenario, supply chains of agricultural commodities, particularly of perishables, have been plagued by inefficiencies such as poor access to marketing channels, inadequate transportation, improper storage, handling and processing, post-harvest losses, and information asymmetry. Such issues can be resolved with smart technologies such as artificial intelligence as well as incentivizing the logistic role played by e-commerce and prompt delivery services.

Most countries are currently focusing on rebuilding economies and then implementing recovery plans that support clean, green, healthy, safe, and a more resilient world-order.

The UN Secretary-General has proposed the following six climate-positive actions to build economies and societies under the current scenario:

1. Green transition (i.e., investments must accelerate the decarbonization of all aspects of our economy)
2. Green jobs, and sustainable and inclusive growth
3. Green economy for making societies and people more resilient through a transition that is fair to all and leaves no one behind
4. Investment in sustainable solutions: Fossil fuel subsidies must end, and the polluters must pay for the pollution they caused
5. Confronting all climate risks
6. Cooperation, as no country can succeed alone

Supporting vulnerable regions through the green climate fund will directly contribute not only to Climate Action, the Goal 13 but also to the other SDGs. These actions must complement other initiatives to integrate disaster risk measures, sustainable natural resource management, and human security into national development strategies. With strong political will, increased investment, and use of appropriate and cost-effective technologies, humanity could achieve limiting global mean temperature rise to 2 °C above pre-industrial levels, or even 1.5 °C above that benchmark.

1.3 Sustainable Development: Global Scenario

It was in the 1972 United Nations Conference on the Human Environment, held in Stockholm, where the world's attention was drawn for the first time to the urgent need to respond to the problem of environmental deterioration. Sustainability was defined by the 1987 Brundtland Commission, as "*meeting the needs of the present generation without compromising the ability of future generations to meet their own needs.*"

In the 1992 United Nations Conference on Environment and Development, held in Rio de Janeiro, all 175 participating nations agreed that protection of the environment, as well as social and economic development, are fundamental to sustainable development. To achieve this, participating countries adopted Agenda 21 and the Rio Declaration, which served as a significant milestone to set a new agenda for sustainable development.

Ten years later, in the 2002 Johannesburg Summit, there was a sharp divide between the views of developed and developing nations. The developed countries emphasized reducing carbon emissions, while the developing nations still struggled to deal with poverty, hunger, hygiene, and shelter for their people. For them, reducing carbon emissions had little meaning in comparison to their more immediate issues of life and survival. This divide led summit participants to shift their focus to the indivisibility of human dignity, and the concept of sustainability was amended to include the dimension of intra-generational equity along with inter-generational equity, thereby targeting increased access to basic requirements such as clean water, sanitation, adequate shelter, energy, health care, food security, and the protection of biodiversity. At the same time, the parties agreed to support access to financial resources, benefits from open markets, capacity building, the use of modern technology to spur development, technology transfer, human resource development, and education and training to eliminate underdevelopment. Thus, worldwide focus was directed toward meeting difficult challenges, including improving people's lives and conserving our natural resources in a world with growing population, with ever-increasing demands for food, water, shelter, sanitation, energy, health services, and economic security.

Consequently, 8 Millennium Development Goals (MDGs) were framed by the UN. Those goals ranged from removing extreme poverty to increasing environmental sustainability, among others. The 2015 UN Millennium Development Goals Report (2017) on ensuring environmental sustainability states the following:

1. Ozone-depleting substances have been drastically reduced since 1990, and the ozone layer is expected to recover by the middle of this century.
2. Terrestrial and marine protected areas in many regions have increased substantially since 1990. In Latin America and the Caribbean, coverage of terrestrial protected areas rose from 8.8% to 23.4% between 1990 and 2014.
3. In 2015, 91% of the global population is using an improved drinking water source, compared to 76% in 1990. Over half of the global population now enjoys this higher level of service of safe, piped water.
4. Globally, 147 countries have met the MDG drinking water target, 95 countries have met the MDG sanitation target and 77 countries have met both.
5. Worldwide, 2.1 billion people have gained access to improved sanitation.
6. The proportion of urban populations living in slums in the developing regions fell from approximately 39.4% in 2000 to 29.7% in 2014.
7. The number of people now living in extreme poverty has declined by more than half, falling from 1.9 billion in 1990 to 836 million in 2015.
8. The proportion of undernourished people in the developing regions dropped by almost half since 1990.
9. The mortality rate of children under-five was cut by more than half since 1990, while maternal mortality fell by 45% worldwide.
10. Over 6.2 million malarial deaths have been averted between 2000 and 2015. New HIV infections fell by approximately 40% between 2000 and 2013.
11. Worldwide 2.1 billion people have gained access to improved sanitation.

In 2015, all United Nations Member States adopted the 17 Sustainable Development Goals (SDGs), as a universal call for action to end poverty, protect the planet, and ensure that all people enjoy peace and prosperity by 2030. The SDGs are the blueprint for achieving a better and more sustainable future for all by addressing global challenges related to poverty, inequality, safe water for all, alternate and affordable energy, climate change, environmental degradation, peace, and justice. The seventeen SDGs are so integrated that action in one area will affect outcomes in others, and the resulting development will balance social, economic, and environmental sustainability.

1.4 Transition to Alternate Energy Resources

Renewable energy can play a significant role in mitigating climate change. Heavy dependence on fossil fuels and old, inefficient coal-fired power plants contribute approximately 60% to global greenhouse gas emissions. Besides global warming, fossil fuels also adversely affect air quality and human health. Indoor air pollution from the use of combustible fuels for household energy caused 4.3 million deaths in 2012, with women and girls accounting for 6 out of every 10 of these deaths (IHME 2018). Adopting renewable energy technologies supports the Paris Agreement on climate change as well as the Sustainable Development Goals (SDGs). Global consumption of renewable energy started increasing since 2006 and in 2016, renewables reached 13.7% of the total energy consumption in 2017. Biofuels account for 9.5%, while hydro energy is about 2.5%. Renewable energies have become increasingly popular around the world as technologies like solar and wind have become cheaper and more advanced (Statista 2020).

The United Nations Development Programme (UNDP) supports the development of on- and off-grid renewable energy technologies and delivery services to transform renewable energy markets and implement policies that catalyze investment in renewable energy technologies. These renewable energy solutions should focus on integrated approaches that benefit climate and development. For example, UNDP's integrated approach and focus on zero-carbon, risk-informed, sustainable development, helps achieve climate targets, reduce disaster risks (associated with rising temperatures), and build back better following a disaster event. Access to electricity in poorer countries has increased, energy efficiency continues to improve, and renewable energy is making impressive gains in the electricity sector.

Still, more focused attention is needed to improve access to clean and safe cooking fuels and technologies. Bioenergy sources is one approach that may prove useful at local levels. Energy from wastewaters using microbial fuel cell technology is an innovative technique that has the dual benefits of wastewater treatment and low-carbon bioelectricity production.

1.5 Sustainability Principles, Practices, and Challenges

While working toward environmental sustainability, it is important to understand the principles of sustainability that have been supporting millions of species over millennia. The three basic principles of Earth's sustainability are:

1. Solar energy is the ultimate energy source for all life forms.
2. Ecosystem biodiversity imparts enormous stability to the ecosystems as complex webs through which exchange of energy and matter takes place.
3. Ecosystem cybernetics operates based on homeostatic properties, which work according to the system's carrying capacity and assimilation capacity and negative (deviation counteracting) and positive (deviation accelerating) feedback loops that determine the system's response to a stress.

The above principles of nature provide the guidelines as we plan our actions for sustainable development. Besides solar energy, many alternate energy sources, such as wind energy, tidal energy, biomass energy, and biofuel energy are solar driven. Biodiversity helps build ecosystem resilience and absorb environmental disturbances. Microbial diversity provides enormous opportunities for boosting sustainable agricultural productivity or eco-friendly bioremediation approaches to degrading various toxic pollutants and detoxifying contaminated environments. In nature, nothing is waste, as the waste of one process is used in another process. Based on the same principle, we must adopt ways of making our industrial and technological processes cyclic, so that waste management focuses on efficient resource recovery, reuse, and recycling.

We must be aware there are proposals to add massive amounts of aerosols to the stratosphere to shade the earth (i.e., stratospheric geoengineering) to counteract greenhouse gas warming. In the absence of coincident reduction in carbon dioxide emissions, the acidification of the oceans would continue with drastic negative effects on marine productivity. If all greenhouse gas emissions were not coincidently reduced, the total amount of aerosols added to the stratosphere would have to be continually increased to counteract increasing greenhouse gas warming. This in turn, would reduce the amount of solar radiation available for crop production. If the global stratospheric aerosol injections were terminated, the pent-up greenhouse gas global warming would snap back within a few years.

In summary, to make life on Earth sustainable, humans must develop alternative energy resources, adopt cleaner technologies, build resilience in our agricultural ecosystems, enhance climate change adaptations in our communities, seek sustainable management of waste generated by industrial processes, and conserve our traditional knowledge. SDGs focus on global human development, but the challenges differ regionally. Thus, "*Think globally and act locally*" is the key to developing action plans in different regions or nations. We must tap local resources and traditional knowledge to address the challenges of sustainable growth. We can formulate innovative action plans for education, health care, and hygiene by adopting new models suited to local conditions.

References

Dockrill, P (2019) It is official: Atmospheric CO_2 just exceeded 415ppm for the first time in human history. Science alert, Environment (https://www.sciencealert.com)

FAO, IFAD, UNICEF, WFP and WHO (2018) The state of food security and nutrition in the world 2018. Building climate resilience for food security and nutrition. Rome, FAO

Institute for Health Metrics and Evaluation (IHME) (2018) Global Burden of disease collaborative network. Global Burden of disease study 2017. Seattle, United States. http://ghdx.healthdata.org/gbd-results-tool

IPCC (Intergovernmental Panel on Climate Change) (2013) Climate Change 2013: The Physical Science Basis. Working group I Contribution to the IPCC fifth assessment report. Cambridge University Press, Cambridge. www.ipcc.ch/report/ar5/wg1

Statista (2020) Global share of energy consumed from renewables 1990–2017, published by N. Sönnichsen, May 28, 2020 www.statista.com/topics/6148/global-energy-industry

The Millennium Development Goals Report 2015/UNDP (2017). www.undp.org/content/undp/en/home/librarypage

Part I

Climate Change, Mitigation, and Sustainable Agriculture

This section includes articles dealing with climate change projection models, impacts, mitigation, adaptation, climate services, and sustainable agriculture.

Climate change is a major challenge across the world, and it is considered a big risk to most countries in Southeast Asia consisting of 11 countries, which vary widely in development status. Robust climate change projections are needed to know the risks and impacts on different sectors that would help in various decision-making processes in this region, which has already faced several climate disasters. A critical analysis of regional climate downscaling simulations (RCDS) over Southeast Asia, both at country and regional levels, shows that despite several projection models, there is still a need for advancing modeling activities, and based on the projections, mitigation measures must be taken to reduce emissions, and countries must adopt climate-resilient pathways for development.

Signatures of climate change are also becoming visible in the isolated icy continent of Antarctica with its unique ocean-ice-air-radiation system that is responsible for its sustenance. Rising temperature in Antarctica is a matter of great concern as the resulting sea level rise will have a direct impact on humans in coastal countries. The Schirmacher Oasis in East Antarctica ideally located between the polar ice and the shelf ice has been found to be an ideal site to study the impact of global warming, and long-term plans are needed to stop melting of ice in this coldest continent that will otherwise lead to deadly sea level rise.

Reducing emissions from deforestation and degradation (REDD) is a mechanism of mitigating climate change. The transformation of the strategy to REDD+ is found to have improved potential of addressing climate change and rural poverty. A critical analysis of the current status of REDD+ efforts in India shows how it could be an effective strategy and tool for climate change mitigation and adaptation, improving ecological and environmental services, conserving biodiversity, and enhancing forest-based livelihood of communities.

Impacts of climate change and vulnerabilities in various developing countries in the Indian Ocean littorals have been found to be quite significant in the twenty-first century due to high GHG emissions and slow transition to renewable energy alternatives. The Indian Ocean Region (particularly the Western Indian Ocean) has been warming faster than any other tropical oceans and the countries while developing have to bear the sustainability goals in mind, though economic constraints are a major hurdle, which needs to be addressed by framing a cost-effective climate policy for the region.

Climate change and variability have a profound influence on agroecosystems, posing serious threats to food security and human health. There is a need for improved assessment of risks associated with variable and uncertain environmental conditions and promoting agriculture adaptation to the same. The National Meteorological and Hydrological Services (NMHSs) have a major role to play in this by making weather and climate information available to the farmers and developing early warning systems that could facilitate both strategic and tactical decisions in increasing and sustaining agricultural production.

Climate services are found to be a positive development for the agriculture sector from the perspective of enabling the farmers to adapt to climate change, develop sustainable and economically viable agricultural systems, increase efficiency in the use of water, labor, and energy, conserve natural resources, and decrease pollution due to the use of agrochemicals that may degrade the environment. Case studies from Tamil Nadu and Bihar reveal that user-aligned information and services, a blended modality of active and passive information delivery system, and investing in inclusive agricultural development will be useful to translate the climate service information to practice that will enable the farmers amidst a changing climate.

As changing climate has led to uncertainty in crop yields and food security, suitable mitigation and adaptation strategies to achieve climate-resilient agriculture has become extremely important. Effective weather and climate services have opened new avenues for sustainable farming with improved income of farmers by providing advisory at different temporal and spatial scales with minimal cost. The agrometeorological services need to be further upgraded incorporating technology-driven crop responses to different conditions using artificial intelligence and promoting collaboration among all stakeholders.

Increased and sustained crop productivity under different environmental stresses and climate variations is very important for ensuring food security. Using multipurpose consortia of beneficial soil microbes isolated from the soil of different agroclimatic conditions has been found to act as bio-stimulants that help increase crop yield without using synthetic agrochemicals. Rhizospheric engineering with inoculated microbial consortia for multiple agronomic benefits can, however, modify the composition and function of the rhizosphere, which can be designed carefully as per the emerging needs of the new agroclimate situations and new cropping systems.

A new strategy for sustainable agriculture is emerging in recent times in the form of vertical farming system. This high-tech approach makes use of methods such as hydroponics, aquaponics, and aeroponics, which replace the requirements for soil-based plantation system. Vertical farming helps water and land conservation at a rapid pace. This approach with strategic development and design has in-built mobility features providing easier market access to the investors or the commercial growers for their produce and helps in significantly reducing transportation cost and storage cost, thus increasing sustainability of the system.

Progress in Climate Change Downscaling Simulations in Southeast Asia

2

Fredolin Tangang, Jing Xiang Chung, Supari, Sheau Tieh Ngai, Ester Salimun, Faye Cruz, Gemma Narisma, Thanh Ngo-Duc, Jerasorn Santisirisomboon, Liew Juneng, Ardhasena Sopaheluwakan, Mohd Fadzil Akhir, and Mohd Syazwan Faisal Mohd

Abstract

This chapter reviews the progress in regional climate downscaling simulations (RCDS) over Southeast Asia, both at country and regional levels. The need to advance RCDS stems from the fact that robust climate change projections are needed in decision-making processes involved in adopting climate-resilient pathways, wherein adaptation, mitigation, and sustainable development aspects

F. Tangang (✉)
Department of Earth Sciences and Environment, Faculty of Science and Technology, Universiti Kebangsaan Malaysia, Bangi, Selangor, Malaysia

Ramkhamhaeng University Center of Regional Climate Change and Renewable Energy (RU-CORE), Bangkok, Thailand
e-mail: tangang@ukm.edu.my

J. X. Chung
Institute of Oceanography and Environment, Universiti Malaysia Terengganu, Kuala Terengganu, Terengganu, Malaysia

Faculty of Science and Marine Environment, Universiti Terengganu Malaysia, Kuala Terengganu, Terengganu, Malaysia

Supari
Center for Climate Change Information, Agency for Meteorology Climatology and Geophysics (BMKG), Jakarta, Indonesia

S. T. Ngai
Department of Earth Sciences and Environment, Faculty of Science and Technology, Universiti Kebangsaan Malaysia, Bangi, Selangor, Malaysia

School of Atmospheric Sciences, Sun Yat-sen University, Zhuhai, China

E. Salimun
Department of Earth Sciences and Environment, Faculty of Science and Technology, Universiti Kebangsaan Malaysia, Bangi, Selangor, Malaysia

A. Kaushik et al. (eds.), *Climate Resilience and Environmental Sustainability Approaches*, https://doi.org/10.1007/978-981-16-0902-2_2

must also be considered. At the country level, notable variation in the progress in RCDS among countries in Southeast Asia exists wherein some countries have not conducted any RCDS, while others show some levels of activities in both dynamical and statistical downscaling. At the regional level, RCDS activities are also limited. However, the establishment of the Coordinated Regional Climate Downscaling Experiment (CORDEX) Southeast Asia (CORDEX-SEA) and the completion of high-resolution multi-model simulations are game changers. Furthermore, through the establishment of the Southeast Asia Regional Climate Change Information System (SARCCIS), a data portal where CORDEX-SEA simulation outputs are archived and linked to the Earth System Grid Federation (ESGF) for worldwide accessibility, the availability of robust climate change projections is expected to spur the research and development activities in the vulnerability, impact, and adaptation aspects in the region. Lastly, the remaining challenges, such as on bias and uncertainties, are discussed, which indicate the way forward concerning RCDS in Southeast Asia.

F. Cruz
Regional Climate Systems Laboratory, Manila Observatory, Quezon City, Philippines

G. Narisma
Regional Climate Systems Laboratory, Manila Observatory, Quezon City, Philippines

Atmospheric Science Program, Physics Department Ateneo de Manila University, Quezon City, Philippines

T. Ngo-Duc
REMOSAT Laboratory, University of Science and Technology of Hanoi Vietnam Academy of Science and Technology, Hanoi, Vietnam

J. Santisirisomboon
Ramkhamhaeng University Center of Regional Climate Change and Renewable Energy (RU-CORE), Bangkok, Thailand

L. Juneng
Department of Earth Sciences and Environment, Faculty of Science and Technology, Universiti Kebangsaan Malaysia, Bangi, Selangor, Malaysia

Marine Ecosystem Research Centre, Faculty of Science and Technology, Universiti Kebangsaan Malaysia, Bangi, Selangor, Malaysia

A. Sopaheluwakan
Center for Research and Development, Agency for Meteorology, Climatology and Geophysics (BMKG), Jakarta, Indonesia

M. F. Akhir
Institute of Oceanography and Environment, Universiti Malaysia Terengganu, Kuala Terengganu, Terengganu, Malaysia

M. S. F. Mohd
National Hydraulic Research Institute of Malaysia (NAHRIM), Seri Kembangan, Selangor, Malaysia

Keywords

Climate models · Climate resilience · CORDEX-SEA · Regional climate downscaling simulations

2.1 Introduction

Southeast Asia is a unique region consisting of 11 developing and least developed countries situated on a part of mainland Asia and archipelagos of the Maritime Continent, where exposure and risk of climate change impacts are considered high (Hijioka et al. 2014). This region has been experiencing recurring climate-related disasters in the past few decades, ranging from the costliest flood events in 2011 in Thailand with a total loss of USD46.5 billion and 815 deaths (Supharatid et al., 2016); typhoon Linda, which was the worst typhoon ever recorded in southern Vietnam in 1997 that killed over 3000 people, destroyed about 200,000 houses and left about 383,000 people homeless (Anh et al. 2019); the century worst flood in northeastern Peninsular Malaysia in December 2014 (Hai et al. 2017); the deadliest cyclone Nargis in Myanmar in 2008 that killed more than 138,000 people (Fritz et al. 2010); typhoon Haiyan in 2013, which is the strongest typhoon recorded in the Philippines in recent history that caused an estimated 6300 deaths and economic losses of USD1.8 billion (Hernandez et al. 2015; Lagmay et al. 2015); to the 2015 forest fires in Indonesia that caused transboundary haze affecting Singapore, Malaysia, and Brunei (Lohberger et al. 2018) and caused USD 16 billion economic losses to Indonesia (World Bank 2016). With the current population of nearly 700 million, where most live in coastal cities and river basins, projected to be above 750 million by 2050 (Hirschman and Bonaparte 2012), hundreds of millions of people could be at greater risk to climate change impacts in the decades to come. Significant variations in precipitation patterns (Tangang et al. 2020), weather and climate extremes (Supari et al. 2020), and sea level rise (IPCC 2019) have been projected in future warmer climates in Southeast Asia, possibly exerting negative impacts on various sectors, such as agricultural yields, biodiversity, forest harvests, health, and water resources (Hijioka et al. 2014). With the region having about 115 million hectares of agricultural land of various crops including rice, maize, oil palm, rubber, and coconut (Asian Development Bank Institute 2009), a significant reduction in yields may occur due to climate change resulting from increasing water and heat stresses, frequent weather and climate extremes, and increased climate-related pests and diseases. Increases in floods and drought would affect water resources and infrastructures in the region. Moreover, increased frequency and severity of future droughts would also provide favorable conditions for forest fires and haze episodes, especially in the Indonesian regions of Sumatra and Kalimantan, which have been occurring in the current climate due to El Niño (Juneng and Tangang 2005; Supari et al. 2018; Supari et al. 2020; Tangang et al. 2020). In addition, the threat of El Niño-induced drought is likely to exacerbate in the future given the projected increase in frequency of extreme El Niño episodes due to climate change (Cai et al. 2014).

The current knowledge strongly points to the fact that climate change is indeed a key risk to most countries in the region. However, as highlighted in a report by the Asian Development Bank (ADB), the region's aggressive and unsustainable development pursuits in turn could further exacerbate the risks of climate change in the coming decades (Asian Development Bank Institute 2009). Hence, countries in the region must embrace sustainable development in economic pursuits by implementing mitigation and adaptation strategies to fight climate change and increase climate resilience. While mitigation is important to ensure sustained reduction in greenhouse gas (GHG) concentrations in the atmosphere by lowering emissions and adding carbon sinks, adaptation measures are needed to reduce vulnerability and impacts of future climate by regulating development activities (Zhao et al. 2018). However, the implementation of adaptation strategies is usually at a local scale, which requires robust and scientifically based future climate change information at sufficient resolution for the assessment of future climate change impacts. Without such information, any attempt in regulating development activities may lead to failed adaptation strategies or even maladaptation (Barnett and O'Neill 2010; Magnan et al. 2016; McMullen et al. 2019). For example, a new climate resilient infrastructure that incorporates robust climate projection in its design can reduce the risk of climate-related disruptions, e.g., flooding, even though it may not fully eliminate the risk (OECD 2018). However, in Southeast Asia, there is a lack of detailed information on the impacts of future climate change on many critical sectors, including water resources and food security (Hijioka et al. 2014). This knowledge gap could be attributed to the lack of robust climate change projections that can be reliably used for climate change impact and vulnerability assessment research.

Generating climate projections involves the use of a Global Climate Model (GCM), which is forced by different GHG emission scenarios (Giorgi 2019). Since GCMs can potentially produce different responses, incorporating future climate projections from all available GCMs has been the common practice so far. For example, previous assessment reports of the Intergovernmental Panel on Climate Change (IPCC) utilize climate information from available GCMs that have been implemented following a common experimental protocol of the Coupled Model Intercomparison Project (CMIP) (Meehl et al. 2000; Eyring et al. 2016). However, due to its coarse spatial resolution (100–300 km), a GCM tends to ignore and smooth out local features, e.g., complex terrains and coastlines, rendering the outputs to be less suitable for impact, vulnerability, and adaptation assessments at local scales. Thus, GCM outputs are required to be "refined" to much finer resolutions, which can be carried out either by dynamical or statistical downscaling (Tangang et al. 2020). Dynamical downscaling requires "nesting" of a regional climate model (RCM) at a much higher resolution within a GCM. The procedure is not just merely refining the scale but potentially incorporating added values to the regional simulations (e.g., Giorgi 2019). However, for robustness and uncertainty estimation, multiple GCMs, RCMs, and emission scenarios are required (Valle et al. 2009), making dynamical downscaling very expensive and time consuming to implement. Hence, the progress of regional climate downscaling research and studies around the world has been less uniform, wherein the advancement in developed regions far outpace those in developing countries. Nevertheless, regional and international collaborations have been undertaken in the past few years to overcome such challenges. This chapter reviews the progress and advancement in regional climate downscaling in Southeast Asia. The rest of the paper is organized as follows: Section 2.2 briefly describes climate models and regional climate downscaling. In Sect. 2.3, progress at the country and regional levels is reviewed. In Sect. 2.4, knowledge gaps and the way forward are discussed. Conclusions are presented in Sect. 2.5.

2.2 Climate Models and Regional Climate Downscaling

The climate system is enormously complex and inherently nonlinear. Energized by solar radiation, it has five major components, namely the atmosphere, hydrosphere, geosphere, biosphere, and cryosphere (Peixoto and Oort 1992). A GCM is a mathematical representation of the climate system and the processes that take place within. It is a tool used to investigate the responses of the climate system to various forcings, to predict climate from seasonal to decadal time scales, and to project future climate, forced by different GHG emission scenarios. In a climate model, Earth is divided into (horizontal) grids and (vertical) layers (Fig. 2.1). For each grid, the model numerically solves mathematical equations representing the laws of physics operating within components of the climate system. However, there are some processes that cannot be explicitly resolved in the model, and are mathematically approximated using parameterizations. In a GCM, multi-way interactions across different modules that represent processes within the components of the climate system are allowed to mimic actual processes in the real world. For example, the ocean and atmosphere modules are coupled to form the so-called Atmosphere-Ocean General Circulation Model (AOGCM) (Fig. 2.1).

Climate models evolved from weather prediction models since the 1940s. After the development of the first coupled climate model in the 1960s (Manabe and Bryan 1969), the level of complexity in the model continued to increase leading to the development of Earth System Models (ESM), which incorporate components and processes related to atmospheric chemistry, land carbon and ocean biogeochemistry. Many climate models have been developed by numerous institutions around the world, and intercompared throughout the years under the framework of CMIP (e.g., CMIP3, CMIP5, and CMIP6) (Meehl et al. 2000; Eyring et al. 2016). A comprehensive review can be found in the recent IPCC Assessment Reports (Randall et al. 2007; Flato et al. 2013) and other relevant literature (Stocker 2014). It is noted that no institution in Southeast Asia has yet developed a GCM, and participated in the CMIP experiments, at this time.

A GCM is without a doubt very useful in terms of providing information regarding future global climate change (Flato et al. 2013). However, the model's coarse horizontal resolution (~100–300 km) limits its ability to represent smaller-

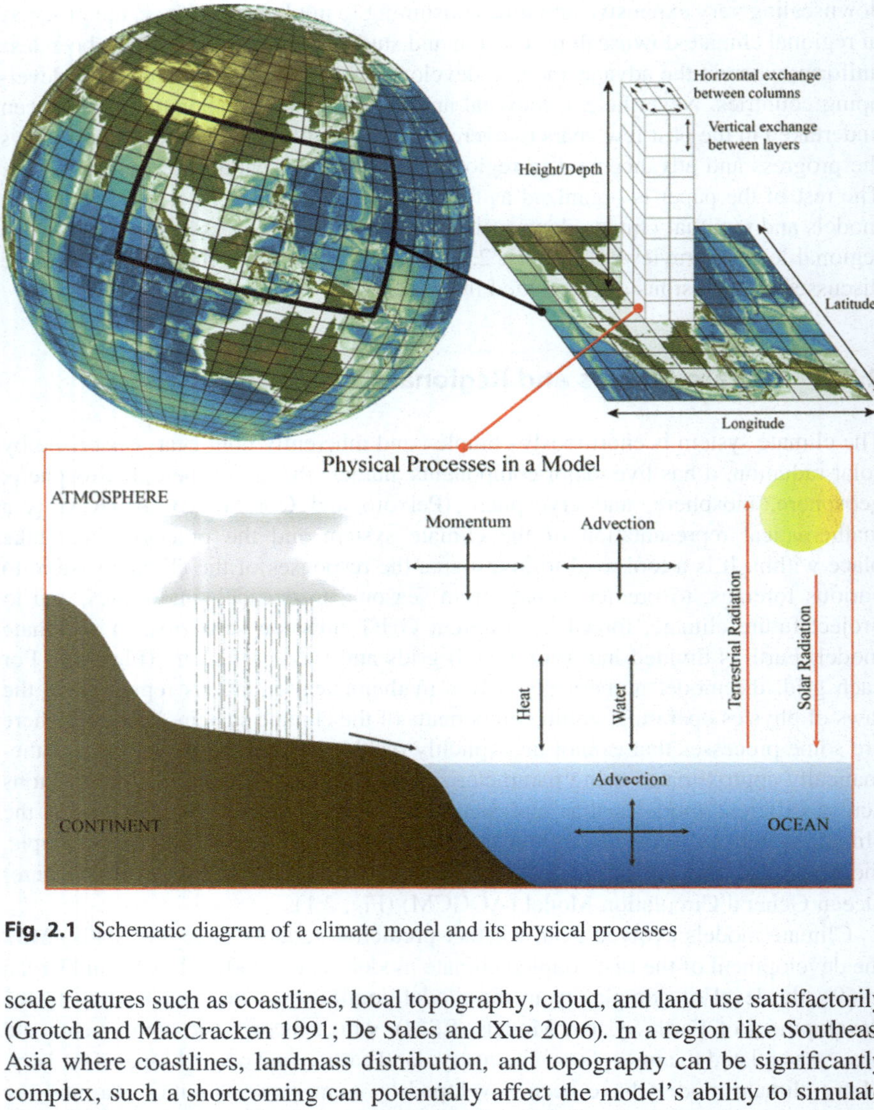

Fig. 2.1 Schematic diagram of a climate model and its physical processes

scale features such as coastlines, local topography, cloud, and land use satisfactorily (Grotch and MacCracken 1991; De Sales and Xue 2006). In a region like Southeast Asia where coastlines, landmass distribution, and topography can be significantly complex, such a shortcoming can potentially affect the model's ability to simulate the complex monsoonal climate and related variability in the region, leading to model biases (Fig. 2.2a). Increasing the grid resolution may overcome this problem but tends to be impractical and computationally expensive for long-term climate simulations at very high grid resolution (e.g., <20 km). A solution to this issue is by nesting a limited area model at a higher grid resolution (i.e., RCM) within a GCM. RCM usually resolves landmasses, topography, and coastlines better (Fig. 2.2b), and simulates climate processes at finer grid scales while constrained by the large-scale information at its boundaries. This approach is better known as dynamical

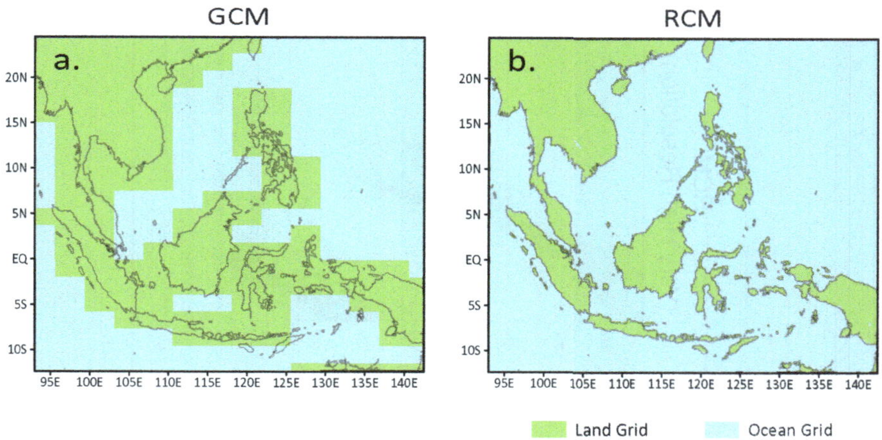

Fig. 2.2 Comparison of land and ocean grid between a typical GCM and an RCM

downscaling (Giorgi 2019). Multiple nesting is also possible where the same RCM is further nested within itself but at a much higher resolution (Fig. 2.3).

Both GCM and RCM operate on the same physical laws but may have different ways of approximating the sub-grid processes, which are often referred to as parameterization schemes. Convective process is one example, which is approximated by a cumulus parameterization scheme that can be formulated in many different ways (Kuo 1974). Likewise, there can be many types of parameterization schemes for other sub-grid processes. Hence, for a particular region and a particular RCM, experiments must be conducted to determine the best combination of schemes for different sub-grid processes (Juneng et al. 2016; Chung et al. 2018). These experiments can be carried out by forcing the RCM with a "perfect" boundary condition, i.e., reanalysis data. The best combination of schemes can then be adopted for the long-term climate simulations where the RCM is forced by the GCM, which is in turn forced by emission scenarios.

Dynamical downscaling has been shown to be beneficial as RCM can add value not just in terms of resolution to suit the needs of data applications, e.g., hydrological modelling, but also better performance than GCM in simulating regional or local climate features, e.g., extremes (Giorgi 2019). The advantage of RCM is its ability to deliver comparable performance to that of GCM even with relatively low computing resources. A recent study by Demory et al. (2020) showed that the performance of RCM in simulating daily precipitation is on par with a GCM of similar resolution. However, a common feature of both GCM and RCM is model biases. Different GCMs tend to produce different climate sensitivities and biases. These arise from different formulations of climate processes in each GCM. This is also the case for RCMs. Apart from poor representation of climate processes in the model, biases in the boundary forcing from GCM can also influence the performance of RCM. These

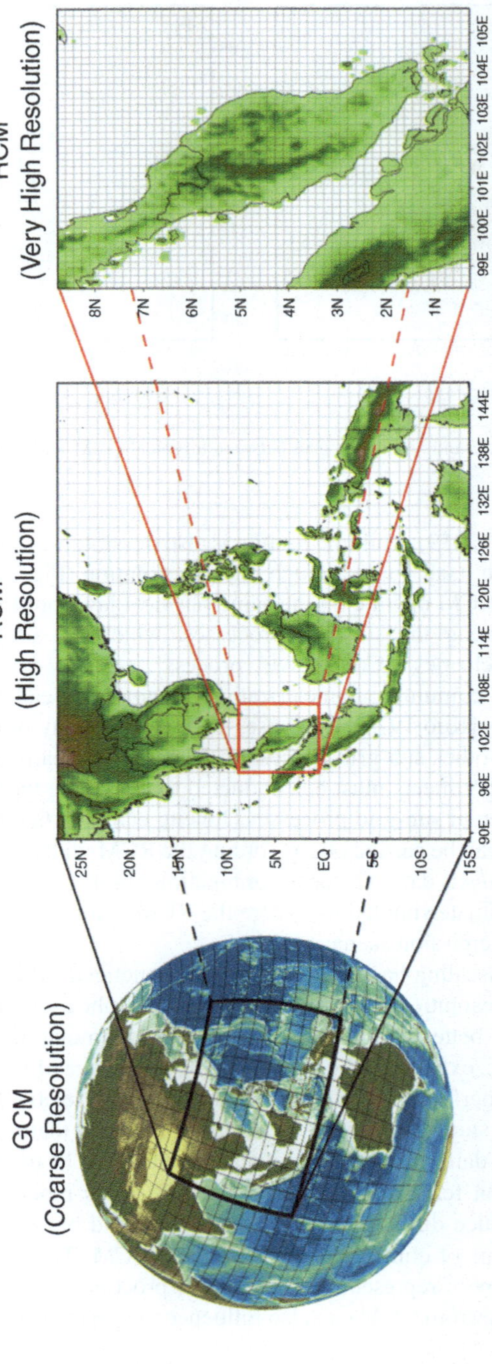

Fig. 2.3 Schematic diagram of regional climate dynamical downscaling of multiple nestings

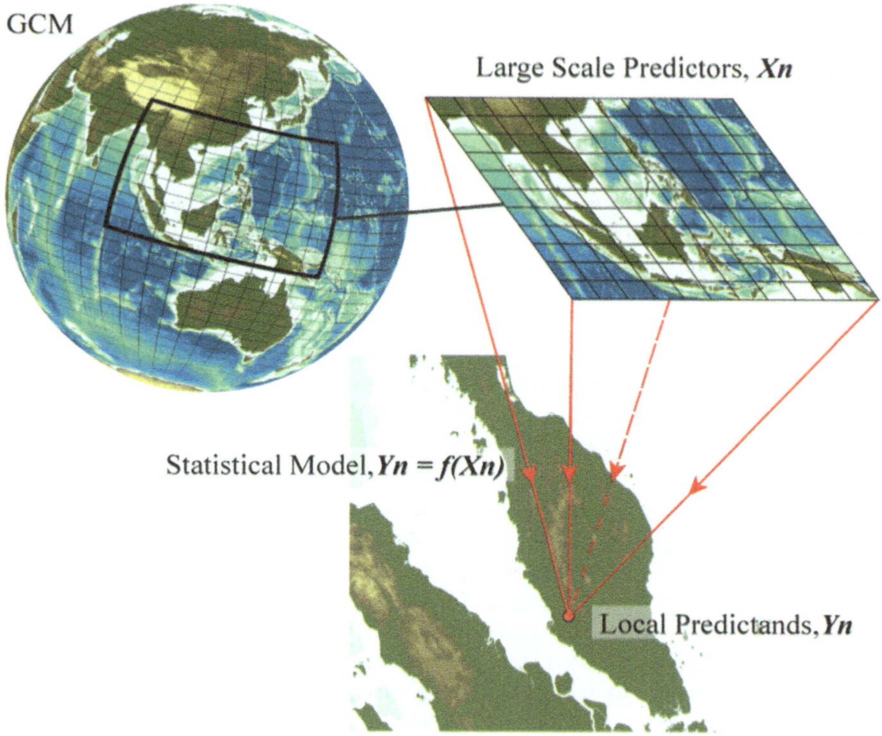

Fig. 2.4 Schematic diagram of regional climate statistical downscaling

biases from both GCM and RCM contribute to the uncertainties in future projections, which need to be adequately sampled for robustness. This requires the inclusion of multiple GCMs and RCMs as well as greenhouse gas emission scenarios (Valle et al. 2009), making dynamical downscaling very expensive and time consuming to carry out. To help this endeavor, the Coordinated Regional Climate Downscaling Experiment (CORDEX) was established under the World Climate Research Programme (WCRP). Much like CMIP, CORDEX allows coordination of regional climate modelling activities around the world but over different regional domains (Giorgi et al. 2009; www.cordex.org). There are 14 simulation domains in CORDEX with CORDEX Southeast Asia (CORDEX-SEA) being the latest one established (Tangang et al. 2020).

An alternative to the dynamical downscaling approach is the statistical downscaling (SD) technique, which assumes that regional climate is influenced by two factors, namely the large-scale climatic conditions and regional/local physiological features such as topography, land-ocean distribution, and land use. SD technique establishes a statistical relationship between large-scale climate information (predictors) of GCM and regional/local variables (predictands) during the historical period (Fig. 2.4). Potential predictor variables include mean sea level pressure, geopotential height, wind field, relative humidity, or temperature while predictand variables can be daily rainfall, mean temperature (Wilby et al. 2004), among others. While there are several methods of estimating the relationship, e.g., regression models (multiple regression, canonical correlation analysis, artificial neural networks, kriging), weather classification (analog, Monte Carlo methods, hybrid approaches), and weather generators (Markov chains, stochastic models, spell length methods), all rely on the availability of observed data (Wilby et al. 2004). Limited studies were found on the application of SD for the whole of Southeast Asia (Raghavan et al. 2018; Ngai et al. 2017). Overall, while SD is much cheaper to implement, its limitation is the existence of long-term high quality archived observed data.

2.3 Progress in Regional Climate Simulation

2.3.1 Country Level

Due to the highly technical nature and high computing cost requirement of regional climate downscaling simulations (RCDS), not all countries in Southeast Asia have been able to carry out the exercise thus far. A review of peer-reviewed literature indicates an apparent lack of regional climate simulations conducted by scientists or institutions from countries such as Myanmar, Cambodia, Laos, Brunei, and Timor-Leste. However, publications related to regional climate downscaling can be found with authors or institutions from countries such as Vietnam, Thailand, Malaysia, Philippines, Singapore, and Indonesia, although the level of activity can vary from one country to another. This section gives an overview of the regional climate downscaling studies done in selected countries in Southeast Asia.

2.3.1.1 Vietnam

Vietnam has conducted a number of RCDS studies. In 2009, the Ministry of Natural Resources and Environment (MONRE) of Vietnam published the report on "the climate change, sea level rise scenarios for Vietnam" (MONRE 2009) based on statistical downscaling of CMIP3 products. The report was first updated in 2012 where MONRE used a combination of statistical and dynamical approaches (MONRE 2012), and later updated in 2016 where a multi-model dynamical downscaling approach was applied with an ensemble of 16 experiments under Representative Concentration Pathway (RCP) 4.5 and 8.5 (MONRE 2016). The first peer-reviewed English paper related to climate simulations in Vietnam was Phan-Van et al. (2009) where the authors conducted an experiment using the Regional Climate Model version 3.0 (RegCM3) with the initial and lateral boundary conditions from the ERA-40 reanalysis data. Since then, a number of publications related to climate simulations in Vietnam have been published, showing remarkable progress in this direction during recent years.

Several studies showed that seasonal climate in Vietnam could be well captured by the downscaling experiments with different regional models (Phan-Van et al. 2009; Ngo-Duc et al. 2014; Van Khiem et al. 2014; Kieu-Thi et al. 2016; Katzfey et al. 2016). Multi-model RCDS have also been conducted (Ngo-Duc et al. 2014;

Van Khiem et al. 2014; Katzfey et al. 2016; Nguyen-Thi et al. 2019). It was shown that the ensemble average experiment generally outperformed an individual downscaling one over Vietnam in simulating temperature and rainfall (Ngo-Duc et al. 2014; Nguyen-Thi et al. 2019). Ability of RCDS in representing extreme climate events was also reported (Ngo-Duc et al. 2014; Kieu-Thi et al. 2016; Ngo-Duc et al. 2017; Herrmann et al. 2020). Added values of regional downscaling experiments over Vietnam compared to their respective host GCMs were investigated (Ngo-Duc et al. 2017; Trinh-Tuan et al. 2019). A simple bias correction technique such as the quantile mapping method can significantly improve the added values of the downscaled products (Trinh-Tuan et al. 2019). To date, future climate over Vietnam has been investigated and discussed using RCDS under the IPCC Special Report on Emission Scenarios (SRES) scenarios (MONRE 2009, 2012; Ho et al. 2011; Ngo-Duc et al. 2014; Kieu-Thi et al. 2016), and RCP scenarios (MONRE 2016; Trinh-Tuan et al. 2019; Nguyen-Thi et al. 2019).

2.3.1.2 Thailand

The level of activity in RCDS in Thailand is somewhat limited. Most studies thus far relied on the use of a single RCM to downscale a single GCM. The earliest RCDS in Thailand was that of Chinvanno et al. (2008) where PRECIS was used to downscale ECHAM4 under SRES A2 and B2 scenarios. This was followed by another study where CCSM model was downscaled with MM5 RCM for SRES A1B and A2 (Kreasuwun et al. 2009). Subsequently, there have been a few more RCDS but still limited to the use of a single RCM to downscale a single GCM (Chotamonsak et al. 2011; Manomaiphiboon et al. 2013). Lack of multi-model RCDS prompted some researchers to resort to coarse GCM products (Supharatid et al. 2016). A number of statistical downscaling initiatives have also been conducted due to its relatively low cost compared to dynamical downscaling (Vu et al. 2016; Saengsawang et al. 2017). A truly multi-model projection of future climate in Thailand was based on CORDEX-SEA simulations (Tangang et al. 2019).

2.3.1.3 Malaysia

There have been some activities of regional climate downscaling in Malaysia. Driven by the requirement to have downscaled climate projections in the Malaysia Initial Communication report, the National Hydraulic Research Institute Malaysia (NAHRIM) has carried out a regional climate downscaling projection study over Peninsular Malaysia (NAHRIM 2006). This projection was implemented on a 9-km grid using the Regional Hydro-Climate Model for Peninsular Malaysia (RegHCM-PM), a coupled model of a regional climate model, and a hydrological model. A similar model was applied to East Malaysia (Sabah and Sarawak) (NAHRIM 2010a). NAHRIM had also conducted other regional climate downscaling studies using the Hadley Centre, UKMO's system called PRECIS at 25 km resolution (NAHRIM 2010a, b, 2013). In addition, Shaaban et al. (2011) used the RegHCM-PM model to downscale Canadian Center for Climate Modeling and Analysis (CCCMA). Follow-up work by NAHRIM on RegHCM-PM model brought improvement in terms of spatial resolution from 9 km to 6 km over Peninsular Malaysia, which allowed refinement of topographic and land characteristics at local and watershed scales. NAHRIM was also able to develop the Watershed Environmental Hydrology-Hydroclimate Model of Peninsular Malaysia (WEHY-HCM-PM) to better understand the hydrological processes at watershed scale (NAHRIM 2014).

In addition to the government's agency initiatives, several studies have been conducted by researchers at universities (Kwan et al. 2013; Loh et al. 2016; Ngai et al. 2020a, b). Kwan et al. (2013) examined the possible changes of future temperature and rainfall extremes over Malaysia based on the SRES A1B emission scenario using HadCM3 and downscaled with PRECIS at 25 km resolution. Loh et al. (2016) also conducted a similar study but using PRECIS to downscale the HadAM3P and HadCM3Q0 GCM to 25 km horizontal grid spacing based on the SRES A2, A1B, and B2 scenarios. In a more recent study by Ngai et al. (2020b), the Non-hydrostatic Regional Climate model (NHRCM) was used to downscale the MRI AGCM3.2 model simulation from 20 km to 5 km resolution. A number of statistical downscaling studies have also been implemented in Malaysia (Hussain et al. 2017; Noor et al. 2019; Ngai et al. 2020a).

2.3.1.4 Indonesia

Occupying most of the Maritime Continent, Indonesia is a nation featured with thousands of islands with complex topography and coastlines. A typical landmass in GCM as in Fig. 2.2a is likely not sufficient to resolve the climate complexity within the country. Yet a number of studies have utilized bias corrected GCM outputs directly (Ministry of Public Works and Housing Indonesia 2017). A number of statistical downscaling projections have also been carried out in Indonesia (Inas et al. 2019; Kuntoro et al. 2018; Daksiya et al. 2017). Daksiya et al. (2017) studied the projected characteristics of daily rainfall over Jakarta. Meanwhile, Kuntoro et al. (2018) statistically downscaled 15 GCMs in projecting future rainfall in the upper Citarum River Basin, Java. Inas et al. (2019) also statistically downscaled a single GCM to project future temperature and rainfall over Komodo Island.

There have been a number of dynamic downscaling carried out in Indonesia, albeit limited. Jadmiko and Faqih (2014) conducted a RCDS of EH5OM Global model using RegCM3 for the Indramayu district, Central Java. McGregor et al. (2016) downscaled a CSIRO conformal-cubic atmospheric model (CCAM) 200 km uniform-grid global into 14 km resolution for the islands of Lombok and Sumbawa which the simulated outputs were used by Kirono et al. (2016) to investigate the impacts of climate change on agriculture and water resources in Lombok and Sumbawa Islands. Chandrasa and Montenegro (2020) conducted a regional climate downscaling of the entire country of MPI-ESM GCM using WRF3.9 and RCP8.5 scenario. The most comprehensive simulations, however, come from CORDEX-SEA. There have been a number of studies on future climate change in Indonesia using CORDEX-SEA simulations including Amsal et al. (2019), Supari et al. (2019), and Hidayat et al. (2019).

2.3.1.5 The Philippines

In the Philippines, there has been a growing number of regional climate downscaling studies done, particularly in the recent decade, to better understand climate processes and climate change in the country (Daron et al. 2018). Dynamical downscaling has been used to understand climate at the local scale, such as to examine potential impacts of interannual variations in local sea surface temperatures (Dado and Takahashi 2017) and urban expansion (Dado and Narisma 2019) on summer monsoon rainfall in some areas in the Philippines, as well as orographic effects on rainfall associated with typhoons (Racoma et al. 2016), among others. Sensitivity studies have also been conducted with dynamical and statistical models, which show the applicability of these models for the Philippines but also the sensitivity of these climate simulations to choices such as lateral boundary forcing and parameterization schemes (Cruz et al. 2016).

In 2011, the Philippine Atmospheric, Geophysical and Astronomical Services Administration (PAGASA), the national weather bureau, released a report describing climate projections for the country produced from dynamically downscaled simulations from ECHAM4 GCM for the A2 and B2 SRES scenarios, and HadCM3Q0 version 3 GCM for the A1B SRES scenario using the PRECIS model (PAGASA 2011). Through the collaboration of local and international research centers, and with the goal of producing more robust climate information, these climate projections have been updated using multiple regional climate models to downscale projections from multiple CMIP5 GCMs under the RCP4.5 and RCP8.5 scenarios to examine future changes in temperature and rainfall, and tropical cyclones affecting the country (Gallo et al. 2019; Villafuerte et al. 2020). The next report to be released will include simulations from CORDEX-SEA and discuss future changes in climate extremes. Furthermore, other studies have conducted downscaling of climate projections for sectoral applications using dynamical (e.g., for assessment on wind energy (Silang et al. 2014) and flood risk (Ushiyama et al. 2016)) and statistical approaches (e.g., for climate change adaptation planning (Basconcillo et al. 2016)).

2.3.2 Regional Level

Much like at the country level, activities in regional climate downscaling at the regional scale is still limited. The resources and coordination needed to carry out climate downscaling at the regional level are much greater, hence hampering its advancement in Southeast Asia. Furthermore, technical capabilities of institutions in most countries in the region are still behind those in East Asia, more so in Europe and North America where the progress in regional climate downscaling has advanced tremendously.

The first coordinated regional climate downscaling activity in Southeast Asia was conducted under the Southeast Asia Climate Analysis and Modeling (SEACAM) Project. From 2012 to 2014, SEACAM was championed by the Centre for Climate Research Singapore (CCRS) of Meteorological Service Singapore and Hadley Centre, UK Met Office with the involvement of many national meteorological and hydrological agencies and universities in the region (Rahmat et al. 2014). A single RCM with a horizontal grid spacing of 25 km was used driven by multiple GCM forcings, including five simulations of HadCM3Q' QUMP and ECHAM5 for the A2, A1B, and B2 SRES scenarios. Hence, the RCM uncertainty in the projections could not be estimated. Another RCM activity was carried out and coordinated by the climate modelling group at the Singapore-MIT Alliance for Research and Technology (SMART), Center for Environmental Sensing and Modeling (CENSAM), Singapore (Kang et al. 2019). Focusing over the Maritime Continent, this study employed three GCMs but was limited to a single RCM. A truly multi-model regional climate downscaling simulation was conducted under the CORDEX Southeast Asia project (e.g., Tangang et al. 2020; Supari et al. 2020). Initiated in 2013 under the Southeast Asia Regional Climate Downscaling (SEACLID), a project funded by the Asia-Pacific for Global Change Research Network (APN; https://www.apn-gcr.org/resources/items/show/1886), this project was a bottom-up initiative by a group of climate scientists from Southeast Asia who have assembled at Hanoi University of Science in Hanoi, Vietnam in 2012 to discuss how to coordinate regional climate downscaling in the region. Subsequently, after its inception, SEACLID was incorporated into CORDEX and became its fourteenth domain. Since then the project is referred to as SEACLID/CORDEX Southeast Asia, although CORDEX-SEA is more frequently used (Tangang et al. 2020).

Streamlining the SEACLID initiative into the globally coordinated CORDEX is a game changer as far as regional climate downscaling in Southeast Asia is concerned. Under SEACLID, a limited number of GCMs and a single RCM were used. However, the involvement of institutions outside the region was made possible under CORDEX-SEA. As described in Tangang et al. (2020), 11 GCMs have been downscaled using 7 RCMs for two RCP scenarios, making CORDEX-SEA simulations the most comprehensive set of projections for the region thus far. Overall, there were 24 simulations covering historical and future periods up to 2100 for middle and worst-case emission scenarios of RCP4.5 and RCP8.5, respectively. Although the simulations can still be considered as an "ensemble of opportunity" (Daron et al. 2018), the progress achieved in CORDEX-SEA is a step in the direction toward a more robust sample of simulations. It has contributed to the advancement of knowledge on projected future climate change in Southeast Asia with the publication of several papers, including Tangang et al. (2020) and Supari et al. (2020), on mean and extreme precipitation, respectively, Tangang et al. (2018) on extremes at 2.0 °C global warming, and Nguyen-Thi et al. (2021) on the climate analog and appearance of novel climate in Southeast Asia. The publications from CORDEX-SEA are expected to contribute to the Sixth IPCC Assessment Report (AR6), which is currently underway.

While it is crucial to analyze CORDEX-SEA simulations and publish findings, it is equally, if not more, important that the data be used by the vulnerability, impact, and adaptation (VIA) communities in the region to assess the risk and impacts of future climate change on critical sectors. Knowledge on this aspect needs to be advanced beyond what has been assessed in Hijioka et al. (2014), which would be the basis for countries in the region to strengthen their climate resilience and minimize climate-related risk and impacts. To facilitate this, and in line with the objectives of SEACLID and funder requirement, a data portal called the Southeast Asia Regional Climate Change Information System (SARCCIS) has been established and launched in 2018 as a venue for CORDEX-SEA data archiving. SARCCIS is hosted by the Ramkhamhaeng University Center for Regional Climate

Change and Renewable Energy (RU-CORE) in Bangkok, Thailand (https://www.rucore.ru.ac.th/SARCCIS) and is linked to the Earth System Grid Federation (ESGF) for worldwide accessibility, making it the first and only ESGF data node in Southeast Asia.

2.4 Challenges and Ways Forward

Although the implementation of CORDEX-SEA is generally successful, a number of issues remain that need to be addressed and should be a basis moving forward. Faced with increasing needs for RCDS in the years to come, improving the achievements under CORDEX-SEA thus far must be a priority. A notable issue in the CORDEX-SEA simulations is biases and uncertainty. As discussed in Sect. 2.2, while biases from GCMs can be reduced by selecting "good" GCMs that are capable of simulating mean climate and variability in the region, participating RCMs must be tuned to ensure parameterization schemes are optimized. In CORDEX-SEA, RegCM4 was extensively configured for its cumulus and air-sea flux parameterizations (Juneng et al. 2016; Ngo-Duc et al. 2017; Cruz et al. 2017). Nevertheless, the simulations still tended to have wet and cold biases even after using the best combination of schemes. To illustrate, Fig. 2.5 indicates the ensemble mean precipitation over three selected areas in Southeast Asia from the CORDEX-SEA simulations (Tangang et al. 2020). In Myanmar, the ensemble mean has approximated the observed annual cycle very well. Biases or inter-model variation is relatively small during boreal winter. However, larger biases in the summer reflect the poor skill in most of the models, which suggests further fine-tuning of RCMs is needed. Climate modellers may need to conduct an intercomparison of RCMs in simulating climate in Southeast Asia, which could be implemented under CORDEX as a flagship pilot project (www.cordex.org).

Unlike in Myanmar, the ensemble means precipitation was not able to capture the observed annual cycle in the Philippines and in Sulawesi (Fig. 2.5). Both areas are characterized by complex coastlines and topography, which may still not be well resolved even at 25 km resolution and would affect the quality of simulations. Increasing the resolution may improve the simulations. Ngai et al. (2020a) indicated that increasing the resolution to 5 km provided added values, especially in the simulation of extreme precipitation. However, the 5-km resolution falls into the so-called "gray zone" as far as convective permitting scale is concerned (Jeworrek et al. 2019). Convective Permitting Models (CPM), where the resolution needs to be <4 km (Prein et al. 2015) to resolve the convective processes, could improve simulations in a complex region such as the Maritime Continent. In addition, given its geographical setting being surrounded by regional seas, climate simulation in Southeast Asia could be better implemented by using a regional atmosphere–ocean–land coupled model compared to an uncoupled model where the atmosphere component is forced by the sea surface temperature. A number of these coupled model simulations indicated improvement and reduced bias in simulated climate (Primo et al. 2019). However, the application of either a CPM or a regional

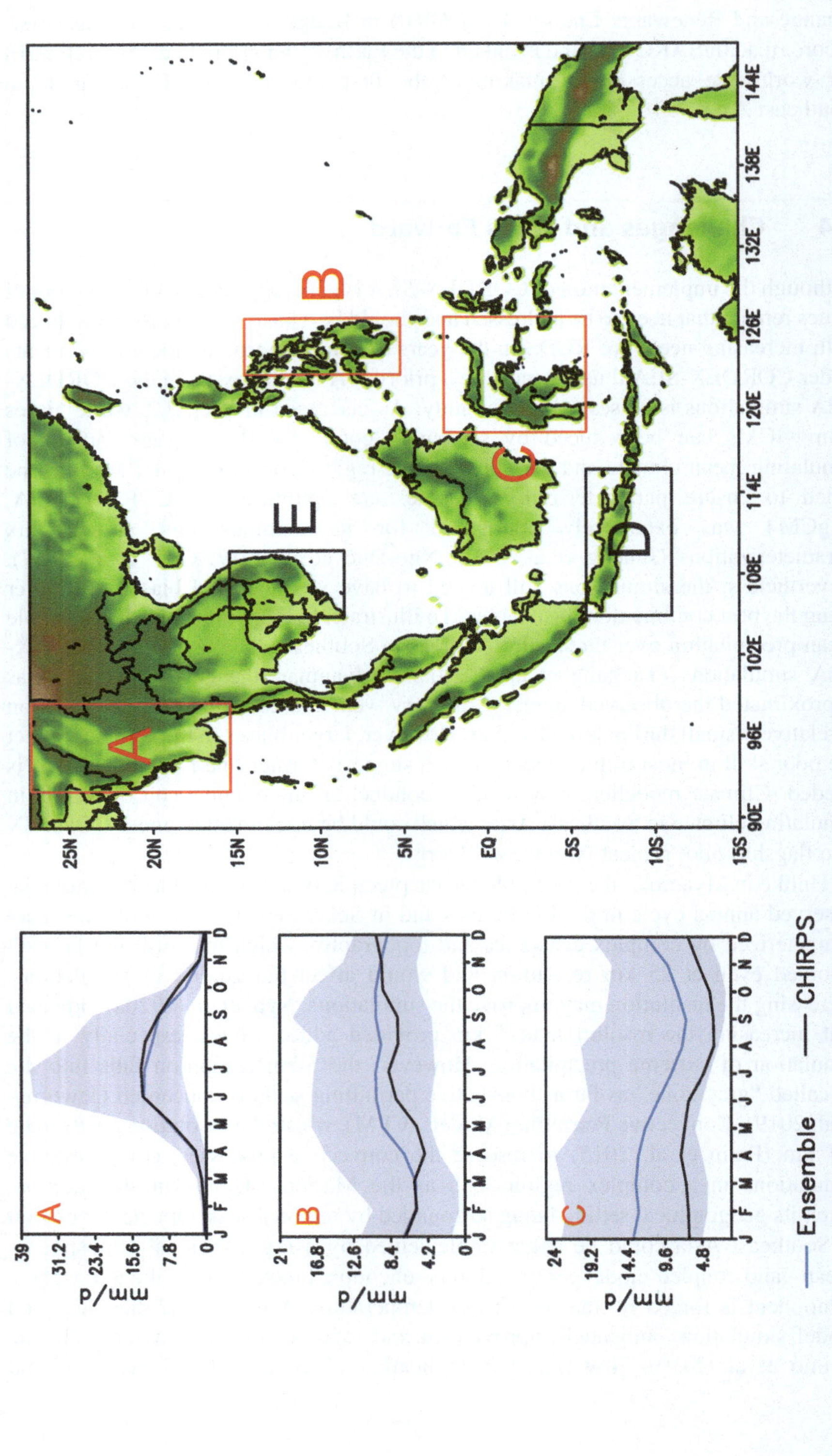

Fig. 2.5 Comparison of the ensemble mean rainfall from CORDEX-SEA simulations and observed GPCC rainfall over three areas indicated in the red boxes (a, b, and c) in the map. The shaded area indicates the ranges of the lowest and highest simulated values

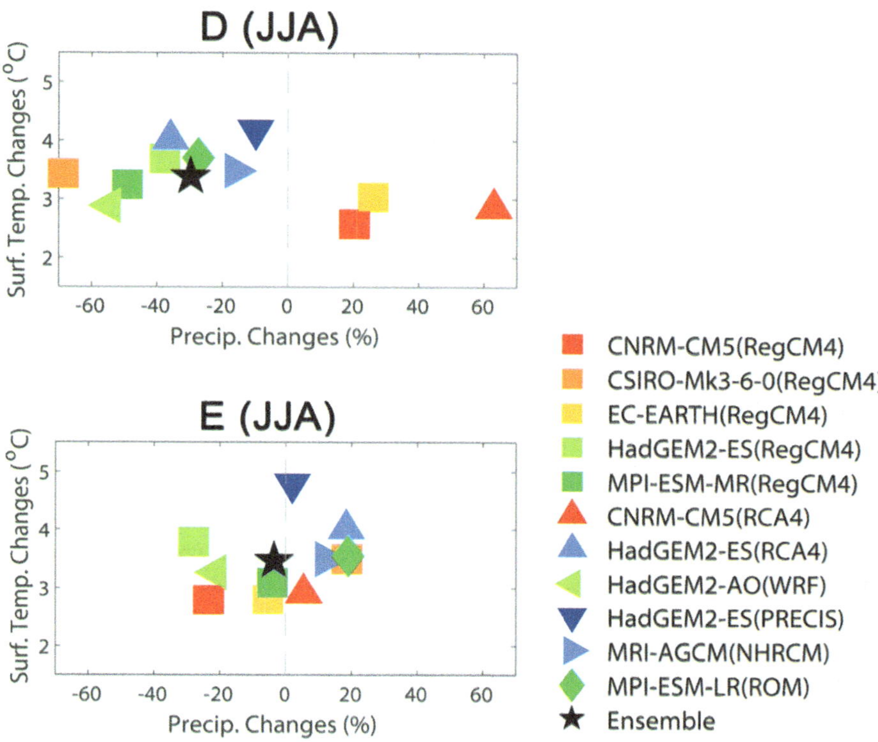

Fig. 2.6 The projected changes in mean rainfall of individual models of the CORDEX-SEA simulations, plotted against the projected mean surface temperature during June to August (JJA) at the end of the twenty-first century for RCP8.5 scenario for two areas: Java (Box D in Fig. 2.5) and Cambodia–southern Vietnam (Box E in Fig. 2.5)

atmosphere–ocean coupled model comes with huge challenges in terms of computing resources.

Biases in the model can translate into uncertainty in the projection. Figure 2.6 shows the projected mean rainfall of CORDEX-SEA simulations for areas over Cambodia–southern Vietnam and Java island at the end of the twenty-first century for RCP8.5 during June to August. Over the Cambodia–southern Vietnam area, the food basket region, the ensemble mean projected change indicates ~5% reduction in rainfall, but the range of projection or uncertainty can be large, i.e., within −25% to 20%. Similarly, for Java despite the ~30% mean rainfall reduction indicated in the ensemble mean, three models projected increases in mean rainfall by as much as 60%. Even though the robustness in climate change signals can be computed in terms of model agreement in the sign of changes, a large uncertainty in the projections makes decision-making on climate change adaptation difficult. By incorporating projections from statistical downscaling, the uncertainty can be even larger (Daron et al. 2018). With the upcoming release of the CMIP6 simulations, a new round of RCDS using these simulations will be required for Southeast Asia in the near future. Hence, there needs to be continued efforts to address these uncertainties in the RCM projections.

2.5 Conclusions

To most countries in Southeast Asia, climate change is a key risk, which is projected to increase in the coming decades. In pursuit of economic progress, countries must embrace climate-resilient pathways for development, in which adaptation, mitigation, and sustainable development must be considered. However, actions in terms of "what to do" and "how to do" must incorporate sound and robust knowledge of climate change projections. At a local level, this requires implementation of high-resolution multi-model regional climate downscaling, which is expensive and time consuming to implement. Based on the review carried out in this study, the level of activities in RCDS in most countries is low, far from being able to provide sufficient information needed for robust assessment of impacts of climate change. The successful implementation of CORDEX-SEA regional climate simulations is a game changer, in terms of how the region and each country should progress in RCDS. With the availability of freely accessible CORDEX-SEA climate projection outputs from SARCCIS data portal, it is hoped that the VIA community would utilize these projections to generate robust information on risk and impacts of climate change on important sectors. Further modelling activities in CORDEX-SEA need to be advanced to address some issues, including bias reduction, and the need for much higher resolution RCDS. Already the second phase of CORDEX-SEA with 5 km resolution on a number of subdomains is progressing well (https://www.apn-gcr.org/resources/items/show/2048). Implementation of CPM and regional atmosphere–ocean coupled models in RCDS must also be pursued in the future. Collaboration of scientists from within and outside the region, particularly the developed countries, will enhance the capacity of scientists in the region and advance the progress of RCDS even further.

Acknowledgments This research was funded by the Malaysian Ministry of High Education through the Long-term Research Grant Scheme LRGS/1/2020/UKM/01/6/1 Project under the LRGS/1/2020/UKM/01/6 Programme. CORDEX-SEA was funded by the Asia-Pacific for Global Change Research (APN) (ARCP2015-04CMY-Tangang, ARCP2014-07CMY-Tangang, ARCP2013-17NMY-Tangang).

References

Amsal F, Harsa H, Sopaheluwakan A, Linarka UA, Pradana RP, Satyaningsih R (2019) Bias correction of daily precipitation from downscaled CMIP5 climate projections over the Indonesian region. IOP Conf Ser: Earth Environ Sci 303:012046

Anh LT, Takagi H, Thao ND (2019) Storm surge and high waves due to 1997 Typhoon Linda: uninvestigated worst storm event in southern Vietnam. J Jpn Soc Civil Eng Ser B3 (Ocean Engineering) 75(2):I_73–I_78

Asian Development Bank Institute (2009) The economics of climate change in Southeast Asia: a regional review. Asian Development Bank, 253 p

Barnett J, O'neill S (2010) Maladaptation. Global Environ Change 20:211–213

Basconcillo J, Lucero A, Solis A, Sandoval JR, Bautista E, Koizumi T, Kanamaru H (2016) Statistically downscaled projected changes in seasonal mean temperature and rainfall in Cagayan Valley, Philippines. J Meteorol Soc Japan Ser II 94A:151–164

Cai W, Borlace S, Lengaigne M, Van Rensch P, Collins M, Vecchi G, Timmermann A, Santoso A, Mcphaden MJ, Wu L, England MH, Wang G, Guilyardi E, Jin F-F (2014) Increasing frequency of extreme El Niño events due to greenhouse warming. Nat Clim Chang 4:111–116

Chandrasa GT, Montenegro A (2020) Evaluation of regional climate model simulated rainfall over Indonesia and its application for downscaling future climate projections. Int J Climatol 40:2026–2047

Chinvanno S, Laung-Aram V, Sangmanee C, Thanakitmetavut J (2008) Simulation of future climate scenario for Thailand and surrounding countries. A final report (in Thai) submitted to the Thailand Research Fund

Chotamonsak C, Salathé EP, Kreasuwan J, Chantara S, Siriwitayakorn K (2011) Projected climate change over Southeast Asia simulated using a WRF regional climate model. Atmos Sci Lett 12 (2):213–219

Cruz FT, Sasaki H, Narisma GT (2016) Assessing the sensitivity of the non-hydrostatic regional climate model to boundary conditions and convective schemes over the Philippines. J Meteorol Soc Japan Ser II 94A:165–179

Cruz F, Narisma G, Dado JB, Singhruck P, Tangang F, Linarka UA, Wati T, Juneng L, Phan-Van T, Ngo-Duc T, Santisirisomboon J (2017) Sensitivity of temperature to physical parameterization schemes of RegCM4 over the CORDEX-Southeast Asia region. Int J Climatol 37 (15):5139–5153

Dado JM, Narisma GT (2019) The effect of urban expansion in metro Manila on the southwest monsoon rainfall. Asia-Pac J Atmos Sci

Dado JMB, Takahashi HG (2017) Potential impact of sea surface temperature on rainfall over the western Philippines. Prog Earth Planet Sci 4:23

Daksiya V, Mandapaka P, Lo EYM (2017) A comparative frequency analysis of maximum daily rainfall for a SE Asian region under current and future climate conditions. Adv Meteorol 2017:2620798

Daron J, Macadam I, Kanamaru H, Cinco T, Katzfey J, Scannell C, Jones R, Villafuerte M, Cruz F, Narisma G, Delfino RJ, Lasco R, Manalo J, Ares E, Solis AL, De Guzman R, Basconcillo J, Tangang F (2018) Providing future climate projections using multiple models and methods: insights from the Philippines. Clim Chang 148:187–203

De Sales F, Xue Y (2006) Investigation of seasonal prediction of the south American regional climate using the nested model system. J Geop Re: Atmos, p 111

Demory M-E, Berthou S, Sørland SL, Roberts MJ, Beyerle U, Seddon J, Haarsma R, Schär C, Christensen OB, Fealy R, Fernandez J, Nikulin G, Peano D, Putrasahan D, Roberts CD, Steger C, Teichmann C, Vautard R (2020) Can high-resolution GCMs reach the level of information provided by 12–50 km CORDEX RCMs in terms of daily precipitation distribution? Geosci Model Dev Discuss, https://doi.org/10.5194/gmd-2019-370, in review

Eyring V, Bony S, Meehl GA, Senior CA, Stevens B, Stouffer RJ, Taylor KE (2016) Overview of the coupled model Intercomparison project phase 6 (CMIP6) experimental design and organization. Geosci Model Dev 9:1937–1958

Flato G, Marotzke J, Abiodun B, Braconnot P, Chou SC, Collins W, Cox P, Driouech F, Emori S, Eyring V, Forest C, Gleckler P, Guilyardi E, Jakob C, Kattsov V, Reason C, Rummukainen M (2013) Evaluation of climate models. In: Stocker TF, Qin D, Plattner G-K, Tignor M, Allen SK, Doschung J, Nauels A, Xia Y, Bex V, Midgley PM (eds) Climate change 2013: the physical science basis. Contribution of working group I to the fifth assessment report of the intergovernmental panel on climate change. Cambridge University Press, Cambridge, pp 741–882. https://doi.org/10.1017/CBO9781107415324.020

Fritz HM, Blount C, Thwin S, Thu MK, Chan N (2010) Cyclone Nargis storm surge flooding in Myanmar's Ayeyarwady River Delta. In: Charabi Y (ed) Indian Ocean tropical cyclones and climate change. Springer, Dordrecht

Gallo F, Daron J, Macadam I, Cinco T, Villafuerte Ii M, Buonomo E, Tucker S, Hein-Griggs D, Jones RG (2019) High-resolution regional climate model projections of future tropical cyclone activity in the Philippines. Int J Climatol 39:1181–1194

Giorgi F (2019) Thirty years of regional climate modeling: where are we and where are we going next? J Geophys Res Atmos 124:5696–5723

Giorgi F, Jones C, Asrar GR (2009) Addressing climate information needs at the regional level: the CORDEX framework. World Meteorol Organ (WMO) Bull 58(3):175

Grotch SL, Maccracken MC (1991) The use of general circulation models to predict regional climatic change. J Clim 4:286–303

Hai OS, Samah AA, Chenoli SN, Subramaniam K, Ahmad Mazuki MY (2017) Extreme rainstorms that caused devastating flooding across the East Coast of peninsular Malaysia during November and December 2014. Weather Forecast 32:849–872

Hernandez JY Jr, Aquino RER, Pacheco BM, Cruz EC (2015) Damage caused by Typhoon Haiyan in the Philippines, review of structural regulations and practice, and research developments in wind engineering. Wind Eng JAWE 40(3):270–274

Herrmann M, Ngo-Duc T, Trinh-Tuan L (2020) Impact of climate change on SEA surface wind in Southeast Asia, from climatological average to extreme events: results from a RegCM4 dynamical downscaling of CNRM-CM5 in the CORDEX-SEA framework. Clim Dyn 54:2101–2134

Hidayat NM, Agustina L, Wibawanty DR, Safril A (2019) Rice plantation projection based on scenario representative concentration pathways (RCP) 4.5 in West Java Province. IOP Conf Series: Earth Environ Sci 303:012021

Hijioka Y, Lin E, Pereira J, Corlett R, Cui X, Insarov G, Lasco R, Lindgren E, Surjan A (2014) Asia. Climate change 2014: impacts, adaptation, and vulnerability. Part B: regional aspects. Contribution of working group II to the fifth assessment report of the intergovernmental panel on climate change. Cambridge University Press, Cambridge and New York, pp 1327–1370

Hirschman C, Bonaparte S (2012) In: Williams L, Philip Guest (eds) Demography of Southeast Asia. (2012) Southeast Asia program. Cornell University, Ithaca, NY

Ho T, Phan V, Le N, Nguyen Q (2011) Extreme climatic events over Vietnam from -observational data and RegCM3 projections. Clim Res 49:87–100

Hussain M, Yusof K, Mustafa M, Mahmood R, Shaofeng J (2017) Projected changes in temperature and precipitation in Sarawak state of Malaysia for selected CMIP5 climate scenarios. Int J Sustain Dev Plan 12:1299–1311

Inas R, Kuswanto H, Fithriasari K (2019) Climate projection using quantile matching bootstrap: a case of temperature and precipitation in Komodo Island, East Nusa Tenggara. AIP Conf Proc 2194:020041

IPCC (2019) IPCC special report on the ocean and cryosphere in a changing climate. In: Pörtner H-O, Roberts DC, Masson-Delmotte V, Zhai P, Tignor M, Poloczanska E, Mintenbeck K, Alegría A, Nicolai M, Okem A, Petzold J, Rama B, Weyer NM (eds) (in press)

Jadmiko SD, Faqih A (2014) Dynamical Downscaling Luaran Global Climate Model (GCM) Menggunakan Model REGCM3 untuk Proyeksi Curah Hujan di Kabupaten Indramayu. Agromet 28:9–16

Jeworrek J, West G, Stull R (2019) Evaluation of cumulus and microphysics parameterizations in WRF across the convective gray zone. Weather Forecast 34(4):1097–1115. https://doi.org/10.1175/WAF-D-18-0178.1

Juneng L, Tangang FT (2005) Evolution of ENSO-related rainfall anomalies in Southeast Asia region and its relationship with atmosphere–ocean variations in indo-Pacific sector. Clim Dyn 25:337–350. https://doi.org/10.1007/s00382-005-0031-6

Juneng L, Tangang F, Chung JX, Ngai ST, Tay TW, Narisma G, Cruz F, Phan-Van T, Ngo-Duc T, Santisirisomboon J, Singhruck P (2016) Sensitivity of Southeast Asia rainfall simulations to cumulus and air-sea flux parameterizations in RegCM4. Clim Res 69(1):59–77

Kang S, Im E-S, Eltahir EAB (2019) Future climate change enhances rainfall seasonality in a regional model of western maritime continent. Clim Dyn 52:747–764. https://doi.org/10.1007/s00382-018-4164-9

Katzfey J, Nguyen K, Mcgregor J, Hoffmann P, Ramasamy S, Nguyen HV, Khiem MV, Nguyen TV, Truong KB, Vu TV, Nguyen HT, Thuc T, Phong DH, Nguyen BT, Phan-Van T, Nguyen-Quang T, Ngo-Duc T, Trinh-Tuan L (2016) High-resolution simulations for Vietnam – methodology and evaluation of current climate. Asia-Pac J Atmos Sci 52:91–106

Kieu-Thi X, Vu-Thanh H, Nguyen-Minh T, Le D, Nguyen-Manh L, Takayabu I, Sasaki H, Kitoh A (2016) Rainfall and tropical cyclone activity over Vietnam simulated and projected by the non-hydrostatic regional climate model – NHRCM. J Meteorol Soc Japan Ser II 94A:135–150

Kirono DGC, Butler JRA, Mcgregor JL, Ripaldi A, Katzfey J, Nguyen K (2016) Historical and future seasonal rainfall variability in Nusa Tenggara Barat Province, Indonesia: implications for the agriculture and water sectors. Clim Risk Manag 12:45–58

Kreasuwun J, Chodamornsuk C, Ratjiranukool P, Wirunvedchayan O, Amnauylawjarurn T, Phansuke P (2009) Climate change simulations for Thailand by the MM5-regional climate model. A final report (in Thai) submitted to the Thailand Research Fund

Kuntoro AA, Cahyono M, Soentoro EA (2018) Land cover and climate change impact on river discharge: case study of upper Citarum River basin

Kuo HL (1974) Further studies of the parameterization of the influence of cumulus convection on large-scale flow. J Atmos Sci 31:1232–1240

Kwan MS, Tangang FT, Juneng L (2013) Projected changes of future climate extremes in Malaysia. Sains Malaysiana 42:1051–1059

Lagmay AMF, Agaton RP, Bahala MAC, Briones JBLT, Cabacaba KMC, Caro CVC, Dasallas LL, Gonzalo LAL, Ladiero CN, Lapidez JP, Mungcal MTF, Puno JVR, Ramos MMAC, Santiago J, Suarez JK, Tablazon JP (2015) Devastating storm surges of Typhoon Haiyan. Int J Disaster Risk Reduct 11:1–12

Loh JL, Tangang F, Juneng L, Hein D, Lee D-I (2016) Projected rainfall and temperature changes over Malaysia at the end of the 21st century based on PRECIS modelling system. Asia-Pac J Atmos Sci 52:191–208

Lohberger S, Stängel M, Atwood EC, Siegert F (2018) Spatial evaluation of Indonesia's 2015 fire-affected area and estimated carbon emissions using Sentinel-1. Glob Chang Biol 24:644–654

Magnan AK, Schipper ELF, Burkett M, Bharwani S, Burton I, Eriksen S, Gemenne F, Schaar J, Ziervogel G (2016) Addressing the risk of maladaptation to climate change. WIREs Climate Change 7:646–665

Manabe S, Bryan K (1969) Climate calculations with a combined ocean-atmosphere model. J Atmos Sci 26:786–789

Manomaiphiboon K, Octaviani M, Torsri K, Towprayoon S (2013) Projected changes in means and extremes of temperature and precipitation over Thailand under three future emissions scenarios. Clim Res 58:97–115. https://doi.org/10.3354/cr01188

McGregor JL, Nguyen KC, Kirono DGC, Katzfey JJ (2016) High-resolution climate projections for the islands of Lombok and Sumbawa, Nusa Tenggara Barat Province, Indonesia: challenges and implications. Clim Risk Manag 12:32–44

Mcmullen C, Downing T, Patt A, Troni J, Resurrección B (2019) Maladaptation to climate change: avoiding pitfalls on the Evolvability pathway-Frontiers 2018/19: emerging issues of environmental concern chapter 5. Frontiers 2018/19: emerging issues of environmental concern

Meehl GA, Boer GJ, Covey C, Latif M, Stouffer RJ (2000) The coupled model Intercomparison project (CMIP). Bull Am Meteorol Soc 81:313–318

Ministry of Public Works and Housing Indonesia (2017) The project for assessing and integrating climate change impacts into the water resources management plans for Brantas and Musi River basins. Jakarta. Available at https://openjicareport.jica.go.jp/pdf/12353090_01.pdf

MONRE (2009) Climate change and sea level rise scenarios for Vietnam. MONRE, Hanoi
MONRE (2012) Updates on the climate change and sea level rise scenarios for Vietnam. MONRE, Hanoi
MONRE (2016) Updates on the climate change and sea level rise scenarios for Vietnam. MONRE, Hanoi
NAHRIM (2006) Study of the impact of climate change on the hydrologic regime and water resources of peninsular Malaysia
NAHRIM (2010a) Climate projection downscaling for peninsular Malaysia and Sabah- Sarawak using Hadley Centre PRECIS model
NAHRIM (2010b) Study of the impact of climate change on the hydrologic regime and water resources of Sabah and Sarawak
NAHRIM (2013) Hydroclimate downscaling projection for malaysia using hadley centre PRECIS model (Moses 2.2)
NAHRIM (2014) Extension study of the impact of climate change on the hydrologic regime and water resources of peninsular Malaysia
Ngai ST, Tangang F, Juneng L (2017) Bias correction of global and regional models simulated daily precipitation and surface temperature over Southeast Asia. Glob Planet Chang 149:79–90
Ngai ST, Juneng L, Tangang F, Chung JX, Salimun E, Tan ML, Amalia S (2020a) Future projections of Malaysia daily precipitation characteristics using bias correction technique. Atmos Res 240:104926
Ngai ST, Sasaki H, Murata A, Nosaka M, Chung JX, Juneng L, Supari S, E. & Tangang, F. (2020b) Extreme rainfall projections for Malaysia at the end of 21st century using the high resolution non-hydrostatic regional climate model (NHRCM). SOLA 16:132–139. https://doi.org/10.2151/sola.2020-023
Ngo-Duc T, Kieu C, Thatcher M, Nguyen-Le D, Phan-Van T (2014) Climate projections for Vietnam based on regional climate models. Clim Res 60:199–213
Ngo-Duc T, Tangang FT, Santisirisomboon J, Cruz F, Trinh-Tuan L, Nguyen-Xuan T et al (2017) Performance evaluation of RegCM4 in simulating extreme rainfall and temperature indices over the CORDEX-Southeast Asia region: performance evaluation of RegCM4 over the CORDEX-Southeast Asia region. Int J Climatol 37:1634–1647
Nguyen-Thi T, Ngo-Duc T, Phan-Van T (2019) Performance of SEACLID/CORDEX-SEA multi-model experiments in simulating temperature and rainfall in Vietnam. Vietnam J Earth Sci 41 (4):374–387. https://doi.org/10.15625/0866-7187/41/4/14259
Nguyen-Thi T, Ngo-Duc T, Tangang FT, Cruz F, Juneng L, Santisirisomboon J, Aldrian E, Phan-Van T, Narisma G (2021) Climate analogue and future appearance of novel climate in Southeast Asia. Int J Climatol 41:E392–E409
Noor M, Ismail TB, Ullah S, Iqbal Z, Nawaz N, Ahmed K (2019) A non-local model output statistics approach for the downscaling of CMIP5 GCMs for the projection of rainfall in peninsular Malaysia. J Water Clim Chang 11:944–955
Oecd (2018) Climate-resilient Infrastructure. OECD environment policy paper No.14
Pagasa (2011) Climate change in the Philippines. Available online at https://dilg.gov.ph/PDF_File/reports_resources/DILG-Resources-2012130-2ef223f591.pdf
Peixoto JP, Oort AH (1992) Physics of climate. Springer, Berlin
Phan-Van T, Ngo-Duc T, Ho TMH (2009) Seasonal and interannual variations of surface climate elements over Vietnam. Clim Res 40:49–60
Prein AF, Langhans W, Fosser G, Ferrone A, Ban N, Goergen K, Keller M, Tölle M, Gutjahr O, Feser F, Brisson E (2015) A review on regional convection-permitting climate modeling: demonstrations, prospects, and challenges. Rev Geophys 53(2):323–361
Primo C, Kelemen FD, Feldmann H, Ahrens B (2019) A regional atmosphere-ocean climate system model (CCLMv5.0clm7-NEMOv3.3-NEMOv3.6) over Europe including three marginal seas: on its stability and performance. Geosci Model Dev 12:5077–5095. https://doi.org/10.5194/gmd-12-5077-2019

Racoma BAB, David CPC, Crisologo IA, Bagtasa G (2016) The change in rainfall from tropical cyclones due to orographic effect of the Sierra Madre Mountain range in Luzon, Philippines. Philipp J Sci 145:313–326

Raghavan SV, Hur J, Liong S (2018) Evaluations of NASA NEX-GDDP data over Southeast Asia: present and future climates. Clim Chang 148:503–518. https://doi.org/10.1007/s10584-018-2213-3

Rahmat R, Archevarahuprok B, Kang CP, Soe W (2014) A regional climate modelling experiment for Southeast Asia. Centre for climate research Singapore, meteorological service Singapore Rep, p 127

Randall DA, Wood RA, Bony S, Colman R, Fichefet T, Fyfe J, Kattsov V, Pitman A et al (2007) Climate models and their evaluation. In: Solomon S, Qin D, Manning M, Chen Z, Marquis M, Averyt KB, Tignor M, Miller HL (eds) Climate change 2007: the physical science basis. Contribution of the working group I to the fourth assessment of report of the intergovernmental panel on climate change. Cambridge University Press, Cambridge

Saengsawang S, Pankhao P, Kaprom C, Sriwongsitanon N (2017) Projections of future rainfall for the upper Ping River basin using regression-based downscaling. Adv Clim Chang Res 8 (4):256–267. https://doi.org/10.1016/j.accre.2017.11.001

Shaaban AJ, Amin MZM, Chen ZQ, Ohara N (2011) Regional modeling of climate change impact on peninsular Malaysia water resources. J Hydrol Eng 16:1040–1049

Silang A, Uy SN, Dado JM, Cruz FA, Narisma G, Libatique N, Tangonan G (2014) Wind energy projection for the Philippines based on climate change modeling. Energy Procedia 52:26–37

Stocker T (2014) Introduction to climate modelling, spring semester 2014. University of Bern, Bern, p 174

Supari F, Tangang E, Salimun E, Aldrian E, Sopaheluwakan A, Juneng L (2018) ENSO modulation of seasonal rainfall and extremes in Indonesia. Clim Dyn 51:2559–2580

Supari S, Sopaheluwakan A, Linarka UA, Rizal J, Satyaningsih R, Chung JX (2019) Indonesian climate under 2°C and 4°C global warming: precipitation extremes. IOP Conf Series: Earth Environ Sci 303:012048

Supari TF, Juneng L, Cruz F, Chung JX, Ngai ST, Salimun E, Mohd MSF, Santisirisomboon J, Singhruck P, Phanvan T, Ngo-Duc T, Narisma G, Aldrian E, Gunawan D, Sopaheluwakan A (2020) Multi-model projections of precipitation extremes in Southeast Asia based on CORDEX-Southeast Asia simulations. Environ Res 184:109350

Supharatid S, Aribarg T, Supratid S (2016) Assessing potential flood vulnerability to climate change by CMIP3 and CMIP5 models: case study of the 2011 Thailand great flood. J Water Climate Change 7(1):52–67. https://doi.org/10.2166/wcc.2015.116

Tangang F, Supari S, Chung J, Cruz F, Salimun E, Ngai S, Juneng L, Santisirisomboon J, Santisirisomboon J, Ngo-Duc T, Phan-Van T, Narisma G, Singhruck P, Gunawan D, Aldrian A, Sopaheluwakan A, Nikulin G, Yang H, Remedio ARC, Sein D, Hein-Griggs D (2018) Future changes in annual precipitation extremes over Southeast Asia under global warming of 2 °C. APN Sci Bull 8(1). https://doi.org/10.30852/sb.2018.436

Tangang F, Santisirisomboon J, Juneng L, Salimun E, Chung J, Supari CF, Ngai ST, Ngo-Duc T, Singhruck P, Narisma G, Santisirisomboon J, Wongsaree W, Promjirapawat K, Sukamongkol Y, Srisawadwong R, Setsirichock D, Phan-Van T, Aldrian E, Gunawan D, Nikulin G, Yang H (2019) Projected future changes in mean precipitation over Thailand based on multi-model regional climate simulations of CORDEX Southeast Asia. Int J Climatol 39:5413–5436. https://doi.org/10.1002/joc.6163

Tangang F, Chung JX, Juneng L, Supari S, Salimun E, Ngai ST, Jamaluddin AF, Mohd MSF, Cruz F, Narisma G, Santisirisomboon J, Ngo-Duc T, Van Tan P, Singhruck P, Gunawan D, Aldrian E, Sopaheluwakan A, Grigory N, Remedio ARC, Sein DV, Hein-Griggs D, Mcgregor JL, Yang H, Sasaki H, Kumar P (2020) Projected future changes in rainfall in Southeast Asia based on CORDEX–SEA multi-model simulations. Clim Dyn 55:1247–1267

Trinh-Tuan L, Matsumoto J, Tangang FT, Juneng L, Cruz F, Narisma G, Santisirisomboon J, Phan-Van T, Gunawan D, Aldrian E, Ngo-Duc T (2019) Application of quantile mapping Bias correction for mid-future precipitation projections over Vietnam. SOLA 15:1–6
Ushiyama T, Hasegawa A, Miyamoto M, Iwami Y (2016) Dynamic downscaling and bias correction of rainfall in the Pampanga River basin, Philippines, for investigating flood risk changes due to global warming. Hydrol Res Lett 10:106–112
Valle D, Staudhammer CL, Cropper WP Jr, Gardingen PR (2009) The importance of multimodel projections to assess uncertainty in projections from simulation models. Ecol Appl 19:1680–1692
Van Khiem M, Redmond G, Mcsweeney C, Thuc T (2014) Evaluation of dynamically downscaled ensemble climate simulations for Vietnam. Int J Climatol 34:2450–2463
Villafuerte MQ II, Macadam I, Daron J, Katzfey J, Cinco TA, Ares ED, Jones RG (2020) Projected changes in rainfall and temperature over the Philippines from multiple dynamical downscaling models. Int J Climatol 40:1784–1804
Vu MT, Aribarg T, Supratid S, Raghavan SV, Liong SY (2016) Statistical downscaling rainfall using artificial neural network: significantly wetter Bangkok? Theor Appl Climatol 126:453–467
Wilby R, Charles S, Zorita E, Timbal B, Whetton P, Mearns L (2004) Guidelines for use of climate scenarios developed from statistical downscaling methods. 27 p. Available online at http://www.narccap.ucar.edu/doc/tgica-guidance-2004.pdf
World Bank (2016) The cost of fire an economic analysis of Indonesia's 2015 fire crisis
Zhao C, Yan Y, Wang C, Tang M, Wu G, Ding D, Song Y (2018) Adaptation and mitigation for combating climate change – from single to joint. Ecosystem Health Sustain 4:85–94

Climate Change Signatures over Schirmacher Oasis, Antarctica

3

H. N. Dutta

Abstract

Antarctica is an isolated dome-shaped icy continent surrounded by the stormiest oceans in the world. It holds the world's most efficiently coupled ocean-ice-air-radiation system that is responsible for its sustenance and growth. However, global warming is knocking at the periphery of Antarctica and the Southern Ocean around it, leading to melting of polar ice. This is a matter of great concern as the resulting sea level rise will have a direct impact on human endeavors in almost all the coastal countries.

The Indian Antarctic station Maitri (70.7666S, 11.7314E) is located over a small rocky region (35 sq. km area) in the Schirmacher Oasis, situated along the periphery of East Antarctic region. Schirmacher Oasis consists of exposed weathered rocks in the forms of hilly undulations formed due to receding polar glaciers. Due to weathering of rocks over thousands of years, many lakes and ponds have formed over this oasis and it offers a convenient site to understand various processes that may help in taking steps to minimize the impact of global warming in Antarctica.

Keywords

Schirmacher Oasis · Indian Antarctic station Maitri · Radiative properties · Thermal stressors

H. N. Dutta (✉)
CSIR-National Physical Laboratory, Delhi, India

A. Kaushik et al. (eds.), *Climate Resilience and Environmental Sustainability Approaches*, https://doi.org/10.1007/978-981-16-0902-2_3

3.1 Introduction

Antarctica holds world's most efficiently coupled ocean-ice-air system (Singh and Dutta 2011) that has led to the formation of dome shaped icy continent, which is surrounded by the Southern Ocean from all around. Since it is a coupled system, the role of radiative transfer of heat, katabatic winds and the formation of cyclones over the ocean, topography, formation of sea ice etc. becomes equally important and any change in any system, will be reflected in all the associated components (Ganajanda et al. 2011). On February 02, 2002, Dutta et al. (2011) observed ocean around east Antarctic periphery to be warmer continuously for 24 hours. With global warming, the coupled system is affecting the Antarctic ice mass balance (Turner et al. 2002). In 2006, Rémy and Frezzotti (2006) had concluded that the West Antarctica ice sheet seems to be more sensitive to current warming but the East Antarctica ice sheet is more or less in balance. However, now there is no doubt that the ice sheet is sensitive to recent climate change and surface ice mass is decreasing (Lenaerts et al. 2016). The most pronounced atmospheric circulation changes over the twentieth and twenty-first centuries to date have occurred in the Southern Hemisphere (Screen et al. 2018).

Rignot et al. (2019) have calculated the grounding line ice discharge of 176 basins draining the Antarctic Ice Sheet from 1979 to 2017. It has been estimated that the East Antarctica is a major participant in the mass loss. During the entire period, the mass loss concentrated in areas closest to warm, salty, subsurface, circumpolar deep water (CDW), that is, consistent with enhanced polar westerlies pushing CDW toward Antarctica to melt its floating ice shelves, destabilize the glaciers, and raise sea level. Bronselaer et al (2018) have projected that meltwater from the Antarctic Ice Sheet can cause up to one metre of sea-level rise by 2100 but meltwater would delay the exceedance of the maximum global-mean atmospheric warming targets of 1.5 and 2 degrees Celsius by more than a ecade

With this scientific background, this chapter examines the basic reasons of ice melting over Antarctica on the basis of several investigations carried out over the Indian Antarctic station Maitri (75.2081°S, 1.2501°E), situated over the Schirmacher Oasis (SO), which is a small, ice-free rocky region with about 35 km^2 located at the periphery of East Antarctica (Bormann and Fritzsche 1995).

3.2 Formation of Schirmacher Oasis

On the basis of numerous minette dykes intersect the Precambrian crystalline basement of SO, the age of whole-rock–biotite isochrons suggests an age of $\sim$455 Ma for emplacement of the minette dykes and the area has experienced both weathering and erosion (Hoch et al. 2001). Dharwadkar et al. (2018) have concluded that the recessional regime in SO was set in at $\sim$22 Ka and subsequently the area remained ice free. The recession of snout of Dakshin Gangotri Glacier in SO was physically confirmed by Chaturvedi et al. (1999). The continuous recession of ice front since eighties, points toward an increase in the area of the SO in future as well (Dharwadkar et al. 2018).

Fig. 3.1 Photograph of a portion of the SO (marked as 1). It shows totally exposed rocky undulations with deep, shallow depressions and the Indian Antarctic station, Maitri (marked as 2), a part of frozen Priyadarshini lake (marked as 3) and a small hill named as Shivling (4) situated over the polar ice. It also shows weathered rocks (5) having different textures, sizes and appearances. In summer, this area normally remains ice free but occasional snowfall/drifting ice deposit incidences are recorded

In the oases regions situated at the periphery of Antarctica, apart from solid ice and fresh snow, water gets accumulated in lakes, ponds, and is drained out toward the shelf ice/ocean in various forms (Phillips 1998). Moreover, many of these lakes and ponds form even over polar and shelf ice (Banwell et al. 2019) and such formations can be physically seen from the imageries in Google Earth (Google Earth). At times, it may rain and an incidence of unusual warming was reported by Kaur et al. (2013) over SO and it rained so much that the whole SO was full of water and first time, we personally realized the havoc of rain for the stability of Antarctic ice.

Figure 3.1 presents photographic display of a small portion of the SO; it shows hilly undulations, presence of boulders, partially frozen Priyadarshani lake in front of the Indian Antarctic station, Maitri. It also shows polar ice and a small hill named as Shivling (Composite Gazetteer of Antarctica n.d.) surrounded by polar ice. There is a general downward slope from polar ice toward the SO. This area gets covered by ice during local winter and in summer season, it is more or less ice free but occasional snowfall or katabatic winds dumping snow/ice are common. Once this type of snowfall is there, the snow gets melted due to the presence of sunlight and the meltwater has many roles in the weathering of rocks.

3.3 Types of Rocks in SO

Formation of SO started on geological scales and presently, it has heterogeneous weathered rocks totally exposed to all the weathering agents present in Antarctica (Dasgupta et al. 2001). The rocks consist of charnockites, enderbites, garnet–sillimanite gneisses, garnet–biotite gneisses, quartzo-feldspathic augen gneisses with minor amounts of foliated lamprophyres, amphibolites, dolerites, metagabbros, and metabasalts (Musta and Tahir 2012). Figure 3.2 shows a closer look of rocks from the SO. It is important to note that within the rocks, since chemical compositions are different, their thermal conductivity, porosity, expansion rates, chemical reaction rates, corrosion properties vary and make them susceptible to weathering at different levels and over different time periods. The erosion and weathering of rocks lead to fragmentation into smaller pieces of all sizes, finally gravel, sand, and fine dust become the end products. The entire SO offers a convenient site to study weathering of rocks and their influence on the ice melt within the SO and the adjoining areas.

3.4 Formation of Ponds and Lakes Over SO

SO has about 150 freshwater lakes, the size of the lakes varies from small, shallow ponds to big, deep natural reservoirs (Gillieson et al. 1990). These lakes are sensitive to radiative properties and other atmospheric parameters for their maintenance and growth in size and volume over a period of time. The basic formation of ponds and lakes is due to undulations over the rocky regions but once these undulations were exposed to various weathering agents, over thousands of years; it has led to the formation of soil (Bockheim 1997), fragments of rocks, modification of size, and

Fig. 3.2 Almost all the rocks in SO are weathered by various weathering agents (radiative, mechanical, chemical and biological) and these fragmented/crushed weathered materials are enhancing melting of ice within and around SO. Many rocks display signatures of shear stresses by the receding thick polar ice. In Antarctica, one form of weathering is complemented or supplemented by another kind of weathering and all of them work together with different erosional capacities and time scales

volume of ponds and lakes. Since the formation of ponds and lakes is linked to weathering of rocks, we shall focus on these aspects in order to realize the power of natural forces in helping ponds and lakes to grow in their sizes, volume, and shapes.

3.5 Impact of Radiative Properties and Thermal Stresses

From an academic pursuit, it was believed that for the Antarctic ice sheet to be stable it must always transfer heat to the atmosphere and the icy surface must remain cooler than the air aloft. This process of heat transfer results in the development of surface-based temperature inversion close to the icy surface (Connolley 1996; Gajananda et al. 2004; Naithani and Dutta 1995) and inversions are extremely stable in the interior of the continent (Baas et al. 2019). To understand the interactive dynamics of atmosphere over SO, an acoustic radar was established by the author's group at Maitri in the year 1990–1991 and it was observed that the atmosphere remains stable for most of the months in a year. However, for 5.63% of the time, atmospheric convection has been observed (Naithani and Dutta 1995).

Figure 3.3 shows thermal convection and stable layers over Maitri, Antarctica as observed by the acoustic radar (Naithani et al. 1995; Gajananda et al. 2004). During thermal convection, if snowfall is there, convection melts snow and water so formed, seeps through the cracks, holes, pores of various types of rocks. Once snowfall ceases, rocks cool immediately and inversion starts forming, the temperature of rocks falls below the freeing temperature of water leading to Freeze-thaw weathering

Thermal convection also shows rocks being subjected to warming events in which different chemical constituents present in rocks expand differently, leading to the mechanical stresses at micro-granular levels. Thermal expansion and contraction makes rocks as brittle. In the case of Antarctica, thermal stability causes winds to flow as katabatic winds which transport snow and ice from the interior of the

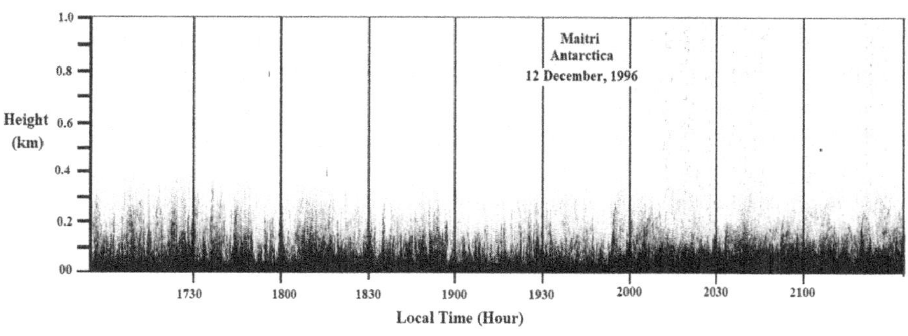

Fig. 3.3 Thermal convection recorded an acoustic radar over SO. Under thermal convection, hot air parcels from the surface of Earth move up in the form of thermal plumes. These plumes carry fine dust and microorganisms from the surface in the upper regions of the air so as to mix and deposit them over a larger portion around SO

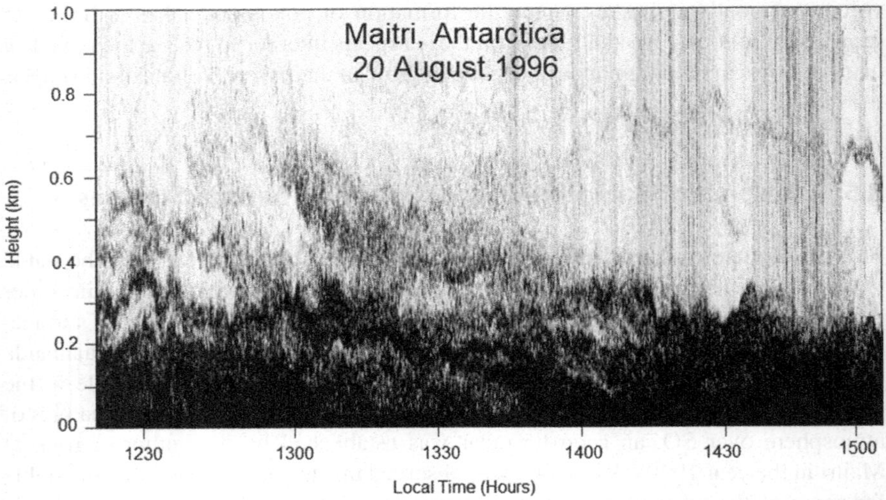

Fig. 3.4 Surface based layers and multiple layers are the most common atmospheric feature prevailing over SO indicating surface to be much cooler than the air aloft. For the stability of Antarctica, presence of inversions is the most important

continent towards the periphery (Naithani et al. 1995). Such events cover the entire rocky region with snow/ice which may remain intact until summer season starts

Figure 3.4 shows stable layers recorded over SO. In winter, the flow of cooler air from higher plateau towards the periphery of Antarctica dominates, it results in the formation of surface based and elevated layers on the acoustic radar

By using an acoustic radar, Argentini et al.(2005) have observed thermal convection at Dome C, Antarctica. This means, for some intermittent period, thermal convection can be observed even over icy surfaces. Matsuoka (1995) has concluded that the humid weathering regime that occurs only locally around the margin of the present sheet, frequent diurnal freeze–thaw cycles in summer cause relatively rapid rock fragmentation (Hall et al. 2002). Most of the mountains are situated in the arid weathering regime, under which rock breakdown is very slow unless the rock contains plenty of salts. Salt weathering becomes more intensive and extensive with exposure age, as a result of salt accumulation in rock, eventually producing soils as small as fine silt size.

In the polar ice, so long a rock remains buried deep below the ice, it has no major role to play except grinding, but the moment, it gets exposed, it drastically lowers the albedo of ice from 90% to ~30% (Gusain and Negi (2016) and the rocks with higher thermal conductivity values show a higher average annual recession of polar ice than that of the lower thermal conductivity values (Swain 2019). Moreover, due to weathering, rocks get fragmented into small fragments/pebbles, sand, and dust. Sand and dust are blown away by the katabatic winds, making ice dusty; thereby it enhances its melting rate to form water. Lamp et al. (2017) have studied thermal stress weathering and spalling of rocks over Mullins Glacier, McMurdo Dry Valleys, Antarctica. The results suggest that thermal stresses induced at the base of thin alteration rinds ~2 mm thick, common on rocks exposed for ~10^5 years, maybe sufficient to cause existing cracks to propagate under present-day meteorological forcing, eventually leading to rind detachment.

Gusain and Negi (2016) have measured albedo over rocky regions of SO and the adjoining area and it was concluded that if similar meteorological observations persist for a longer duration in coming years then it will contribute to increasing ice-free areas near SO. Ganju and Gusain (2017) observed albedo of polar ice near SO as 0.64, which is an indicator of the influence of SO around the entire icy area.

3.5.1 Impact of Thermal Convection

Thawing Process During summer, there are days when SO experiences snowfall (from northerly oceanic winds) or snow/ice deposited by katabatic wind events. In both cases, the moment solar heating is there, snow/ice deposited over rocks, pebbles, silt, etc. melts to form water. This water has low quantity but it is enough to fill small pits. The water in these pits freezes at night and thus the thawing forces weathering of rocks.

Aerosols in Atmosphere Thermal convection also supports the transport of dust/microorganisms in the atmosphere over SO, thus it helps spreading the influence of SO by depositing fine dust over ice (Ganajanda et al. 2004). Kavan et al. 2018 have found that over the Ulu Peninsula, James Ross Island to be an important active high latitude dust source. The highest aerosol concentrations of 57μg m^{-3} for PM_{10} were detected during high wind speed events that exceeded 10 m s^{-1}, which is also a threshold level for activating local mineral material sources. Deshpande and Kamra (2004) have measured aerosol 800–1200 particles cm^{-3} over SO during January–February, 1997. Total aerosol concentration increases by approximately an order of magnitude whenever a low-pressure system passes over the station. This shows that the aerosols over SO are of local origin as well as, these are inducted from the adjoining oceanic winds. In fact, Gajananda et al. (2007) have reported fog incidence over SO. Surface deposition of dust particles can have significant environmental impacts such as changes in properties of atmosphere or enhanced snow melting.

Aerosols in the atmosphere and coarse dust blown by katabatic winds get deposited on ice and they lower the albedo of the ice. When ice with lower albedo starts melting, the fine dust flows down in the ponds/lakes leaving behind coarse dust particles to assist further melting. Figure 3.5 shows dust/bioaerosols deposited on ice and it reduces albedo so that dusty ice melts faster. It is important to note that lacustrine sediments of the Big Lachman Lake (James Ross Island, Antarctica) are dominated by clay- and silt-sized particles of varied petrographic composition (Vašinka et al. 2020).

Fig. 3.5 Dust/coarse sand deposited over ice enhance formation of meltwater by lowering albedo of the ice. Dust deposited on ice is marked as "1", clear fresh ice is marked as "2" and the meltwater is marked as "3". Once meltwater forms over the ice, it lowers the albedo and starts acquiring thermal energy, resulting in enhancement of its temperature. As temperature of meltwater enhances, it melts more ice to form a small pond

3.6 Mechanical Weathering by Katabatic Winds

Antarctica is the windiest continent and therefore, the study of weathering of rocks by winds has been investigated by many groups (Matsuoka et al. 1996). Gillies et al. 2009 presented well-defined evidence of wind abrasion and ventifaction of rocks. Due to the dome shape of the continent, Katabatic winds are the most persistent, surface winds blowing from the interior of the continent toward periphery and are responsible for snow transport (Kaur et al. 2012) and lead to extreme wind–ice interactions (Das et al. 2015). Such winds lead to drifting of snow and blizzards in Antarctica (Palm et al. 2011). Due to these processes, snow and ice eroded from polar ice get deposited over the rocky regions of SO and the most significant part is that since snow or ice crystals are like dry sand, they get deposited inside the cracks and holes. Now when it melts during sunny period, it results in the formation of water leading to chemical weathering. Drifting or blowing snow can accumulate several meters of ice on the rocks; it leads to physical weight as well. Sabacka et al. (2012) have shown aeolian flux of biotic and abiotic materials in the Taylor Valley, Antarctica. Due to aeolian processes of fine soil, it gets deposited over the polar cap ice, thereby it reduces its albedo and the probability of it getting melted becomes much higher (Fig. 3.4). In fact, snow-laden katabatic wind results in frost weathering and wind erosion are only locally effective where plenty of moisture or an abrasive material is available, while salt weathering and removal of the waste by wind play a major role in constructing erosional landforms over the mountains.

Katabatic winds sweep a great mass of snow off the surface of the ice sheet and carry it outward to the periphery of the ice sheet (Kobayashi 1978). This snow deposits over SO and is important for water and weathering of rocks. Strong winds that transfer sediment of sand and dust are an important feature of McMurdo Dry Valleys environment (Nicholas et al. 2010; Nylen et al. 2004). Similarly, evidence for aeolian processes and sediment deposits are widespread over the SO (Sedov et al. 2019). Katabatic winds blowing across the surface of a lake interact with lake water and cause the water to move in a downward direction. Since the speed of katabatic winds is extremely high, lake water is pitched against rocks, resulting in severe weathering of rocks and expansion of lakes over SO. Katabatic winds also mix air in the lake water for the survival and growth of microorganisms.

3.7 Chemical Weathering

During local summer, meltwater and pieces of ice from polar glaciers flow toward the Schirmacher Oasis (SO) in various forms. This water gets collected in lakes and ponds over the hilly regions of SO and at the periphery; it also goes toward shelf-ice. When water and ice rush, they grind rocks with their momentum, result in mechanical and chemical reactions and over a period of time, may consequently break the rocks (Balke et al. 1991). Flowing water deposits fine dust/microorganisms in ponds and lakes and removes chemicals from active debris layer of SO (Balke et al. 1991). Water chemistry in lakes offers an insight into the role of lithology, weathering, and precipitation (Asthana et al. 2019) and fine soil transport is carried out by water channels and winds (Srivastava et al. 2013).

Once water gets added to the surface, chemical weathering of rocks starts (Dold et al. 2013). However, Acid rock drainage and rock weathering in Antarctica has been found to be important sources for iron cycling in the Southern Ocean and the same process is important in the contribution fine fragments of rocks in lakes (Dold et al. 2013). Due to the availability of liquid water, common soils on the Earth are made up largely of chemically weathered minerals (Wilson 1995). Over Taylor Valley polar desert located in the McMurdo Dry Valleys, glacial meltwater streams flow through unconsolidated boulders and pebbles to perennially ice covered closed-basin lakes for 6–10 weeks in a year (Nezat et al. 2001). Because there is no overland flow or groundwater input, most chemical weathering reactions are restricted to the stream channels. These high rates in Taylor Valley suggest that temperature and precipitation are not the primary controls on chemical denudation. In this case, high stream discharges, high rates of physical weathering (e.g., frost action and salt weathering), and/or the interaction between the stream and hyporheic zone contribute to the high chemical denudation rates in the Taylor Valley stream channels.

3.8 Biological Weathering

Due to abundance of water and soil nutrients, SO harbors a diversified and adapted biodiversity (Gajananda and Dutta 2005, Nayaka and Upreti 2005). In fact, moss is seen as a dark black carpet covering (green-yellowish portion underneath) almost entire soil around major lakes. Saxicolous species of lichens are able to induce and accelerate weathering of their rock substrate, and effects of lichens on substrate can be attributed to both physical and chemical causes (Chen and Blume 2002). Chen et al. (2000) discussed that the weathering of minerals can be accelerated by the growth of at least some lichen species. Since lichens and mosses thrive on soil nutrients, bioweathered rocks have been studied by various investigators (Wierzchos and Ascaso 1996; Jie and Blume 2002). As a result of the weathering induced by lichens, many rock-forming minerals exhibit extensive surface corrosion (Chen et al. 2000).

Dold et al. (2013) examined three sites during three expeditions during 2009–2011 on the South Shetland Islands in Antarctica. Two of them displayed intensive sulfide mineralization and generated acidic (pH 3.2–4.5), iron-rich drainage waters (up to 1.78 mM Fe), which infiltrated as groundwater (as Fe^{2+}) and as superficial runoff (as Fe^{3+}) into the sea, the latter with the formation of schwertmannite in the sea ice. The increased iron input correlates with increased phytoplankton production close to the source. This might even be enhanced in the future by a global warming scenario, and could be a process counterbalancing global warming. It is seen that even moss when it becomes black and dry, a part of this is blown by wind to be dispersed over SO and the adjoining areas (Gajananda et al. 2004). Interaction of clay minerals has been discussed in details by Cuadros (2017). It has been emphasized that researchers are missing out important information if they do not investigate the influence and modifications that minerals have on microbial activity and evolution. Singh et al. (2018) have concluded that the changes in temperature, pattern of cyclonic activity, availability of nutrients, water, high UV-B radiation along with other stresses poses obstacles for survival and distribution of Antarctic flora.

Sedov et al. (2019) performed micromorphological investigation of several thin soil profiles developed under mosses and lichens on the sediments deposited by meltwater in a lake depression of Schirmacher Oasis and concluded that the studied soil bodies are synsedimentary: fluvial (and possibly aeolian) deposition of mineral material that occurs simultaneously with the growth of primitive plants. The mineral grains are trapped between plant tissues, which form a framework of the soil matrix. Figure 3.6 shows photos of various terrestrial communities over SO, each community has its own radiative, physical, chemical, and biological reaction rates rendering the formation of variety of weathered rocks.

Fig. 3.6 Texture and quantity of terrestrial vegetation community is highly variable over SO, it adds to the variability in weathering of rocks by chemical and biological reactions and when transported by water channels and winds over ice, these microorganisms change the albedo of ice

3.9 Rock–Ice Interaction

Figure 3.7 shows a close up of a small hill called Shivling (see Fig. 3.1), it is a small rocky hill with base area 1460 m^2 (calculated from Google Earth) and the area of influence on ice calculated from the Google Earth is 26125 m^2. Figure 3.7 shows one side of this hill, it has created a pit on the polar ice by the meltwater falling from the hill and by the thermal conduction/radiation. At the same time, we can also see that there is a well demarcated area on ice which is influenced by this hill. This hill also pours fragmented/weathered sand/dust particles and other biological material (that grows on the lower side of the hill) on the icy surface, resulting in lowering its albedo. The rocks in the hill are also weathered by all the weathering agents. Since meltwater forms a pit around the hill, meltwater starts cutting the polar ice much below the surface of ice and may form a meltwater cave below the icy surface

Figure 3.8 shows a portion of shelf ice along SO, large undulations in the shelf ice are created by the flowing water and the katabatic winds. It is important to note that a shelf ice is the floating tongue of ice that extends from grounded polar ice on the ground and the existence of shelf ice is extremely important for the stability of the polar ice. The most dangerous is the addition of meltwater from polar ice on shelf ice via SO. When water flows through SO, it adds heat, dissolved chemicals, and fine particulate matter. Polar ice glacier movement in the eastern and western side of the SO transports meltwater as well during local summer. Also, formation of meltwater ponds, caves, lakes, and rivers on the shelf ice itself is highly dangerous for the stability of shelf ice (Silvano et al. 2018).

Fig. 3.7 Evidence of direct influence by Shivling hill (marked as 1) South of SO in melting the polar ice. It is a group of weathered rocks (marked as 2) and the meltwater from the hill forms a pit in the polar ice (marked as 3). Polar ice is marked as "4" and the area within which, polar ice is directly affected by the hill is marked as "5"

The ground reality is that Shepherd et al. (2018) have found a mass balance of Antarctic ice sheet declining from 1992 to 2017 and Alexander et al. (2019) have concluded that the speed at which Larsen B Ice Shelf collapsed was caused by a season of anomalously high surface meltwater production. Bothale et al. (2015) using snowmelt signatures from 2001 to 2014 using OSCAT and QuikSCAT scatterometer data calculated that the East Antarctic shelves experienced large melt during the study period.

It is important to note that in January 1993, many of us (members of the 12th Indian Scientific Expedition to Antarctica) could venture over the shelf ice but in January 2006, the same area was full of flowing river with serious undulations and now, when we see the same area through Google Earth (between 70.75S, 11.745E and 70.75S, 11.755E), it is full of ponds, lakes, rivers, cracks, and undulations, indicating weathering and warming processes actively disintegrating shelf ice.

Fig. 3.8 A small portion of the shelf ice adjoining the SO (marked as 1). The flow of water is marked as "2" and the undulations in the shelf ice are marked as "3". Shelf ice is the portion of ice that floats over the ocean and water on the surface of this ice, lowers its stability

3.10 Conclusion

SO offers an excellent site to study the impact of global warming as it is ideally located in between the polar ice and the shelf ice. Weathering processes are expanding ponds and lakes over SO and at the same time, SO itself is contributing toward warming the adjoining areas. There is a discussion that artificial snow deposit could save Antarctic ice melt, hills like Shivling offer an experimental facility to dump artificial snow in the pit and at least on the lower portion of the hill so as to gain the experience of combating global warming at the micro level. The energy for artificial snow formation must be generated through wind and the meltwater is to be converted into snow. If we compare Shivling with the SO, both are having almost the same processes in melting the ice around. It is important to note that global warming is an interdisciplinary subject and India must induct a long-term plan of measurements so that steps are taken to stop melting of most dangerous water resulting in a silent deadly sea level rise.

Acknowledgments I thank CSIR-NPL authorities for giving me an opportunity to be part of various Indian Scientific Expeditions to Antarctica and to lead the PBL program at Maîtri. I would like to thank Dr. Jaswant Singh and Dr. Sanjay Bist for sharing photographs from Maitri, Antarctica.

References

Alexander A et al (2019) A speed limit on ice shelf collapse through Hydrofracture. Geophys Res Lett. https://doi.org/10.1029/2019GL084397

Argentini S, Viola A, Sempreviva AM et al (2005) Summer boundary layer height at the plateau site dome C Antarctica. Bound-Layer Meteorol 115(3):409–422

Asthana R, Shrivastava PK, Srivastava HB et al (2019) Role of lithology, weathering and precipitation on water chemistry of lakes from Larsemann Hills and Schirmacher oasis of East Antarctica. Adv Polar Sci. https://doi.org/10.13679/j.advps.2019.1.00035

Baas P, van de Wiel BJH, van Meijgaard E et al (2019) Transitions in the wintertime near-surface temperature inversion at dome C Antarctica. Q J R Meteorol Soc 145(720):930–946

Balke J et al (1991) Contribution to the weathering-controlled removal of chemical elements from the active debris layer of the Schirmacher oasis, East Antarctica. Z Geol Wiss 19:153–158

Banwell AF, Willis IC, Macdonald GJ et al (2019) Direct measurements of ice-shelf flexure caused by surface meltwater ponding and drainage, Nat. Commun 10(1):730

Bockheim JG (1997) Properties and classification of cold desert soils from Antarctica. Soil Sci Soc Amer J 61(1):224–231

Bormann P, Fritzsche D (1995) The Schirmacher oasis, Queen Maud Land, East Antarctica, and its surroundings. Peterm Geogr mitt erg -h 289, Perthes, Gotha

Bothale RV, Rao PVN, Dutt CBS, Dadhwal VK, Maurya D (2015) Spatiotemporal dynamics of surface melting over Antarctica using OSCAT and QuikSCAT scatterometer data (2001–2014). Curr Sci 109(4):733–744

Bronselaer B, Winton M, Griffies SM et al (2018) Change in future climate due to Antarctic meltwater. Nature. https://doi.org/10.1038/s41586-018-0712-z

Chaturvedi A, Singh A, Gaur MP et al (1999) A confirmation of polar glacial recession by monitoring the snout of Dakshin Gangotri glacier in Schirmacher range. Sci report fifteenth Indian expedition to Antarctica, Dept Ocean dev., Govt. of India, New Delhi Tech Pub No 13, pp 321–336

Chen J, Blume HP (2002) Rock-weathering by lichens in Antarctic: patterns and mechanisms. J Geogr Sci 12:387–396

Chen J, Hans PB, Lothar B (2000) Weathering of rocks induced by lichen colonization – a review. Catena. https://doi.org/10.1016/S0341-8162(99)00085-5

Composite Gazetteer of Antarctica (n.d.). https://data.aad.gov.au/aadc/gaz/scar/display_name.cfm?gaz_id=112248

Connolley WM (1996) The Antarctic temperature inversion. Int J Climatol 16(12):1333–1342

Cuadros J (2017) Clay minerals interaction with microorganisms: a review. Clay Miner 52 (2):235–261. https://doi.org/10.1180/claymin.2017.052.2.05

Das I, Scambos TA, Koenig LS et al (2015) Extreme wind-ice interaction over recovery ice stream East Antarctica. Geophys Res Lett 42(19):8064–8071

Dasgupta S, Sengupta S, Bose S et al (2001) Polymetamorphism in the Schirmacher Hills granulites East Antarctica: implications for tectonothermal reworking of an isobarically cooled deep continental crust. Gondwana Res 4(3):337–357

Deshpande CG, Kamra AK (2004) Physical properties of aerosols at Maitri, Antarctica. J Earth Syst Sci 113:1–25. https://doi.org/10.1007/BF02701995

Dharwadkar A, Shukla SP, Verma A et al (2018) Geomorphic evolution of Schirmacher oasis, central Dronning Maud land. East Antarctica Polar Sci. https://doi.org/10.1016/j.polar.2018.06.006

Dold B, Gonzalez-Toril E, Aguilera A et al (2013) Acid rock drainage and rock weathering in Antarctica: important sources for iron cycling in the southern ocean. Environ Sci Technol. https://doi.org/10.1021/es305141b

Dutta HN, Sharma PK, Deb NC et al (2011) An insight into the ocean-ice-air interactions over the East Antarctic marine boundary layer: unique phenomena. In: Singh J, Dutta HN (eds) Antarctica: the Most interactive ice-Air-Ocean environment. NOAA Publisher, USA, pp 13–38

Gajananda K, Dutta HN (2005) Terrestrial vegetation community structure and biomass of the Schirmacher oasis ecosystem East Antarctica. Int J Ecol Dev 3(1):39–64

Gajananda K, Kaushik A, Dutta HN (2004) Thermal convection over East Antarctica: potential microorganism dispersal international. Aerobiologia 20(1):21–34

Gajananda K, Dutta HN, Lagun VE (2007) An episode of coastal advection fog over. East Antarctica Curr Sci 93(5):654–659

Gajananda K et al (2011) Land-ice-Air-Ocean interactions in the Schirmacher oasis, East Antarctica. In: Singh J, Dutta HN (eds) Antarctica: the Most interactive ice-Air-Ocean environment. NOAA Publisher, USA, pp 39–88

Ganju A, Gusain HS (2017) Six years observations and analysis of radiation parameters and surface energy fluxes on ice sheet near 'Maitri' Research Station, East Antarctica. Proc INSA. https://doi.org/10.16943/ptinsa/2017/48953

Gillies JA et al (2009) Ventifacts and wind-abraded rock features in the Taylor Valley, Antarctica. Geomorphology 107(3–4):149–160

Gillieson D, Burgess J, Spate A et al (1990) An atlas of the lakes of the Larsemann Hills, Princess Elizabeth land, Antarctica. ANARE res notes 74. Kingston, Tasmania, Australia: the pub off, Aust ant div. 174

Google Earth, https://earth.google.com/web/search/schirmacher+oasis, Imagery and data attributions CNES/ Airbus (accessed on June 20, 2020)

Gusain HS, Negi HS (2016) Estimation of surface energy fluxes of Antarctic snow/ice media near Schirmacher oasis and Larsemann Hills. Sci Rep Twenty Sixth Ind Ant Expe Tech Pub No 24:261–278

Hall K et al (2002) Weathering in cold regions some thoughts and perspectives. PPGEEC 26 (4):577–603

Hoch M et al (2001) Sr Nd Pb and O isotopes of minettes from Schirmacher oasis East Antarctica: a case of mantle metasomatism involving subducted continental material. J Petrol 42 (7):1387–1400

Jie C, Blume H (2002) Rock-weathering by lichens in Antarctic: patterns and mechanisms. J Geogr Sci 12:387–396. https://doi.org/10.1007/BF02844595

Kaur R, Dutta HN, Deb NC et al (2012) Role of coreless winter in favoring survival of microorganisms over the schirmacher oasis, East Antarctica. J Ecophysiol Occup Hlth 12:21–25

Kaur R, Dutta HN, Deb NC et al (2013) Investigation of unusual atmospheric warming over the Schirmacher oasis, East Antarctica. IJST 2:550–559

Kavan J, Waldhauserova PD, Renard JB et al (2018) Aerosol concentrations in relationship to local atmospheric conditions on James Ross island. Antarctica Front Earth Sci. https://doi.org/10.3389/feart.2018.00207

Kobayashi S (1978) Snow transport by katabatic winds in Mizuho camp area, East Antarctica. J Meteorol Soc Jpn 56(2):131–139

Lamp JL, Marchant DR, Mackay SL et al (2017) Thermal stress weathering and the spalling of Antarctic rocks. J Geophys Res Earth Surf. https://doi.org/10.1002/2016JF003992

Lenaerts JTM, Vizcaino M, Fyke J et al (2016) Present-day and future Antarctic ice sheet climate and surface mass balance in the community earth system model. Clim Dyn. https://doi.org/10.1007/s00382-015-2907-4

Matsuoka N (1995) Rock weathering processes and landform development in the sor Rondane Mountains, Antarctica. Geomorphology 12(4):323–339

Matsuoka N et al (1996) Field experiments on physical weathering and wind erosion in an Antarctic cold desert. ESPLDB 21(8):687–699

Musta B, Tahir SH (2012) Geochemical study of weathered gneiss rock from Schirmacher oasis Antarctica, scientific report of twenty fourth indian antarctic expedition. Tech Pub No 22:187–197

Naithani J, Dutta HN (1995) Acoustic sounder measurements of the planetary boundary layer at Maitri Antarctica. Bound-Layer Meteorol 76(1–2):199–207

Naithani J, Dutta HN, Pasricha PK et al (1995) Evaluation of turbulent fluxes over Maitri Antarctica. Bound-Layer Meteorol 74(1–2):195–208

Nayaka S, Upreti DK (2005) Schirmacher oasis, East Antarctica, a lichenologically interesting region. Curr Sci 89(7):1069–1071

Nezat CA et al (2001) Chemical weathering in streams of a polar desert (Taylor Valley, Antarctica). Geol Soc Amer Bull. https://doi.org/10.1130/0016-7606(2001)113<1401:CWISOA>2.0.CO;2

Nicholas L et al (2010) Sand transport by wind on complex surfaces: field studies in the McMurdo dry valleys, Antarctica. J Geophys Res Earth Surf. https://doi.org/10.1029/2009JF001408

Nylen TH et al (2004) Climatology of katabatic winds in the McMurdo dry valleys, southern Victoria land. Antarctica J Geo Res. https://doi.org/10.1029/2003JD003937

Palm SP, Yang Y, Spinhirne JD et al (2011) Satellite remote sensing of blowing snow properties over Antarctica. J Geophys Res. https://doi.org/10.1029/2011JD015828

Phillips HA (1998) Surface meltstreams on the Amery ice shelf East Antarctica. Ann Glaciol 27:177–181

Rémy F, Frezzotti M (2006) Antarctica ice sheet mass balance. CR Geoscience. https://doi.org/10.1016/j.crte.2006.05.009

Rignot E, Mouginot J, Scheuchl B et al (2019) Four decades of Antarctic ice sheet mass balance from 1979–2017. PNAS. https://doi.org/10.1073/pnas.1812883116

Sabacka M, Priscu JC, Basagic H et al (2012) Aeolian flux of biotic and abiotic material in Taylor Valley Antarctica. Geomorphology. https://doi.org/10.1016/j.geomorph.2011.12.009

Screen JA et al (2018) Polar climate change as manifest in atmospheric circulation. Curr Clim Change Rep 4:383–395. https://doi.org/10.1007/s40641-018-0111-4

Sedov S, Zazovskaya E, Dmitriy Fedorov-Davydov D et al (2019) Soils of East Antarctic oasis: interplay of organisms and mineral components at microscale. Bol Soc Geol Mex 71(1):43–63

Shepherd A et al (2018) Mass balance of the Antarctic ice sheet from 1992 to 2017. Nature. https://doi.org/10.1038/s41586-018-0179-y

Silvano A, Rintoul SR, Peña-Molino B et al (2018) Freshening by glacial meltwater enhances melting of ice shelves and reduces formation of Antarctic bottom water. Sci Adv. https://doi.org/10.1126/sciadv.aap9467

Singh J, Dutta HN (eds) (2011) Antarctica: the Most interactive ice-Air-Ocean environment. NOVA Science Publisher Inc, New York

Singh J et al (2018) Influence of climate change on Antarctic flora. Pol Sci. https://doi.org/10.1016/j.polar.2018.05.006

Srivastava A et al (2013) Mineralogical and geochemical studies of glacial sediments from Schirmacher oasis. East Antarctica Quatern Int. https://doi.org/10.1016/j.quaint.2012.07.028

Swain AK (2019) Influence of thermal conductivity of rocks on polar ice sheet recession near Schirmacher Oasis, East Antarctica. J Geol Soc India. https://doi.org/10.1007/s12594-019-1200-2

Turner J, Lachlan-Cope TA, Marshall GJ et al (2002) Spatial variability of Antarctic peninsula net surface mass balance. JGR: Atmospheres https://doi.org/10.1029/2001JD000755

Vašinka M, Krmíček L, Všianský D et al (2020) Chemical weathering in Antarctica: an example of igneous rock particles in big Lachman Lake sediments. James Ross Island Environ Earth Sci. https://doi.org/10.1007/s12665-020-08926-3

Wierzchos J, Ascaso C (1996) Morphological and chemical features of bioweathered granitic biotite induced by lichen activity. Clay Miner 44:652–657

Wilson MJ (1995) Interactions between lichens and rocks: a review. Cryptogram Bot 5:299–305

REDD+ in the Indian Context: Planning and Implementation Scenario

4

Prodyut Bhattacharya and Swapan Mehra

Abstract

The basic idea behind Reducing Emissions from Deforestation and Degradation (REDD) pertains to countries that are willing and able to reduce emissions from deforestation and should be financially compensated for doing so. Previous approaches to curb global deforestation have so far been unsuccessful. However, REDD provides a new framework to allow deforesting countries to break this historical trend. The "+" in REDD+ has drawn increasing attention towards the activities related to the conservation and enhancement of carbon stocks. A future REDD mechanism has the potential for implementing the Sustainable Forest Management (SFM) concept. REDD could simultaneously address climate change and rural poverty, while conserving biodiversity and sustaining vital ecosystem services. Green India Mission has various components to ensure support for REDD+ strategy.

In India, there are around 300 million forest dependent people including around 87 million tribals who derive their livelihood and a substantial part of their income from forests. Joint Forest Management (JFM), a partnership between local communities and the Forest Department started in India in the nineties has been a successful model under the "Care and Share" principle. India is world's 4th largest economy and 5th largest greenhouse gas (GHG) emitter, accounting for about 5% of global emissions. Its emissions increased to 65% between 1990 and 2005 and are projected to grow another 70% by 2020. On a per capita basis, India's emissions are 70% below the world average. Forests neutralize 11% of

P. Bhattacharya (✉)
University School of Environment Management, GGS Indraprastha University, Delhi, India

S. Mehra
IORA Ecological Solutions, Delhi, India
e-mail: swapan@ioraecological.com

A. Kaushik et al. (eds.), *Climate Resilience and Environmental Sustainability Approaches*, https://doi.org/10.1007/978-981-16-0902-2_4

India's GHG Emissions. India added around 3 million ha of forests during 1997–2007. There is significant scope for improving quality of forest cover by addressing drivers of degradation, as a significant part of the country's forest cover falls in the open to medium categories owing to various drivers of degradation. As estimated, REDD+ programme could provide for capture of around 1 billion tonnes of additional CO_2 over the next 3 decades and significant financial incentives as carbon services under REDD+ including flow of positive incentives to local communities. This chapter analyses the current status of REDD+ efforts of India and about how it could be an effective strategy and tool for mitigation and adaptation of climate change, improving ecological and environmental services, biodiversity conservation as well enhancing forest-based livelihood of forest-dependent communities.

Keywords

Sustainable forest management · Climate change · Biodiversity conservation · Community · REDD+

4.1 Concept of REDD+

Forest ecosystem provides multiple benefits to humankind. Forests also play an important role in the global carbon (C) cycle because they store 80% of the global above-ground carbon of the vegetation and about 40% of the soil Carbon and interact with atmospheric processes through the absorption and respiration of CO_2. Global forests cover is about 30% of the earth's surface and it is spread over about 4 billion ha of the landmass. Forests are considered a dominant sink of carbon after the ocean. The total carbon content of forests has been estimated at 638 Gt in 2005 by FAO, which is more than the amount of carbon in the entire atmosphere. Emissions from deforestation and forest degradation in developing countries constitute about 17.4% of global greenhouse gas emissions or approximately 5.8 Gt of carbon dioxide equivalent (CO_2eq), per year (IPCC 2007). This has led to aggravation of the impacts of climate change across the globe. The small and marginal food cultivators who establish a major chunk of the farming community are worst affected and vulnerable due to this changing climate regime.

The developmental path of India and other agrarian countries will become more difficult if this trend continues a little longer. There is a global understanding on the sustainable use and management of forest resources, which has led to the framing of the idea of Reducing Emissions from Deforestation and Degradation (REDD+), wherein the "+" in the REDD has drawn increasing attention towards the activities related to conservation and enhancement of carbon stocks. Checking deforestation and forest degradation has been the agenda of the United Nations Framework Convention on Climate Change (UNFCCC) for quite long. UN-REDD was launched in 2008 in 29 countries with $51.4 million. REDD is an effort to create a financial value for the carbon stored in forests, offering incentives for developing countries to

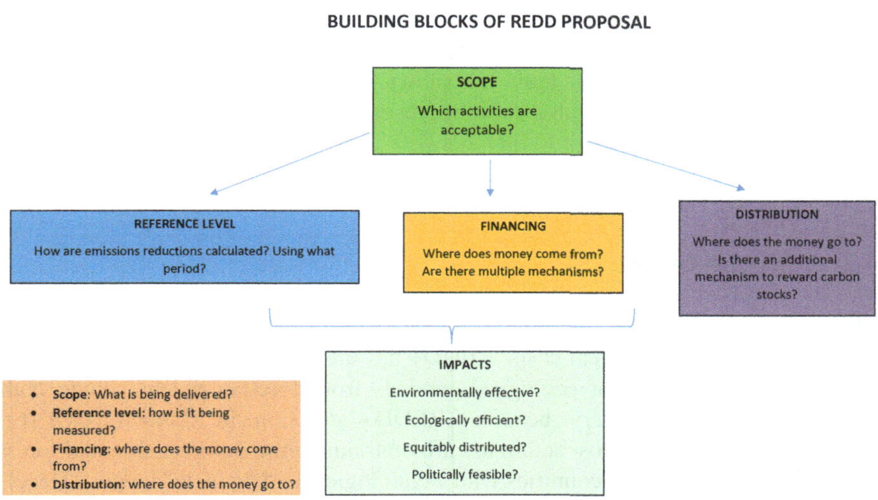

Fig. 4.1 Understanding REDD proposal

	LOW FOREST COVER (< 50%)	HIGH FOREST COVER (> 50%)
HIGH DEFORESTATION RATE (> 0.22%/yr)	**Quadrant I** e.g. Guatemala, Thailand, Madagascar No. of Countries: 44 Forest area: 28% Forest carbon total: 22% Deforestation annual 48%	**Quadrant III** e.g. Papua New Guinea, Brazil, Congo (DR) No. of Countries: 10 Forest area: 39% Forest carbon total: 48% Deforestation annual 47%
LOW DEFORESTATION RATE (< 0.22%/yr)	**Quadrant II** Dominican Republic, Angola, Vietnam No. of Countries:15 Forest area: 20% Forest carbon total: 12% Deforestation annual 1%	**Quadrant IV** e.g. Suriname, Belize, Gabon, No. of Countries: 11 Forest area: 13% Forest carbon total: 18% Deforestation annual 3%

Fig. 4.2 A matrix to split countries by their forest cover and historical rate of deforestation

reduce emissions from forested lands and invest in low-carbon paths to sustainable development. The REDD+ mechanism comprises scope, reference level, financing, and distribution as the building blocks as illustrated in Fig. 4.1.

The forest cover of a country and the rate of deforestation have crucial implications in overall reduction of emissions from deforestation and forest degradation (Fig. 4.2). A larger number of countries lie in quadrant I with low forest cover and high deforestation rate, while much lesser countries show high forest cover and low deforestation rate. The countries have to take up appropriate stand to adopt REDD+. The Government of India has shown its readiness to implement REDD+ mechanism in the country with a set of representations that have been submitted to UNFCCC execution approach (TERI 2013).

4.2 Scope of REDD+

REDD provides a unique opportunity to achieve large-scale emission reduction at comparatively low abatement costs. REDD+ has the additional benefits, along with the production function of ecological services from forests, which include both monetary and nonmonetary benefits. REDD+ decision in Cancun (COP16), Governments agreed to boost action to curb emissions from deforestation and forest degradation in developing countries with technological and financial support. Developing country Parties, following their respective capabilities and national circumstances are encouraged to contribute to mitigation actions in the forest sector by undertaking various activities (Fig. 4.3), as deemed appropriate by each Party

I. Reducing emissions from deforestation.
II. Reducing emissions from forest degradation.
III. Conservation of forest carbon stocks.
IV. Sustainable management of forest.
V. Enhancement of forest carbon stocks.

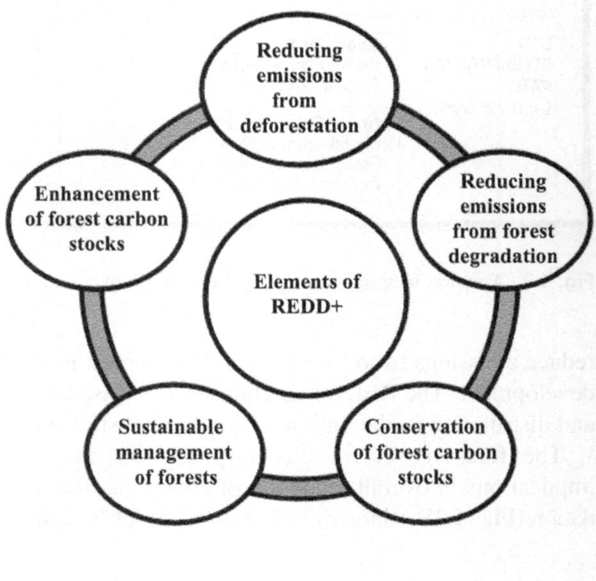

Fig. 4.3 Crucial elements of REDD+ (Source: COP 16)

4.3 India's Stand on REDD+

India is a massive country with rich biological diversity. Forest is the second largest land use in India after agriculture. About, 275 million rural people in India depend on forests for at least part of their sustenance and livelihood (World Bank 2006). As per the India State of Forest Report (FSI 2017), the forest cover of the country stood at 708,273 km^2, while it was 701,495 km^2 in 2015 updated assessment (FSI 2017), recording an increase of 6778 km^2 within two years. The National Forest Policy of India envisages a goal of 33% of its geographical area under forest and tree cover. The total forest and tree cover of the country is 24.4% of its geographical area. The issues related to forest and environment management have been given adequate importance in the overall policy and planning for balanced development of the country. The basic approach is development without destruction of natural resources, which is also well-fitted in the REDD + concept (MoEF&CC 2014).

India is the world's 4th largest economy and 5th largest GHG emitter, accounting for about 5% of global emissions (Busby and Shidore 2017). Its emissions increased to 65% between 1990 and 2005 and are projected to grow another 70% by 2020. On a per capita basis, India's emissions are 70% below the world average. Forests neutralize 11% of India's GHG Emissions (Busby and Shidore 2017). India added around 3 MHA of forests in the decade of 1997–2007.

India is one of the few tropical countries where forest cover has stabilized over the period from 1947 to 2007; and between 1997 and 2007, it further increased by 3.13 m ha (FSI 2009). Forest resources support livelihoods of approximately 200 million people in India. In addition to the livelihood dependence of the local communities, prevalence of widespread poverty, shrinking common property resource base, and increasing human and livestock population are factors responsible for the unsustainable harvest of forest produce and major causes of forest degradation in the country (WB 2000). Since India's forest policy gives priority to the rights of local communities on forest resources for meeting their subsistence needs, it is, therefore, a challenge to enhance forest carbon stock without undermining the dependence of the local forest-dependent communities.

There is significant scope for improving quality of forest cover by adopting drivers of degradation, since a significant part of the country's forest cover falls in the open to medium categories owing to various drivers of degradation. As estimated, REDD+ programme could provide for capture of around 1 billion tonnes of additional CO_2 over the next 3 decades and significant financial incentives as carbon services under REDD+ including the flow of positive incentives to local communities (Sarkar 2011). REDD+ can be a part of an effective strategy and tool for mitigation and adaptation of climate change, improving ecological and environmental services, biodiversity conservation as well as enhancing the forest-based livelihood of forest-dependent communities.

The objectives of the National REDD+ Policy of India is given below:

Objectives of the National REDD+ Policy

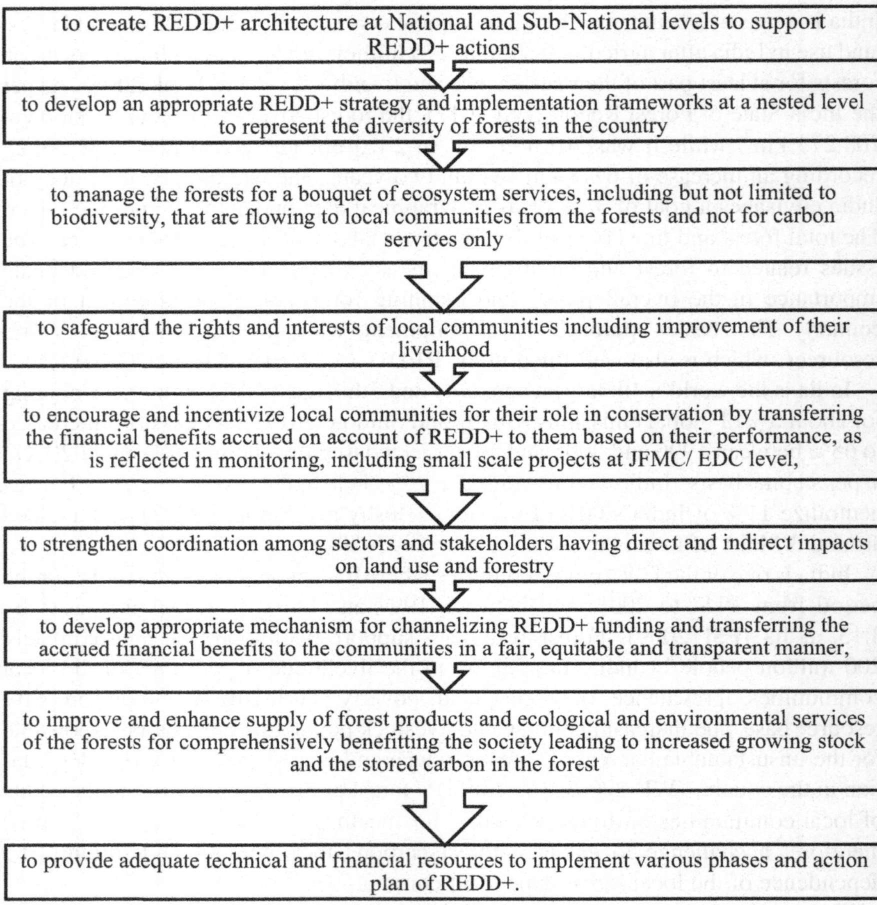

There are around 300 million forest-dependent people including 87 million indigenous communities who are deriving their livelihood and major income from forests. India is one of the few countries in the world, which has shown the successful implementation of Joint Forest Management (JFM), a partnership between local communities and the forest department, a successful model under "Care and Share" principle. There are around 1,20,000 JFM committees involving around 20 million people managing approximately 20 million hectares of forest primarily to address the control of degradation and deforestation (Bhattacharya et al. 2010).

4.4 India's Preparedness for REDD+

At the national level, India has a strong policy and legal framework in the form of the Wildlife Protection Act 1972; Forest Conservation Act 1980; National Forest Policy 1988; Biodiversity Act 2002; National Environmental Policy 2006; and Forest Rights Act 2006 for the sustainable management of its forests. India's National Forest Policy formulated 4 years before the Earth Summit (UNCED 1992), embodies all elements—social, environmental, and economic—of Sustainable Forest Management. The principal aim of India's National Forest Policy is to ensure environmental stability and maintenance of ecological balance, including atmospheric equilibrium, which is vital for the sustenance of all life forms—human, animal, and plant. The derivation of direct economic benefit is subordinate to this principal aim. India's National Forest Policy is in alignment with the four global objectives agreed in 2006 by all member countries of the United Nations under the umbrella of the United Nations Forum on Forests (UNFF).

These global objectives aim to:

1. Reverse the loss of forest cover worldwide through sustainable forest management, including protection, restoration, afforestation, and reforestation; and increase efforts to prevent forest degradation.
2. Enhance forest-based economic, social, and environmental benefits, including improving the livelihoods of forest-dependent people.
3. Increase significantly the area of protected forests worldwide and other areas of sustainably managed forests, as well as the proportion of forest products from sustainably managed forests.
4. Reverse the decline in official development assistance for sustainable forest management and mobilize significantly increased, new, and additional financial resources from all sources for the implementation of sustainable forest management.

A range of policies and programmes have been initiated at the national level to address the impacts of climate change in the context of sustainable development. The National Mission for a Green India (GIM) is one of eight Missions prepared under the aegis of the National Action Plan on Climate Change (NAPCC) (MoEF&CC 2014). Its objective is to increase forest cover by over five million ha and improve the quality of forests over another five million ha during the next 10 years; enhance annual CO_2 sequestration by 50–60 million tonnes by the year 2020, and increase forest centred livelihood income of about three million households living in and around the forest. The National Afforestation Programme (NAP) is another such effort, which aims to restore degraded forests with the active participation of local communities. For further strengthening, this programme has been reviewed by the Government of India based on feedback from implementing States and other stakeholders (GoI 2011). The MoEF&CC is executing a centrally sponsored scheme, namely "Intensification of Forest Management Scheme," which aims at the creation of infrastructure for the development, protection, and conservation of forest resources in the country. For India, carbon services from forests and plantations are one of the co-benefits and not the primary objective of forest management. Therefore, as India is poised to pilot REDD+ projects in the country, it is important to re-evaluate some of the issues underpinning the design of an appropriate national architecture. In the past three years, MoEF&CC has carried out 12 regional capacity building programmes in Northern, Southern, Central and North-Eastern zone of India regarding the REDD+ strategy and their implementation framework (MoEF Undated). The JICA sponsored forestry projects in several states has incorporated the REDD+ model in the projects for the field level implementation.

4.5 Implementation Framework

Restating India's commitment to Paris Agreement on Climate Change, the then Union Minister for Environment, Forest and Climate Change, Dr. Harsh Vardhan has emphasized that the cooperation and involvement of the tribal, other forest-dwelling people and the society as a whole, is crucial for the implementation of the REDD+ strategy. The Minister has further emphasized that REDD+ activities will help in the sustainable livelihood of local communities and in the conservation of biodiversity. India has communicated in its Nationally Determined Contribution (NDC) under the Paris Agreement, that it will capture 2.5–3 billion tonnes of Carbon dioxide through additional forest and tree cover by 2030. India's first biennial update report to UNFCCC has revealed that forests in India capture about 12% of India's total GHG emissions. Thus, the forestry sector in India is making a positive cost-effective contribution to climate change mitigation.

Complying with the UNFCCC decisions on REDD+, India has prepared its National REDD+ Strategy in accordance with the global trend. The Strategy builds upon existing national circumstances that have been updated in line with India's National Action Plan on Climate Change, Green India Mission, and India's Nationally Determined Contribution (NDC) to UNFCCC.

- *Conformity with UNFCCC Decisions*: National REDD+ Policy will follow the necessities laid down by all the relevant UNFCCC decisions. Most prominent among these decisions is the Paris Agreement, the Warsaw Framework for REDD +, and Cancun Agreements. These three decisions form the main source of the implementation guidance for REDD+ in developing countries. Cancun Agreements have a special place amongst all UNFCCC decisions as this provides the scope of REDD+ implementation by developing countries.
- *Conformity with National Policy and Legal Framework*: National REDD+ Policy will cover all types of forest areas including TOF (Trees outside Forest) irrespective of ownership or control, but meeting the requirement of the definition of forest as adopted by the Forest Survey of India. The strategy aims at optimization of all forest ecosystem services, including the sequestration of carbon thereby adding to the existing forest carbon stocks, and reducing pressure on forests by addressing drivers of deforestation and forest degradation. Performance-based

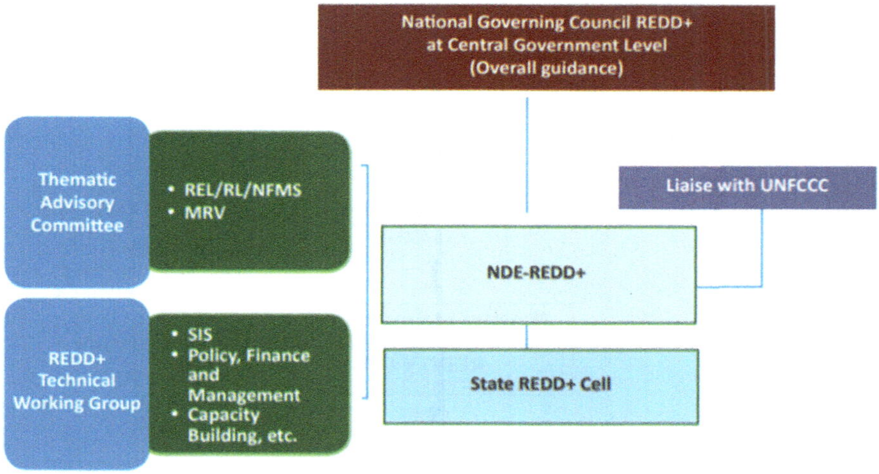

Fig. 4.4 Institutional mechanism for REDD+ implementation in India

incentives will be paid to stakeholders particularly local communities in tune with their contribution to enhancing carbon stocks. All actions according to the NRPS will be implemented with due observance of all relevant national laws, policies, rules, and regulations. This legal and policy framework provides guidance for implementation of REDD+ and also constitutes the boundary of applicable environmental, social, and economic standards which all the actions, their consequences and implications should conform to (Fig. 4.4).

- *Institutional Mechanism: Roles and Responsibilities of Stakeholders*: The broad institutional framework for implementing REDD+ is already in place in India. However, for REDD+ to be functional, a system capable of forest carbon stocks accounting, Measuring, Reporting, and Verification (MRV) system, and enforcing social and for environmental safeguards, will be required. The institutional arrangement for the implementation of REDD+ in India is shown in Fig. 4.5.

4.6 Apportioning Targets

To facilitate the implementation of REDD+, and to channelize it towards achieving the NDC target of achieving the additional CO_2 sink of 2.5–3.0 billion t CO_2 equivalent by 2030, the Government of India in the MoEF&CC may in future consider undertaking an exercise in consultation with the State/UT Governments to work out the appropriate targets of afforestation and reforestation (A&R) for each State/UT that would enable the country meet the objectives of Green India Mission

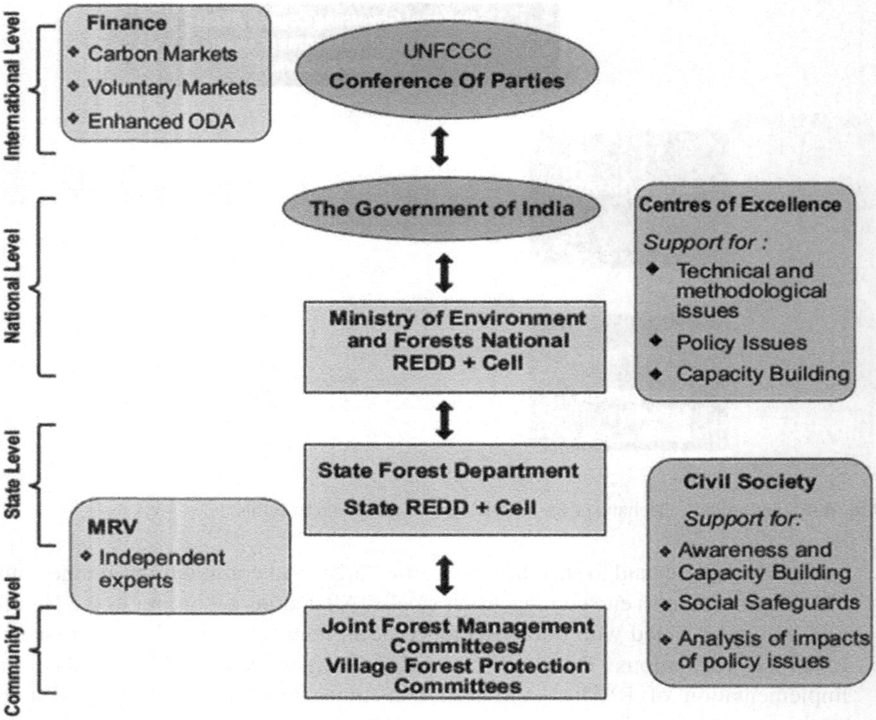

Fig. 4.5 Institutional Framework for REDD+ in India

and NDC target, CAMPA fund may be utilized for the same by the government (National REDD+ Strategy India 2018). The State targets will also include the attainments of tree plantation schemes and projects being implemented by other Ministries and Departments of Central Government (excluding NHAI), like Ministry of Rural Development, Ministry of Agriculture and Farmers' Welfare, Ministry of Panchayati Raj etc.

4.7 Infrastructure Required

Supplementary infrastructure in the shape of trained human capital and state of the art equipment for carrying out forest and carbon measurements at all levels is required. FSI (Forest Survey of India) is charged with the overall responsibility of measuring forest goods and services including carbon. In its pivotal role of measuring, FSI will be reinforced by SFDs and local communities. Building capacity of the forest staff at all levels and that of the local community will not suffice, and it will need to be complemented by creating modern measuring capability with the latest remote sensing and GIS equipment's in each State. The existing space application centres and GIS facilities in the States will be strengthened and upgraded to enable these to carry out measurements of forest carbon stocks of the State under the guidance of the FSI.

4.8 Addressing Gender Equity

Gender equity is integral in the relevant laws, policies, rules, and regulations dealing with joint management of forests, and local governance organizations like Panchayats and Gram Sabhas influencing the management and health of forests. Reservation for women in JFMCs and EDCs functioning to the extent of 50% in Panchayati Raj Institutions (PRIs) are good indicators of the government's intention to ensure meaningful participation of women in the local decision-making mechanisms in all local matters including management and development of forests. REDD+ implementation intrinsically being the core of the existing and future management and developmental initiatives and activities in the forest, will ensure adherence to gender centred sensitivity and transparency in forest governance.

4.9 State Governments

The strategy devolves major responsibility for the execution of REDD+ activities and measurement of their performance on the State Forest Departments. It places a high priority on capacity building of all levels of the SFDs, the line departments (such as Rural Development, panchayat, horticulture, tribal development agriculture etc.) and the local communities to enable the proper implementation of REDD+ and accurate assessment and measuring of REDD+ performance. Each State will create a REDD+ Cell in the State Forest Department and appoint a Nodal Officer to coordinate the activities of the State REDD+ Cell. Strategies prescribing role and mandate of REDD+ Cell, and flow of incentives will be issued by the Government of India. States will be encouraged to develop their State Action Plan for REDD+. The proposed composition and terms of reference of the State REDD+ Cell is given in Table 4.1.

Table 4.1 Proposed constitution and terms of reference of the State REDD+ Cell

Constitution of State REDD+ Cell: Composition of the State REDD+ Cell shall be as follows:		
1.	Principal Chief Conservator of Forests & HoFF	Chair
2.	Principal Chief Conservator of Forests (Planning/Budget)	Member
3.	PCCF/APCCF (nominated by Chairman)	Member
4.	APCCF/CCF (Monitoring)	Member
5.	Regional APCCF, MoEFCC or his representative	Member
6.	Two REDD+ Experts (Nominated by Chairman)	Member
7.	Representative of prominent NGO	Member
8.	APCCF/CCF/CF (In-charge of Afforestation)	Nodal Officer

(Source: National REDD+ Strategy India 2018)

4.10 How Community-Based Monitoring May Contribute to a National MRV System for REDD+?

Community Monitoring and National MRV Development have mutually supportive benefits (Table 4.2.) In India under JFM institution, the National REDD+ MRV can benefit from community monitoring variously, particularly in terms of obtaining data on local level stock changes and impacts of REDD+ activities. This data will be valuable in carbon accounting, especially for reduced degradation and forest enhancement; it may sharpen the estimates of rates of change and increase the level of confidence in these estimates, and provide excellent information on the non-carbon impacts of REDD+.

For community monitoring to function well as an integral element within the national MRV system, however, there is a need for governments at the national level to formally define the role of community forest monitoring within the REDD+ MRV

Table 4.2 Matrix of proposed community-based monitoring dimensions within national MRV system

<table>
<tr><th colspan="2">Forest change activity</th><th>Monitoring options at national level</th><th>Potential contribution of community based monitoring</th></tr>
<tr><td colspan="2">Reforestation</td><td>• Remote sensing
• National forest inventory
• Monitoring through forestry companies</td><td rowspan="3">• Acquiring/signaling the location, time, area and type of change events (in near real time)
• Ground level measurements for local implementation (i.e. of reforestation plots)
• Independent local reference for national/other data sources</td></tr>
<tr><td colspan="2">Deforestation</td><td>• Remote sensing
• National forest inventory</td></tr>
<tr><td rowspan="3">Forest degradation</td><td>Commercial activities, incl. selctive logging</td><td>• National forest inventory
• Commercial companies (i.e. harvest estimates)
• Remote sensing</td></tr>
<tr><td>Wild fire</td><td>• Remote sensing, national forest inventory</td><td>• Acquiring/signaling date, area and type of change event (near real time)</td></tr>
<tr><td>Subsistence forest use incl. fuel wood, charcoal, community forest management etc.</td><td rowspan="2">• Very limited historical data
• Rough estimates using gain-loss methodology
• National forest inventory data not sufficiently detailed</td><td rowspan="2">• Regular ground level measurements and reporting of forests and carbon stocks
• Tracking growth/decrease of local activities (drivers)</td></tr>
<tr><td>Forest enhancement</td><td>Increases in biomass due to REDD+ activities at local level</td></tr>
</table>

system and to develop the necessary institutions, training, and protocols to set it in motion. Before communities commit themselves, likely, they will also need information on future benefit-sharing mechanisms under REDD+, so that they can make rational choices as to whether or not to participate. All protocols for monitoring should be based on IPCC Good Practice Guidance and be internationally acceptable, but need to be presented in a way that can be easily understood at the local level. The protocols should include instructions for mapping, sampling, measurement, data storage, and reporting.

The reporting structure should be well defined, with clear communication channels so that the data can be entered into national databases, and there should be a clear code of ethics regarding who can access the data for what purposes, since it may in some cases be confidential (for example, information on land tenure and rights).

4.11 Finance/Funding

The arrangement of funds is crucial to the timely implementation of REDD+ strategy adopted by India. Finance will be organized internally by an allocation through GIM, CAMPA, Namami Gange Programme, Green Highways Policy etc. The deficit in finance to meet the NDC commitment on REDD+ implementation will be sourced through external funding from GCF under the UNFCCC mechanism. The budget deficit on this count will be estimated separately and communicated to UNFCCC and GCF in due course of time. In the ongoing Indian context, the financing avenue could be explored in the following way:

(a) *Finance Commission*: Fourteenth Finance Commission has recommended for devolution of funds from the federal pool to the States attaching a weightage of 7.5% of the State's forest cover. A part of this devolution is expected to be ploughed into the forestry sector.
(b) *Compensatory Afforestation Fund Act, 2016*: This Act has been enacted by the government to systematically channelize the funds accrued from a levy on user agencies on account of diversion of forest land under the Forest (Conservation) Act, 1980. The Compensatory Afforestation Fund Management and Planning Authority (CAMPA) funds are currently utilized by Indian States for raising plantations, undertaking natural regeneration, protection of forests, Green India Programme, wildlife protection, and other related activities. CAMPA funds at present account for an amount of around Rs. 500 billion which mainly will be used by the State Forest Departments for undertaking aforesaid activities.
(c) *Green India Mission and National Afforestation Programme*: The GIM is one of the eight Missions under the National Action Plan on Climate Change (NAPCC) and was launched in 2014. It aims to enable the forestry sector to play an important role in the protection of the environment while enhancing the country's green cover. India needs to revisit the Green India Mission objectives and timeframe in the light of new developments under global climate change

regime, especially India's NDCs to UNFCCC. The actual implementation period of the Mission was to spread over 10 years, coinciding with the 12th and 13th Five Year Plan periods. The Mission has a preparatory phase (2011–2012), a first phase (five years) and a second phase (five years). GIM activities need to be revisited and carry forwarded till 2030 to coincide with achieving NDC forestry targets.

(d) *Green Climate Fund*: Green Climate Fund (GCF) is the funding window of the UNFCCC established in 2010 to limit or reduce the GHG emissions in developing countries, and to help adapt vulnerable societies to adverse impacts of climate change. The developed economies have agreed to mobilize USD 100 billion per year by 2020. A part of this collection will be routed through GCF. The Board of GCF in its eighteenth meeting adopted the terms of reference for the REDD+ Results-Based Payments pilot programme, as well as a scorecard to evaluate countries' submissions to the GCF. GEF also supports REDD+ readiness programmes for phase I (Preparation of National Strategy or Action Plans) and phase II (implementation of national Strategy, Capacity building, and demonstration activities) of REDD+. MoEF&CC will work out the amount of gap funding required for full implementation of REDD+, and communicate the same to UNFCCC for funding support by the GCF.

(e) *Mobilizing Finance for Adaptation*: The Green Climate Fund (GCF) is investing in developing countries' efforts to adapt to the effects of climate change. GCF aims to deliver a 50:50 balance between mitigation and adaptation allocations in its portfolio. Forestry activities/actions supporting enhanced livelihoods of the most vulnerable people, communities, and regions by effectively engaging local communities in the forest sector adaptation programmes, shall be developed. Skill development of community youths for various forest sectors adaptation activities like Assisted Natural Regeneration (ANR), soil and moisture conservation, fire protection, weed management, management of forest insects and pests, agroforestry, tree fodder production, NTFP management, bioenergy production, and biodiversity and ecotourism management activities shall be provided. This will ensure approximately 1 million jobs to the communities and in turn, help forest ecosystems to adapt to climate change. Local communities will also be trained to make them capable of assisting the SFDs in carrying out forest-related measurements. Such actions will also complement various REDD+ activities that have been taken in Nepal and Indonesia.

(f) *Other External Sources of Funding*: COP Decision 9/CP.19 reaffirms that the progression of developing country Parties towards results-based actions occurs in the context of the provision of adequate and predictable support for all phases of REDD+ implementation. The result-based finance provided to developing country Parties that is new and additional and predictable may come from a wide variety of sources referred to in decision 2/CP.17, paragraph 65 (public and private, bilateral, and multilateral, including alternative sources), and decision 9/CP.19 (including Green Climate Fund in a key role) to collectively channel adequate and predictable results-based finance in a fair and balanced manner. Finance for REDD+ shall be mobilized per the UNFCCC decisions on REDD+

finance. State REDD+ Cells in collaboration with NDE-REDD+ shall explore the relevant and appropriate financing entity for the support of REDD+ actions following the laid down principles of the financing entities.

4.12 Some Case Studies

There are several success stories of REDD implementation, out of which Nepal and Indonesia have been the prominent Asian countries to implement the REDD+ programmes, to counter-attack deforestation and forest degradation.

Nepal

In 2010, Nepal had 4.80 Mha of natural forest, extending over 33% of its land area. In 2019, it lost 2.08 kha of natural forest, equivalent to 913 kt of CO_2 of emissions (GFW, 2020). In Nepal from 2001 to 2018, 12% of tree cover loss occurred in areas where the dominant drivers of loss such as wildfire, shifting agriculture, urbanization, and commodity-driven deforestation have resulted in deforestation (GFW, 2020). Nepal has participated in the REDD+ processes since 2008, its readiness activities have been financially and technically assisted by World Bank's Forest Carbon Partnership Facility, UN and some bilateral and multilateral partners. The REDD implementation centre under MoEF is the lead organization to carry out the readiness activities, implement emission reduction programmes, and demonstrate results related to REDD+ in Nepal.

In Nepal, the performance-based Emissions Reduction (ER) Program covered 12 contiguous districts of Nepal's Terai Arc Landscape (TAL) with the potential to recover up to US$45 million instead of 9.16 million tons of CO_2 sequestered over six years ending 2024. This represents a benchmark achievement for Nepal in its efforts to support local communities and end forest loss in the TAL. The ER Program is a model for the implementation of performance-based activities to address drivers of deforestation and degradation in Nepal to enable full implementation of REDD+ in alignment with the Paris Agreement, which was an appropriate strategy as they had experience in community forestry for the last 25 years.

The vision of the Nepal National REDD+ strategy is the enhanced Carbon and non-carbon benefits of forest ecosystem contributing to the prosperity of the people of Nepal.

The approval of the ERPD, which aims to reduce emissions through enhanced sustainable management of forests and community-based forest management, is a milestone for Nepal's REDD+ Program. The ER Program, under the leadership of the Ministry of Forests and Environment, has covered community-based forest management regimes (community forestry and collaborative forestry), and enhance the benefits of localized forest management with increased knowledge and application of Sustainable Forest Management principles. Nepal has experienced substantial deforestation and forest degradation resulting in the conversion of Himalayan forest land into shrub and agricultural land. While obvious factors such as overharvesting, encroachment, fire, and invasive species have exacerbated deforestation, it is also affected by deep-rooted issues such as poverty, inequality, the struggle for control over resources, and overdependency and degradation of resources. Deforestation is embedded in a complex sociopolitical dynamic involving multiple actors, relations of power between them, and a range of social-ecological factors. Proper documentation, regular monitoring, and in-depth analysis of the drivers and dynamics of deforestation are lacking. Two major forest management regimes in Terai that are important in the context of REDD+, CF (Community Forestry), and government managed forests, have experienced different governance challenges.

Nepal is preparing for REDD+ with support from FCPF (Forest Carbon Partnership Facility) and several other REDD+ policies, institutional, and methodological initiatives. These initiatives have largely adopted a multi-stakeholder approach, which was regularly active during the early phase of the programme but appears to be meeting with decreased enthusiasm recently, in the government.

CSOs (Civil Society Organisations) are concerned over the adoption of a rather technical and peripheral process in REDD+ and argue that the core issues of tenure, governance, and institutional reform have been overlooked. Overall, prospects remain good for REDD+ implementation in Nepal, yet immense challenges exist. A robust policy, legal, and institutional foundation for Community Forestry and well-functioning community institutions are required for sustainable forest management in Nepal. The performance of CF and other CBFM (Community-Based Forest Management) modalities give grounds for optimism that they can deliver carbon sequestration and other co-benefits (Paudel et al., 2013).

Indonesia

Indonesia is one of the most forested countries in the Asian region, spanning 90 million hectares, Indonesian forests constitute 10 percent of the world's remaining tropical forests and provide people with a variety of benefits or ecosystem services. Since 1990, approximately 25 percent of the country's forests about 28 million hectares have been cleared (WRI 2008). According to the Food and Agriculture Organization (FAO), the country has the second highest amount of annual forest loss of any nation, losing approximately 1.9 million hectares per year.

The Indonesian government released its official deforestation numbers in May 2018, showing that the rate of forest loss has been declining from 2015 to 2018. The data reported deforestation of 440,000 hectares in 2018, slightly lower than the 2017 number of 480,000 hectares (WRI 2019). Global Forest Watch released similar numbers showing a 40 percent decrease in deforestation in Indonesia's primary forests in 2018, compared to the average annual rate of loss from 2002 to 2016.

The Government of Indonesia is committed at national and international levels to addressing the challenges of climate change and to using forest carbon offsets to consolidate its forestry sector reforms. Indonesia has pledged to reduce its emissions from land use, land-use change, and forestry (LULUCF) by at least 26% by 2020. One way the country plans to meet this target is by reducing its emissions from deforestation and forest degradation, through the REDD+ mechanism.

By implementing REDD+, Indonesia will become eligible to receive financial payments based on forest carbon credits. REDD+ offers the potential for innovative and stable forms of financing for local governments and community-based resource management. REDD+ could also assist in the distribution of benefits for community development aligned with national and local government policies. In some parts of Indonesia, particularly in Java, the forest cover has been shown to increase by up to 4% in the past decade. Despite a slightly declining trend, however, deforestation rates in most parts of Indonesia remain alarmingly high. Deforestation is intricately linked to the structure of the Indonesian economy, which remains reliant on revenues from natural resources. This dependence has shaped the country's political economy and institutional landscape and, naturally, affects the forestry sector and, in turn, REDD+. Indonesia is improving several regulations related to REDD+. The current development of REDD+ is highly influenced by the content of some regulations conflicting with its REDD+. Existing policies do not fully support emissions reductions. Indonesia is still in the process of development. The government needs to reconsider policies related to the plantation and mining sectors that have an impact on deforestation and forest degradation, particularly involvement of private sector for long-term forestry lease.

Besides, policy implementation is also largely determined by the existing institutions at the central and regional levels. Several institutional arrangements and regulations related to REDD+ architecture such as national strategy, REL, MRV, safeguards and financial systems, and incentive distribution mechanisms have been developed that reflect the readiness of REDD+ implementation. However, there are some challenges related to this, such as unclear REDD+ institutional systems at the local level, financial issues, and the regulations on market and carbon incentive distribution that can be obstacles in achieving the effectiveness of REDD+ (Ekawati et al. 2019).

4.13 Key Issues and Challenges

(a) *An integrated approach and interdepartmental coordination*: Critical interlinkages between forest resources, rural livelihoods, as well as competing land-use demands, proliferate deforestation, and forest degradation (Davidar et al. 2010). The forestry sector alone cannot bear the burden of supporting livelihoods of a large section of the population living in and around forest areas. Linkages with other sectors, especially agriculture, health, rural development, and energy programmes need to be studied in-depth, to develop a synergistic approach to address the challenge, by dove-tailing various poverty alleviation programmes, prioritizing the development of forest-based enterprises, and increasing agricultural productivity in lands adjoining forests for yielding food, fodder, and fuelwood. The forestry sector can play a pivotal role in this due to its large-scale presence in interior areas, through JFM Committees. Ministries such as the Ministry of Rural Development, Ministry of New and Renewable Energy among others must come forward to provide alternate livelihood opportunities, as there is a great scope of convergence of developmental programmes of Rural Development, Tribal and Social Welfare, Health and Education, and Power Departments, in areas within the forests and fringe forest areas.

(b) *Funds for Sustainable Forest Management (SFM)*: The Rio Principles adopted at the 1992 United Nations Conference on Environment and Development underscore that States have common but differentiated responsibilities regarding collective global interests and concerns related to forests, and the sovereign right to utilize their resources to meet their national policy objectives. At the same time, it was also emphasized that international cooperation should focus on building human and institutional capacity in developing countries to conserve and manage their forests. The recently concluded United Nations Conference on Sustainable Development at Rio in 2012 recognizes the need for strengthening cooperation arrangements for REDD+ in the areas of finance, trade, transfer of environmentally sound technologies, capacity building and governance, and benefit sharing, among other key areas, which will be under national legislation and priorities. The Government of India may establish linkage with the Forest Carbon Partnership Facility (FCPF) of the World Bank and UN-REDD programmes to access adequate financial resources necessary to build the capacity of communities and forest officials for the implementation of REDD +. The Government of India may also explore the possibility of attracting funds from voluntary markets and carbon trading under REDD+. The money generated through additional carbon added or carbon saved will supplement incentives already derived by JFMCs through harvesting of Non-Timber Forest Products (NTFPs) and would help support livelihoods of communities and contribute to overall socioeconomic development. Incentives received from REDD+ would be passed to local communities involved in the conservation and management of the forests to ensure sustained protection of India's forests.

(c) *The clarity in methodology*: The progress on the REDD+ negotiations is hampered due to ambiguity on key definitions like forest degradation, sustainable management of forests, deforestation, and sustainable harvest. The concept of Sustainable Forest Management (SFM) was first articulated in the Forest Principles at the Earth Summit in Rio in 1992. As the concept of SFM evolved, the 14 members of the Collaborative Partnership on Forests (consisting of international organizations and secretariats) recognized that "SFM provides an effective framework for forest-based climate change mitigation and adaptation". It was emphasized that without a comprehensive approach to forests (recognizing that countries manage their forest for multiple socioeconomic, productive, and environmental functions) and without sound policy, legislative, and governance frameworks, forestry-related climate change mitigation and adaptation efforts on the ground will not be successful. The Bali Action Plan (COP-13) introduces the term Sustainable Management of Forests (SMF) in the context of REDD+. It defines REDD+ as: "*Policy approaches and positive incentives on issues relating to reducing emissions from deforestation and forest degradation in developing countries; and the role of conservation, sustainable management of forests, and enhancement of forest carbon stocks in developing countries*". Therefore, to take the process forward there is a need to bring clarity on the usage of the terms, SMF, SFM, and forest degradation. Also, while REDD+ emphasizes compensation of carbon sequestration services, more clarity is required on the compensation mechanism of other ecosystem services, particularly intangible services.

(d) *COVID-19 Pandemic situation*: Many climate stories these days give various takes on how to view the issue through the lens of the coronavirus and vice versa. When lawmakers are convinced a crisis is a real and a mortal threat, most of them can find the will to put everything into stopping it. Some evidence has marvelled at the emissions reductions the virus has brought. *"There's an unlikely beneficiary of coronavirus: The planet"*, a CNN headline proclaimed. The Guardian too went ahead: *"Coronavirus could cause a fall in global CO_2emissions"*. 2020 is meant to be a pivotal year in global efforts to turn the tide on climate change, but the coronavirus is threatening to throw a spanner in the works. Just 18 months ago, the UN Intergovernmental Panel on Climate Change warned that the world was way off course in its efforts to limit warming to below 2 degrees Celsius (3.6 Fahrenheit) and avoid the worst impacts of global warming. Paradoxically, the coronavirus catastrophe has given the world an idea of how quickly pollution could be reined in with extreme measures. There is a misunderstanding that nature is "getting a break" from humans during the COVID-19 pandemic. Instead, many rural areas in the tropics are facing increased pressure from land grabbing, deforestation, illegal mining, and wildlife poaching. People who have lost their employment in cities are returning to their rural homes, further increasing the pressure on natural resources while also increasing the risk of COVID-19 transmission to rural areas. Meanwhile, there are reports of increased deforestation in Asia, Africa, and Latin America. Illegal miners and loggers are encroaching on indigenous territories, which could expose remote indigenous communities to the virus. Areas that are economically dependent on tourism face reduced resources as tourism has come to a halt, resulting in a rise in bushmeat (wild meat) consumption in Africa. Meanwhile, illegal mining for gold and precious stones in Latin America and Africa is on the rise, as prices spike and protected areas are left unguarded (DW 2020).

Governments must stop rolling back legal protections for the world's protected areas, as this can accelerate the pace of climate change, eliminate an important source of sustainable livelihoods, and contribute to biodiversity loss and deforestation—two significant drivers of disease outbreaks. Instead of scaling back protected areas, the government should seize the opportunity to scale them up. Governments in countries experiencing a rise in deforestation, illegal mining, and poaching urgently need to maintain enforcement efforts, even during the COVID-19 pandemic. Equally important, countries must start planning for rebuilding their economies in a way that fosters green structural transformation, including through long-term commitments to public spending and pricing reforms.

There is a recent trend wherein various scientific organizations have mentioned that the funding for conservation efforts, deforestation, degradation, and climate change issues, are diverted to the health sector due to the current

COVID-19 pandemic. After restrictions are lifted, governments and development financing institutions should prioritize stimulus efforts that have high economic multiplier effects and reduce carbon emissions. Such investments would have additional benefits for biodiversity and reduce the risk of zoonotic disease outbreaks, too, thereby addressing an important root cause of the current pandemic. It is equally important that climate and biodiversity stay at the top of the agenda in 2020 and beyond, and that leaders leverage every opportunity to maintain the momentum (Conservation International 2020).

4.14 Conclusion and Future Coverage

The REDD+ architecture is evolving in various parts of the globe, and so in India. It is a global endeavour towards supporting developing countries to safeguard their forest. However, the initial enthusiasm for REDD+ implementation and support through financial provision by developed countries has reduced to a great extent after the Paris Convention.

(a) *Centre and State government dialogue*: A regular and effective consultation process between the Centre and State governments is a prerequisite for successful implementation of REDD+. The State level REDD+ Cell should actively coordinate with and act as a link between the National level REDD+ Cell and the village-level forest governance unit. To take on board views of all stakeholders, pre-negotiation consultation, and post-policy formulation briefing on REDD+ issues need to be institutionalized.

(b) Present scope under the REDD+ mechanism is to cover the forests. Research efforts are underway to assess the potential of carbon sequestration by grasslands, coastal seagrasses etc., and to develop methodologies to monitor the change in carbon stocks in these ecosystems. Therefore, with an eye on future, grasslands, blue carbon, and phytoplankton are included as potential sinks of carbon.

Grasslands: Presently, pristine natural grasslands are not eligible to qualify as a REDD+ activity as these do not meet the eligibility criterion of the definition of a REDD+ forest. However, since these are a potential sinks of biomass carbon, sequestering more and more carbon and adding to the pool of soil organic carbon, it is considered worthwhile to include these as a land use with high mitigation potential. Grasslands are being flagged in the NRPS to work on and develop methodologies in future for MRV of the carbon accumulation these lands. India with its vast natural grasslands in the mountains, plains, and deserts of the country can contribute substantially to enhancing its land carbon sinks by investing technological, financial, and human resources in the effective management of its grasslands resource.

Blue Carbon: Mangroves, coastal sea grasslands, and salt marshes are also important sinks of carbon of biomass origin. These ecosystems are not only important for their contribution in addressing climate change mitigation, but also for the valuable ecosystem services they provide to the local communities. Presently, some of these systems are also not covered under REDD+ for the same reason as natural grasslands as these do not qualify the definition of a REDD+ activity. However, if these are protected and managed scientifically, these have the potential of capturing and locking huge quantities of CO_2 from the atmosphere in their vegetative parts and organic soil.

Phytoplankton: Phytoplanktons are huge sinks of atmospheric CO_2. Efficient scientific management of phytoplanktons has the potential of substantially enhancing the mitigation capability of this resource. Presently, such ecosystems do not qualify to be a REDD+ activity, but seeing their large mitigation potential, it is considered apt to flag these for future consideration as a potential sink.

Others Forest or tree cover in less than one-hectare area and having less than 10% canopy growth will also be considered for REDD+ being a potential sink of carbon.

(c) *Support for one pilot in each State:* As the REDD+ mechanism is being negotiated by member countries of the UNFCCC and the intricacies worked out, undertaking pilot REDD+ projects would help generate valuable experiences before REDD+ is fully operationalized at the national level. Pilot projects will also help to field-test the methodologies and provide important learnings, which can address research gaps, help refine SFM methodology and provide policy inputs for the large-scale design of REDD+ projects. In the process, this would help build technical capability and also help identify barriers to operationalizing REDD+. It is, therefore, imperative that at least one pilot study be undertaken in each State of the country where feasible. The pilot projects will be undertaken with the help of village-level Forest Resource Committees at Gram Sabhas, Van Panchayats, and JFMCs, among others. These could be the implementing institutions for undertaking REDD+ pilots in India, with support from the State Forest Department.

(d) *Building capacities*: Capacity building is a key requirement for the REDD+ readiness process. Capacity has to be built of officials at all tiers of forest governance, of the supporting institutions and the local forest-dependent communities, on various issues ranging from general awareness about forest policies and programmes to the benefit-sharing mechanism under REDD+, MRV mechanism, and social and environmental safeguards. The role of local communities is vital in not only addressing the drivers of forest degradation but also enhancing the carbon stock through forest conservation, protection, and reforestation. The carbon stock of Indian forests in 2004 has been estimated by the Forest Survey of India (FSI 2011) to be 6663 m tones, which can be increased significantly with the involvement of local communities. There is also a need to build capacity on methodologies for carbon assessment and procedures for project preparation under REDD+. The National REDD+ Cell will be responsible to build the capacity of the State REDD+ Cell and State Forest Departments (SFDs) on these issues. The SFDs would further sensitize the JFMCs and local-level institutions on REDD+.

References

Bhattacharya P, Pradhan L, Yadav G (2010) Joint forest management in India: experiences of two decades. Resour Conserv Recycl 54:469–480

Busby JW, Shidore S (2017) When decarbonization meets development: the sectoral feasibility of greenhouse gas mitigation in India. Energy Res Soc Sci 23:60–73. https://doi.org/10.1016/j.erss.2016.11.011

Conservation International (2020). https://www.conservation.org/stories/impact-of-covid-19-on-nature. Accessed 4 August 2020

Davidar P, Sahoo S, Mammen PC, Acharya P, Puyravaud J-P, Arjunn M, Garrigues JP, Roessingh K (2010) Assessing the extent and causes of Forest degradation in India: where do we stand? Biol Conserv 143:2937–2944

DW (2020). https://www.dw.com/en/corona-stimulus-plans-overlook-historic-chance-for-climate crisis/a-52876341. Accessed 5 August 2020

Ekawati S, Subarudi KB, Sari GK, Muttaqin MZ (2019) Policies affecting the implementation of REDD+ in Indonesia (cases in Papua, Riau and Central Kalimantan). Forest Policy and Economics 108:101939

FSI (2009) India state of Forest report 2009 Forest survey of India (FSI), Ministry of Environment, and forests (MoEF), Government of India

FSI (2011) India state of Forest report 2011 Forest survey of India (FSI), Ministry of Environment, and forests (MoEF), Government of India

FSI (2017) India state of Forest report 2017, Forest survey of India, Dehradun. Ministry of Environment, Forest and Climate Change, Government of India

GFW (2020) Nepal: land cover and forest change. Retrieved from Global Forest Watch: https://www.globalforestwatch.org/dashboards/country/NPL/?category=land-cover&dashboardPrompts=eyJzaG93UHJvbXB0cyI6dHJ1ZSwicHJvbXB0c1ZpZXdlZCI6WyJ2aWV3TmF0aW9uYWxEYXNoYm9hcmRzIiwiZG93bmxvYWREYXNoYm9hcmRTdGF0cyIsInNoYXJlV2lkZ2V0Iiwid2lkZ2V0U2V0dGluZ3MiXSwi. Accessed 15 September 2020

IPCC (2007) In: Edenhofer O, Pichs-Madruga R, Sokona Y, Farahani E, Kadner S, Seyboth K, Adler A, Baum I, Brunner S, Eickemeier P, Kriemann B, Savolainen J, Schlömer S, von Stechow C, Zwickel T, Minx JC (eds) Climate change 2007: mitigation of climate change, Contribution of working group III to the fifth assessment report of the intergovernmental panel on climate change. Cambridge University Press, Cambridge and New York

MoEF (Undated) India's forests and REDD+. Ministry of environment and forests, government of India

MoEF&CC (2014). Reference document for REDD+ in India. Ministry of Environment and Forests and Climate change, Government of India. New Delhi, India

National REDD+ Strategy INDIA (2018) Ministry of environment, forest and climate change government of India

Paudel NS, Khatri DB, Khanal DR, Karki R (2013) The context of REDD+ in Nepal drivers, agents and institutions. Occasional paper 81

Reference Document for REDD+ in India (2014) Ministry of environment, forests and climate change New Delhi, India

Sarkar J (2011) REDD, REDD+ and India. Curr Sci 101(3):265

TERI (2013) Sustainable Forest management and REDD+ in India. TERI Press, New Delhi

World Bank (2000) India: alleviating poverty through forest development. Washington, DC: The World Bank. Details Available at www.worldbank.org/html/oed

World Bank (2006) India: unlocking opportunities for forest dependent people in India. Report No. 34481 - IN, World Bank: South Asia Region, p 85

WRI (2008, September) Forest cover, 1990–2005, Indonesia. Retrieved from World Resource Institute. https://www.wri.org/resources/charts-graphs/forest-cover-1990-2005-indonesia#:~:text=Forest%20Cover%2C%201990-2005%2C%20Indonesia%20Spanning%2090%20million%20hectares%2C,forests%20for%20food%2C%20medicine%2C%20freshwater%2C%20and%20building%20material. Accessed 15 September 2020

Climate Change Impacts, Vulnerability, and Mitigation in the Indian Ocean Region: Policy Suggestions

5

R. S. Aswani, Mohammad Younus Bhat, and Shambhu Sajith

Abstract

Climate Change is considered as the mother of many global concerns in the twenty-first century and is a significant driver of biodiversity loss leading to major threats to species, ecosystems, and people's livelihood around the globe, in general, and the Indian Ocean in particular. Warming of oceans and ocean acidification can prove critical and challenging for the present and future generations. Oceans are one major realm that is impacted due to the natural climatic fluctuations and other anthropomorphic activities. The capacity of oceans to absorb carbon dioxide, a significant ecosystem service, is declining over time continuously with increasing carbon emissions. Thus, a reduction in greenhouse gases calls for alternate energy resources that are more secure, efficient, and environmental friendly. Moreover, the scenario of climate change vulnerabilities in developing countries is studied extensively. The unanimous call by a great majority of the scientific community is for a drastic reduction in greenhouse gasses particularly CO_2 as well as significant measures aiming at conservation of resources. The Indian Ocean climatic conditions and its variations are studied to arrive at climate mitigation measures. It also attempts to deliver

R. S. Aswani (✉)
School for Life, University of Petroleum and Energy Studies, Dehradun, Uttarakhand, India
e-mail: aswani.rs@ddn.upes.ac.in

M. Y. Bhat
Department of Economics and International Business, School of Business, University of Petroleum and Energy Studies, Dehradun, India
e-mail: younus.bhat@ddn.upes.ac.in

S. Sajith
Department of Energy Management, School of Business, University of Petroleum and Energy Studies, Dehradun, India
e-mail: ssajith@ddn.upes.ac.in

A. Kaushik et al. (eds.), *Climate Resilience and Environmental Sustainability Approaches*, https://doi.org/10.1007/978-981-16-0902-2_5

constructive recommendations toward policy development. A good policy should account for alternative risk reduction strategies-mitigation, adaptation, and insurance.

Keywords

Climate change · Indian ocean region · Carbon trading

5.1 Introduction

Climate change is inarguably the biggest challenge the world is facing today, as it comes with a range of costs to life on earth. Climate change is generally not considered as regional distress, but a global one, majorly because Green House Gases (GHGs) are highly permeable in the atmosphere and no matter where they are emitted, they are absorbed into the atmosphere and dispersed equally, thereby arousing the need for a cooperative mechanism of climate change adaptation. According to the reports by International Energy Agency, China is the major emitter of CO_2 gases with 28% of the world share, followed by the USA at 14%, and Europe including the whole of Russian Federation) at 12%. Among the low-middle income countries, India tops the polluter list accounting for 7% of global emissions. Though a global phenomenon, it has evolved to become a major concern with stark consequences to the developing countries, due to the sociopolitical volatilities of the region. Until recently, international discourses surrounding climate change aimed to establish a multilateral framework to regulate GHG emissions from the developed countries. Despite mitigating regulations on emissions and pollutants, historically disadvantageous and recently industrializing developing countries, especially in the Indian Ocean Region, lack an effective implementation strategy to achieve climate friendly actions. This chapter aims to study the impacts and vulnerabilities of climate change in the Indian Ocean Region, with a specific focus on the maritime climatic challenges.

Responding to climate change is a profound challenge, especially on the Oceanic front. However, climate adaptation measures need to be embraced for ensuring the continuing existence of humans on the planet. If the present conditions persist, by the end of 2030, GHGs are predicted to increase by 25–90%, increasing the global temperatures, leading to extreme climatic calamities impacting billions of people living in the low lying littorals of Indian Ocean. The Indian Ocean comprises 20% of the World's surface water and the scope of the Indian Ocean Region for this study includes the area bounded by the African continent in the East, Asian landmass (including entire West Asia, Central Asia, and India) to the North, and certain Southeast Asian countries, Sri Lanka, and Australia to the West. Thus, mitigating climate change should become a policy priority for the countries, as it is not a phenomenon that could be addressed in isolation. It requires an international level of cooperative remedial action to assess these issues and to arrive at corrective measures. Fighting climate change is a global responsibility. However, low-income countries will need financial and technical assistance, along with political support from the developed world, to shift to a low-carbon economy, while addressing the issues within their domestic economies and politics. Hence, this chapter also aims to provide constructive suggestions toward policy creation, to mitigate these wider concerns of humanity.

Global warming and climatic change are caused due to natural physiological fluctuations but are accelerated due to anthropogenic interventions, such as industrialization and pollution. Climate change is known as the "mother of all issues" and cannot be considered in isolation. Any measures to address this global menace needs to accommodate and comprehend its interlinkages to science, technology, trade, commerce, economy, diplomacy, and politics. This unique feature of climate change and the devastating impacts it yields make it yet another issue in this complicated world of proliferating geopolitical concerns.

5.2 Climate Change Vulnerabilities in the Indian Ocean Region

Rapidly changing climatic patterns have threatened the Indian Ocean and its littoral nations, posing physiological, economic, and geopolitical repercussions to its ecosystem, flora and fauna, and the livelihood of people around the region. Global warming, rise in sea levels, and ocean acidification might raise many unforeseen challenges to this generation, and the ones ahead. It is identified that the surface level seawater temperatures of the tropical Indian Ocean have shown noticeable warming since the 1950s (Bader and Latif 2003). Substantiating this, have studied the SST (Sea Surface Temperature) in the Indian Ocean Region, with a particular focus on the western tropical Indian Ocean. This study has further identified that Indian Ocean Region (particularly the Western Indian Ocean) has been warming faster than any other tropical oceans—such as the Pacific or Atlantic for more than a century and the Indian Ocean has been identified as the "largest contributor" toward the rise in overall global mean SST. Additionally, some studies analyzed the results of 163 climate model simulations, which has expressed that in the 20th and 21st centuries, the relatively higher SST in the Indian Ocean had influenced the climatic changes in the Pacific Region (Luo et al. 2012). This study has also stated that such increased warming up of the tropical Indian Ocean causes a La Nina-like situation with stronger trade winds in Western Pacific.

For centuries, oceans have safeguarded civilizations from the adverse effects of climate change by absorbing heat and carbon dioxide from the atmosphere. This capability of oceans to absorb CO_2, a noteworthy ecosystem service, is diminishing due to the increase in the volume of GHG emissions. Within the Indian Oceanic Region, the Western Ocean is recorded to have experienced unusual warming of 1.2 °C in summers, while the whole of Indian Ocean SST increases only 0. 7 °C (in the period from 1901 to 2012). This region comprises ecologically sensitive island chains including Lakshadweep, Madagascar, Mauritius, and Seychelles. Much like other global oceans, this increase in SST is observed to be the impact of "anthropogenic forcing". The major consequence of climate change and rising surface level temperature of the Indian Ocean is the rising sea levels in the region. As the ocean is warming up, it expands and the increasing global temperatures melt the ice sheets and polar glaciers, leading to higher sea levels, which are destructive for the low-lying islands of the Indian ocean and the low-lying countries such as Bangladesh.

Alternatively, these increases in CO_2 emissions have adverse impacts in the Indian Ocean, in the form of "evil twin of climate change," i.e., Ocean Acidification. Ocean acidification is the chemical processes that happen inside the ocean, while it absorbs the CO_2 emissions. This process is so widespread that more than ¼ of the world's CO_2 emissions from traditional energy sources such as coal and oil are dissolved into the ocean, and acknowledging that Asia accounts for 60% of world GHG emissions, the vulnerability of the Indian Ocean spikes up the radar. In the past 200 years alone, ocean water has become 30% more acidic—faster than any known change in ocean chemistry in the last 50 million years. Since the beginning of the industrial era, the ocean has absorbed some 525 billion tons of CO_2 from the atmosphere, presently around 22 million tons per day. The role of oceans in mitigating climate change is huge, and rapid increase in ocean acidification creates imbalances in the chemical components of seawater, making it acidic and affecting the marine ecosystems. Perhaps the most adverse impact of ocean acidification for the Indian Ocean is the change in ocean chemistry that leads to erosion of corals. This is because of the presence of numerous coral island reefs in the region such as those in Lakshadweep, Maldives, Mauritius, and Seychelles. These GHG emissions induced by human activity dissolves into the oceans, rivers, and lakes, and hinders the natural growth of microorganisms like marine plants/algae or corals and shellfish to develop their skeletons and shells.

The impact of climate change in the oceans, in general, is huge, mostly due to the two-way relationship between oceans and climate. On the one hand, oceans largely decide the climatic conditions and variations and on the other, the oceans are at the receiving end of climatic changes. In addition to environmental and economic bearings, it has far-reaching geopolitical repercussions on the oceanic waters. The Indian Ocean, which is in the realm of geopolitical competition in the twenty-first century, with the likely rivalry between the rising powers, India and China, and the Asian Pivot by the USA is key to impacts of climate change. Unlike other places, the Indian Ocean is largely sensitive due to the volatile political situation of its littoral countries.

5.3 Case Study: Maldives

The Maldives, the exotic island nation in the Indian Ocean was among the 49 LDCs (Least Developed Countries) of the world and managed to graduate to UMIC (Upper Middle Income Country) in 2011 January. Have detailed the vulnerability of the LDCs based on its location and "low capacities to adapt to these changes." They indicate that domestic politics and various developmental factors retard the capacities of these nations to embrace climatic adaptations as a policy priority.

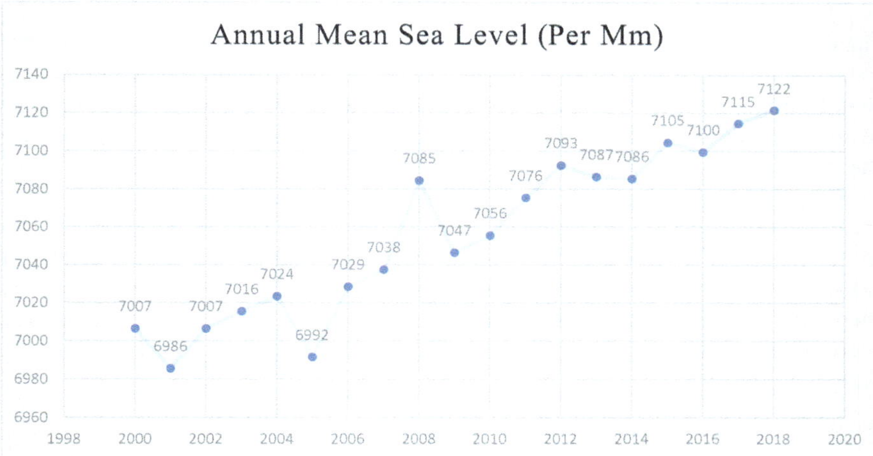

Fig. 5.1 Annual mean sea level data (Source: Permanent Services for Mean Sea Level 2019)

The Maldives consists of 1190 islands roughly out of which 188 is inhabited. Climate change is a major threat to the Maldives, as rising sea levels can sink many atolls underwater. Maldivian government's National Adaptation Programme of Action (NAPA) initiated plans for quick response to climate-related impacts. To build the Maldives, which is suffering from the rising sea level and other climate-related impacts, the Government has also set up a Climate Change Policy Framework, which addresses them and builds capacity to ensure the safety, sustainability, and adaptability toward the future.

In the case of Maldives, the effect of climate change and global warming is calamitous as the reports state that if the sea level rises by around 3 feet (1 meter) in the Indian Ocean, Maldives could virtually and entirely be drowned by about 2085. In 2018, *Science Advances* has reported that many atolls will be dilapidated by 2050 due to colossal rise in sea level that possibly will surpass 2.0 m directly above 2000 levels, with a weighty bearing on coastal and low-lying areas, particularly the tropical ones such as Maldives. The Maldives is noted as the most vulnerable to the effects of global warming and is also called the, "flattest nation". IPCC has also predicted that 85% of Maldives would be inundated by the year 2100, under the GHG (Green House Gases) emission scenarios.

Figure 5.1 shows the average increase in tidal levels across all Maldives between 2000 and 2018. One can comprehend that though the mean sea level has been intensifying for quite some time now, recent years have witnessed a faster rise in sea levels. The year 2008 has shown a sudden increase of 20 million meters and this year has been a watermark in the history of Maldives due to its sudden transition from an authoritarian to a democratic nation. However, it faced an equal reduction in the coming 2 years, ever since the political turmoil in the Maldives, the annual mean sea level is on a consistent increase.

Though a rise in the mean sea levels the most penetrating indicator of the climatic changes, increase in average sea surface level temperature is no less of a major peril to this Island nation. It has led to large-scale ocean acidification and coral degradation, leading to a grim risk of an ecosystem, which has led the Maldives to become the poster child of climate change in the recent years. The hopes of resurrection for the Maldives is clear through its government-funded geoengineering projects of building new high altitude islands to relocate population from low lying ones. If the most vulnerable place in the Indian Ocean, to climate change, is the Maldives, the most vulnerable species are the Corals. Ocean acidification is a major factor that leads to coral inundation in the Maldives, raising graveyards for the coral reefs. The increase in GHG emissions in the atmosphere has led to an increase in its oceanic absorptions, making the waters more acidic, thereby leading to acidification of the reefs in Maldives.

5.3.1 Maldives: Path to Climate Resilience

There are multiple internationally funded projects in Maldives to address these issues of climate changes such as the United Nations Development Programme, which has identified the action plan for assimilation of climate change threats into the Maldives Safer Island Development Program. In addition to it, the *USAID Project Regenerate* aims to generate resilience in the atolls around the Maldives to climate change and implement adaption techniques. Various bilateral arrangements of climate adaptation funding are in vogue between Maldives and major countries such as India, Norway, Germany, USA and EU. The utilization of these climate financing is an essential query that would dictate the future of this eco-sensitive coastal nation.

5.4 Mitigation Policies and Challenges among Littorals of Indian Ocean

The International Panel for Climate Change was established in 1988, to consolidate the world scientists working on the issues of climate change. This was the first effective step toward operative global action in the reduction in the GHGs. Further efforts toward addressing the issue began with the UNFCC (United Nations framework convention on Climate Change) in 1992, which held periodic Conference of Parties and through the third Conference, emerged the Kyoto Protocol on '97 (in force on 2005), which drew out emission targets for each country in the world, to regulate global GHG emissions. An ideal climate mitigation policy should aim to identify alternative strategies to reduce climatic risks and provide effective strategies for accommodation, adaptation and mitigation of the impacts of climate change. Below is a careful analysis of the existing climatic policies of major countries in the periphery of the Indian Ocean, better attributed as the Indian Ocean Littorals.

Bangladesh has made tremendous progress in climate change adaptation practices, backed by appropriate planning at its national as well as local level. At the local level, it is critical to have policies that could safeguard the population in low-lying coastal areas, which are more prone to climate change-induced catastrophes and rise in sea levels. There are ongoing efforts to divert the coastal population from crop cultivation into livestock raring, to save them from climate change-induced flood damage to crops, however, this would need institutional support in the form of credit (Alam et al. 2017). At the national level, the targeted 10-year strategy Bangladesh Climate Change Strategy and Action Plan (BCCSAP) was rolled out. BCCSAP has vision 2021 goals toward capacity building and resilience to climate change. The Sixth Five Year plan (2011–2015) took climate change as a central issue; it pointed out multiple adaptation and mitigation strategies. To protect people and property from adverse effects of climate change Perspective Plan (2010–2021) was set up, however, the plan did not prove to be successful since the priorities have not shifted to projects or programs to counter climate change (Fatemi et al. 2020).

The climate change curtailment policies of Bangladesh government through promoting the production and consumption of Renewable Energy, not only provides long-term gains from an economic point of view, but also ensures energy security by undermining the use of coal and forest biomass. The existing policies in Bangladesh have proved to be ineffective as GHG emission for every unit of energy has increased over the last 25 years adding to the same period the contribution from Renewable Energy to the Energy mix has decreased (Hasan et al. 2020).

National Action plan on Climate Change has eight national missions by which India's growing demand for Energy could be achieved in a sustainable and eco-friendly manner. To achieve the objectives of adaptation and mitigation various steps were outlined by this plan. It is crucial to have a public–private partnership catering to the objectives of each of the eight individual mission backed by the use of new advanced technologies (Pandve 2009).

Recently Indian Prime Minister has shown his interest in climate change and with the attempt to gracefully revive the council and set agenda to deal with climate change. This was with the help of concerned ministries to develop policies to counter climate change along practical lines. A new component was added to climate change by PM Modi on his address to the United Nations Educational, Scientific and Cultural Organization (UNESCO) when he said, "Too often, our discussion is reduced to an argument about emission cuts. However, we are more likely to succeed if we offer affordable solutions, not simply impose choices". India with her fast growth economically has claimed her seat in the global table to discuss climate change policies. India herself has a major contribution to the global emission of carbon but aims to carry the responsibility to hold Asia together with values and lifestyle changes that can place the climate change mitigation at a better place.

Sri Lanka's National Climate Change policy largely depends on the adaptation toward the target of sustainability with National Adaptation plan in 2013 on Climate Change Impacts. However, Sri Lanka did not allocate any substantial funds to the National budget for supporting Climate Change project over the last 25 years; these policies will not find its true objectives without funds in support (Hewawasam and Matsui 2019).

Owing to the Paris Agreement 2015, Indonesia has stated on its agreement on Nationally Determined Contribution (NDC) to set a target of emissions of GHG, coming under an unconditional category to 29% and conditional with international support to 41% by 2030. This huge responsibility needs to be carried out through a well-drafted policy which aims to slow down the occurrence of climate change induced catastrophes, if not to actively prevent them. Indonesia identified the sectors, which will act towards mitigation: they are agriculture, forestry, and land (AFOLU), Energy, Industrial Processes, and Product and Waste Management. For a Nation, which is spread over 1700 islands, the communication regarding climate adaptation and mitigation needs to be effective (Patrianti et al. 2020).

The Malaysian Climate Change Policy (MCCP) covers Adaptability, Vulnerability, and Mitigation as its primary pillars. Adaptation options can be at Sub-national level, from a city level the measures can focus on urban development and governance at multiple levels. By using a top-down and Bottom-up approach designed to adapt to climate change Maleka City has closed its gap between the city and national authority. This is an effective method of Multi-Level Governance. Multi-Level Governance was placed on the framework of Horizontal and Vertical understanding and executions of the climate initiatives bringing the regional and national objectives to the same page (Zen et al. 2019).

5.4.1 Path to Resolution: Policy Suggestions

Mitigation efforts toward addressing the rise in global SST comprises not only of national governments but also of the scientific, citizen, and political communities, along with collaborative determinations of national governments. A major problem in achieving this is the challenge of international coordination. Despite the common call by the policymakers and the world governments, many countries find it hard to participate in the measures to reduce GHGs, as in the contest to economic growth; no country wants to be drawn backward, by curtailing its emissions. However, the threat of climate change is real, and the developing countries are at the receiving end of this big menace, and the attempts to reduce GHGs calls for alternative renewable energy sources that are clean, efficient, secure, and environment friendly.

To study the challenges in arriving at international coordination toward addressing climate change, certain tools of game theory such as "prisoner's dilemma" have been in vogue (Hanley et al. 2013). This tool of game theory shows that even for their best interests, two rationally thinking individuals might not cooperate, due to lack of trust. Drawing parallel to this study, in the context of mitigating global climate change, each country has to strictly adhere to the allowed GHG emissions, trusting that every other country has also been following the guidelines. However, it is seen that there is a huge lack of trust among countries in approving and ratifying these guidelines. It is seen from the non-ratification of the Kyoto Protocol by USA (Hovi et al. 2012), the subsequent withdrawal of Canada

(Lao 2015), and the exemptions to India and China (Fedon 2007). This tool of prisoner's dilemma helps in understanding why nations have not ratified or distrusted such international cooperative mechanisms, despite the warnings of large-scale climatic catastrophes (Hassler 2019).

The aim of effective climate policy should be its cost-effective climate risk reduction strategies, such as carbon taxes and carbon emission trading. Carbon taxes allows fixing the cost of carbon emissions and allows for specific quantify of emissions. This tax may be fixed and collected by an agency in several ways. For example, depending on the domestic production and consumption of fossil fuels, a severance tax would be imposed and similarly, on fossil fuel imports and transportation specific taxation schemes can be laid out. However, the argument of economic efficiency is a major hindrance and thus there is a need to address climate change from the perspective of political economy (Spash 2010). In this regard, the CO_2 cap and trade permit system are significant and such permits ensure mutual gains between partners and guarantee a fixed level of reductions in CO_2 emissions (Hanley et al. 2013). A market-based climate policy instrument would ensure that the firms would reduce their tax exposure, by targeting reductions in CO_2 emissions. An internationally harmonized carbon tax over a permit system, in which the carbon tax would likely be a better competent incentive, simply because of its simplicity (Nordhas 2008). Such a taxation regime would be advantageous over licenses and permits if the goals were to maintain the flexibility needed to ensure political economy.

5.5 Conclusion

The scenario of climate change vulnerabilities in the developing countries is studied extensively in the chapter and policy challenges are identified in a specific relation with Indian Ocean littorals, as the delinquent effects of climate change have spread excessively in this region in the twenty-first century. The policy analysis suggests a massive spread in GHG emissions in developing countries and a slow transference to the use of renewable energy. Through this study, we suggest the sustainable use of renewables, as they are secure, efficient, environmentally justifiable, and economically viable. The increase in population in these developing countries is another impetus to accomplish environmental sustainability at the earliest. If we replace the traditional petro-based fuels with biofuels, it has the potential to avoiding the dangerous impacts of GHGs. IRENA data indicates that, for instance, renewables from India has sidestepped 180.1 million tonnes CO_2e in 2016 compared to 73.8 million tonnes CO_2e in 2000. Besides, renewable electricity generation in India in 2016 has replaced the fossil fuel emission by 181.8 million tonnes CO_2e. Total emissions avoided in India from bioenergy use were 2.405 million tonnes CO_2e in 2000, which increased to a maximum of 34.24 million tonnes CO_2e in 2014.

This chapter attempted to deliver constructive recommendations toward policy creation in IOR, to mitigate the concerns of sea level rise, ocean acidification, etc. This study is significant for both developing and developed countries and it especially focuses on mitigating climate change challenges in achieving sustainable development goals such as the SDG (13) that aims to achieve "climate action." The need is to understand and study the climatic impacts of this region, as the lives of millions of people and their resources are impacted in case the lack of apathy toward addressing these major issues continue. Economic growth should be achieved without triggering environmental destruction and planned and sustainable use of the natural resources ensures a bright future for generations to come.

References

Alam GMM, Alam K, Mushtaq S (2017) Climate risk management climate change perceptions and local adaptation strategies of hazard-prone rural households in Bangladesh. Clim Risk Manag 17:52–63. https://doi.org/10.1016/j.crm.2017.06.006

Bader J, Latif M (2003) The impact of decadal-scale Indian Ocean Sea surface temperature anomalies on Sahelian rainfall and the North Atlantic oscillation. Geophys Res Lett 30(22):2169

Fatemi N, Okyere SA, Diko SK, Kita M (2020) Multi-level climate governance in Bangladesh via climate change mainstreaming: lessons for local climate action in Dhaka City, pp 9–13

Feldon J (2007) The black hole in the Kyoto protocol: was the exclusion of black carbon regulation a "fatal flaw"? Sustain Dev Law Policy Winter 60–62:85

Hanley N, Shogren J, White B (2013) Introduction to environmental economics. Oxford University Press, Oxford

Hasan A, Rimi I, Masiur S, Aina YA, Islam M, Khondaker AN (2020) The synergy between climate change policies and national development goals : implications for sustainability. J Clean Prod 249:119369. https://doi.org/10.1016/j.jclepro.2019.119369

Hassler M (2019) What the "prisoner's dilemma" tells us about climate change. Schroders perspective. https://www.schroders.com/en/insights/economics/what-the-prisoners-dilemma-tells-us-about-climatechange/#:~:text=In%20the%20context%20of%20global,warnings%20of%20the%20risks%20ahead

Hewawasam V, Matsui K (2019) Historical development of climate change policies and the climate change secretariat in Sri Lanka. Environ Sci Pol 101:255–261. https://doi.org/10.1016/j.envsci.2019.09.001

Hovi J, Sprinz DF, Bang G (2012) Why the United States did not become a party to the Kyoto protocol: German, Norwegian, and US perspectives. Eur J Int Rel 18(1):129–150

Lao E (2015) Canada and the Kyoto protocol. Altern J https://www.alternativesjournal.ca/community/blogs/current-events/canada-and-kyoto-protocol

Luo JJ, Sasaki W, Masumoto Y (2012) Indian Ocean warming modulates Pacific climate change. Proc Natl Acad Sci USA 109(46):18701–18706

Nordhas W (2008) A question of balance: weighing the options on global warming policies. Yale University Press, New Haven, CT

Pandve H (2009) Indian. J Occup Med 13(1):17–19. https://doi.org/10.4103/0019-5278.50718

Patrianti T, Bajari A, Agustin H, Bakti I (2020) Climate change communication on mitigation policy and its challenges towards. Sustainable Development Goals (SDGs) 29(6):4054–4062

Spash CL (2010) The brave New World of carbon trading. New Political Economy 15(2):169–195. https://doi.org/10.1080/13563460903556049

Zen IS, Al-Amin AQ, Doberstein B (2019) Mainstreaming climate adaptation and mitigation policy: towards multi-level climate governance in Melaka, Malaysia. Urban Clim 30:100501. https://doi.org/10.1016/j.uclim.2019.100501

Climate Change, Agriculture Adaptation, and Sustainability

6

Mannava Sivakumar

Abstract

Climate change is widely accepted as the single most pressing issue facing society on a global basis due to the increasing greenhouse gas (GHG) emissions. GHG emissions growth between 2000 and 2010 has been larger than in the previous three decades. Hence, each of the past 3 decades has been successively warmer than the preceding decades since 1850. Continued emissions of greenhouse gases will cause further warming and changes in the climate system. Agriculture is one of the important economic sectors of the global society and climate change/variability is having a profound influence on the agroecosystems, posing serious threats to food security, human health, and protection of the environment. Of the total annual crop losses in world agriculture, many are due to increasing frequency and magnitude of weather and climate extremes such as droughts, flash floods, untimely rains, frost, hail, and severe storms. Agricultural impacts from natural events and disasters most commonly include alteration of ecosystems, contamination of water bodies, loss of harvest or livestock, increased susceptibility to disease, and destruction of irrigation systems and other agricultural infrastructures. Between 2003 and 2013, natural hazards and disasters in the developing regions affected more than 1.9 billion people and resulted in nearly 500 million US dollars in estimated damage. Effective decision-making to limit climate change and ensure economic development of agriculture sector can be made by a wide range of analytical approaches for evaluating expected risks and benefits; recognizing the importance of governance, ethical dimensions, equity, value judgments, economic assessments, and diverse perceptions and responses to risk and uncertainty. Comprehensive agrometeorological adaptation policy

M. Sivakumar (✉)
27 Chemin des Corbillettes, Cointrin, Geneva, Switzerland
e-mail: mannavas@gmail.com

A. Kaushik et al. (eds.), *Climate Resilience and Environmental Sustainability Approaches*, https://doi.org/10.1007/978-981-16-0902-2_6

guidelines, focusing on preparedness and adaptation measures to support sustainable agricultural development are needed to cope with the impacts of climate change. The major international agreements in 2015, such as the Sustainable Development Goals (SDGs) and Paris Agreement, to which almost all countries are now committed, emphasize that countries should implement policies aimed at climate change adaptation, and greater sustainability in the agricultural sector.

Keywords

Climate change · Agriculture adaptability · Agrometeorological adaptation · SDGs

6.1 Introduction

Earth's climate system is controlled by a complex set of interactions among the atmosphere, oceans, continents, and living systems. The United Nations Framework Convention on Climate Change (UNFCCC), in its Article 1, defines climate change as "a change of climate which is attributed directly or indirectly to human activity that alters the composition of the global atmosphere and which is in addition to natural climate variability observed over comparable time periods." Evidence is pointing to the fact that human-driven changes in Earth's energy balance are driving a warmer and wetter atmosphere, with this trend superimposed on and magnifying natural variability. Over the last three decades, climate change has emerged as one of the most crucial issues for humankind, with serious implications for sustainable development.

The Intergovernmental Panel on Climate Change (IPCC), is the leading international body for the assessment of climate change, established by the United Nations Environment Programme (UNEP) and the World Meteorological Organization (WMO) in 1988, to provide the world with a clear scientific view on the current state of knowledge in climate change and its potential environmental and socioeconomic impacts. Governments participate in the review process and the plenary sessions, where main decisions about the IPCC work program are taken and reports are accepted, adopted, and approved. According to the Fifth Assessment Report of the WMO/UNEP Intergovernmental Panel on Climate Change (IPCC), *each of the last three decades has been successively warmer at the Earth's surface than any preceding decade since 1850* (IPCC 2013). The globally averaged combined land and ocean surface temperature data as calculated by a linear trend show a warming of 0.85 °C over the period 1880–2012. More frequent and intense extreme weather events and increasing uncertainty in rainy season patterns are already having significant impacts on food production, food distribution infrastructure, livelihood assets, human health, and food emergencies, in both rural and urban areas (FAO 2008).

Rapidly rising population growth and diminishing arable land, particularly in developing countries, has increased the stress on the natural resource base. Lal (1991) calculates that the per capita arable land will progressively decline from about 0.3 ha in 1990 to 0.23 ha by the year 2000, 0.15 ha by 2050, and 0.14 ha by 2100. Combined with the growing concerns regarding the decline in the nonrenewable sources of energy and the degradation of environment, it is certainly time that the world is taking a hard look at the way agricultural adaptation to climate change and improved sustainability. As Swindale (1988) explained, sustainability conveys the idea of a balance between human needs and environmental concerns.

Observed Climate Change

According to IPCC's Fifth Assessment Report (AR5), in recent decades, changes in climate have caused impacts on natural and human systems on all continents and across the oceans. Human influence on climate has been the dominant cause of observed warming since the mid-twentieth century, while global average surface temperature warmed by 0.85 °C between 1880 and 2012, as reported in the IPCC Fifth Assessment Report, or AR5 (IPCC 2013). According to the IPCC Special Report in 2018 on the impacts of global warming of 1.5 °C, human activities are estimated to have caused approximately 1.0 °C of global warming above pre-industrial levels, with a likely range of 0.8 °C–1.2 °C. Global warming is likely to reach 1.5 °C between 2030 and 2052 if it continues to increase at the current rate. The observed global mean surface temperature (GMST) for the decade 2006–2015 was 0.87 °C (likely between 0.75 °C and 0.99 °C) higher than the average over the 1850–1900 period (IPCC 2018). In the Northern Hemisphere, 1983–2012 was likely the warmest 30-year period of the last 1400 years.

Evidence from observations of the climate system has led to the conclusion that human activities are contributing to the warming of the Earth's atmosphere. According to Smil (2017), between 1850 and 2000, the human world's energy use increased by a factor of 15 or so. Over its course, the mixture of fossil fuels used changed quite dramatically. Atmospheric concentrations of carbon dioxide (CO_2), methane (CH_4), nitrous oxide (N_2O), and other greenhouse gases are increasing steadily, due to human activities. According to IPCC (2013), the atmospheric concentrations of CO_2, CH_4, and N_2O have all increased since 1750 due to human activity and these concentrations now substantially exceed the highest concentrations recorded in ice cores during the past 800,000 years. Developing country emissions comprised more than half of global emissions in 2010, and grew during the preceding decade by an amount that accounted for the total global emissions rise (JRC/PBL 2013). Carbon dioxide concentrations have increased by 40% since pre-industrial times, primarily from fossil fuel emissions and secondarily from net land use change emissions. The ocean has absorbed about 30% of the emitted anthropogenic CO_2, causing ocean acidification.

Warming from such anthropogenic emissions from the pre-industrial period to the present will persist for centuries to millennia and will continue to cause further long-term changes in the climate system, such as sea level rise, with associated impacts.

The attribution chapter of the AR5 of IPCC (Bindoff et al. 2013) and recent papers (Sun et al. 2016; Wan et al. 2018) assessed that, over every continental region

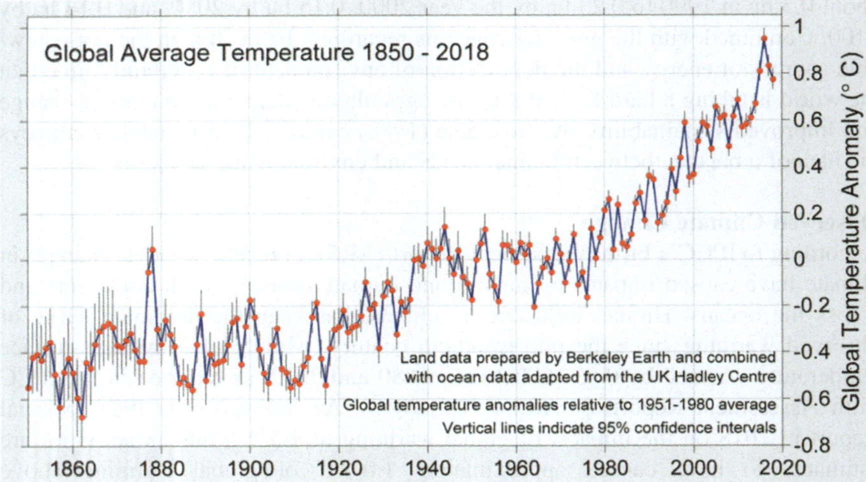

Fig. 6.1 Change in global average temperature 1850–2018 (Source: Berkeley Earth 2019)

and in many subcontinental regions, anthropogenic influence has made a substantial contribution to surface temperature increases since the mid-twentieth century. The change in global average temperature relative to 1880–1900 is shown in Fig. 6.1.

Human-induced global warming has already caused multiple observed changes in the climate system (high confidence). Changes include increases in both land and ocean temperatures, as well as more frequent heatwaves in most land regions (Hoegh-Guldberg et al. 2018). There is also high confidence that global warming has resulted in an increase in the frequency and duration of marine heatwaves. Most of the problems people have with weather and climate come from extremes. Extreme weather events—floods, violent storms, droughts, and forest fires (Fig.6.2)—occurred on all inhabited continents in 2019. Annual losses from weather disasters such as hurricanes, hailstorms, or wildfires frequently run into hundreds of billions of dollars. In total, weather-related natural disasters have caused losses of some US $4200 bn since 1980 and killed nearly a million people (Munich Re2019).

Over the past 100 years, fifteen of the hottest summers have occurred since 2000 and the world is already facing climate change-induced impacts, such as rising sea levels, changing rainfall patterns, increased droughts, and more erratic storms. Furthermore, higher temperatures along with decreased precipitation have been associated with observations of more intense and longer droughts over wider areas since the 1970s. According to IPCC (2013), changes in many extreme weather and climate events have been observed since about 1950. It is very likely that the number of cold days and nights has decreased and the number of warm days and nights has increased on the global scale. These stresses have exacerbated existing economic, political, and humanitarian issues around the world.

Increases in the frequency of droughts and floods are projected to negatively affect local crop production, especially in subsistence sectors at low latitudes. Some

Fig. 6.2 Extreme weather events: (**a**) Hurricanes, (**b**) Storms, (**c**) Floods, and (**d**) Wild fires (Source: Munich Re 2019)

regions are more vulnerable than others. There is substantial evidence that human-induced global warming has led to an increase in the frequency, intensity, and/or amount of heavy precipitation events at the global scale (medium confidence), as well as an increased risk of drought in the Mediterranean region. On November 21, 2016, a line of thunderstorms passed through the Australian state of Victoria. Late that afternoon, a peculiarly downdraft generated by the storm front pushed a layer of cold air thick with pollen, dust, and other particles through Melbourne. By the end of the next day, it sent 3000 people to the hospital, caused by asthma.

El Niño-Southern Oscillation (ENSO) is a recurring cycle that refers to year-to-year variations in sea surface temperatures, convective rainfall, surface air pressure, and atmospheric circulation that occur across the equatorial Pacific Ocean. The warm phase, in which the central and east-central equatorial Pacific Ocean warms, occurs at intervals of two to seven years (El Niño) and alternates with an opposite cold phase (La Niña) (Guilyardi et al. 2009). ENSO is known to affect weather events globally (teleconnection) such as depending on location, warmer/colder or drier/wetter climate than normal conditions (potential droughts or floods), monsoon rainfall changes, and intensity and frequency of tropical cyclones. El Niño conditions tend to bring slightly lower hurricane activity in the North Atlantic, drier weather in the northern USA and Canada, stronger precipitation in the southern

USA, and more droughts in southeast Asia and Australia. La Niña has the opposite effect in many cases.

According to Munich Re NatCatService (2020), there was a worldwide increase in natural catastrophes, and proportions of insured losses, in recent decades. About half of the world's population lives within 200 km of a coastline, with absolute numbers increasing, and a number of coastal regions are under threat of flooding.

Tropical cyclones can form over a sea or ocean with a surface temperature of 27 °C or more. The area where such temperatures are possible will definitely increase with warming. For example, on May 20, 2020, Supercyclone Amphan made landfall, the most powerful storm to do so in the Bay of Bengal region in 20 years. Winds gusting up to 185 km/hr pounded the coast of the Indian state of West Bengal. Huge waves swept over the Indian and Bangladeshi coasts. Trees were lifted out of the ground, city streets turned into rivers and tens of thousands of people lost their homes. As of May 27, 2020, the number of deaths were about 100 overall.

A recent study (Hansen and Stone 2016) shows that most of the detected temperature-related impacts that were reported in the Fifth Assessment Report of IPCC (Cramer et al. 2014) can be attributed to anthropogenic climate change, while the signals for precipitation-induced responses are more ambiguous.

In 2020, during the Coronavirus Pandemic crisis, because of the abandonment of planes, trains, and automobiles, the emissions of greenhouse gases decreased. Between January and March 2020, demand for coal dropped by 8% and oil by 5%, compared with the same period in 2019. According to International Energy Agency (IEA), by the end of 2020, the global industrial greenhouse gas emissions could be 8% lower than they were in 2019.

6.2 Future Climate Change

Continued emissions of greenhouse gases will cause further warming and changes in all components of the climate system. The magnitude of warming toward the end of the twenty-first century depends highly on GHG emissions for the next decades, which are driven by many socioeconomic and technological factors, and climate policy.

Projections of changes in the climate system are made using a hierarchy of climate models ranging from simple climate models, to models of intermediate complexity, to comprehensive climate models, and Earth System Models (IPCC 2013). These models simulate changes based on a set of scenarios of anthropogenic forcing. In all models, atmospheric CO_2 concentrations are higher in 2100 relative to the present day as a result of further increase of cumulative emissions of CO_2 to the atmosphere during the twenty-first century.

According to IPCC (2013), global surface temperature change for the end of the twenty-first century is likely to exceed 1.5 °C relative to 1850–1900 for all model scenarios except one. The increase in temperature will be larger on the land than over the ocean and larger than the mean. It will be larger in the Arctic (IPCC 2014b). There will be more frequent hot temperature extreme episodes over most land areas

(IPCC 2014b). An increase in temperature will trigger increased demand for water for evapotranspiration by crops and natural vegetation and will lead to more rapid depletion of soil moisture (FAO 2013). Climate change is projected to reduce renewable surface water and groundwater significantly in most dry subtropical regions (Jiménez Cisneros et al. 2014). This will intensify competition for water use.

Average precipitation will very likely increase in high-latitudes and parts of the mid-latitudes, and the frequency and intensity of heavy precipitation will also likely increase on average. ENSO will very likely continue to be the dominant mode of interannual variability in the future (Christensen et al. 2013). It is not well understood how ENSO will change in the twenty-first century, but the associated precipitation variability on regional scales is likely to increase due to larger moisture availability in the atmosphere. The contrast in precipitation between wet and dry regions and between wet and dry seasons will increase.

Short-duration precipitation events will shift to more intense individual storms and fewer weak storms are likely as temperature rises. As a result of climate change, freshwater availability increases in regions in the temperate zones but decreases in regions in the low latitudes, including prominent agricultural and heavily irrigated areas in India, China, and Egypt (Elbehri and Burfisher 2015). Constraints on freshwater availability in heavily irrigated areas, however, may lead to large reductions in the irrigated share of overall agricultural production, amplifying direct climate change impacts and increasing weather-induced variability in these regions. Globally averaged, maximum windspeed and rates of precipitation from tropical cyclones will likely increase in the long run. Warming will continue beyond 2100 under all model scenarios. Warming will continue to exhibit interannual-to-decadal variability and will not be regionally uniform.

Extreme precipitation events over most of the mid-latitude land masses and over wet tropical regions will very likely become more intense and more frequent by the end of this century, as global mean surface temperature increases (IPCC 2013). The global ocean will continue to warm during the twenty-first century. Heat will penetrate from the surface to the deep ocean and affect ocean circulation (IPCC 2013). As seawater continues to warm and glaciers and ice sheets are lost, global average sea level will rise during the twenty-first century faster than in the past decades. In 2046–2065 (relative to 1986–2005), global average sea level rise is likely in the range of 0.17–0.32 m and 0.22–0.38 m for the lowest and highest GHG concentration pathways, respectively (Church et al. 2013).

Climate change will also reduce glaciers, which often play a key role to provide river flows in summer. It is very likely that the Arctic sea ice cover will continue to shrink and thin and that Northern Hemisphere spring snow cover will decrease during the twenty-first century as global mean surface temperature rises. Under all model scenarios, the rate of sea level rise will very likely exceed that observed during 1971 to 2010 due to increased ocean warming and increased loss of mass from glaciers and ice sheets.

6.3 Impacts of Climate Change on Agriculture

According to the International Fund for Agricultural Development (IFAD), at least 70% of the very poor live in rural areas, most of them depending partly or completely on agriculture for their livelihoods. It is estimated that 500 million smallholder farms in the developing world are supporting almost 2 billion people, and in Asia and sub-Saharan Africa these small farms produce about 80% of the food consumed (IFAD 2011).

Agriculture is a complex system, within which changes are driven by the joint effects of economic, environmental, political, and social forces (Olmstead 1970; Bryant and Johnston 1992). Agriculture is the sector most vulnerable to climate change due to its high dependence on climate and weather. Climate change and increasing climate variability affect food security in all of its 4 dimensions—availability, accessibility, utilization, and stability (FAO 2008). More frequent and intense extreme weather events and increasing uncertainty in rainy season patterns are already having significant impacts on food production, food distribution infrastructure, livelihood assets, human health, and food emergencies, in both rural and urban areas (FAO 2008). Battisti and Naylor (2009) concluded that projected changes in the frequency and severity of weather and climate events, including higher growing season temperature, have significant consequences for food production, farm income, and food security.

Climate change can have both direct and indirect impacts on agricultural production systems. The direct impacts are those that are directly caused by a modification of physical characteristics such as temperature levels and distribution along the year and water availability on a specific agricultural production. Indirect effects are those that affect production through changes in other species such as pollinators, pests, disease vectors, and invasive species.

In their article on climate change in South Asia, Sivakumar and Stefanski (2011) enlisted the following climate change impacts in South Asia: impacts of enhanced temperatures; impacts of precipitation variability and water resources; impacts of increased frequency of extreme events and natural disasters; impacts of crop, pasture, and forest productivity; impacts on crop pests and diseases; impacts on fisheries; and impacts of sea level rise.

Impacts in the mid-to-high latitudes are expected to be more mixed, especially at lower levels of warming (IPCC 2014a). Some high-latitude regions are expected to benefit—sometimes substantially—from warmer temperatures and longer growing seasons; however, other environmental conditions, such as soil quality issues in the far north, will likely constrain expansion (Porter et al. 2014). Spatial differences are also observed at regional and subregional scales, particularly where there are substantial differences in elevation. Contrasted impacts between high- and low-latitude regions indicate that climate change is likely to exacerbate existing imbalances between the developed and developing world (Elbehri et al. 2015). Overall climate change will also increase variability in crop yields in many regions.

The risks of climate change impacts on agricultural production directly translate into additional risks for the food security and nutrition of the people who directly depend on agriculture for their food and livelihood. They can also have an impact on the food security and nutrition of distant populations through price volatility and disrupted trade. As shown in Fig. 6.3, there is thus a cascade of risks from climate changes to agroecosystems, to agricultural production, to economic and social consequences, and finally to food security and nutrition (FAO 2016).

The effects of climate change on the ecosystems are already severe and widespread, and ensuring food security in the face of climate change is among the most daunting challenges facing humankind. Extreme weather and climate events such as severe droughts, floods, or temperature shocks often strongly impede sustainable farming development, particularly in the tropics and subtropics. Analysis of drought trends in different locations confirms that there are an increasing frequency and intensity of droughts, which can lead to increased land degradation. Factors such as climate variability and change contribute to the vulnerability of individual farms, as well as on whole rural communities.

Developing countries are likely to face the highest reductions in agricultural potential in the world due to climate change. As a result, climate change will place an additional burden on efforts to meet long-term sustainable development goals. Slow agricultural productivity growth, declining income growth, and problems of maintaining food security already pose challenges to many developing countries in the world. The impacts on the territories of indigenous people and marginalized populations are both early and severe due to their location in vulnerable environments, including low rainfall areas, drylands, and desert margins.

Climate change affects livestock production in multiple ways, both directly and indirectly. The most important impacts are experienced in animal productivity, yields of forages and feed crops, animal health, and biodiversity, as summarized in Table 6.1.

Consistent warming trends and more frequent and intense extreme weather events have been observed across the different regions in the world in recent decades. Of the total annual crop losses in world agriculture, many are due to direct weather and climatic effects such as droughts, flash floods, untimely rains, frost, hail, and severe storms (Hay 2007). Chattopadhyay and Lal (2007) estimated that around 28% of the land in India is vulnerable to droughts, 12% to floods, and 8% to cyclones. But in the year 1918, which was ranked as the worst drought year of the last century in India, about 68.7% of the total area of the country was affected by drought (Chowdhury et al. 1989).

Farmers and farming communities throughout the world have, in most instances, survived and developed by mastering the ability to adapt to widely varying weather and climatic conditions. However, the dramatic growth in human population is imposing enormous pressure on existing farming production systems. Agricultural production systems are expected to produce food for a global population that will amount to 9.1 billion people in 2050 and over 10 billion by end of the century (UNFPA 2011). Most of the additional billions of people will live in developing countries. At the same time, more people will be living in cities (70% against the current 50%). Population and income increase, as well as urbanization, are driving increased and changing food and feed demand. Given these trends, FAO estimates

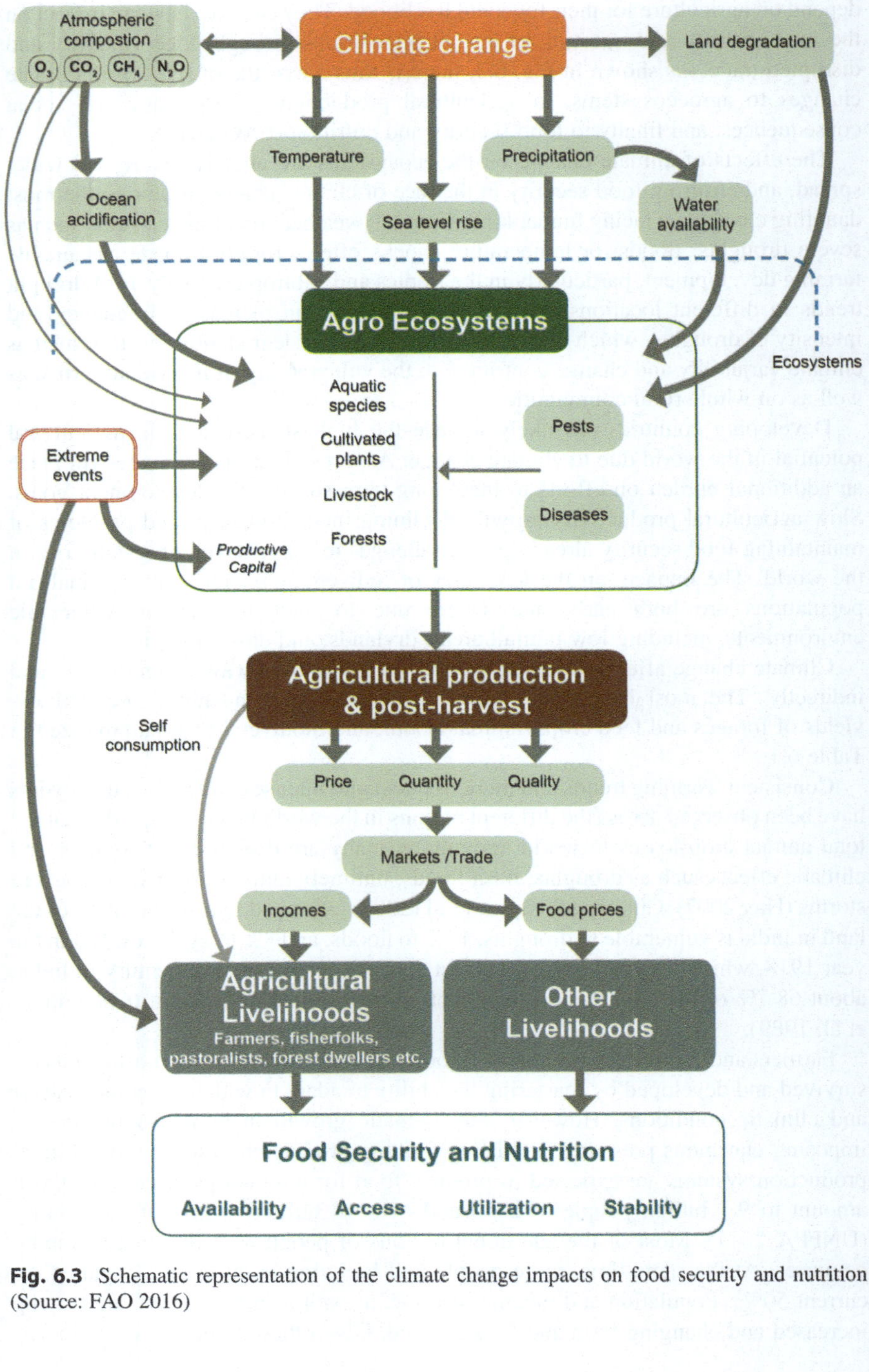

Fig. 6.3 Schematic representation of the climate change impacts on food security and nutrition (Source: FAO 2016)

Table 6.1 Pathways of impacts of climate change on livestock (Source: Thornton et al. (2009), IUCN (2010), Niang et al. (2014))

	Animals	Forage and feed crops	Labor force and capital
Variability in rainfall	– Shortages in drinking and servicing water. – Diseases • Increased pathogens, parasites, and vectors • Changed distribution and transmission • New diseases	– Decreased yields. – Decreased forage quality. – Changes in pasture composition (species, communities). – Changes in production system (e.g., from mixed crop-livestock to rangelands)	– Altered human health and resources allocation to livestock. – Decreased productivity. – Migrations. – Conflicts.
Temperature	– Heat stress • Decreased feed intake and livestock yields • Decreased conception rates • Altered metabolism and increased mortality – Diseases • Distribution and transmission through pathogens, parasites, and vectors • Decreased resistance of livestock • New diseases – Domestic biodiversity losses	– Decreased yields – Decreased forage quality – Change in pasture composition.	
CO_2 in the atmosphere		– Partial stomata closure and reduced transpiration. – Change in pasture composition.	

that production will have to increase by 60% by 2050 to satisfy the expected demands for food and feed (FAO 2015).

The extent of these impacts depends not only on the intensity and timing of the changes but also on their combination, which are more uncertain, and on local conditions.

Slow agricultural productivity growth, declining income growth, and problems of maintaining food security already pose challenges to many developing countries in the world. In future, the increase of mean temperature, changes in rainfall patterns, increased variability both in temperature and rainfall patterns, changes in water availability, the frequency and intensity of extreme events, sea level rise, and salinization and perturbations in ecosystems will have intense impacts on agriculture. Developing countries are likely to face the highest reductions in agricultural potential in the world due to climate change.

Climate change is adding significant uncertainty to the availability of water in many regions in the future. It will affect precipitation, runoff, and snow/ice melt, with effects on hydrological systems as well as on water quality, water temperature, and groundwater recharge. According to the IPCC (2012), there is "medium confidence" that "droughts will intensify in the twenty-first century in some seasons and areas, due to a combination of more variable precipitation and/or increased evapotranspiration." Climate change will also significantly impact sea level with potential impacts on the salinity of surface and groundwater in coastal areas.

The observed effects of past climate trends on crop production are evident in several regions of the world (Porter et al. 2014), with negative impacts more common than positive ones, including several periods of price spikes following climate extremes in key producing regions. There is evidence that climate change has already negatively affected wheat and maize yields in many regions and also at global level (Lobell et al. 2008). The increased frequency of unusually hot nights in most regions is damaging for most crops, with observed impacts on rice yields and quality.

6.4 Agriculture Adaptation

According to Allen et al. (2018), climate adaptation refers to the actions taken to manage impacts of climate change by reducing vulnerability and exposure to its harmful effects and exploiting any potential benefits. Adaptation takes place at international, national, and local levels. The countries party to the United Nations Framework Convention on Climate Change (UNFCCC) took on the commitment to a global adaptation framework and set up the Green Climate Fund, with the goal of raising $100 billion annually by 2020 to help developing nations reduce their emissions and establish plans for adaptation.

6.4.1 Framework for Adaptation Strategies

The development of adaptation policy must begin by identifying both the hazards and the vulnerabilities, which then can be used to identify particular disaster scenarios (Motha 2009). The disaster scenarios must account for all scales, ranging from actions to managing the local manifestations of global climate disasters, through global measures to reduce hazards, and to reduce vulnerability. Ways and means of dealing with integrated climate disasters need to include the following elements: adaptation to ensure that future developments reduce that increase potential disasters; actions to mitigate the losses associated with disasters; and measures to ensure that disasters do not reoccur after a disaster event. These measures should take into account potential impacts on socioeconomic and environmental systems in the local area.

6.4.2 Adaptation Strategies

According to IPCC (2014b), adaptation is place- and context-specific, with no single approach for reducing risks appropriate across all settings. Effective risk reduction and adaptation strategies consider vulnerability and exposure and their linkages with socioeconomic processes, sustainable development, and climate change. According to Thomas (2008), the world's drylands will face not only increasing temperatures with climate change but more importantly also disruptions to their hydrological cycles resulting in less and more erratic rainfall that will exacerbate the already critical state of water scarcity and conflicts over water allocation.

The overall risks of climate change impacts can be reduced by limiting the rate and magnitude of climate change. With the adoption of the Paris Agreement at the 21st Session of the Conference of the Parties (COP-21) to the United Nations Framework Convention on Climate Change (UNFCCC) in December 2015, the global community took a significant step toward tackling climate change, and a much more ambitious one than many had expected before the Conference.

The Paris Agreement represents a major milestone in international efforts to combat climate change and it sets out a global action plan to put the world on track to avoid dangerous climate change by limiting global warming to well below 2 °C above pre-industrial levels and pursue efforts to limit the temperature increase to 1.5 °C, which will require countries to transition to zero net GHG emission economies by the end of the century. The Paris Agreement commits signatories to support vulnerable communities as they seek to adapt and become more resilient. In response, the Least Developed Countries (LDC) Group—during Ethiopia's chairmanship—launched the LDC Initiative for Effective Adaptation and Resilience (LIFE-AR). This LDC-led initiative seeks to develop effective longer-term climate adaptation interventions, and in so doing, strengthen the case for domestic, donor, and private investment that can be steered towards LDCs priorities. As part of the LIFE-AR, nature-based solutions were implemented across different ecosystems and landscapes—for example, integrated river basin management in Uganda, coastal mangrove restoration in the Pacific and ecosystem-based adaptation in South Africa, and climate-resilient value chains in agriculture, forestry, energy, and livestock that promote SME development and create links between households, governments, and the private sector.

According to FAO (2007), the two main types of adaptation are autonomous and planned adaptation. Autonomous adaptation is the reaction of, for example, a farmer to changing precipitation patterns through changing crops or planting dates. Planned adaptation measures, on the other hand, are conscious policy options or response strategies, often multi-sectoral in nature, aimed at altering the adaptive capacity of the agricultural system, or facilitating specific adaptations. Judicious use of water using supplementary irrigation systems, more efficient irrigation practices, and the adaptation and adoption of existing and new water harvesting technologies have been suggested as appropriate strategies to cope with these problems.

FAO (2016) identified broad adaptation themes (Table 6.2). These adaptation practices at the farm level can be complemented and supported by measures in other

Table 6.2 Options for adaptation to climate change at farm level (Source: FAO 2016)

Risk	Response
Changing climate conditions and climate variability and seasonality	Participate in monitoring schemes when available. Optimization of planting schedules such as sowing dates (including for feedstocks and forage). Plant different varieties, species, or cultivars of crops. Use of short duration cultivars. Varieties or breeds with different environmental optima may be required, or those with broader environmental tolerances. The use of currently neglected or rare crops and breeds should be considered. Early sowing enabled by improvements in sowing machinery or dry sowing techniques. Increased diversification of varieties or crops to hedge against the risk of individual crop failure. Use intercropping. Make use of integrated systems involving livestock and/or aquaculture to improve resilience. Change post-harvest practices, for example, the extent to which grain may require drying and how products are stored after harvest. Consider the effect of new weather patterns on the health and Well-being of agricultural workers.
Change in rainfall and water availability	Participate in monitoring schemes when available. Change irrigation practices. Adopt enhanced water conservation measures. Use marginal and wastewater resources. Make more use of rainwater harvesting and capture. In some areas, increased precipitation may allow irrigated or rain-fed agriculture in places where previously it was not possible. Alter agronomic practices. Reduced tillage to lessen water loss, similarly the incorporation of manure and compost, and other land use techniques such as cover cropping increase soil organic matter and hence improve water retention.
Increased frequencies of droughts, storms, floods, wildfire events, sea level rise	Participate in monitoring schemes with available General water conservation measures are particularly valuable at times of drought. Use flood, drought, and/or saline resistant varieties. Improved drainage, improved soil organic matter content, and farm design to avoid soil loss and gullying. Consider (where possible) increasing insurance cover against extreme events.
Pest, weed and diseases, disruption of pollinator ecosystem services	Participate in risk monitoring and preventing schemes when available. Use expertise in coping with existing pests and diseases. Build on natural regulation and strengthen ecosystem services.

sectors, such as agricultural R&D and innovation, and at landscape levels. They very often require farmers to engage with other food producers to share best practices and experience, and participate in risk monitoring systems (inform and be informed), so as to enhance community-based adaptation.

According to IPCC (2014b), adaptation options for agriculture include technological responses, enhancing smallholder access to credit and other critical production resources, strengthening institutions at local to regional levels, and improving market access through trade reform (medium confidence). Responses to decreased food production and quality include: developing new crop varieties adapted to changes in CO_2, temperature, and drought; enhancing the capacity for climate risk management; and offsetting economic impacts of land use change. Improving financial support and investing in the production of small-scale farms can also provide benefits. Expanding agricultural markets and improving the predictability and reliability of the world trading system could result in reduced market volatility and help manage food supply shortages caused by climate change.

The available adaptation strategies in five sectors (Yohe et al. 2007) are described as follows:

Agricultural cropping: Choice of crop and cultivar (Use of more heat/drought-tolerant crop varieties in areas under water stress; Use of more disease and pest tolerant crop varieties; Use of salt-tolerant varieties; Introduce higher yielding, earlier maturing crop varieties in cold regions); Farm management Altered application of nutrients/fertilizer; Altered application of insecticide/pesticide; Change planting date to effectively use the prolonged growing season and irrigation; and Develop adaptive management strategy at farm level.

Livestock production: Breeding livestock for greater tolerance and productivity; Increase stocks of forages for unfavorable time periods; Improve pasture and grazing management including improved grasslands and pastures; Improve management of stocking rates and rotation of pastures; Increase the quantity of forages used to graze animals; Plant native grassland species; Increase plant coverage per hectare; Provide local specific support in supplementary feed and veterinary service.

Fisheries: Breeding fish tolerant to high water temperature; Fisheries management capabilities to cope with impacts of climate change must be developed.

Development of agricultural biotechnologies: Development and distribution of more drought, disease, pest, and salt-tolerant crop varieties; Develop improved processing and conservation technologies in livestock production; and Improve crossbreeds of high productivity animals.

Improvement of agricultural infrastructure: Improve pasture water supply; Improve irrigation systems and their efficiency; Improve use/store of rain and snow water; Improve information exchange system on new technologies at national as well as regional and international level; Improve sea defense and flood management; Improve access of herders, fishers, and farmers to timely weather forecasts.

6.5 Sustainability

A Multidimensional Poverty Index for 2005–2015 reported nearly 1.5 billion people in developing countries as multidimensionally poor, and notes that deprivation also occurs in high-income countries (UNDP 2016). The majority of the multidimensionally poor live in South Asia (53.9%) and sub-Saharan Africa (33.5%), and more in rural than in urban areas; projections (which do not include climate change) suggest that poverty will increase by 2030 if development goals are not met (UNDP 2016).

The idea of sustainability was a key theme of the United Nations Conference on the Human Environment in Stockholm in 1972. There was a tangible need for a developmental concept that would allow reconciling economic development with environmental protection. In December 1983, the Secretary-General of the United Nations appointed the Brundtland Commission, or more formally, the United Nations World Commission on Environment and Development (UNWCED). The term sustainable development was coined in the paper *Our Common Future* (UNWCED 1987). Sustainable development was defined as "development that meets the needs of the present generation without compromising the ability of future generations to meet their own needs."

In September 2015, international community endorsed a universal agenda entitled "Transforming our World: the 2030 Agenda for Sustainable Development," known as the Sustainable Development Goals (SDGs). The 17 goals and 169 targets to be met by 2030 were developed with widespread participation and were adopted in 2012 for the overarching goals of sustaining people, prosperity, peace, partnerships, and the planet. The preamble to the SDGs announces "to take the bold and transformative steps which are urgently needed to shift the world onto a sustainable and resilient path." With their explicit aim to "leave no one behind," the SDGs provide a promising basis for addressing inclusive growth, shared prosperity, and multidimensional equalities (UNRISD 2016).

Over the last 2 decades, climate change has emerged as one of the most crucial issues for humankind, with serious implications for sustainable development. In order to meet the multifaceted challenges of climate variability and climate change, there is a need for wise and well-informed decision-making at every level from households, communities, countries, and regions, to international fora, including the UN Framework Convention on Climate Change. Those decisions will require, directly or indirectly, access to the best possible climate science and information, and effective application of this information through climate services (Sivakumar et al. 2011). Climate information delivers economic value by providing users whose activities are sensitive to climate conditions with a basis for making decisions.

The AR5 reported with high confidence that sustainable development is strongly connected to climate change and that disruptive levels of climate change would preclude sustainable development including reducing poverty (Denton et al. 2014; Fleurbaey et al. 2014). The AR5 concluded that risks from climate change are "unevenly distributed and generally greater for disadvantaged people and communities in countries at all level of development" (IPCC 2014a) and that multidimensional inequalities, often produced by uneven development processes, shape differential vulnerabilities to, and risks from climate change (IPCC 2014b).

The AR5 introduced the notion of "climate-resilient pathways," defined as "sustainable-development trajectories that combine adaptation and mitigation to reduce climate change and its impacts; they include iterative processes to ensure that effective risk management can be maintained" (IPCC 2014b). These pathways are best seen as "future trajectories of development" with a variety of alternative pathways at any possible scale, each of which requires an evaluation of associated risks and benefits regarding climate resilience (Denton et al. 2014). Climate-resilient pathways rely on flexible, innovative, and participatory problem solving and, especially in the context of severe effects of climate change, require transformational change, including transformation in social processes (Denton et al. 2014).

The outcomes of climate policy options can generally be observed in the three spheres related to the three pillars of sustainable development: the economic, the social, and the environmental sphere (Fleurbaey et al. 2014). Sustainability in the economy refers to the preservation of standards of living and the convergence of developing economies toward the level of developed countries. Sustainability in the social sphere refers to fostering the quality of social relations and reducing causes of conflicts and instability, such as excessive inequalities and poverty, lack of access to basic resources and facilities, and discriminations. Sustainability in the environmental sphere refers to the conservation of biodiversity, habitat, natural resources, and to the minimization of ecosystem impacts more generally. The way the three spheres (and pillars) flourish can be viewed as contributing to sustaining well-being for humans as well as for other living creatures. Human well-being depends on economic, social, and natural goods, and other living beings depend on the quality of the ecological system. It may therefore be convenient to summarize the multiple relevant considerations by saying that the ultimate end result, for sustainability assessment, is the wellbeing of all living beings.

As policymakers are concerned with the broader issues of SD, it is important to reflect on how climate risks and policies fit in the general outlook (Fleurbaey et al. 2014). Many factors determine the development pathway. Among the main factors that can be influenced by policy decisions, one can list governance, human and social capital, technology, and finance. Population size, behaviors, and values are also important factors. Managing the transition toward sustainable development also requires taking account of path dependence and potential favorable or unfavorable lock-ins (e.g., via infrastructures), and attention to the political economy.

End hunger, achieve food security, and improve nutrition is at the heart of the sustainable development goals (FAO 2016). The growing awareness of the impacts of climate change on agriculture is forcing decision makers to refocus on the sustainability of agricultural production. Broad concepts in sustainable agriculture encompass ecological, economic, and social parameters, whereas more narrowly defined concepts are mostly concerned with environmental issues such as optimal resource and environmental management (McCracken and Pretty 1990). The notion that sustainable agricultural systems "maintain output in spite of major disturbance, such as caused by intensive stress or large perturbation" (Conway 1985) is of particular relevance in the current concerns with the possible impacts of climate change on the agroecosystem.

Sustainable agriculture has offered considerable scope for achieving resilience over the years, and now these efforts need to be further enhanced. Agronomic management practices have a major role to play to enhance the resilience in addition to production enhancement. One of the crucial issues in this context is farm decision-making which is seen as an ongoing process, whereby producers are continually making short- and long-term decisions to manage risks emanating from a variety of climatic and non-climatic sources.

Sivakumar et al. (2000) highlighted some of the priorities for agrometeorologists to address sustainable agriculture in the twenty-first Century. These include improvement and strengthening of agrometeorological networks; development of new sources of data for operational agrometeorology; improved understanding of natural climate variability, promotion and use of seasonal to interannual climate forecasts; establishment and/or strengthening of early warning and monitoring systems and promotion of geographical information systems and remote sensing applications; and agroecological zoning for sustainable management of farming systems, forestry, and livestock. Other priorities include use of improved methods, procedures, and techniques for the dissemination of agrometeorological information; development of agrometeorological adaptation strategies to climate variability and climate change; mitigation of the effects of climate change; more active applications of models for phenology, yield forecasting, etc.; active promotion of tactical applications such as response farming at the field level and promoting a better understanding of the interactions between climate and biological diversity.

The opportune use of agrometeorological information in crop planning and management not only results in higher crop yield but also lower the production costs with a host of other side benefits such as better product quality, environmental protection, efficient water management, and combat climate change. It also makes land management far more efficient through appropriate microclimate modification, crop, and soil management (Rathore 2020). The decisions farmers make have a significant impact on the returns to their investments and on their overall family welfare. The climate-based decisions that farmers make are mainly strategic in nature, e.g., choice of a crop/cropping system, allocation of acreage, purchase of inputs such as seed and fertilizer ahead of the cropping season. In contrast, the weather-based decisions are tactical in nature and affect operational activities such as sowing, fertilizer application, irrigation, weeding, and harvesting. Farm-level risk management strategies have to deal with both the changing and the variable climatic conditions as well as the weather conditions.

6.6 Conclusions

Agriculture is the most weather-dependent and climate-sensitive sector. There are many vulnerable regions in the world to climate change in view of the high population, the large number of people facing food insecurity and projected impacts of climate change on the agriculture sector. Climate change adaptation strategies are now a matter of urgency, especially for the most vulnerable communities. In order to increase and stabilize agricultural production to meet the increasing demands for food, it is important to emphasize the necessity to make more efficient use of natural resources while ensuring at the same time the conservation of the environment. There is a need for better assessment of risks associated with variable and uncertain environmental conditions. The National Meteorological and Hydrological Services (NMHSs) have much to contribute to addressing this priority and the timely availability of weather and climate information and early warnings could facilitate both strategic and tactical decisions in increasing and sustaining agricultural production.

Urgent steps are needed to increase the adaptive capacity to face current as well as future climate risks. The adaptation strategies will need to simultaneously consider the background of changing demand due to population growth as well as the socioeconomic and environmental consequences of possible adaptation options. Comprehensive profiling of the user communities in collaboration with social scientists and regular dialogue with the users could help identify the important needs and opportunities to help the user communities effectively deal with climate change impacts.

A universally preferable solution is to start with adaptation strategies that are anyway needed for sustainable development. Strategies that maximize synergies between adaptation, food production, and sustainable development would be most appropriate.

References

Allen MR, Dube OP, Solecki W, Aragón-Durand F, Cramer W, Humphreys S, Kainuma M, Kala J, Mahowald N, Mulugetta Y, Perez R, Wairiu M, Zickfeld K (2018) Framing and Context. In: Masson-Delmotte V, Zhai P, Pörtner H-O, Roberts D, Skea J, Shukla PR, Pirani A, Moufouma-Okia W, Péan C, Pidcock R, Connors S, Matthews JBR, Chen Y, Zhou X, Gomis MI, Lonnoy E, Maycock T, Tignor M, and Waterfield T (eds) Global warming of 1.5°C. An IPCC Special Report on the impacts of global warming of 1.5°C above pre-industrial levels and related global greenhouse gas emission pathways, in the context of strengthening the global response to the threat of climate change, sustainable development, and efforts to eradicate poverty

Battisti DS, Naylor RL (2009) Historical warnings of future food security with unprecedented seasonal heat. Science 323:240–244

Berkeley Earth (2019) Global temperature report for 2018. Berkeley Earth, Berkeley, California

Bindoff NL et al. (2013) Detection and attribution of climate change: from global to regional – supplementary material. In Stocker TF, Qin D, Plattner G-K, Tignor M, Allen SK, Boschung J, Nauels A, Xia Y, Bex V, Midgley PM (eds) Climate change 2013: the physical science basis. Contribution of working group I to the fifth assessment report of the intergovernmental panel on climate change Cambridge University Press, Cambridge and New York, NY, pp 1–25

Bryant CR, Johnston TRR (1992) Agriculture in the city's countryside. University of Toronto Press, Toronto

Chattopadhyay N, Lal B (2007) Agrometeorological risk and coping strategies – perspective from Indian subcontinent. In: Sivakumar MVK, Motha R (eds) Managing weather and climate risks in agriculture. Springer, Berlin Heidelberg, pp 83–98

Chowdhury A, Dandekar MM, Raut PS (1989) Variability of drought incidence over India: a statistical approach. Mausam 40:207–214

Christensen JH, Krishna Kumar K, Aldrian E, An S-I, Cavalcanti IFA, de Castro M, Dong W, Goswami P, Hall A, Kanyanga JK, Kitoh A, Kossin J, Lau N-C, Renwick J, Stephenson DB, Xie S-P, Zhou T (2013) Climate phenomena and their relevance for future regional climate change. In Stocker TF, Qin D, Plattner G-K, Tignor M, Allen SK, Boschung J, Nauels A, Xia Y, Bex V, Midgley PM (eds) Climate change 2013: the physical science basis. Contribution of working group I to the fifth assessment report of the intergovernmental panel on climate change, Cambridge and New York, Cambridge University Press

Church JA, Clark PU, Cazenave A, Gregory JM, Jevrejeva S, Levermann A, Merrifield MA, Milne GA, Nerem RS, Nunn RD, Payne AJ, Pfeffer WT, Stammer D, Unnikrishnan AS (2013). Sea level change. In Stocker TF, Qin D, Plattner G-K, Tignor M, Allen SK, Boschung J, Nauels A, Xia Y, Bex V, Midgley PM (eds). Climate change 2013: the physical science basis. Contribution of working group I to the fifth assessment report of the intergovernmental panel on climate change. Cambridge and New York, Cambridge University Press

Cramer W et al. (2014) Detection and attribution of observed impacts. In Field CB, Barros VR, Dokken DJ, Mach KJ, Mastrandrea MD (eds) Climate change 2014: impacts, adaptation, and vulnerability. Part a: global and sectoral aspects. Contribution of working group II to the fifth assessment report of the intergovernmental panel on climate change. Cambridge University Press, Cambridge and New York, NY, pp 979–1037

Denton F et al. (2014) Climate-resilient pathways: adaptation, mitigation, and sustainable development. In Field CB, Barros VR, Dokken DJ, Mach KJ, Mastrandrea MD (eds) Climate change 2014: impacts, adaptation, and vulnerability. Part A: global and sectoral aspects. Contribution of working group II to the fifth assessment report of the intergovernmental panel of climate change, Cambridge University Press, Cambridge and New York, NY, pp 1101–1131

Elbehri A, Burfisher M (2015) Economic modelling of climate impacts and adaptation in agriculture: a survey of methods, results and gaps. In: Elbehri A (ed) Climate change and food systems: global assessments and implications for food security and trade. Food and Agriculture Organization of the United Nations, Rome

Elbehri A, Elliott J, Wheeler T (2015) Climate change, food security and trade: an overview of global assessments and policy insights. In: Elbehri A (ed) Climate change and food systems: global assessments and implications for food security and trade. Food and Agriculture Organization of the United Nations, Rome

FAO (2007) The state of food and agriculture 2007. Paying farmers for environmental services. Food and Agriculture Organization of the United Nations, Rome

FAO (2008) Climate change adaptation and mitigation: challenges and opportunities for food security. Information document prepared for the high level conference on World Food Security: the Challenges of Climate Change and Bioenergy, 3–5 June 2008. Food and agriculture Organization of the United Nations, Rome

FAO (2013) Report of the first meeting of the plenary assembly of the global soil partnership, (Rome, 11–12 June 2013). Hundred and forty-eighth session, 2–6 December 2013. Food and Agriculture Organization of the United Nations, Rome

FAO (2015) The impact of natural hazards and disasters on agriculture and food and nutrition security: a call for action to build resilient livelihoods. Food and Agriculture Organization of the United Nations, Rome

FAO (2016) Climate change and food security: risks and responses. Food and Agriculture Organization of the United Nations, Rome

Fleurbaey M, Kartha S, Bolwig S, Chee YL, Chen Y, Corbera E, Lecocq F, Lutz W, Muylaert MS, Norgaard RB, Okereke C, Sagar AD (2014) Sustainable development and equity. In Edenhofer O, Pichs-Madruga R, Sokona Y, Farahani E, Kadner S, Seyboth K, Adler A, Baum I, Brunner S, Eickemeier P, Kriemann B, Savolainen J, Schlömer S, von Stechow C, Zwickel T, Minx JC (eds) Climate change 2014: mitigation of climate change. Contribution of working group III to the fifth assessment report of the intergovernmental panel on climate change. Cambridge University Press, Cambridge and New York

Guilyardi E, Wittenberg A, Fedorov A, Collins M, Wang C, Capotondi A, van Oldenborgh GJ, Stockdale T (2009) Understanding El Niño in ocean-atmosphere general circulation models: progress and challenges. Bull Amer Meteor Soc 90:325–340

Hansen G, Stone D (2016) Assessing the observed impact of anthropogenic climate change. Nature Clim Change 6:532–537

Hay J (2007) Extreme weather and climate events, and farming risks. In: Sivakumar MVK, Motha R (eds) Managing weather and climate risks in agriculture. Springer, Berlin Heidelberg, pp 1–19

Hoegh-Guldberg O, Jacob D, Taylor M, Bindi M, Brown S, Camilloni I, Diedhiou A, Djalante R, Ebi KL, Engelbrecht F, Guiot J, Hijioka Y, Mehrotra S, Payne A, Seneviratne SI, Thomas A, Warren R, Zhou G (2018) Impacts of 1.5°C global warming on natural and human systems. In Masson-Delmotte V, Zhai P, Pörtner H-O, Roberts D, Skea J, Shukla PR, Pirani A, Moufouma-Okia W, Péan C, Pidcock R, Connors S, Matthews JBR, Chen Y, Zhou X, Gomis MI, Lonnoy E, Maycock T, Tignor M, Waterfield T (eds) Global warming of 1.5°C. An IPCC special report on the impacts of global warming of 1.5°C above pre-industrial levels and related global greenhouse gas emission pathways, in the context of strengthening the global response to the threat of climate change, sustainable development, and efforts to eradicate poverty. IPCC Secretariat, WMO, Geneva

IFAD (2011) Rural poverty report 2011. New realities, new challenges: new opportunities for tomorrow's generation. International Fund for Agricultural Development, Rome

IPCC (2012) In: Field CB, Barros C, Stocker TF, Qin D, Dokken DJ, Ebi KL, Mastrandrea MD, Mach KJ, Plattner G-K, Allen SK, Tignor M, Midgley PM (eds) Managing the risks of extreme events and disasters to advance climate change adaptation. Cambridge University Press, Cambridge, p 582

IPCC (2013) Climate change 2013: the physical science basis. In Stocker TF, Qin D, Plattner G-K, Tignor M, Allen SK, Boschung J, Nauels A, Xia Y, Bex V, Midgley PM (eds) Contribution of working group I to the fifth assessment report of the intergovernmental panel on climate change. Cambridge University Press, Cambridge and New York, NY, p 1535

IPCC (2014a) Climate change 2014: synthesis report. In Core Writing Team, Pachauri RK, Meyer LA (eds) Contribution of working groups I, II and III to the fifth assessment report of the intergovernmental panel on climate change. IPCC, Geneva, Switzerland, p 151

IPCC (2014b) Summary for policymakers. In: Field CB, Barros VR, Dokken DJ, Mach KJ, Mastrandrea MD (eds) Climate change 2014: impacts, adaptation, and vulnerability. Part A: global and sectoral aspects. Contribution of working group II to the fifth assessment report of the intergovernmental panel on climate change. Cambridge University Press, Cambridge and New York, NY, pp 1–32

IPCC (2018) Summary for policymakers. In: Masson-Delmotte V, Zhai P, Pörtner H-O, Roberts D, Skea J, Shukla PR, Pirani A, Moufouma-Okia W, Péan C, Pidcock R, Connors S, Matthews JBR, Chen Y, Zhou X, Gomis MI, Lonnoy E, Maycock T, Tignor M, Waterfield T (eds) Global warming of 1.5°C. An IPCC special report on the impacts of global warming of 1.5°C above pre-industrial levels and related global greenhouse gas emission pathways, in the context of strengthening the global response to the threat of climate change, sustainable development, and efforts to eradicate poverty. IPCC, Geneva, Switzerland

IUCN (2010) Building climate change resilience for African livestock in sub-Saharan Africa. World initiative for sustainable pastoralism (WISP): a program of IUCN eastern and southern Africa regional office, Nairobi, March 2010, pp viii + 48

Jiménez Cisneros BE, Oki T, Arnell NW, Benito G, Cogley JG, Döll P, Jiang T, Mwakalila SS (2014) Freshwater resources. In: Field CB, Barros VR, Dokken DJ, Mach KJ, Mastrandrea MD (eds) Climate change 2014: impacts, adaptation, and vulnerability. Part A: global and sectoral aspects. contribution of working group II to the fifth assessment report of the intergovernmental panel on climate change. Cambridge University Press, Cambridge and New York, NY, pp 229–269

JRC/PBL (2013) Emission database for global atmospheric research (EDGAR), Release Version 4.2 FT2010. European Commission, Joint Research Centre (JRC)/PBL Netherlands Environmental Assessment Agency. Available at: http://edgar.jrc.ec.europa.eu

Lal R (1991) Soil structure and sustainability. J Sustain Agr 1:67–92

Lobell DB, Burke MB, Tebaldi C, Mastrandrea MD, Falcon WP, Naylor RL (2008) Prioritizing climate change adaptation needs for food security in 2030. Science 319:607–610

McCracken JA, Pretty J (1990) Glossary of selected terms in sustainable agriculture. Gatekeeper series no. SA6, International Institute for Environment and Development, London

Munich Re (2019) Natural disasters overview. Munich Re, Munich, Germany. https://www.munichre.com/topics-online/en/climate-change-and-natural-disasters/natural-disasters.html

Munich Re NatCatservice (2020) Relevant natural loss events worldwide 1980–2019. Munich Re, Munich, Germany

Niang I, Ruppel OC, Abdrabo MA, Essel A, Lennard C, Padgham J, Urquhart P (2014) Africa. In: Barros VR, Field CB, Dokken DJ, Mastrandrea MD, Mach KJ, Bilir TE, Chatterjee M, Ebi KL, Estrada YO, Genova RC, Girma B, Kissel ES, Levy AN, MacCracken S, Mastrandrea PR, White LL (eds) Climate change 2014: impacts, adaptation, and vulnerability. Part B: Regional aspects, Contribution of working group II to the fifth assessment report of the intergovernmental panel on climate change. Cambridge University Press, Cambridge and New York, NY, pp 1199–1265

Olmstead CW (1970) The phenomena, functioning units and systems of agriculture. Geogr Pol 19:31–41

Porter JR, Xie L, Challinor AJ, Cochrane K, Howden SM, Iqbal MM, Lobell DB, Travasso MI (2014) Food security and food production systems. In: Field CB, Barros VR, Dokken DJ, Mach KJ, Mastrandrea MD, Bilir TE, Chatterjee M, Ebi KL, Estrada YO, Genova RC, Girma B, Kissel ES, Levy AN, MacCracken S, Mastrandrea PR, White LL (eds) Climate change 2014: impacts, adaptation, and vulnerability. Part A: global and sectoral aspects, Contribution of working group II to the fifth assessment report of the intergovernmental panel on climate change. Cambridge University Press, Cambridge and New York, pp 485–533

Rathore LS (2020) Challenges and opportunities in agrometeorology. J Agrometeorol 22:3–6

Sivakumar MVK, Stefanski R (2011) Climate change in South Asia. In: Lal R, Sivakumar MVK, Faiz SMA, Mustafizur Rahman AHM, Islam KR (eds) Climate change and food security in South Asia. Springer, New York and London, pp 13–30

Sivakumar MVK, Gommes R, Baier W (2000) Agrometeorology and sustainable agriculture. Agric For Meteorol 103:11–26

Sivakumar MVK, Bessemoulin P, Peterson TC, Asrar G (2011) Changing climate and demands for sustainable development. Clim Res 47:3–4

Smil V (2017) Energy and civilization: a history. MIT Press, Cambridge, MA, p 568

Sun Y, Zhang X, Ren G, Zwiers FW, Hu T (2016) Contribution of urbanization to warming in China. Nat Clim Chang 6:706–709

Swindale LD (1988) The impact of agricultural development on the environment: an IARC point of view. Consultative Group on International Agricultural Research (CGIAR) Mid-Term Meeting, 15–19 May 1988, West Berlin, Germany

Thomas RJ (2008) Opportunities to reduce the vulnerability of dryland farmers in central and West Asia and North Africa to climate change. Agr Ecosys Environ 126:36–45

Thornton PK, Van de Steeg J, Notenbaert A, Herrero M (2009) The impacts of climate change on livestock and livestock systems in developing countries: a review of what we know and what we need to know. Agric Syst 101:113–127

UNDP (2016) Human development report 2016: human development for everyone. United Nations Development Programme, New York, p 286

UNFPA (2011) The state of world population 2011. United Nations Population Fund, New York

UNRISD (2016) Policy innovations for transformative change: implementing the 2030 agenda for sustainable development. United Nations Research Institute for Social Development, Geneva, p 248

UNWCED (1987) Our common future. Report of the United Nations World Commission on Environment and Development. Oxford University Press, Oxford, p 383

Wan H, Zhang X, Zwiers F (2018) Human influence on Canadian temperatures. Clim Dyn 52:479–494

Yohe G, Burton I, Huq S, Rosegrant MW (2007) Adaptation, mitigation and climate risk. A synthesis with emphasis on Asia. Prepared for Policy Forum Agricultural and Rural Development for Reducing Poverty and Hunger in Asia: In Pursuit of Inclusive and Sustainable Growth. 9–10 August 2007, Asian Development Bank, Manila, Philippines

Enhancing Climate Service Delivery Mechanisms in Agriculture Sector to Cope with Climate Change

7

Anshul Agarwal, G. Srinivasan, Mitesh V. Sawant, and Kareff Rafisura

Abstract

A review of available weather and climate-related information products and services for the agriculture sector and their delivery mechanism is presented. This study draws upon available literature and author's experience from the project implemented in South Asia, particularly in India, by Regional Integrated Multi Hazard Early Warning Systems (RIMES). One of the highlights of this chapter is a case study that demonstrates the advantage of involving field-based NGOs and farmers to improve the delivery mechanism of agro-advisory services. As climate services involve a wide variety of forecasts, climate outlooks, and projections, it is imperative to understand and link them with appropriate decision-making institutions and communities for building long-term climate resilience. Climate services for the agriculture sector are a positive development from the perspective of enabling farmers to adapt to climate change by building their experience to deal with the present-day climate. Climate services can also help develop sustainable and economically viable agricultural systems, increase efficiency in the use of water, labor and energy, conserve natural resources, and decrease pollution by agricultural chemicals or other agents that contribute to the degradation of the environment. Field-level work demonstrates that more information does not automatically translate into better farm services unless farmers are empowered to use it in farm-level decisions.

A. Agarwal (✉) · G. Srinivasan · M. V. Sawant
Regional Integrated Multi Hazard Early Warning Systems, Klong Luang, Pathumthani, Thailand
e-mail: anshul@rimes.int

K. Rafisura
United Nations Economic and Social Commission for Asia and the Pacific, Bangkok, Thailand

A. Kaushik et al. (eds.), *Climate Resilience and Environmental Sustainability Approaches*, https://doi.org/10.1007/978-981-16-0902-2_7

Keywords

Climate services · Agriculture · Resilience · Agro-advisory

7.1 Introduction

Every aspect of the agricultural system from food production to distribution and food storage is affected by the weather and climate. Knowledge of climate and forthcoming weather can enhance preparedness to cope with climate hazards even if the predictions have limitations of spatial and temporal accuracy. Agriculture in a location or region has evolved over centuries as a combination of human needs and the physical environment controlled by soil, climate, and ecology of the place. As technology and industries developed at a rapid rate since the beginning of the twentieth century, human needs started dictating agricultural development, many times pushing the limits of the physical environment which exposes the agriculture sector even more to weather and climate variabilities. The growing realization to put agriculture into an environmentally sustainable path is now beginning to drive changes in agricultural systems. Climate knowledge is an important and primary driver that is likely to lead this change. Communities in various parts of the world have displayed complex adaptation strategies to deal with climatic variability by building water harvesting structures, water-regulating structures, and by resorting to agricultural strategies such as crop diversification (Shanmugasundaram et al. 2017).

Climate change is expected to result in warmer temperatures, increased rainfall variability, and extreme weather that may pose substantial additional risk to food security. Climate change is also accompanied by increased weather uncertainty (Lobo et al. 2017). This may possess a serious threat to agriculture production in future if current agriculture practices are not adapted to changing climate. Nelson et al. (2009) reported that in South Asia rice production is set to fall by 14%, wheat production by up to 49%, and maize production by up to 19% by 2050. Sustaining high water consumption cropping systems such as the paddy systems on a large scale may become increasingly challenging due to increasing climate stress. This could undermine food security and the livelihoods of millions of people.

The agriculture sector in India employs more than 50% of the total workforce in India and contributes around 17–18% to the country's GDP (Gopalakrishnan and Subramanian 2020). However, the unpredictability of weather negatively affects Indian agriculture, and climate change will further worsen this condition. The increased incidence of climate-related disasters such as droughts and floods, high incidences of pests and diseases, increased demand for water along with the need to increase agricultural production to satisfy the increasing demand necessitate the effective utilization of available climate information and weather and climate forecasts. Hence it is imperative to understand the requirements of farmers related to weather and climate and enhance its application in their decision-making to minimize the risk in agriculture.

The India Meteorological Department (IMD) generates weather and climate forecasts and agrometeorological bulletins with the support of Agro Meteorological Farm Units (AMFUs) on a regular basis for all the districts in India. Many international research centers like the Global Producing Centres (GPCs) of the World Meteorological Organization (WMO) and regional institutes such as Regional Integrated Multi Hazard Early Warning Systems (RIMES), support national agencies to produce state-of-the-art forecasts for different timescales. The forecasts and bulletins are disseminated regularly to users, which include irrigation agencies, disaster and management agencies, local government units, farmers, agricultural planners, hydrologists, and researchers.

Despite the availability of state-of-the-art forecasts and outlooks at different lead-times, ranging from 3-days up to a season in advance (3–6 months), using such information to make decisions is still a challenge for the farmers (Agarwal et al. 2018). Underutilization of forecast information is due to many reasons: information does not reach at a time when it could have made a difference; it is not clearly understood; and users do not trust the information enough to make critical decisions that involve considerable financial consequences. For example, it is difficult for a smallholder farming household to change its usual cropping practice unless it has confidence in the information. Thus, it is required to build the capacity of farmers to utilize the available weather and climate information products and apply the wealth of traditional knowledge to overcome the challenges due to climate variability and change. To build the confidence of farmers, there is a need to capacitate them to better manage resources and risks through the utilization of climate information of different timescales.

Another factor that limits the use of climate information and services to agriculture is weak institutional linkages and integration of decisions across timescales. Lack of awareness and appreciation of climate information leads to poor demand. Capacities for service provision are also at a very rudimentary level because of scarce resources, untrained manpower, and sparse climate data. The available information and data do not link to key user sectors in agriculture owing to the following reasons (Srinivasan et al. 2019):

- Besides farm-level decision-making for each growing season, there is a need to consider climate information for agricultural long-term planning. As small land in a climatically suitable location can produce the same amount of food grain as from a larger area from an unsuitable area. Climate knowledge can be useful for land use planning and thereby bring balance between fallow land and land used for the Urbanization and development of industries.
- Most climate information is focused on responding to the needs of crop production. The inclusion and development of sub-sectors like animal husbandry, fisheries, and agroforestry, through climate information/climate knowledge can help increase the income of farmers and can make agriculture an attractive business.

- Crop diversification and organic farming are the future directions that a large part of the sector is moving toward, and tailored climate information is required to support these efforts.

Effective weather and climate information and advisory services can inform the decision-making of farmers and improve their management of agricultural risks. Such services can help develop sustainable and economically viable agricultural systems, improve production and quality, reduce losses and risks, decrease costs, increase efficiency in the use of water, labor, and energy, conserve natural resources, and decrease pollution by agricultural chemicals or other agents that contribute to the degradation of the environment. For better management of risk, the farmers need to meet the following requirements:

- Knowledge about weather and climate, variability in climate, and its impact on farming activities.
- Understanding the causes of climate variabilities and the processes involved in weather and climate.
- Historical knowledge about the extreme weather events of the locality where the operations are performed.
- Skills for analyzing the forecasts and bulletins in their farming context.
- Ability to apply advanced knowledge about the weather extremes in decision-making.

On-farm adaptation measures, such as changing cropping dates or varieties, will continue to be necessary to adapt to weather variability and to shifts in temperature and precipitation, but investments in various on-farm infrastructure measures, such as more efficient irrigation systems or water storage. Because the benefits of most adaptation measures are local and directly captured by farmers, self-interest should be a sufficient incentive for individuals to adapt, i.e., farmers will act when the private benefits they generate outweigh the costs.

Thus, it is required to build the capacity and confidence of farmers for utilization of the available weather and climate information products and apply the wealth of traditional knowledge to overcome the challenges brought by climate variability and change. To build the confidence of farmers, there is a need to capacitate them to better manage resources and risks through the utilization of climate information of different timescales. The big challenge, also, is to develop a mechanism that will effectively disseminate the information in a language that could be easily understood at different levels to obtain the maximum benefits from the information they receive.

This study presents a review of available weather and climate-related information products and services in the agriculture sector and their delivery mechanism, drawing from both the literature and the authors' observations and firsthand experiences in South Asia. A case study is presented to demonstrate how the involvement of field-based NGOs and farmers improves the delivery mechanism of agro-advisory services.

7.2 Emerging Information Products

The importance of forecast is realized even in traditional societies and many traditional indigenous ways exist in every community for this purpose. Indigenous agrometeorological knowledge of farmers has always been very useful in managing environmental risks. This indigenous knowledge is a product of keen observations, centuries of trial and error, and natural selection and thus is very effective for long and will remain effective in future. In Asia, the farmers usually have a certain degree of knowledge about astrology to understand the weather and decide their farming activities linked with the movement of planets. Rainfall was mostly forecasted on the basis of predetermined position of the planet. The first formal agrometeorological service developed in India which is still in use was the classical Hindu almanac called "Panchanga" (Vanadeep et al. 2012). It is in the form of a book containing the details about astronomical phenomena with all details about calendar days, weeks, and months. Based on this system, cropping calendar, types of crops, and all farming activities are decided for a state or country.

With the development of science and availability of modern computing facilities, various state-of-the-art products are emerging for forecasting weather and climate for agriculture. The establishment of numerical weather prediction has dramatically improved the weather prediction process in the twentieth century. Since the establishment of the first numerical integration experiment in the 1950s, the possibility to access the weather forecast at a large temporal and spatial scale has increased (Harper et al. 2007; Lynch 2008). General forecasts are classified into various categories (Fig. 7.1) such as, nowcast (with a lead time of 1–6 h for a specific geographic area), very short-range forecast (up to 12 h for local weathers), short-range forecast (12–72 h) medium-range forecast (7–10 days), monthly extended forecast and long-range or seasonal forecast with a certain degree of accuracy. The forecast could be either deterministic or probabilistic depending on the lead time. Increased lead time is generally linked with decreasing quality of forecast and thus limits the applications at farm scale. An extensive review of agrometeorological advisory services in India is conducted by Gopalakrishnan and Subramanian (2020).

In recent years, national meteorological services and many private companies have started more specific types of prediction for the farmers. To become relevant for agriculture, climate information need to meet the following (Srinivasan et al. 2019):

- *Climate data*: Good coverage and quality long-term climate data are essential to building robust climate services. While agencies external to the NMHSs can contribute substantially toward the collection of climate data, an effective collaboration from the very beginning is key to sustained sector-specific monitoring systems. Observed data needs to be well organized and accessible through appropriate Climate Data Management Systems (CDMS). NMHSs worldwide are now implementing such systems.
- *Operationalizing* systematic use of such information requires a well-coordinated end-to-end institutional system that begins with monitoring of weather and climate events and ends with a community-level response. The main challenge

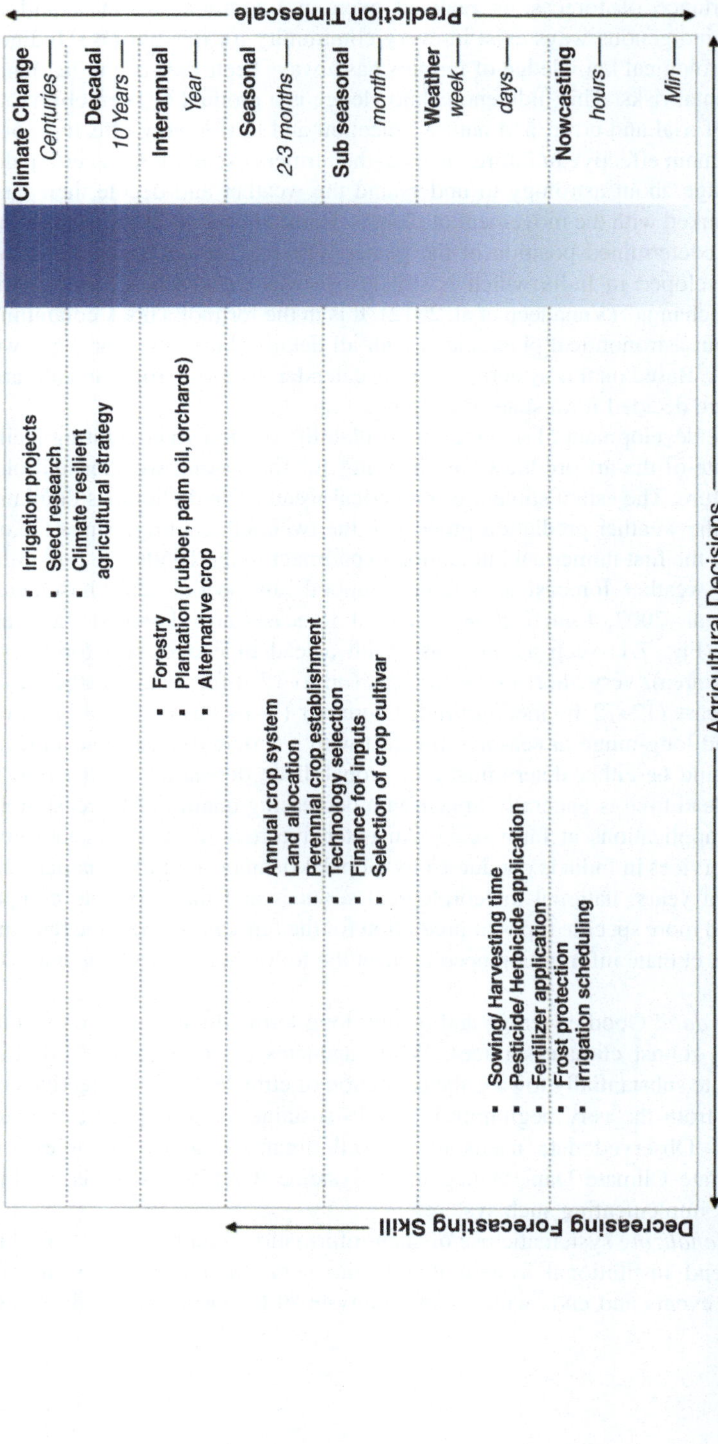

Fig. 7.1 Probable agricultural decisions based on different timescales of forecasts

of such a system would also be the customization of information to climate-sensitive points of decision-making. This needs to be evolved iteratively through an interactive process of stakeholder engagement.

- *Climate as resource:* Climate from an applications perspective needs to be viewed at a risk as well as a resource. Effective interface with users can deliver climate information that is clear, precise, and relevant. Working with stakeholders involves knowledge and understanding of a wide range of issues that run across the traditional science domains within which NMHSs generally function. Continued communications through National Climate Forums (NCFs) with updates through e-mails and website, videoconferencing, social networks, and text messages through cellular networks are crucial. Partnership with interface institutions and agencies working directly with target communities are essential to render climate services.
- *Impact-based forecasting:* There is currently a movement toward "impact-based forecasting" that is producing the impacts of weather and climate for the variables that are of interest to stakeholders. Impact-based forecasting suites that link climate forecasts to impact models can be operationalized by scaling up successful demonstration projects.
- *Integration with weather scale:* From a sector user perspective, climate information cannot be viewed separately from the information being provided by NMHSs on the weather time scales, typically short- and medium-range covering periods from 3 to 10 days ahead. This is because the sector generally tends to use climate and weather information "telescopically," with short lead-time information for responsive actions and longer lead-time information for strategic planning and preparedness. DSS, therefore, needs to bring both weather and climate information together, in a customized platform to address specific decision contexts.

The Agromet Advisory Services of the India Meteorological Department (IMD) are built on these principles and are aimed at "weatherproofing" farm production by providing location and crop-specific actionable weather and climate services. The most important products developed for agriculture are the following: Daily weather forecasts; Monthly climate outlooks; Seasonal climate outlooks; Observed climate change signals; Climate change vulnerability assessment maps; Advisories/Alerts on extreme climate events such as droughts and floods; Climate data, maps, and atlas; Agro-ecological maps; Dekadal Agrometeorological Bulletins; Seasonal distribution maps of onset and cessation of rains; Length of growing period maps; Maps on chances of dry and wet spell of given lengths; Maps on rainfall variability during different seasons. These products and services developed by IMD link in available technologies, best practices and aimed to go the last mile to reach all farmers (Rathore and Chattopadhya 2016).

7.3 Reach and Application in Existing Contexts and Challenges of Delivering These to Users

7.3.1 Reach and Application in Existing Contexts

The information generated in the climate services should be appropriate, timely, and reliable to the user-sector to make an important livelihood decision that enhanced their resilience to climate shocks (Nkiaka et al. 2019). The poor infrastructure (Sivakumar et al. 2014), limited resources, and illiteracy are a few of the significant issues faced by the rural population in India. In such circumstances finding the appropriate delivery mechanisms for providing climate service is a challenging task.

7.3.2 Delivery of Products and Services

The climate services providers such as NMHS's (National Meteorological and Hydrological Services) employed several delivery mechanisms to push the relevant climate-related information to the agriculture sector targeting the farming community (Sivakumar et al. 2014; Ramakrishna 2013). The various delivery channels could be categorized as interactive communications and push services for information dissemination. The interactive communication channels take into account direct participation of the farmers via face-to-face interaction, village meetings, "Kisan Melas," farmers club, and conversation with extension and private commercial agents and Kisan call centers. Whereas, the push services involve publishing the climate information via local newspapers, websites, village notice boards, communicating through local radio, tv, mobile, SMS, and microphone announcement. The direction of flow of information in case of interactive communication is bilateral, while the latter has only a single direction of information flow.

Climate services delivery mechanisms	Interactive communications (Active channel)	Push services (Passive channel)
Modes of delivery	Face to face, village meetings, kisan melas, farmers club, conversation with extension and private agents, kisan call centers	Local newspaper, TV, radio, website, notice boards, SMS, voice SMS

The interactive communications and push services have their benefits as well as limitations for the user sector and service providers. For example, radio and television broadcasting timings are sometimes not appropriate for most of the farmers (Mittal and Mehar 2013). Whereas, another study concluded that farmers in India rely and trust the information from fellow farmers compared to other sources (Tall et al. 2014a, b; Mittal and Mehar 2013; Jha et al. 2012).

The effectiveness of individual and combined channels has been studied earlier in various researches (Tall et al. 2014a, b; Mittal and Mehar 2013; Jha et al. 2012; Dileepkumar et al. 2014; Vedeld et al. 2019). The studies also showed numerous requirements of the farmers' community were not fulfilled through a single source or channel and encouraged the farming community to rely on multiple sources rather than a single source of Information (Mittal and Mehar 2013; Tall et al. 2014a, b). The situation assessment survey of farmers in India conducted by the National Sample Survey Organization (NSSO) showed that progressive farmers and radio/TV/newspaper/Internet were the two primary sources accessed by the agricultural households for technical advice.

With the advancement of information and communications technology (ICT), the use of mobile phones for information dissemination has increased in the southern parts of Asia and Africa (Mittal and Mehar 2013). However, few studies have also confirmed that it would be difficult for advanced technologies to completely replace the traditional modes of information delivery (Mittal and Mehar 2013; Jha et al. 2012).

7.4 Receptivity and Capacity of Users to Understand and Use

Although several channels used for the dissemination of climate information in the agriculture sector, the receptivity and actual use of climate information is still unclear (Singh et al. 2018).

The good receptivity of climate information is directly attributed to the modes to information delivery, its accessibility, reliability, accuracy, timeliness, and financial benefits (Dileepkumar et al. 2014). The responsiveness to climate information is also determined by the socioeconomic characteristics of the farming community, such as its level of education, age, and farm size. The delivery of the information should be in the local language as most of the users from rural communities have difficulty in reading or typing the message on their mobile phones (Mittal and Mehar 2012).

The level and quality of interaction between the information producer and user decide the usability of the climate information produced, as well as how the information "fits" processes of decision-making. However, there is a very loose connection found between knowledge production and its use (Singh et al. 2016, 2018). The current format of many agro-advisories still do not lend themselves to easy translation for application in decision-making context. As the decision-making is itself an overly complicated procedure involving various factors such as climate, agronomy, economy and sociology, and financial capacity (Singh et al. 2018).

It was also found that the current short-term decisions are being taken without considering their long-term effects. Also use of long-term climate projection in local policies and planning is very limited in South Asia. The reasons may be the lack of reliable historical observations, both to understand the current climate and evaluate climate models and the coarse scale of future climate projections (Singh et al. 2018).

7.4.1 Empowering Agriculture Community to Use Weather and Climate Information in their Decision Context

A large number of farmers in Asia are small landholders (FAO 2010), heavily dependent on rain, and highly sensitive to extreme climate events (Dileepkumar et al. 2014). The farming community usually thinks about which input to use, when, and how much? Where to purchase. All these decisions are not directly related to climate information but could be resolved by supplementing climate services as input (Tall et al. 2014a, b).

The farming community all over SA is losing interest in the farming business. The country like India is losing more than 2000 farmers every single day, which will be a severe implication for food security in Asia (Banu and Mrunalini 2017). As per the survey conducted by the NSSO (National Sample Survey Organization of India), over 40% farmers would ready to quit farming business due to high risk in agriculture occupation, if provided with option to do so (Mehta 2011). The difficulty in the farming business could be eased to some extent by using information created by proper utilization of available climate information. The agriculture in general urgently required to strengthen agro-meteorological services and the understanding and use of these services by the farming community (Ramakrishna 2013).

Key issues with current climate services delivery mechanisms:

1. The interaction between various stakeholders of climate services is multimodal, multi-interface, and involves a multi-tier authorization system. There are a number of departments operating in isolation to serve the varying and different needs of the farmers.
2. Weak existence of agriculture extension system (Mittal and Mehar 2013; Ramakrishna 2013).
3. Weather-based advisories are not currently scaled-down in a robust fashion to meet the requirements of smallholder farmers.
4. The local forecast is not well-tailored or targeted and spatially adjusted to local conditions and changing circumstances (Vedeld et al. 2019). There is a gap between knowledge production and its use (Singh et al. 2018).

To enhance the climate services delivery mechanism, farming communities need to empower in interpreting and using climate information in their decision-making. It is very essential to understand the context of the decision-making by individual, household, and communities (Tall et al. 2014a, b). Decision-making within a household or a community, more broadly, and culturally, can have a robust gendered component (Tall and Njinga 2013). Thus, it also becomes essential for evaluators to define the gendered aspects of decision-making within households and their communities.

Access to the relevant information at the right time through the proper channels enables smallholder farmers to make informed decisions to prepare themselves for adverse climatic conditions, thereby improving their productivity and profitability while managing risks (Dileepkumar et al. 2014). There is a need to develop continuous and multiple ways of eliciting feedback from end-users and intermediary users of the services and utilize this information to continuously improve institutional design and co-creation processes (Vedeld et al. 2019).

Fundamental for the empowering agriculture community to use climate information in their decision-making:

1. Involve farmer the co-design, co-production, and co-evaluation of the climate services. The local indigenous knowledge could be utilized with integration of meteorological information to build the trust among rural users (Tall et al. 2014a, b).
2. Guiding the farming community in the translation and use of climate services through Farmers Field School (FFS) and climate-related training programs.
3. Exploit scalable, innovative, and financially viable model of the communication channel to reach the last miles and sustain itself over a long period (Tall et al. 2014a, b; Mittal and Mehar 2013).

Need to develop continuous and multiple ways of eliciting feedback from end-users and intermediary users of the services and utilize this information to continuously improve institutional design and co-creation processes (Vedeld et al. 2019).

7.5 Case Study of Tamil Nadu: Integrating Products to Users

While a large range of weather and climate information is now available for farmers, some types of information under development or already operational, particularly climate forecasting, formation, may be ill-suited for use by farmers for their decision-making. Stone and Meinke (2006) brought out the important challenges of linking meteorological and climatological information with a wide range of farming decisions many of which remain valid even in the current context. They showed the importance of involving the boundary organizations at district level and participatory approach involving farmers with ownership of the process as success for linking climate and weather information and forecasts to farm decisions.

The Forecast Application in Risk Management (FARM) School in agriculture is one of such processes introduced by RIMES in Tamil Nadu and customized for the context of farmers in Delta districts of Tamil Nadu with help of local partners, including the Indian Meteorological Department (IMD), Department of Agriculture (DoA), and NGO partner Centre for Ecology and Research (CER). FARM School is a bottom-up process designed to bridge the gap in relaying the information generated to the farmers and to make sure that the information is clearly understood in their specific farm context. Designed as a season-long series of interfacing sessions among forecasters, agriculture officers, and farmers, the FARM School is a multi-tier learning-by-doing capacity building program customized for farmers in specific areas.

The objectives of the FARM school is to:

- Provide a mechanism for identification of farmers' information requirements for customizing/tailoring information products and services for the agriculture sector.
- Facilitate regular and better access to enhanced information products and services by building on available channels of information communication/dissemination, and harnessing extension workers–farmers network.

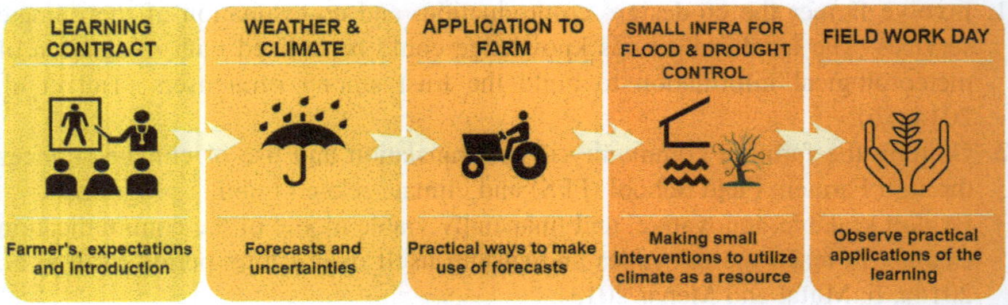

Fig. 7.2 FARM School learning process flowchart

- Enhance understanding, by farmers, of climate information of various timescales and build their capacity for seamless integration of the same into plans and decisions.
- Establish a robust mechanism for regularly obtaining farmers' feedback concerning information products and services amidst evolving capacities and gaps, technologies, risks, and other relevant concerns relating to agriculture.

The FARM School process consists of a training that captures and describes the changes in weather and its effects to crop water demand, water supply availability, crop growth and yield, and incidence of pest and diseases. The training is fully field-oriented, participatory and discovery learning based with the field as primary learning material. A multi-cycle training approach is adopted, which starts from the training of trainers (TOT) with experts from concerned agencies (e.g., India Meteorological Department (IMD)) and local experts identified by RIMES as trainers and agricultural officers at block level as trainees. It is followed by the training of the first batch of farmers with the trained agricultural officers as trainers this time. Emerging "champion farmers" during the training are then utilized as resource speakers to support the agricultural technicians. A "champion farmer" is a trainee who is articulate, smart, and shows authority and leadership during the training session. In the process, the trained agricultural technicians could organize several sessions simultaneously in their respective area. In every session, the learning process outlined in Fig. 7.2 is followed.

A web base platform "Specialized Expert system for Agro-meteorological early warning for climate resilient agriculture (SESAME)" was developed for the selected districts of Tamil Nadu. SESAME is a web portal, developed for providing simplified Agro-meteorological bulletins to farmers for faster and effective decision-making. The tool can generate crop advisory with weather forecast information at four different time scales: (a) seasonal (three months) outlook, (b) Monthly outlooks, (c) 10-day weather forecast (separated into two Pentads), and (d) 3-day high resolution weather forecast. The system works with the idea of mapping the sensitivity of a specific crop to certain weather conditions and comprehend how the weather in general, influencing the growth of a crop during the entire growth cycle. The tool is designed to be farmer-centric and requires very basic inputs to enumerate all the scheduled activities for a selected crop for a season. The system gathers this knowledge, progressively and uses it to devise the advisories for that crop for different weather conditions. The system thus is not limited to any specific crop rather scalable to include any number of crops with some historical knowledge about its characteristics. To meet this requirement the system combines a dynamic crop–weather calendar with sufficient information on the growth stages and sensitivity of the crop to different weather conditions.

The benefit FARM School process and SESAME brought are:

- FARM School process encourages participants to think and analyze what challenges they are facing and how these could be resolved.
- Farmers are now aware that so much information is available for free which could help them to take decisions for day-to-day farming activities.
- Training imparted them with skills to use the weather forecasts and take decisions based on the forecast. This brought a holistic change in approach toward farming because of the new level of understanding of forecasts.
- They are able to relate the agronomic practices to weather which was very nicely packaged in the training.
- Farmers are able to generate the advisories relevant to their own field based on forecast.
- Use of pesticides is significantly reduced as they are applying locally available organic materials to control pests and rodents.

The process does not end with the training but it continues in form of weekly advisories, guidance, and regular support for farmers. It continues until farmers get the full confidence of transforming their traditional farming to climate/weather informed and need based farming. Within the villages farmers now take collective decisions for farming operations and thus support each other during critical periods. It removes the stress from farmer's to decide for farming activity with so much uncertainty around. Now they can better plan each activity well in advance and also have the capacity to adjust for any weather surprises. In case of any difficulty or doubt they know where to look for.

Feedback from the farmers who participated in the FARM schools showed that despite the availability of agro-advisories they were unable to fully use it due to lack of understanding, interpretation and confidence in the information provided. The training helped them build their confidence to apply the information in their farming decisions. Also, involvement of local NGO's like CER in Tamil Nadu where experts were available to resolve any queries related to the forecasts and advisories and also to provide the suggestions for farming decisions built their trust in the process. RIMES introduced a similar mechanism in Bihar state and findings revealed the same—the involvement of local partners and easy access to customized products increase the trust and confidence of farmers.

7.6 Conclusions and Future Directions

The availability of a wide-range of climate information for the agriculture sector is a positive development from the perspective of enabling farmers to adapt to climate change by building their experience to deal with the climate now. However, the discussion above shows that more information does not automatically translate into better services for farmers. Experiences from Tamil Nadu and Bihar examined in this chapter reinforce the following key lessons:

User-aligned information and services: Climate information can more effectively assist farmers in making risk-informed farm decisions if it is aligned with their intended application in terms of the variables being forecasted, format, and timing of delivery. It is therefore imperative to bring farmers on board in the entire process from scoping of information requirements, product development, pilot testing, and evaluation of usability.

Information delivery mechanisms: A blended modality of active and passive information delivery seems to work well in many contexts. Innovations in delivering integrated agrometeorological services to farmers through mobile phone-based applications are useful in enhancing outreach, but traditional modality cannot be written-off just yet, particularly in the context of India's small farmers. Finally, introducing climate information through participatory capacity building processes, such as FARM Schools, are most useful as it provides a platform for discussion among users and for providing feedback to information providers and their intermediaries.

Investing in inclusive agricultural development: As shown in the discussion above, farm-level decision-making is a highly complex process that involves consideration of several factors, including, but not just limited to expectations of how the climate will evolve throughout the cropping season. The success of autonomous efforts by farmers to manage climate risks and opportunities is determined to a large extent by the availability of resources and infrastructure, such as access to advisory/extension services and agricultural infrastructure. Investing in broader agricultural development in a way that benefits smallholder farmers is therefore critical no matter the climate.

Agriculture remains the backbone of food security and is the biggest provider of employment for many countries, including in India. It is also increasingly expected to shift to environmentally sustainable practice while continuing to deliver for economic growth. User-oriented climate services will be an important enabler as the sector endeavors to deliver across all these fronts amidst a changing climate.

References

Agarwal A, Srinivasan G, Dash I (2018) Forecast application for risk management in agriculture: case study from Tamil Nadu, India. In Proceedings and recommendations of 'Regional expert consultation on underutilized crops for food and nutritional security in Asia and the Pacific'. Asia-Pacific Association for Agricultural Research Institutions (APAARI), Bangkok, Thailand, November 13–15, 2017

Banu N, Mrunalini A (2017, April) Empowering young farmers in the context of climate change. Int J Agric Sci Res (IJASR) 7(2), 449–456. ISSN(P): 2250–0057; ISSN(E): 2321–0087

Dileepkumar G, Davis A, Tall A (2014) Reaching the last mile: best practices in leveraging the power of ICTs to communicate climate services to farmers at scale. CCAFS working paper: 70

FAO (2010) A System of integrated agricultural census and surveys: Volume 1: World Programme for the Census of Agriculture (WCA) 2010, FAO Statistical Development Series 11

Gopalakrishnan T, Subramanian K (2020) Agrometeorological advisory Services in India: an assessment. Centre for Science and Environment, New Delhi. Available at: www.cseindia.org

Harper K, Uccellini LW, Kalnay E, Carey K, Morone L (2007) 50th anniversary of operational numerical weather prediction. Bull Am Meteorol Soc 88(5):639–650

Jha A, Bisht H, Tanner T, McGray H, Khan MA, Niazi Z (2012) Addressing the information and knowledge needs of farmers to enable climate change adaptation-a study of Bundelkhand region in Central India. Development Alternatives, New Delhi, India

Lobo C, Chattopadhyay N, Rao K (2017) Making smallholder farming climate-smart. Economic and Political Weekly, LII 53(1):53–58

Lynch P (2008) The origins of computer weather prediction and climate modeling. J Comput Phys 227(7):3431–3444

Mehta R (2011) Statistics on farmers and farm management for furthering synthesis of agricultural development and related policy analysis. In 4th meeting of Wye City group on statistics on rural development and agriculture household income, pp 9–11

Mittal S, Mehar M (2012) How mobile phones contribute to growth of small farmers? Evidence from India. Q J Int Agric 51(3):227–244

Mittal S, Mehar M (2013) Agricultural information networks, information needs and risk management strategies: a survey of farmers in Indo-Gangetic plains of India, vol 10. CIMMYT, Mexico

Nelson GC, Rosegrant MW, Koo J, Robertson R, Sulser T et al (2009) Climate change. Impact on agriculture and costs of adaptation. International Food Policy Research Institute, Washington, DC

Nkiaka E, Taylor A, Dougill AJ, Antwi-Agyei P, Fournier N, Bosire EN et al (2019) Identifying user needs for weather and climate services to enhance resilience to climate shocks in sub-Saharan Africa. Environ Res Lett 14(12):123003

Ramakrishna YS (2013) Current status of agrometeorological services in South Asia, with special emphasis on the indo-Gangetic Plains. CCAFS working paper no. 53. Copenhagen, Denmark: CGIAR research program on climate change, agriculture and food security (CCAFS). Available at: https://hdl.handle.net/10568/33842

Rathore LS, Chattopadhya N (2016) Weather and climate services for farmers in India. Available at: https://public.wmo.int/en/resources/bulletin/weather-and-climate-services-farmers-india

Shanmugasundaram J, Gunnell Y, Hessl AE, Lee E (2017) Societal response to monsoon variability in medieval South India: lessons from the past for adapting to climate change. Anthropocene Rev. https://doi.org/10.1177/2053019617695343

Singh C, Dorward P, Osbahr H (2016) Developing a holistic approach to the analysis of farmer decision-making: implications for adaptation policy and practice in developing countries. Land Use Policy 59:329–343

Singh C, Daron J, Bazaz A, Ziervogel G, Spear D, Krishnaswamy J et al (2018) The utility of weather and climate information for adaptation decision-making: current uses and future prospects in Africa and India. Clim Dev 10(5):389–405

Sivakumar MVK, Collins C, Jay A, Hansen J (2014) Regional priorities for strenghtening climate services for farmers in Africa and South Asia. CCAFS working paper no. 71. Copenhagen, Denmark: CGIAR research program on climate change, Agriculture and Food Security (CCAFS). Available at: https://hdl.handle.net/10568/41599

Srinivasan G, Agarwal A, Sawant M (2019, October) Enhancing climate services in south Asia. MAUSAM 70(4):725–730

Stone RC, Meinke H (2006) Weather, climate, and farmers: an overview. Meteorol Appl 13 (S1):7–20

Tall A, Njinga JL (2013) Developing a methodology to evaluate climate services for farmers in Africa and South Asia workshop report

Tall A, Davis A, Agrawal S (2014a) Does climate information matter? A proposed monitoring and evaluation framework for participatory assessment of the impact of climate services for male and female farmers

Tall A, Hansen J, Jay A, Campbell BM, Kinyangi J, Aggarwal PK, Zougmoré RB (2014b) Scaling up climate services for farmers: mission possible. Learning from good practice in Africa and South Asia

Vanadeep KSS, Murty R, Krishnaiah M (2012) Meteorological predictions preserved in the Panchangam versus real-time observations–a case study over Tirupati region–a semi-arid tropical site in India. Indian J Sci Technol 5(4):2491–2509

Vedeld T, Mathur M, Bharti N (2019) How can co-creation improve the engagement of farmers in weather and climate services (WCS) in India. Climate Services 15:100103

Agrometeorological Services for Climate Resilient Agriculture

8

S. D. Attri and M. Mohapatra

Abstract

Implications of rapidly changing climate on different sectors of socioeconomic activities including agriculture have focussed the attention of global community to devise ways to address the negative consequences. The scenario has been compounded owing to increasing extreme events and resulting in severe impacts on life and livelihood in general and food security in particular. The status of impacts of climate change on agriculture in different regions of the world by global organizations has been presented. Various initiatives at global, regional, and local levels are underway to cope up with the adverse impacts and achieve sustainability in crop production. The chapter attempts to present the current and future status of climate change, impacts on agriculture, and emerging climate and agrometeorological services in the world and India to support climate resilient agriculture.

Keywords

Climate change · GFCS · Agrometeorological Services · WMO

8.1 Introduction

Out of the 10 risks by likelihood over the next 10 years identified by the World Economic Forum (2020), the top 5 are related to environment viz. extreme weather events, failure of climate change mitigation and adaptation, major natural disasters,

S. D. Attri (✉) · M. Mohapatra
India Meteorological Department, Ministry of Earth Sciences, Government of India, Delhi, India
e-mail: sd.attri@imd.gov.in

A. Kaushik et al. (eds.), *Climate Resilience and Environmental Sustainability Approaches*, https://doi.org/10.1007/978-981-16-0902-2_8

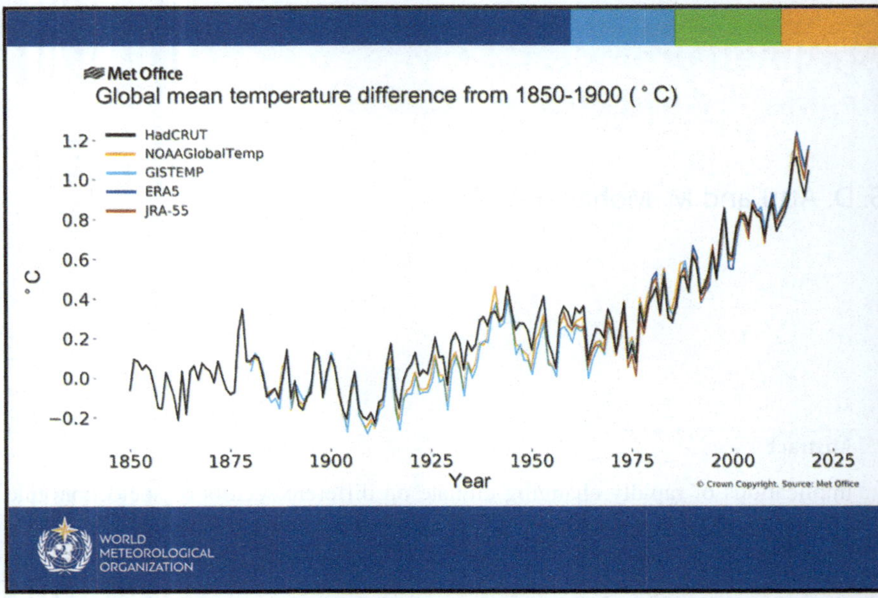

Fig. 8.1 Global annual mean temperature difference from preindustrial conditions (1850–1900)

major biodiversity loss and ecosystem collapse, and human made environmental damage and disasters. Climate change is now defining the issues owing to implications on lives and livelihoods of the people and socioeconomic development across the globe. The shifting weather patterns, rising sea levels, and increasingly extreme weather and climate events are being observed on an unprecedented scale. However, climate science has achieved reasonable degree of robustness to provide evidence of global temperature rise and associated features like sea level rise, shrinking sea ice, glacier mass loss, and extreme events (WMO 2019a). The average global temperature has increased about 1.1 ± 0.1 °C above the pre-industrial period (Fig. 8.1) as per WMO Statement on the State of the Global Climate in 2019 (WMO 2020). The period 2015–2019 is the warmest of any equivalent period on record as per a high-level synthesis report entitled "United In Science" of latest climate science information convened by the Science Advisory Group of the UN Climate Action Summit 2019 in September 2019. The global mean sea level has risen from 3.04 millimeters per year (mm/year) during 1997–2006 to ~4 mm/year during 2007–2016 mainly due to increased rate of ocean warming and melting of the Greenland and West Antarctica ice sheets. The last time Earth's atmosphere contained 400 parts per million CO_2 was about 3–5 million years ago, when global mean surface temperatures were 2–3 °C warmer than today, ice sheets in Greenland and West Antarctica melted, parts of East Antarctica ice had retreated, all causing global sea level rise of 10–20 m compared with today (UN 2019). The continuous growth of CO_2 has been observed during 1985–1995 (1.42 ppm/year), 1995–2005 (1.86 ppm/year), and 2005–2015 (2.06 ppm/year). Observed Concentration of CO_2, CH_4, and

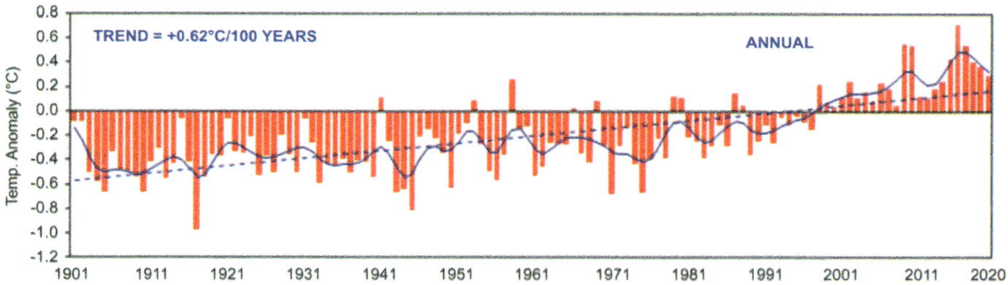

Fig. 8.2 Trends in temperature in India during 1901–2019 (Source IMD)

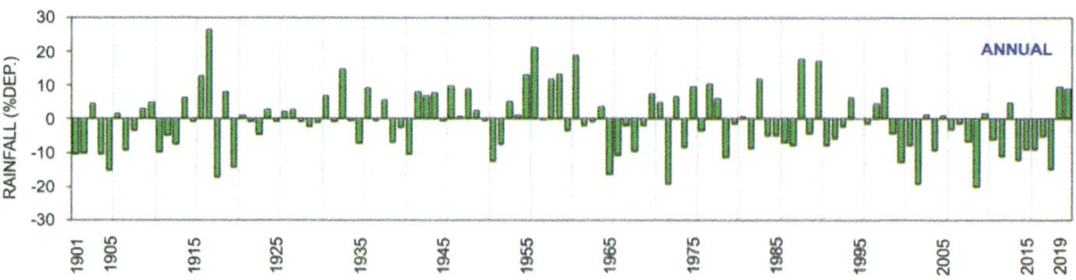

Fig. 8.3 Trends in annual rainfall in India during 1901–2019 (Source IMD)

N_2O were 407.8 ± 0.1 ppm, at 1869 ± 2 ppb, and at 331.1 ± 0.1 ppb during 2018 showing an increase of 146%, 257%, and 122% from pre-industrial levels (WMO 2020).

Three IPCC Special Reports (IPCC 2018, 2019) viz. Global Warming of 1.5 °C, Climate Change and Land and Ocean and Cryosphere in a Changing Climate assess complementary and specific aspects of climate change to support global actions for addressing climate change issues ahead of the Sixth Assessment Report (AR 6) to be released in 2021. It appears that implementing unconditional Nationally Determined Contributions and assuming that climate action continues consistently throughout the twenty-first century, would lead to a global mean temperature rise between 2.9 °C and 3.4 °C by 2100 relative to pre-industrial levels. This level of ambition needs to be roughly tripled to align with the 2 °C limit and must be increased around fivefold to align with the 1.5 °C limit (UNEP 2019).

The annual mean temperature during 1901–2019 in India showed an increasing trend of 0.61 °C/100 years with a significant increasing trend in maximum temperature (1.0 °C/100 years) and relatively lower increasing trend (0.22 °C/100 years (IMD 2020). The 2019 annual mean land surface air temperature for the country was +0.36 °C above the 1981–2010 period average, thus making the year 2019 the seventh warmest year on record since1901 (Fig. 8.2). However, annual monsoon rainfall does not show any trend during 1901–2019 (Fig. 8.3).

There is a growing recognition that climate impacts on various sectors will be harder and sooner than climate assessments indicated even a decade ago. According to the Food and Agriculture Organization of the United Nations (FAO 2018a) report on the State of Food Security and Nutrition in the World, climate variability and extremes are negatively affecting all dimensions of food security—food availability, access, utilization, and stability. The frequency of drought conditions from 2015 to 2017 shows the impact of the 2015–2016 *El Niño* on agricultural vegetation. Large areas in Africa, parts of Central America, Brazil, and the Caribbean, as well as Australia and parts of the Near East, have experienced a large increase in frequency of drought conditions in 2015–2017 compared to the 14-year average. As per International Food Policy Research Institute research (Rosegrant et al. 2014), climate change will put an additional 70 million people at risk of hunger in 2050. It will be impossible to solve the global hunger problem without also addressing the climate problem through adaptation and mitigation. There is a need to follow the Climate Smart Agriculture approach to handle the challenges of a changing climate by sustainably increasing agricultural productivity; helping food systems adapt and building their resilience; and reducing GHG emissions and promoting better knowledge, good practices and available technologies, and innovations.

8.2 Climate Change Impacts on Agriculture

The number of extreme climate-related disasters, including extreme heat, droughts, floods, and storms, has doubled since the early 1990s, with an average of 213 of these events occurring every year during the period 1990–2016. These disasters have significant impacts on agricultural productivity of major crops like wheat, rice, and maize and have caused food price hikes and income losses. 20–80% of the inter-annual variability of crop yields is associated with weather phenomena and 5–10% of national agricultural production losses are associated with climate variability. Climate change may result in increase in number of people at risk of hunger by 10 to 20 percent and malnourished children by 21% by 2050 (FAO, IFAD, UNICEF, WFP and WHO 2019). In India, the estimated countrywide agricultural loss in 2030 may be more than US$7 billion, which will severely affect the income of 10% of the population (IPCC 2014).

Agriculture contributes between 1% and 60% of national GDP in many countries, with a world average of about 4% in 2017 (World Bank 2019), while it accounts for about 14% in India. Under the current food system, FAO has projected the need of producing about 50% more food by 2050 to feed the increasing world population (FAO 2018a). However, crop production is projected to decrease in many areas of the world during the twenty-first century due to rapid changes in climate barring a few areas/crops with favorable conditions that could increase the yields of some crops. Between 2006 and 2016, agriculture (crops, livestock forestry, fisheries, and aquaculture) in developing countries accounted for an estimated 26% of total loss and damage incurred during medium and large-scale climate-related disasters. While two-thirds of loss and damage to crops was associated with floods, almost 90% of loss and damage in the livestock sector was attributed to drought (FAO 2017). In parallel with these trends, yields may decline by up to 30% by 2050 in the absence of

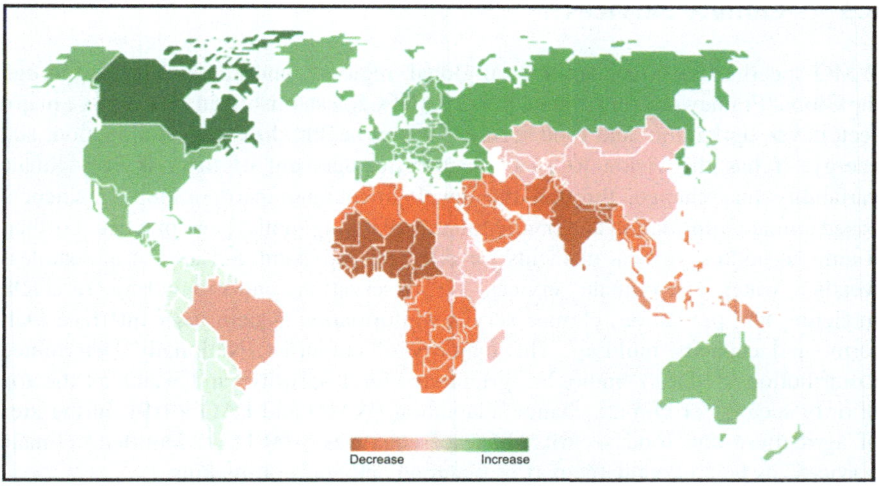

Fig. 8.4 Changes in agricultural production by the year 2050 (Source FAO)

ambitious climate action (Global Commission on Adaptation 2019). The majority of modeling studies agree that climate change impacts on crop yields will be negative from 2030 onwards. Nearly half of the projections beyond 2050 indicate a yield reduction of more than 10%. As per FAO (2018b), climate change is expected to result in declining agricultural production in large parts of Africa, the Middle East, and South and Southeast Asia. The reductions are projected to be more pronounced in West Africa and India, where production could decrease by 2.9% and 2.6%, respectively, due to climate change impacts (Fig. 8.4).

The latest report of WMO has presented regional level impacts in recent years. In Southern Africa, the start of the seasonal rains was delayed and extensive dry periods were observed. Regional cereal output is forecasted to be about 8% below the five-year average with 12.5 million people in the region expected to experience severe food insecurity up to March 2020. Deteriorating situation of food security has been observed in several areas of Ethiopia, Somalia, Kenya, and Uganda due to a poor long/*Gu* rainy season. Somalia was affected by intense flooding between October and November 2019 and parts of Afghanistan in March 2019. The year 2019 has been categorized as the most difficult planting season in history for North American farmers, with over 10 million acres of crops going unplanted due to extreme weather conditions (WMO 2020). At the same time, farmers in Punjab, in India, are experiencing rain showers almost every month and, for the first time in its history, more humid air is leading to greater pest infestations.

8.3 Climate Services

WMO spearheads climate services at global, regional, and national levels through the Global Framework for Climate Services(GFCS) established in 2009 in five major sectors viz. agriculture and food security, water, health, disaster risk reduction, and energy. It has the vision to enable better management of the risks of climate variability and change, through the development and incorporation of science-based climate information and prediction into planning, policy, and practice. Further, it aims to address critical elements for effective provision and uptake of seamless weather, water, and climate services, viz. observations and monitoring; research; modeling and prediction; climate services information system; user interface platform; and capacity building. The majority of countries' Nationally Determined Contributions (NDCs) highlight agriculture, food security, and water as the top priority sectors for climate change adaptation (WMO and FAO 2019). In the area of agriculture and food security, 85% of countries (100/117) identified "climate services" as being a valuable part of planning and decision-making.

"2019 State of Climate Services" brought out WMO in partnership with the Adaptation Fund, the Consultative Group on International Agricultural Research, Research Program on Climate Change Agriculture and Food Security, the Food and Agriculture Organization, Green Climate Fund, Global Environment Facility, the Global Facility for Disaster Reduction and Recovery, the World Bank, and the World Food Programme indicates that climate services investments have a cost–benefit ratio of 1–10 (WMO 2019b). The evidence suggests that the benefits of investing systematically in strengthening the operational global regional-national hydrometeorological system needed for climate services outweigh the costs by about 80 to one. It is estimated that improved weather, climate, water observations, and forecasting could lead to up to USD30 billion per year in increased global productivity and up to USD2 billion per year in reduced asset losses. It also points out that Operational Agrometeorological Advisory Services in India have decreased cultivation costs overall by up to 25% and, increased net returns to farmers up to 83%. The crops that benefited most are paddy, wheat, pearl millet, fruits, and vegetables. The economic benefit is estimated at USD7.575 billion per year.

With rapid urbanization, it is important to use combined land goals (e.g., zero-carbon energy, smart irrigation systems, and climate resilient agriculture) to minimize the negative side effects of climate change while securing quality food for a growing population. As per IPCC Special Report on Climate Change and Land (IPCC 2019), building resilience into productivity and production should be based on simultaneous coordinated actions to improve land, food security, and nutrition; dietary choice that can help reduce emissions and pressure on land, renewable energy, biodiversity conservation, etc. Some of the sustainable actions are delineated as under:

- Closing yield gaps through adapted cultivars, sustainable land management, that combine production and preservation of ecosystems essential functions such as sustainable intensification approaches based on conservation agriculture and

community-based adaptation with functioning support services and market access (Mbow et al. 2014).
- Identifying Sustainable Land Management practices (agroecology, agroforestry, etc.)
- Addressing different ecosystem services (food production, biodiversity, reduction of GHG emissions, soil carbon sequestration) for improved land-based climate change adaptation and mitigation (Sanz et al. 2017; Francis 2016).
- Paying attention to the food–energy–water nexus, especially water use and reutilization efficiency but also management of rainwater (Albrecht et al. 2018).
- Implementing institutional designs focused on youth, women through new economic 10 models that help access credit and loans to support policies that balance cash and food crops.
- Build on and use of local knowledge, culture, and traditions while seeking innovations for food waste reduction and transformation of agricultural products.

At present more than USD3 billion is invested in nearly 200 climate services and early warning systems projects with funding from the Adaptation Fund, the Green Climate Fund, the Global Environment Facility, and the World Bank Group. The World Bank estimates that additional investments beyond those already programmed of up to USD billion are needed to strengthen National Meteorological and Hydrological Services and other national institutions to be capable, and fully equipped, to deliver timely, reliable climate, weather and water information, and services relevant for policy and investment decisions. Africa and Small Island Developing States (SIDS) are facing the largest capacity gaps, especially with regard to the density of the observing network and reporting frequency of observations essential for weather and climate forecasts and services. Overall, investments need to be more focused and holistic. Current investments are ad hoc and piecemeal and there is a need to use the resources more efficiently to strengthen the global–regional–national operational hydrometeorological system that supports country-level service delivery more systematically and in a more integrated way. In particular, the services need to overcome the "last mile" barriers and reach farmers on the ground (WMO and FAO 2019). There is a need to improve the systematic estimation and documentation of the socioeconomic benefits of investments and the resulting services (WMO 2019b).

8.4 Agrometeorological Services

Major challenges to agriculture in India include size and trends of agricultural labor, climate change (not mitigating but evolving coping strategies), issues related to organic farming, conservation of ethnicity and biodiversity, multicropping/Integrated farming/crop diversification, value chain for agricultural produce, developing market/creating market linkage—reforms of wholesale markets, restructuring the architecture of Agricultural supply chain, etc. It has also been recognized that agriculture and food security are facing multiple risks like limited water resources, drought, desertification, land degradation, erosion, hail, flooding, frosts, etc. These risks could be managed by efficient Agrometeorological services which can make a significant difference in crop production by assisting farmers in taking advantage of favorable weather and minimizing the losses/impact of adverse weather conditions. Such services address the real-time needs of farmers and contribute to weather-based crop/livestock management strategies and operations for enhancing crop production and sustainability. At the same time, they can play a significant role in sustainable development and climate change adaptation in different parts of the globe. These services can help in making better policy decisions, improving the use of limited resources and enhance crop, horticulture, and livestock and fisheries production. The advisory bulletins include information on pest management related to relative humidity forecast, rising or falling temperatures or high or low winds, irrigation management depending on rainfall and temperature forecasts, crop protection from extreme high and low temperatures, etc. The farmers take decisions for chemical applications, irrigation scheduling, controlling disease and pest outbreaks, and other weather-related agriculture operations from cultivar selection to harvesting and postharvest operations through sowing, planting, transplanting, and intercultural operations. Without concerted action today, adapting and mitigating adverse impacts on various sectors including agriculture will be very difficult and costly in the future. There is a need to spread good agriculture practices and success stories of farmers across the country. The use of improved drought-tolerant cultivars, optimum fertilizer use, integrated pest management, crop diversification with vegetables, postharvest systems, mechanization, and use of Information Communication Technology in agriculture, Artificial intelligence along with value chain approach and market linkages including efficient Agrometeorological Services is the need of the hour for climate resilient agriculture and sustainability.

Government launched project named "National Initiative on Climate Resilient Agriculture" in 2010–2011 for enhancing the resilience of Indian agriculture covering crops, livestock, and fisheries to climatic variability and climate change through development and application of improved production and risk management technologies; demonstrating site-specific technology packages on farmers' fields for adapting to current climate risks; building up the capacity of scientists and other stakeholders in climate resilient agricultural research and its applications. Climate and early warning information services should underpin decision-making on climate action for adaptation. However, the capacities to deliver and access these services are highly uneven across regions and countries. The challenge is to strengthen the global–regional–national hydrometeorological system needed to operationalize and deliver these products and services at country level, particularly in developing countries, so that everybody benefits.

IMD has a long history of providing weather services for agriculture since 1945 when it initiated "Farmers' Weather Bulletin" and broadcast through All India Radio in regional languages. It started State-level Agrometeorological Advisory Services jointly with State Agriculture Departments in 1976. Agroclimatic zone level Agrometeorological Advisory initiated by NCMRWF in 1991 was merged with IMD's in 2007 and Integrated Agrometeorological Advisory Service at district level was commenced. The existing district level Agrometeorological Advisory Services under Gramin Krishi Mausam Sewa (GKMS) rendered by IMD through a network of existing 130 Agro-Met Field Units (AMFUs) established in State Agriculture Universities, Indian Council of Agricultural Research (ICAR) institutes, Indian Institute of Technology (IITs), etc. are being augmented by addition of another 530 District Agrometeorological Units at Krishi Vigyan Kendras (KVKs) of ICAR to extend services to block level (6500 blocks) with outreach to Panchayat level. The weather-based actionable advisories are generated using quantitative and probabilistic information on past and forecast weather of rain, maximum and minimum temperatures, wind speed and direction, morning and evening relative humidity, and cloud amount in different temporal and spatial domain, viz., nowcast (up to 3 hours), short range (1–3 days), medium range (3–10 days), extended range (10–30 days) at block, district, state, and national levels and state-wise stage of different crops in the country. Such advisories are framed by Expert Panel drawn from IMD, ICAR, State Agriculture Department, and State Agricultural Universities. The agrometeorological advisories prepared at district level on every Tuesday and Friday in regional languages are disseminated to around 40 million farmers in the country through SMS on mobile in regional languages on mKisan portal of Ministry of Agriculture and Farmers Welfare and Public–Private Partnership mode in addition to other modes of communication like Electronic and Print media, Extension Services of Central and State Agriculture Departments, Universities, Kisan Call Centres, WhatsApp, Website, and Toll-free telephone. The advisories are populated through conducting Farmers Awareness programs and feedback mechanisms in villages by AMFUs and DAMUs. State and District level Stakeholders meetings are also organized for this purpose.

IMD has started generating experimental block-level advisory from 2000 blocks since 2019 and plans to cover all 7000 blocks in the next 2–3 years with dissemination to all farming households in the country using all available modes of communication. In addition to biweekly block and district level advisories, weekly State and National level crop weather bulletins are prepared and disseminated to concerned Central and Government Departments, and other stakeholders. IMD also provides weekly All India and State Agrometeorological Advisory Bulletins, Monthly All India Weather and Crop Bulletins, Drought Aridity Anomaly Maps, Standardized Precipitation Index, Agroclimatic Information, and Weekly and Seasonal Rainfall to World Agro-Meteorological Information Service (WAMIS) of World Meteorological Organization.

Information technology is advancing at a very rapid rate, which entitles the current century to be called the century of information technology. Agrometeorological advisories should make use of all contemporary technologies including community radio and social media. IMD/Ministry of Earth Sciences should bring out specific guidelines and support provide financial support, if needed, for Agrometeorological Community Radio Stations in different agro-climatic zones of the country. There is also an urgent need to make effective use of remotely sensed data from Unmanned Aerial Vehicles (UAVs), Satellites, Radars, etc., different temporal and spatial-scale deterministic and probabilistic forecasts, new

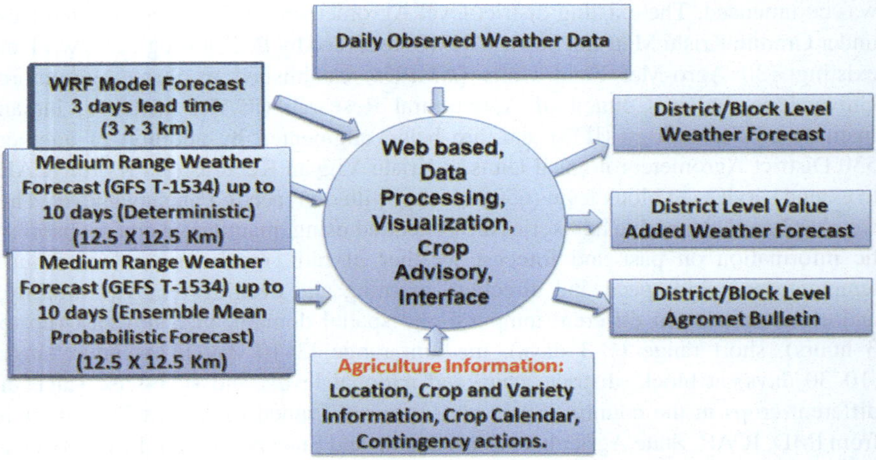

Fig. 8.5 Agrometeorological Decision Support System

agrometeorological products and graphics in agrometeorological services toward achieving the goal of precision farming at field level. Further, specialized efforts are required to systematically measure agrometeorological observations like leaf area, canopy structure, photosynthesis, soil erosion, soil moisture, biological, and related phenomena, pest and disease, direct and indirect damage, phenological observations of crops, and trees, which are generally not available. The service should also address issues of reducing uncertainties, insurance and risk, and provide quantitative information about the weather in different temporal ranges so that the farmers can utilise the information about the likely outcome of alternative or relief management options.

These services have significantly contributed to enhancing the production and income of farming community. A study on impact of these services by the National Council of Applied Economic Research (NACER 2015) has estimated incremental profit of the order of Rs. 38,463 crores in 2010 and Rs. 42,000 crores in 2015 on 4-principal crops viz. wheat, paddy, sugarcane, and cotton with potential of generating net economic benefit up to Rs. 3.3 lakh crores on the 22-principal crops. For achieving the above goal, there is a need to expand the existing services and integrate all available tools, techniques, resources, information, and knowledge. IMD has taken several steps to augment the services like:

- Installation of Agro-Automatic Weather Stations (Agro-AWS) with the capability of measuring weather parameters including soil temperatures and soil moisture at different depths in all District Agrometeorological Units (DAMUs) in the country.
- Issuance of Block level forecast in different temporal scales.
- Development of Agrometeorological–Decision Support System (Agrometeorological DSS) for advisory preparation and dissemination (Fig. 8.5).

- Development of Dynamic Crop Weather Calendars for different crops.
- Development of Expert system for auto-generation of advisory at block level.
- Capacity building of farmers through intensive awareness campaign.
- Developing dynamic feedback mechanism.
- Extensive use of ICT for real-time dissemination of advisories.
- Strengthening the use of remote sensing products and drones.
- Special focus on Agrometeorological services in 115 Aspirational districts in the country.
- Enhancing collaboration with concerned Central and State Government departments and Agricultural Universities, User agencies, Stakeholders, Farmers Organizations, Cooperatives, Public–Private Partnership, etc.
- Specialized products and services toward the aim of doubling the income of farmers by 2022 through integrated farm advice for crop protection and appropriate risk mitigation matrix based on experienced and expected extreme weather events.

IMD still has a long way to go and need to focus on enhancing the accuracy of weather forecasts at different temporal and spatial scales with village/individual farmer as last mile objective and address livestock, poultry, and fisheries issues in its advisories along with capacity building of stakeholders and farming community.

8.5 Conclusion and Future Scope

The projected reduction in crop yields under rapidly changing climate in different parts of the globe has made crop sustainability and food security uncertain. This needs to be overcome by suitable adaption and mitigation strategies to cope with weather and climate extremes toward the goal of climate resilient agriculture. Effective weather and climate services have opened new avenues and provide appropriate advisory at different temporal and spatial scales with minimal cost implications and resulted in increase in income of farmers. The existing block level services need to be augmented to Gram Panchayat level with an increase in accuracy of forecast and contribute to precision farming. All tools and technology will have to be integrated to make use of scarce resources including collaboration among all stakeholders.

Future Agrometeorological services should incorporate:

- Technological driven drought, temperature, and salt-resistant cultivars, varietal response to different domains of weather forecasting, improvement in seed, water and pesticide use efficiency, etc.
- Policy-driven cluster farming, optimization of agriculture land, market linkage, etc.
- Personalized Agrometeorological Advisory for farmers using artificial intelligence and data mining/big data concept, query-based individual farmer services from Decision Support System for precision farming, etc.

References

Albrecht TR, Crootof A, Scott CA (2018) The water-energy-food Nexus: a systematic review of methods for nexus assessment. Environ Res Lett 13:43002. https://doi.org/10.1088/1748-9326/aaa9c6

FAO (2017) The impact of disasters and crises on agriculture and food security. FAO, Rome, pp 3–143

FAO (2018a) The state of food security and nutrition in the world-building climate resilience for food security and nutrition. FAO, Rome, pp 1–181

FAO (2018b) the state of agricultural commodity markets, agricultural trade, climate change and food security. FAO, Rome, pp 1–112

FAO, IFAD, UNICEF, WFP and WHO (2019) The state of food security and nutrition in the world 2019. Safeguarding against economic slowdowns and downturns. Rome, pp 1–212

Francis C (2016) The carbon farming solution: a global toolkit of perennial crops and regenerative agriculture practices for climate change mitigation and food security, by Eric Toensmeier. Agroecol Sustain Food Syst 40:1039–1040. https://doi.org/10.1080/21683565.2016.1214861

Global Commission on Adaptation (2019) The global commission on adaptation seeks to accelerate adaptation action and increase political support for building climate resilience, pp 1–81

IMD (2020) Statement on climate of India during 2019, press release, pp 1–4

IPCC (2014) In: Field CB, Barros VR, Dokken DJ, Mach KJ, Mastrandrea MD, Bilir TE, Chatterjee M, Ebi KL, Estrada YO, Genova RC, Girma B, Kissel ES, Levy AN, MacCracken S, Mastrandrea PR, White LL (eds) Climate change 2014: impacts, adaptation, and vulnerability. Part B: regional aspects. contribution of working group II to the fifth assessment report of the intergovernmental panel on climate change. Cambridge University Press, Cambridge and New York, p 688

IPCC (2018) Summary for policymakers. In: Masson-Delmotte V, Zhai P, Pörtner H-O, Roberts D, Skea J, Shukla PR, Pirani A, Moufouma-Okia W, Péan C, Pidcock R, Connors S, Matthews JBR, Chen Y, Zhou X, Gomis MI, Lonnoy E, Maycock T, Tignor M, Waterfield T (eds) Global warming of 1.5°C. An IPCC special report on the impacts of global warming of 1.5°C above pre-industrial levels and related global greenhouse gas emission pathways, in the context of strengthening the global response to the threat of climate change, sustainable development, and efforts to eradicate poverty. WMO, Geneva

IPCC (2019) Summary for policymakers. In: Climate change and land: an IPCC special report on climate change, desertification, land degradation, sustainable land management, food security, and greenhouse gas fluxes in terrestrial ecosystems Shukla PR, Skea J, CalvoBuendia E, Masson-Delmotte V, Pörtner H-O, Roberts DC, Zhai P, Slade R, Connors S, van Diemen R, Ferrat M, Haughey E, Luz S, Neogi S, Pathak M, Petzold J, Portugal Pereira J, Vyas P, Huntley E, Kissick K, Belkacemi K, Malley J (eds)

Mbow C, Van Noordwijk M, Luedeling E, Neufeldt H, Minang PA, Kowero G (2014) Agroforestry solutions to address food security and climate change challenges in Africa. Curr Opin Environ Sustain 6:61–67. https://doi.org/10.1016/j.cosust.2013.10.014

National Council of Applied Economic Research (NCAER) (2015) Impact assessment and economic benefits of weather and marine services. Ministry of Earth Sciences, New Delhi

Rosegrant MW, Koo J, Cenacchi N, Ringler C, Robertson R, Fisher M, Cox C, Garrett K, Nicostrato D, Perez ND, Pascale Sabbagh P (2014) I food security in a world of natural resource scarcity, the role of agricultural technology, pp 1–77

Sanz MJ, Coauthors (2017) Sustainable Land Management contribution to successful land-based climate change adaptation and mitigation. A report of the science-policy interface, Bonn, Germany, p 178. https://www.unccd.int/sites/default/files/documents/2017-09/UNCCD_Report_SLM_web_v2.pdf

UN (2019) United In Science, High-level synthesis report of latest climate science information convened by the Science Advisory Group of the UN Climate Action Summit 2019, New York. public.wmo.int/en/resources/united_in_science

UNEP (2019) Emissions Gap Report 2019, 1–81
World Economic Forum (2020) The Global Risk Report 2020, 15th edn, 1–102
WMO (2019a) WMO Statement on the State of the Global Climate in 2018
WMO (2019b) 2019 state of climate services report focuses on agriculture and food security, WMO No 1242, 1–43
WMO (2020) WMO Statement on the State of the Global Climate in 2019, 1–44
WMO and FAO (2019) Investments in climate services for agriculture and food security outweigh costs, WMO Press release-2019, 1–5

Microbial Diversity and Multifunctional Microbial Biostimulants for Agricultural Sustainability

9

Pawan Kumar and Rana Pratap Singh

Abstract

Increasing evidences support that the beneficial soil microbes play multiple roles in plant growth and productivity as well as increase in soil microbial carbon biomass. The beneficial soil microbes fix or solubilize plant nutrients, produce plant growth hormones, degrade agro waste, increase soil carbon sequestration, act as biocontrol agents for pests and diseases, and degrade toxic residues of agrochemicals in the agricultural soil. The soil microbial biodiversity and abundance are significant component of its multifunctional activities in nature but due to decline of population and density of specific beneficial strain or genera or their beneficial potential caused by the intensified cropping, excessive use of toxic agrochemicals, climate change-induced drought and flooding stresses, low and high temperature, unfavorable changes in soil pH, and deposition of salts in the upper crust. Due to low soil organic matter and soil quality deterioration, natural multiplication of the native soil microbial population may not occur adequately and some useful species may get lost from the rhizospheric niche. The single potential soil microbe or application of multipurpose microbial consortia of beneficial soil microbes isolated from the soil of different agroclimatic conditions as a biostimulant in the agricultural soil may help to sustain and enhance the crop productivity in a sustainable way without the use of synthetic agrochemicals. The stability, ecological succession, and impact of the inoculated microbes on native beneficial soil microbes in stressed and non-stressed agroecosystems, however, have not been investigated in most of the cases, which will be essential to assess

P. Kumar · R. P. Singh (✉)
Department of Environmental Science, Babasaheb Bhimrao Ambedkar (Central) University, Lucknow, Uttar Pradesh, India
e-mail: rpsingh@bbau.ac.in

A. Kaushik et al. (eds.), *Climate Resilience and Environmental Sustainability Approaches*, https://doi.org/10.1007/978-981-16-0902-2_9

their role in agricultural sustainability and to develop stable formulations of new generation bio-inoculants.

Keywords

Agroecosystem · Bio-inoculant · Carbon sequestration · Plant growth promoting microbes · Sustainable agriculture

9.1 Introduction

Nutrients play an essential role in sustaining the life of both plants and animals. Plants take up mineral nutrients from lithosphere as well as atmosphere (Maathuis and Diatloff 2012) whereas animals depend directly or indirectly upon the plants for their nutrient requirements. All plants take up nutrient in inorganic form from the soil even if it is added or available in the soil as organic nutrient forms as they are capable to assimilate inorganic nutrients into organic forms (Jacoby et al. 2017; Makarov 2019). The plants require only a defined amount of the essential nutrients for their optimal productivity and its growth and productivity often get retarded on insufficient supply of the mineral nutrients. Most of the agricultural soils do not possess sufficient phytoavailable N, P, or K, and the micronutrients to satisfy the requirement of rapid growth and productivity of crops (Chen and Liao 2017). Exogenous supply of the nutrients is required during vegetative and reproductive phases of the plant development because of continuous cycles of the intensified crop cultivation (Gregory et al. 2002; Hati et al. 2006). These elements are supplied as fertilizers in both intensive and widespread agricultural systems. Inorganic fertilizers such as urea, diammonium phosphate (DAP), and NPK (the common synthetic fertilizers) are manufactured chemically (Pan et al. 2016; Jiang et al. 2017). Crop productivity has increased manyfold by the green revolution practices in many countries including India primarily because of the hybrid seeds, exogenously loaded chemical fertilizers and pesticides and flood irrigation, etc. These excessively supplied synthetic fertilizers, however, cause adverse effects on human health, soil biodiversity, and soil fertility in long-term intensified agricultural practices (Liu et al. 2013; Zhu et al. 2016; Pozzer et al. 2017). The excess soluble chemical nutrients that are taken up by the plants but do not get assimilated due to the limited assimilation capacity of the plants may get accumulated in the plant parts as unassimilated inorganic molecules and subsequently cause health hazards to the animal or human consumers (Lea and Miflin 2018). The transformed products of nitrogen fertilizers like Nitrate, nitrite, and other reactive nitrogen species as well as its derivatives like Nitrosamines are poisonous, mutagenic, and carcinogenic for animals and humans if they consume these foodstuff or water contaminated with these chemicals (Takahashi and Morikawa 2019; Ziarati et al. 2018; Takahashi et al. 2007). The synthetic nutrients and agrochemicals used in intensive agricultural practices in excess and not taken up by the plants get leached in the below ground aquifers and get run off with rainwater and floods to the surface water bodies (Adisa et al. 2019; Swartjes and Van der Aa

2019). The inorganic reactive nitrogen and phosphorous species and toxic organic residues of agricultural origin are known to enhance ecological and hazardous risk to the aquatic ecosystems too (Mondal et al. 2018). In addition, emissions of NOx and volatilization of NH_3 are known to build up additional warming potential of the Green House Gases (GHGs) considered as a major cause of global warming and climate change (Jiang et al. 2017). It is believed that the intensive agricultural practices supported by these synthetic fertilizers are no more sustainable and have caused serious threats to the agroecology and human health.

Searching for alternatives against chemical fertilizers, and pesticides that can maintain the productivity of plants without causing any adverse effects to the ecosystem are a major concern of agricultural sustainability today. Farm yard manure (FYM), Zero-Budget Natural Farming (ZBNF), Vermicompost, no-tillage farming, use of biopesticides, Non-Pesticide Management (NPM), and application of biofertilizers are seen as possible alternatives to sustain beneficial soil microbial biodiversity (Kundu et al. 2016; Desmond 2016; Khadse et al. 2017; Li et al. 2019; Chauhan et al. 2020). These beneficial soil microbes commonly known as plant growth promoting microorganisms have been reported to benefit crop productivity by providing nutrients, acting as biocontrol agents, growth stimulators to bioremediate toxic substances, and help to achieve good crop productivity in a sustainable and ecological way. In the ZBNF, crude microbial inocula like Beejamruth and Jeevamruth are added to the soil in order to accelerate the propagation of the native soil microflora beneficial for soil enrichment (Khadse et al. 2017). Using the selected beneficial soil microbes by seed priming or soil amendment as biofertilizers or biopesticide possessing single microbe or microbial consortia is another ecological way to provide additional nutrients to the crops. Due to lack of authenticity and consistent effectivity, the microbial formulations commonly known as biofertilizers and biopesticides are not very popular among farmers and commercial food growers. Lack of adequate understanding of Plant–microbe interactions, changes in rhizospheric microbial flora in the soil of given agroclimatic conditions, changes in land use pattern and contamination of agroecosystems with toxic agrochemicals, etc. are the major reasons which reduce or limit the efficacy of these microbial biofertilizers after application of the new microbial formulations (Rajkumar et al. 2013; Mohanty and Swain 2018; Rodriguez et al. 2019). The ecological succession of the amended bioinoculant in the rhizospheric niche of crop plants and their impacts on the endemic beneficial soil microbes, however, have not been studied extensively and require urgent attention. This chapter analyzes critically the potentials (Fig. 9.1) and limitations of these microbial biostimulants as an alternative to chemical agriculture to achieve a non-contaminated and sustainable food security for a mounting world population with decreasing land holdings for plant agriculture and opens up possibilities of developing new innovative insights and technologies to delimit the bottlenecks of its adoption and popularity.

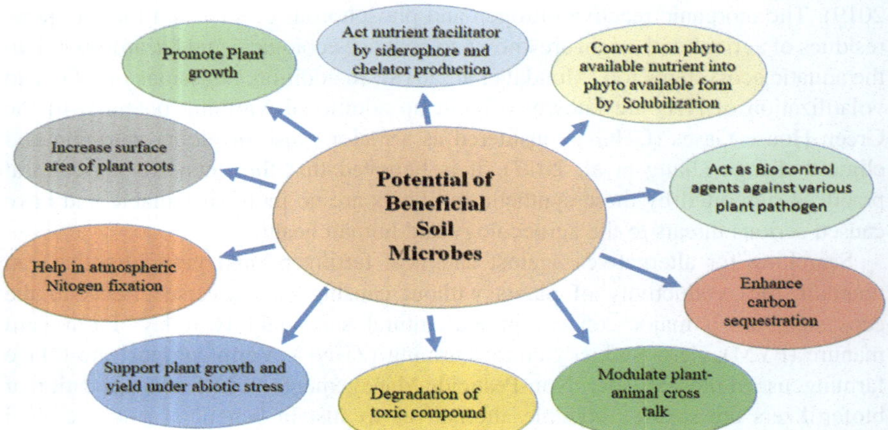

Fig. 9.1 Image showing the potential of beneficial microbe

9.2 Microbial Diversity

Soil consists of a structurally and functionally heterogeneous environment on the Earth (Totsche et al. 2017). In soil, microbes experience soil complex physical structure, geochemistry, and fluctuations in seasonal parameters such as temperature, moisture, and nutrient availability at microscales (Hellmann et al. 2017; Durán et al. 2018). Due to all these parameters, the higher diversity and biomass of microbes can be seen in soil compared with other environments (Singh and Gupta 2018). Certainly, the total number of microbial species and their function in soil remains unknown (Fierer 2017; Baldrian 2019). The recent major advances in cell sorting and DNA probing techniques help in understanding marine microbial ecology but soil's complex matrix of minerals and organic matter made it difficult to extract microbes or their cellular constituents from soil for study (Fierer 2017; Li et al. 2018; Sadler et al. 2018). In addition, for understanding and preserving microbial genome, there is a need to culture soil microbes from outside of its soil environment (Malla et al. 2018; Wilpiszeski et al. 2019). However, recent high-throughput DNA sequencing technologies and advances in isotopic approaches have given us advantage in understanding cultivation-independent assessments of microbial diversity and adaptability of microbes in soil (Slatko et al. 2018; Cheng et al. 2019; Wu et al. 2020). Through this technique, we have found a huge biodiversity of fungal and bacterial genera in the rocky soil of the Plateau of Ralegarh Sidhi in Maharashtra, which is semi-arid. We have isolated 21 total microbes from these soil in 11 are bacterial and 10 are fungal strains. The identification of these isolated microbes showed that *Bacillus* and *Aspergillus* were dominating groups in bacteria and fungi, respectively. Most of the isolated microbes show PGPR activity in characterization

(Pawan kumar, Roli Mishra, Sanjay Dwivedi, Rana Pratap Singh unpublished results).

9.3 Microbes as Nutrient Provider

The microbes or group of microbes available in soil or applied exogenously have the ability to capture or mobilize important nutrients that are essential for plant growth and productivity (Ali et al. 2019). These soil microbes commonly known as plant growth promoting rhizobacteria (PGPR) or more recently plant growth promoting microbes (PGPMs) due to the use of fungi in addition to bacteria as soil inoculant can provide nutrients to the plants through various biological processes such as N_2-fixation, solubilization of macro and micronutrients (e.g., phosphorus, potassium, and zinc), production of siderophores, which chelates Fe^{3+}, production of growth promoting hormones and enzymes for degradation of organic biological substances lignins and proteins (Trimurtulu and Rao 2014; Pal et al. 2015; Shen et al. 2016; Ali et al. 2019). These groups of microbes are gaining popularity to be used as biofertilizers due to its low cost, compatibility with the soil health, and being a biological supplement in place of synthetic chemical fertilizers (Alori et al. 2017; Umesha et al. 2018). Certain microorganisms isolated from the rhizosphere exhibit strong plant growth promoting (PGP) activity are selected and developed to be used in the preparation of biofertilizers (Li et al. 2017; Chinakwe et al. 2019). Soil is the most complex microbial habitat on the earth, which consists of an integrated network of plant roots, soil fauna, diverse microbial consortia, soil particles, and water associated with it (Totsche et al. 2017; Singh et al. 2018). The activities and occurrence of rhizospheric microbes are also influenced by the plants' roots as the release of substances by the plants in its rhizosphere act as signals and affect microbial activity (Verma et al. 2017; Canarini et al. 2018; Zhang et al. 2018). The rhizosphere is characterized as a zone of high microbial dynamics and hence form a unique ecological niche (Garcia and Kao-Kniffin 2018; Nuccio et al. 2020). The activity and microbial population gradually decrease depending upon the distance from the plant roots (Kuzyakov and Razavi 2019; Wood et al. 2019). Hence, the rhizosphere constitutes a niche suitable for obtaining more specific beneficial microbes for the use as soil inoculant for cultivation of plants after selecting the better strain for improved potentiality (Lareen et al. 2016). These associated microorganisms play a significant role in plant productivity by proper supply of nutrients to the plants in order to ensure their growth and productivity being present in the soil or on supply exogenously as a bioinoculant in an adequate amount. The understanding of plant microbiome and their attributes can have various benefits towards the development of area specific, crop specific, and soil specific biofertilizers to achieve the agricultural sustainability without the input of synthetic xenobiotic substances in the agroecosystem. A great emphasis was recently given on decoding of microbial diversity associated with the plants from diverse habitats (Bhardwaj et al. 2014). The rhizosphere beneficial microbes of different groups such as archaea, bacteria, and fungi from the extreme environments (acidic, alkaline, drought, pressure, salinity, and temperatures) and their inoculation are understood to be more climate resilient in the rhizosphere of plants suffering from such stresses can provide a novel niche of climate resilient beneficial microbes to enhance plant productivity in those agroclimatic conditions without application of agrochemicals (Yang et al. 2009; Verma et al. 2017; Lee et al. 2019). In addition, epiphytic and endophytic microbes associated with plants as well as able to grow in the extreme environment may also be very useful in designing a new rhizospheric soil microbe interaction in a planned way to restore the deteriorated' agroecology (Ajar et al. 2017).

These nutrient provider soil microbes are broadly grouped into mineral solubilizing microorganisms, nitrogen-fixing bacteria, and organic matter decomposers.

9.3.1 Nitrogen Fixer

Nitrogen-fixing bacteria are a group of prokaryotes, which have the ability to fix atmospheric nitrogen into ammonium by nitrogenase enzyme that is encoded by *nif* gene present in bacteria (Dos Santos and Addo 2020). On the basis of association with plants they are classified into symbiotic, nonsymbiotic and associative nitrogen fixers (Zaidi et al. 2017). In symbiotic nitrogen-fixing bacteria, a special structure in plants are produced to carry and protect the bacteria. The symbiotic microbes are not able to fix nitrogen, if they are not associated with plant in the close vicinity provided by the nodules (Martínez-Hidalgo and Hirsch 2017). The well-known example of these association is *Rhizobium* and legumes. The bacteria enclosed within the root nodules formed by the plants (Zaidi et al. 2017). In this association, bacteria provide reduced N to plant and plant provide organic substances for energy to the bacteria (Yoneyama et al. 2017). *Rhizobium*, *Mesorhizobium*, *Bradyrhizobium*, *Azorhizobium*, *Allorhizobium*, and *Sinorhizobium* belong to family Rhizobiaceae, which are exploited as symbiotic N_2-fixing bacteria to enhance the productions of legumes (Naik et al. 2020). Different strains of *Rhizobium* were studied in association with different legume crops and trees, for example, *Rhizobium phaseoli* for green gram and black gram, *R. trifolii* for berseem, *R. lupini* for chickpea, *R. japonicum* for soya bean, *R. leguminosarum* for pea and lentil, and *R. meliloti* for Lucerne (Bhattacharjee et al. 2011; Datta et al. 2015; Le et al. 2015; Singh et al. 2018). Similar to the *Rhizobium*, the bacteria of genus *Frankia* fix nitrogen in the nodules of non-leguminous plants (Van Nguyen and Pawlowski 2017). The association of *Frankia* with actinorhizal plants can improve its productivity as these plants used in production of timber and fuel wood and environmental services, e.g., land reclamation and shelterbelts in coastlines and desert area (Schwencke and Caru 2001; Ribeiro-Barros et al. 2018).

In nonsymbiotic N_2 fixation, a close association of plants with bacteria is not required as needed in symbiotic N_2 fixation (Deng et al. 2019). The free-living (nonsymbiotic) bacteria, includes the *Anabaena*, *Nostoc*, *Azotobacter*, *Beijerinckia*, and *Clostridium*. *Azotobacter* is nonsymbiotic nitrogen fixation and rate of fixation depends on amount of carbohydrates utilized by it (Gomare et al. 2013). The increase in population of *Azotobacter* depends on the content of organic matter and complementary, antagonistic or synergistic relationships with other native microorganisms present in the same niche in soil (Ebrahimi et al. 2017; Mohanram and Kumar 2019). About seven species of *Azotobacter* were reported, which gave PGP activity, these are *A. salinestri*, *A. chroococcum*, *A. beijerinckii*, *A. vinelandii*, *A. nigricans*, *A. paspali*, and *A. nigricans* (Chen et al. 2017). *Clostridia* spp. reported as obligatory anaerobic heterotrophs that are capable of fixing N_2 only in the complete absence of oxygen usually they are isolated from soils of rice fields (Brück et al. 2010; Horino et al. 2015). The cyanobacteria (*Anabaena* and *Nostoc*) are most commonly used nonsymbiotic nitrogen fixer as nitrogen (N) fertilizer. Cyanobacteria can be used by small landholder farmers in regions where soil N is depleted and fertilizer availability is unreliable. It can locally produce on-farm as it is easy to grow and have low requirement of land and water for its growth. Twomlow et al. (2008) showed that production and application of as low as 10–17 kg N ha^{-1} anabaena can increase at least 30% crop yield. The nitrogen-fixing bacteria of the genus *Beijerinckia* occur mainly in soils of tropical regions (Oggerin et al. 2009). In these regions, *Beijerinckia* is far more frequently distributed than *Azotobacter* along with similar efficiency of nitrogen fixation to it (Arahal 2016).

Apart from symbiotic and nonsymbiotic relationship with plants some bacteria show associative nitrogen fixation. In these type of fixation, bacteria do not form nodule but live with plants in associated form, i.e., either attached to its root surface or penetrated into the root tissues (Roley et al. 2018). The examples of associative nitrogen fixer which showed association with sorghum, rice, maize, and sugarcane are *Acetobacter*, *diazotrophicus*, *Azospirillum*, *Bacillus*, *Enterobacter*, *Klebsiella*, *Pseudomonas*, *Citrobacter*, and *Herbaspirillum* spp. *Herbaspirillum* can form endophytic colonies in the roots of sugarcane, rice, maize, sorghum, and other cereals (Deng et al. 2019). Baldani et al. (2000) reported that *Herbaspirillum* can fix 31–45% of total plant N in 30-day-old rice seedling from the atmospheric nitrogen. The significant increase of up to 7.5 g $plant^{-1}$ in the yield of rice was reported in greenhouse study when inoculated with *Herbaspirillum* (Mirza et al. 2000). *Herbaspirillum seropedicae* shows endophytic diazotrophic relation with roots of wheat (Kennedy and Islam 2001) and sugarcane plants (Polese et al. 2017) whereas *Herbaspirillum rubrisubalbicans* is obligate endophyte of roots, stems, and leaves (Mehnaz 2011; Polese et al. 2017). *Azospirillum*, an associative nitrogen-fixing organism that are aerobic heterotrophs that fix N_2 under microaerobic conditions and extensively found associated with the rhizosphere of gramineous plants (Eckert et al. 2001; Hayat et al. 2010). The plant–*Azospirillum* association lead to enhanced yield and development of plant mainly due to improved root growth and biological N_2-fixation (Okon and Itzigsohn 1995). The *Azospirillum* forms root–bacterial interactions by chemotaxis and chemokinesis in response to chemoeffectors temporal gradient. The inoculation of *Azospirillum lipoferum* increased the yield of wheat by 30% (Okon and Labandera-Gonzalez 1994) and up to 6.7 g $plant^{-1}$ in rice (Mirza et al. 2000). Bashan (1998) showed a significant increase in height and dry matter of cotton plant when inoculated with *Azospirillum brasilense*. Reinhold-Hurek et al.

(2002) reported intensive colonization of *Azoarcus* sp. BH72 behind the root meristem when it inoculated in a rice field. The 67 species of genus *Burkholderia* are validly published several of these including *Burkholderia phynatum, B. tuberum, B. kururiensis, and B. vietnamiensis* being capable of fixing N_2 (Estrada-delos Station et al. 2001; Vandamme et al. 2002; Eberl and Vandamme 2016). In a field trial, a significant increase of 8 t ha^{-1} in grain yield of rice and reduction of 25–30 kg N ha^{-1} of fertilizer were reported when the crop was inoculated with *B. vietnamiensis* (Tran Vân et al. 2000).

The nitrogen fixation of atmospheric nitrogen into ammonia is mainly done by nitrogenase metalloenzyme that is encoded by different nif genes in microorganisms (Bellenger et al. 2020). Nitrogenase enzymes are present in bacteria and archaea, which give them a competitive advantage in the bioavailable nitrogen-depleted environment (Smith-Moore and Grunden 2018). On the basis of metallic combination present in nitrogenase, it is divided into three different types—molybdenum-iron (MoFe) mostly present in symbiotic nitrogen fixer, vanadium-iron (VFe), and iron-iron (FeFe) (Bellenger et al. 2020). The component of nitrogenase enzyme, i.e., iron, vanadium, and molybdenum in active center are encoded by anfDGK, vnfDGK, or nifDK, respectively, and iron-containing electron transfer proteins are encoded by anfH, vnfH, or nifH (Dos Santos and Addo 2020). Most of the microbes consist of only MoFe encoded gene while *Azotobacter vinelandii* reported to consist of all three type of nitrogenases encoded gene. During nitrogen fixation, 16 molecules of ATP are required to fix per molecule of nitrogen (Bothe et al. 2010). In the fixation process, the iron protein is reduced by electron carriers such as ferredoxin and reduced iron protein subsequently reduces the catalytic component (Rebelein et al. 2016). The dissociation and reassociation of nitrogen molecules required iron and catalytic proteins. The production of powerful reductants and protection of oxygen labile nitrogenase increases the bioenergetic cost for the fixation of nitrogen per molecule (Rebelein et al. 2016). The exposure of oxygen can deactivate nitrogenases, so microbes adopted various mechanisms to protect nitrogenase such as separate nitrogen fixation from photosynthesis either spatial or temporal in oxygenic phototrophs (e.g., *Trichodesmium* spp., *Crocosphaera watsonii*, and *Nodularia* spp.), formation of nod symbiotic nitrogen fixer, conformational changes of nitrogenase, increased oxygen respiration, and detoxification via superoxide dismutase (Rebelein et al. 2016; Bellenger et al. 2020).

9.3.2 Mineral Solubilizing Microbes

Various minerals that are essential for plant growth and productivity are present in soil which are not soluble and hence cannot be utilized by the plants unless they get solubilized, e.g., Phosphate, Potassium, Zinc, and Sulfur (Kumawat et al. 2017). The various groups of soil microbes have been reported to solubilize these insoluble non-phytoavailable forms of nutrients into soluble phyto-available forms (Etesami and Adl 2020). These microbes are categorized further into different groups based on the minerals they solubilize, e.g., Phosphate-solubilizer, Potassium-solubilizer,

Sulfur-solubilizer, and Zinc-solubilizer (Kumawat et al. 2017; Etesami and Adl 2020). The insoluble form of various minerals are as follow, for phosphate its dicalcium phosphate, tricalcium phosphate, rock phosphate and hydroxyapatite; for potassium its muscovite, mica, orthoclase and illite; for Sulfur its gypsum, barite, chalcopyrite, sphalerite, galena, and pyrrhotite; and for Zinc its sulfide ores, complex with organic compound and rock zinc (Kumawat et al. 2017). The organic acids, such as lactic, gluconic, glyoxylic, citric, oxalic, tartaric, acetic, fumaric, and maleic acids, which are produced by the microbes help in converting into solubilized form that can be utilized by the plants (Kumawat et al. 2017; Etesami and Adl 2020). *Pseudomonas*, *Bacillus*, and *Rhizobia* are found to be the most effective phosphate solubilizers over other known bacterial genera (Table 9.1) (Kalayu 2019). Sundara et al. (2002) showed that application of rock phosphate with PSB (*Bacillus megaterium* var. *phosphaticum*) increased sugarcane yield up to 12.6% over without PSB amendment along with improved juice quality. *Bacillus*, *Clostridium*, *Frateuria*, and *Thiobacillus* are some reported potassium solubilizing bacteria (Sindhu et al. 2016). *Frateuria aurantia* is capable of solubilizing potassium from mica into usable form for the plants, now it has fairly been applied in association with other biofertilizers to crops without any antagonistic effects (Mazid and Khan 2014; Raghavendra et al. 2016; Sindhu et al. 2016). Saraf (1988) applied *Azotobacter pasturianam* as a biofertilizer to remove the deficiency of sulfur in pulses and the result showed an increase in yield and quality. Sulfur is essential for the synthesis of some glycoside, disease resistance property, and pungent compound. In oilseed crops, some vegetables (onion, oat, cauliflower, etc.) and some species (ginger, garlic, etc.). Sulfur is one of the major elemental constituents. The growth and yield of crops also get influenced by the deficiency of Zinc as a micronutrient (Singh et al. 2016; Debnath et al. 2015). The availability and cost of zinc fertilizer make zinc solubilizers more important for facilitating adequate supply of zinc. *Bacillus subtilis*, *Thiobacillus thiooxidans*, and *Saccharomyces* sp. are some well-known zinc solubilizers (Raj 2007).

9.3.3 Siderophores as Facilitator

A large amount of Fe is present in the earth's crust but is in the unavailable form to plants. Microorganisms producing siderophore and other chelating substances can play a crucial role to overcome the Fe and Zn deficiency in soil (Schalk et al. 2011; Rana et al. 2012). Siderophores are low molecular weight (500–100 dt) chelating compounds having high affinity toward Fe(III) and form siderophore–Fe (III) complexes which help in plant Fe acquisition (Saha et al. 2012; Hayat et al. 2012; Ganz 2018). Many bacteria like *Pseudomonas*, *Azotobacter*, *Bacillus*, *Enterobacter*, *Serratia*, *Azospirillum*, and *Rhizobium* are reported to synthesize specific ferric ion chelating agents under iron-limited conditions (Thiem et al. 2018). Over 500 biomolecules are revealed as siderophores; hence, on a cellular level biosynthesis of different sets of multiple genes, enzymes and regulators are involved (Rangel et al. 2018). Essen et al. (2007) showed that a facultative aerobic bacterium *Pseudomonas stutzeri* CCUG 36651 can synthesize siderophore but of different type

Table 9.1 Certain beneficial microbes are characterized and studied as Plant Growth Promoting microbes and amended in rhizosphere during the crop cultivation

Nutrient and plant growth promoters	Biological pathways	Microbes	References
Nitrogen	N_2 Fixation	*Azospirillum* spp., *Azotobacter* spp., *Azoarcus* spp., cyanobacteria, *Bacillus* spp., *Enterobacter* spp., *Xanthobacter* spp., *Gluconacetobacter diazotrophicus*, *Achromobacter* spp., *Arthrobacter* spp., *Azomonas* spp., *Clostridium* spp., *Beijerinckia* spp., *Corynebacterium* spp., *Rhodospirillum* spp., *Klebsiella*,spp., *Derxia*, spp., *Herbaspirillium* spp., *Pseudomonas* spp., *Rhodopseudomonas* spp., *Rhizobium* spp., *Mesorhizobium* spp., *Bradyrhizobium* spp., *Azorhizobium* spp., *Allorhizobium* spp., *Sinorhizobium* spp., *Citrobacter*	Hayat et al. (2010), Kaymak (2010), and Prasad et al. (2015)
Phosphorus	Phosphate solubilization	*Alcaligenes* sp., *Aerobactor* sp., *Achromobacter* sp., *Actinomadura oligospora*, *Bacillus* sp., *Pseudomonas* sp., *Penicillium* sp., *Aspergillus* spp. *Bradyrhizobium* sp., *Brevibacterium* sp., *Citrobacter* sp., *Flavobacterium* sp., *Nitrosomonas* sp., *Erwinia* sp., *Achrothcium* sp., *Cephalosporium* sp., *Cladosporium* sp., *Curvularia* sp., *Rhizobium* sp., *Burkholderia* sp., *Agrobacterium* sp., *Micrococcus* sp.,	Hayat et al. (2010), Chen et al. (2006), Kaymak (2010), and Prasad et al. (2015)

(continued)

Table 9.1 (continued)

Nutrient and plant growth promoters	Biological pathways	Microbes	References
		Enterobacter sp., *Serratia* sp., *Proteus* sp., *Klebsiella* sp., *Arthrobacter* sp., *Phyllobacterium myrsinacearum* sp., *Rhodococcus erythropolis* sp., *Delftia* sp., *Morganella* sp.	
Potassium	Potassium solubilization	*Bacillus mucilaginosus, Bacillus, Clostridium, and Thiobacillus*	Chen et al. (2006), Kaymak (2010), and Prasad et al. (2015)
Fe	Siderophore production	*Pseudomonas aeruginosa, Gallionella and Leptothrix species, Frankia, Streptomyces* sp.	Saha et al. (2012) and Zhang et al. (2018)
Zn	Zinc solubilization	*Burkholderia* sp., *Acinetobacter* sp.	Saravanan et al. (2011) and Zhang et al. (2018)
Auxin	Cell enlargement, root initiation, vascular differentiation, apical dominance	*Pisolithus tinctorius, Azospirillum, Rhizobium* sp., *Bradyrhizobium, Azotobacter* sp. *Gluconacetobacter, Herbaspirillum, Cyanobacteria, Nostoc, Kluyvera ascorbata, Azotobacter* sp., *Pseudomonas fluorescens, Streptomyces acidiscabies, Alcaligenes* sp.	Hayat et al. (2010)
Gibberellins	Seed germination, development, and reproduction of plants, floral development	*Gibberella fujikuroi, Azospirillum brasilense, Azospirillum lipoferum*	Hayat et al. (2010)
Cytokinins	Cell division, chloroplast differentiation, photosynthesis, senescence, and nutrient metabolism	*Azospirillum, Pseudomonas syringae* pv, *Agrobacterium tumefaciens, Erwinia herbicola*	Chen et al. (2006), Kaymak (2010), and Prasad et al. (2015)

depends upon state oxygen availability, i.e., aerobic and anaerobic states. According to the oxygen ligands for Fe^{3+} organization, siderophores can be differentiated into three main categories, namely, hydroxamates, catecholates, and carboxylates. Hydroxymate types of siderophores belong to C (=O) N-(OH) R and R is an amino acid and mostly produced by bacteria and fungi. The soil bacteria such as *Erwinia*, *Nocardia*, *Streptomyces*, *Arthrobacter*, *Chromobacterium*, and *Pseudomonas* produce Ferrioxamine-type siderophores whereas fungal species produce ferrichrome (*Ustilago sphaerogena*), ferricrocin (*Aspergillus fumigates*), Coprogens (*Trichoderma* sp.), and Fusigen (*Fusarium* sp.) type siderophores (Wallner et al. 2009). Hydroxamates form 1:1 hexadentate octahedral structure complexes with ferric iron, which show strong absorption between 425 and 500 nm (Pahari et al. 2017). Another important group of siderophore is catecholate mainly produced by bacteria and having a backbone of polyamine (peptide or a macrocyclic lactone) in structure. The chelating group, catecholate or phenolate is attached with the backbone of polymine and upto three group can in one molecule (Dave et al. 2006). The tricatechol siderophore enterobactin was isolated from the fluid culture of *Aerobacter aerogenes*, *E. coli*, and *Salmonella typhimurium*. The most important feature of enterobactin is extremely strong pH dependence, low redox potential for physiological reductants, and high complex formation constant. Carboxylate type of siderophores consist of carboxylate and hydroxyl donor groups and mainly produced by bacteria such as *Staphylococcus* and *Rhizobium* (Dave and Dube 2000).

9.3.4 Organic Matter Decomposer

The dead organic matter is a very good source of nutrient to plant as it consists of most of the essential element in it but presence in complex form it is un-available to plant. So, microbial decomposer plays important role in degrading complex organic matter to simpler form in less time as much as possible. Many fungal species were reported, which act as activator and decomposer plant containing lignocellulosic compounds such as *Phanerochaete chrysosporium*, *Trichoderma viride*, *Penicillium*, *Trichurus spiralis*, and *Aspergillus niger* (Narayanasamy et al. 2012). Microbial decomposing communities are mainly dominated by bacterial and fungal heterotrophs even though they are phylogenetically very distant (Hieber and Gessner 2002; Mille-Lindblom and Tranvik 2003). The metabolic requirements, cellular capabilities, and mode of attachment are different in bacteria and fungi. The bacteria are attached to surface of substrate whereas, filamentous structure of fungus can also penetrate into the substrate. Fungal decomposers have low metabolic nutrient requirements and wider enzymatic capabilities; on the other hand, bacteria have high metabolic activity and shorter turnover (Güsewell and Gessner 2009; Attermeyer et al. 2013; Danger et al. 2016). Due to phylogenetic and functional differences in fungi and bacteria they occupy different niches in organic matter (OM) decomposition (Gessner et al. 2007; Krauss et al. 2011). The presence of energy-rich OM in litter can increase microbial degradation ability for more complex, and low-quality C (Kuehn et al. 2014). The decomposition of organic matter leads to the release of macro- and micronutrients in the soil into simpler forms that can be easily utilized by plants or other microbes further converted from nonavailable form to plant bioavailable form. The health of soil and the plant depends upon the nutrient cycle in which microbial decomposer communities play very important role.

9.3.5 Phytohormone Production

The uptake of nutrient is equally important as its presence in the soil, certain microbes show PGP activity by producing different compounds which trigger the growth of plants roots. These microbes use different mechanisms for promoting plant growth but generally influence via production of plant growth hormone and nutrient uptake enhancement (Richardson 2001; Canarini et al. 2018). Dal Cortivo et al. (2017) showed a 65% increase in the number of root tips and ramifications compared to non-inoculated controls when wheat plants were treated with a consortium of *Azospirillum* spp., *Azoarcus* spp., and *Azorhizobium* spp. Similarly, Bertrand et al. (2000) reported enhancement in the root hair and length when rapeseed treated with rhizobacterium. The phytohormones produced by *Pseudomonas* and *Bacillus* showed an increase in the absorptive surface of plant roots by increasing fine roots that enhance uptake of water and nutrients by plant. The PGPMs inoculants promote growth by producing phytohormones such as Auxin, Gibberellins, Cytokinins, and inhibitors of ethylene. Indole-3-acetic acid is an auxin phytohormone involved in cell division, cell enlargement, and root initiation. Some plant-associated bacteria regulate plant hormonal level such as ethylene, which is considered an inhibitor of plant growth and also known as senescence hormone but in several plants such as *Arabidopsis thaliana* at low level promote plant growth (Pierik et al. 2006). Fuentes-Ramirez and Caballero-Mellado (2005) showed elevation in root IAA and dihydroxyze riboside when pole pine plantlets inoculated with *Paenibacillus polymyxa* and *Pseudomonas fluorescens*, respectively. Similarly, different microbes that are producing plant growth hormones are given in Table 9.1. In general, the direct effect is on plant growth promotion and indirect effect on improving uptake of plant nutrition and water via the development of the roots.

9.4 Root Development

The roots are regarded as a crucial part of plant owing to their support in physical standing and transportation of nutrients and water from soil to plant. Numerous studies exhibited that plant growth promoting microbes (PGPMs) possess potential in modulating root structure (Ogawa et al. 2014; Ning et al. 2014). Alterations in root system architecture by PGPMs implicate the synthesis of phytohormones and other signal molecules that result in elevated lateral root branching and root hair development. PGPMs also enhance plant nutrition, amend root functioning, and affect the physiology of the whole plant. These microbial communities in the rhizosphere rely on the kind and composition of root exudates released including sugars, organic acids, amino acids, fatty acids, enzymes, phenolics, and flavonoids (Canarini et al. 2018). The capability of bacteria to flourish in the rhizosphere relies on their ability of attraction toward the plant-derived carbon sources.

9.4.1 Role of Rhizobacteria in Root Architecture

The roles of PGPR initiated root phenotypes have been reported in some literature. One such example includes inhibition of primary root growth conjugated with the propagation of lateral roots and root hairs resulting in elevated shoot biomass (Ryu et al. 2005; Chagas et al. 2018). Another phenotype involves elevation in plant biomass in conjunction with enhancement of primary root growth (Sati et al. 2020). Bacterial density and distance between bacteria and plant roots are common determinant factors for these phenotypes (Naylor and Coleman-Derr 2018; Hassani et al. 2018). While growth promotion phenotypes have been well pronounced, yet a limited number of studies have been performed to analyze the underlying mechanism for these effects. Presently, numerous PGPRs have been reported to persuade root developmental variations in cell division and differentiation processes at both root meristem and sites of lateral root formation (Small and Degenhardt 2018). These changes at the cellular level modify the overall root structure of the plants. Colonization of the *Arabidopsis* root by different PGPR species, such as *Pseudomonas simiae* WCS417 (formerly *Pseudomonas fluorescens* WCS417) or *Bacillus megaterium* UMCV1 affect the transition from proliferation to differentiation in the root (Zúñiga et al. 2013; Berendsen et al. 2015). In the meristematic zone, *P. simiae* enhances cell division whereas *B. megaterium* lowers the cell division rate resulting in lowered primary root length on direct application 40% and 70%, respectively (López-Bucio et al. 2007; Zamioudis et al. 2013). Owing to early differentiation, root hairs appear closer to the root tip in the colonized plants. Additionally, root hair density enhances upon colonization owing to a higher number of cortical cells around the radial axis resulting in increased number of root hair-forming cells as root hair formation is accelerated by another yet undefined mechanism (Khan et al. 2016).

Alongside the primary root development, PGPRs are also reported to affect the lateral root formation in various studies (Khan et al. 2016). The lateral root formation can be divided into two steps, firstly the pairs of cells in the pericycle (the cell layer in the vasculature neighboring the endodermis), also known as lateral root founder cells, must acquire competency to form a lateral root. Then lateral root founder cells are activated, divide multiple times, and further differentiate to form the lateral root primordia and they ultimately emerge from the primary root in the differentiation zone (Péret et al. 2012). Endodermal cells superimposing the lateral root primordia disperse to lodge the evolving lateral root, by change in shape, size reduction, and formation of small pores in the Casparian strip. Further, division and differentiation of the lateral root cells after advent result in a lateral root anatomy similar to the anatomy of the primary root, involving a meristematic zone that regulates the lateral root's growth rate (Verbon and Liberman 2016). *P. simiae* WCS417 and *B. megaterium* UMCV1 enhance the density of both lateral root primordia and lateral roots, representing that the number of lateral root-competent sites and lateral root outgrowth depend on colonization. It is still unclear whether PGPR also affects lateral root founder-cell speciation. Interestingly, initiation of lateral root formation and shoot growth can happen without the aforementioned observed primary root growth inhibition. Volatile organic entities synthesized by *P. simiae* WCS417 stimulate lateral root formation but do not inhibit primary root growth (López-Bucio et al. 2007; Zamioudis et al. 2013). Root-associated bacteria help in deciding the root architecture as it can modulate the growth of vertical and lateral roots.

9.4.2 Plant Fungus Interaction

The effects of PGP microbes on plant development are not only limited to plant–bacteria interactions as similar phenotypes are induced among the roots on their exposure to fungi. The ectomycorrhizal fungus *Laccaria bicolor* S238N, the truffles *Tuber borchii* strains ATCC96540 and 43BO, and *Tuber melanosporum* strains Bal1 and Rey_t, *Trichoderma virens* Gv. 29-8, and *Trichoderma atroviride* (IMI 206040) have been reported to enhance lateral root growth (Contreras-Cornejo et al. 2009; Verbon and Liberman 2016). Also, *T. atroviride* is also reported to enhance root hair length and density. Additionally, the truffles hinder primary root growth. Therefore, it is alluring to speculate on the preserved effects of biotic factors. Like PGPR, fungi also affect endogenous plant mechanisms to affect root development. The elevated development rate of lateral roots in response to *T. virens* is escorted by augmented expression of the auxin-inducible marker in the primary root tip, developing lateral roots and depends on functional auxin transport and effects in the plant. Since *T. virens* synthesizes the auxin precursor IAA, it leads to a change in auxin signaling in the plant, thus leading to the identified root phenotype (Contreras-Cornejo et al. 2016). Plants also retort to interaction with *L. bicolor* with an augmented auxin response. The auxin produced by *L. bicolor* is not the only trigger to induce lateral root formation in *Arabidopsis*, instead, a volatile compound synthesized by the fungus persuades the phenotype, potentially results in elevated auxin biosynthesis resulting in the identified root phenotype (Felten et al. 2010). This volatile compound produced might be ethylene, produced by the truffles and alongside fungal auxin, it initiated the mentioned root phenotype on application to the roots (Contreras-Cornejo et al. 2016). The possible relationship between auxin and ethylene in response to beneficial fungi is supported by studies in *T. atroviride* as it is reported to produce both auxins and ethylene. Ethylene signaling within the plant is crucial for the fungus-induced root hair phenotype whereas the lateral root induction is ethylene independent (Contreras-Cornejo et al. 2016). Lateral root induction might depend on auxin signaling in the plant, that is, affected by the bioactive metabolite 6-pentyl-2H-pyran-2-one(6-PP) synthesized by *T. atroviride* (Garnica-Vergara et al. 2015). 6-PP increases auxin responsiveness in the lateral root primordia, as exhibited by elevated DR5: GFP expression in the primordia, possibly by changing the expression of PIN proteins acting as auxin transporters (Garnica-Vergara et al. 2015). The signals from the two hormones might be integrated into a positive feedback loop by MPK6, although the precise is still uncovered (Contreras-Cornejo et al. 2016).

9.5 Microbes as Biocontrol Agent for Plants

The protection of plant from various diseases caused by different pathogens is one of the major concerns to achieve the goal of food security. From the era of the green revolution, the use of chemical pesticides started under various names as; weedicide, insecticide, rodenticide, and fungicide to protect crops from various pests and diseases. These chemicals were produced by multinational chemical companies and aggressively marketed throughout the world. The pesticides are effective for certain pests. Although, the continuous use of these chemical pesticides eventually developed dose tolerance in the pest, results in an increase of dose at every successive use. The unlimited use of pesticide has some serious consequences, i.e., the evolution of pest's species, those can tolerate the higher doses of pesticides, also many species of microbes, plants, and animals are affected by these factors, consequently brings on the edge of extinction and bioaccumulation of pesticides induced various diseases in live stocks and human (Chattopadhyay et al. 2017). In addition to that, the contamination of agroecosystem, water, and air with pesticides and its tiny particles and toxic heavy metals have fatal consequences on the microphytes, earthworms, beneficial insects, and birds and badly affect biodiversity of the particular area with predominance of only a few tolerant species. The application of Biocontrol using natural antagonistic microorganisms is a potentially effective and environmentally safe method available to prevent plant diseases.

During the coevolution of plants and microbes, a diverse form of mutualistic interactions are formed between them (Germida and Siciliano 2001; Cardinale et al. 2015). The microbes acting as biocontrol agents show facultative mutualistic or mutualistic relationship with plants. The mycorrhizal interaction shows mutualistic relation with plants, which prevents root infection by stimulating plant defenses. The pathogenic disease suppression is mainly caused due to competition between biocontrol agents (BCA) and pathogens for space, nutrients, and growth stimulators. Some microbes form Parasitic and Hyperparasitic interaction with pathogens, which can be used as BCA such as *Trichoderma viride* which is parasitic to pathogen *Rhizoctonia solani*. The BCA can act on pathogen by using more than one mechanism such as, competition, parasitism, plant growth promotion, inducing host resistance, antibiosis, and signal interference (quorum sensing) (Amer and Utkhede 2000; Collins and Jacobsen 2003; Jataraf et al. 2005; Jorjani et al. 2011; Mansoori et al. 2013). Both bacterial and fungal groups are used as BCA for different type of pathogen along with a single group of BCA i.e. either bacterial or fungal groups can act on both fungal and bacterial pathogens. Vascular wilts can be caused by either bacterial pathogens such as *Ralstonia*, *Erwinia*, *Pseudomonas*, and *Clavibacter* or fungal pathogens such as *Fusarium and Verticillium*. Vascular wilts can cause major food losses as they can affect both the woody perennials and the annual crops (Yadeta and Thomma 2013). The vascular wilt pathogen obstructs the transportation of water and minerals by invading xylem vascular tissues. The endophytic bacteria such as *Enterobacter*, *Agrobacterium*, *Bacillus*, *and Pseudomonas* can inhibit the harmful effects of vascular wilts pathogen. Mercado-Blanco et al. (2004) reported that *Pseudomonas fluorescens* PICF7 reduced the incidence and severity of the

Verticillium wilt disease by 82% and 96%, respectively, in olive trees under greenhouse condition. Among the various fungal BCA, *Trichoderma* is most extensively used for suppression of various pathogens. It has the ability to produce cell wall lytic enzymes, antibiosis, and secondary metabolites that can induce resistance in plants (Woo et al. 2014). The first time biocontrol capability of *Trichoderma* was reported by Weindling (1932, 1934) against the fungal pathogen *Rhizoctonia solani* in which *T. lignorum* was used. Later it was found that *T. lignorum* is also effective against other common fungal pathogens such as *Sclerotium rolfsii*, *Pythium*, *Phytophthora*, and *Rhizopus*. Currently, many strains of *Trichoderma* were reported, which are effective against various pathogenic diseases (Table 9.2).

9.6 Role of Microbes in Bioremediation of Toxic Pesticide Compounds

The microbes play a crucial role in curbing various diseases and promoting the growth of plants without showing any negative impact on nature and health. On the other hand, the applied chemical pesticides and their residues present in the agroecosystems can deteriorate the quality of food by accumulating but they can be degraded by microbes. The complete biodegradation of pesticides is the oxidation of parent compounds to produce CO_2 and H_2O.

For degradation process, various stages may be taken and at every stage a particular enzyme is needed to be catalyzed (Jacquet et al. 2016). The degradation of organic compounds may get cease if a suitable enzyme does not exist at any stage of degradation. The main reason for the persistence of pesticides in the environment is the nonexistence of suitable enzymes or microbes that can produce them (Van Dyk and Pletschke 2011). The diversity of microorganisms plays a crucial role in deciding metabolic, genetic, and morphologic features of that community. Huang and Wang (2013) demonstrated the unbelievable diversity and distribution of microbes in the environment on the basis of small subunit rRNA and rDNA sequencing systems. The presence of microbial community activity decides the fate of pesticides in the environment. Biodegradation of pesticides is also depending on ecological factors along with the availability of degradative microbes. Different groups of bacterial species such as *Alcaligenes*, *Pseudomonas*, *Bacillus*, *Rhodococcus*, and *Flavobacterium*, and fungal species such as *Aspergillus*, *Penicillium, and Fusarium* metabolize pesticides as shown in Table 9.3. To utilize these pesticides as a source of energy and nutrient, several microbes develop different catabolic degradative mechanisms (Wang et al. 2014; Liu et al. 2016; Azadi et al. 2017; Alvarenga et al. 2018). The effective degradation of organochlorines was shown by an aerobic fungal species *Chrysosporium* strain by using several pathways. *Fusarium verticillioides* strain was able to utilize lindane as a carbon and energy source by degrading it under aerobic conditions (Quintero et al. 2008). The white-rot fungus *Phanerochaete chrysosporium* is more effective than other fungi in degrading poisonous or insoluble compounds (Mori et al. 2017). Several other white-rot fungal species, e.g., *Pleurotus tuberregium*, *P. ostreatus*,

Table 9.2 Exogenously applied microbes helping in disease and pest management

Crop	Disease	Pathogen	Possible biocontrol agents
Rice	Blast	*Pyricularia oryzae*	*P. fluorescens Trichoderma* spp.
	Bunt	*Neovossia indica*	*T. harzianum, T. viride, T. virens, T. deliquescens*
	Sheath blight	*Rhizoctonia solani*	*P. fluorescens* and *P. putida, T. harzianum, T. viride, T. virens, A. niger* AN27
	Brown spot	*Drechslera oryzae*	*T. viride*
	Bacterial leaf blight	*Xanthomonas oryzae*	*Bacillus* spp.
Wheat	Karnal bunt	*Neovossia indica*	*T. viride, T. harzianum, T. pseudokoningii*
	Loose smut	*Ustilago segetum*	*T. viride, T. harzianum, T. Koningii, T. lignorum*
	Root rot	*S. rolfsii, F. oxysporum*	*T. harzianum*
	Take-all	*Gaeumannomyces graminis var. tritici*	*T. harzianum*
Maize	Charcoal rot, Banded Blight *R. solani*	*Macrophomina phaseolina*	*Trichoderma* spp.
Sorghum	Charcoal rot	*M. phaseolina*	*A. niger* AN27
Pigeon pea	Wilt	*Fusarium udum*	*T. viride, T. hamatum, T. harzianum, T. koningii, B. subtilis*
	Seed-borne disease	*Xanthomonas campestris pv. Vinae*	*T. viride, T. harzianum*
Chickpea	Wilt	*F. oxysporum f.* sp. *Ciceri*	*T. viride, T. harzianum, T. virens, B. subtilis, A. niger AN27*
	Root rot	*Rhizoctonia solani/M. phaseolina*	*T. viride, T. harzianum*
	Collar rot	*Sclerotium rolfsii*	*T. viride, T. harzianum, P. fluorescens*
	Gray mold	*B. cinera*	*Trichoderma* sp.
	Stem rot	*S. sclerotiorum*	*T. harzianum*
Cowpea	Wilt	*F. oxysporum f.* sp. *ciceris*	*T. viride*
	Charcoal rot and wilt	*M. phaseolina, F. oxysporum, f.* sp. *Tracheiphilum*	*T. viride, T. harzianum, T. koningii, T. pseudokoningii*
Soybean	Dry root rot	*M. phaseolina*	*T. viride, T. harzianum*
Mungbean	Root rot	*M. phaseolina*	*T. harzianum, T. viride*
Oil Seed Crops			
Groundnut	Crown rot	*Aspergillus niger*	*T. viride, T. harzianum, B. subtilis*
	Stem and pod rot	*Sclerotium rolfsii*	*T. harzianum, Rhizobium*

(continued)

Table 9.2 (continued)

Crop	Disease	Pathogen	Possible biocontrol agents
	Late leaf spot	*Phaeoisariopsis personata*	*Penicillum islandicum, P. fluorescens, T. harzianum, B. subtilis*
	Root and stem rot	*R. solani*	*T. virens, T. longibrachiatum*
	Rust	*Puccinia arachidis*	*Verticillium lecanii, T. harzianum*
Castor	Wilt	*Fusarium oxysporum, f. sp. ricini*	*T. viride, A. niger* AN27
	Gray mold	*Botrytis cinerea*	*T. viride, P. fluorescens*
Mustard	Damping-off	*Pythium aphanidermatum*	*T. harzianum, T. viride*
Vegetables			
Bottle gourd	Wilt	*F. oxysporum*	*A. niger* AN27
	Root rot	*R. solani*	*A. niger* AN27
	Collar rot	*Sclerotinia sclerotiorum*	*T. viride, T. virens, B. subtilis*
Cauliflower	Damping-off	*Rhizoctonia solani, P. aphanidermatum*	*T. harzianum, A. niger* AN27
	Stalk rot	*S. sclerotiorum*	*A. niger* AN27
Potato	Black-scurf	*R. solani*	*T. viride, T. viride, B. subtilis*
	Charcoal rot	*M. phaseolina*	*A. niger*
	Late blight	*P. infestans*	*Trichoderma* sp.
Tomato	Damping-off and wilt	*F. oxysporum, B. cinerea f.* sp. *Lycopersici*	*T. harzianum, P. fluorescens*
	Gray, *B. cinerea*	*Meloidogyne incognita*	*T. harzianum*
	Root Knot-Meloidogyne incognita	*M. javanica*	*T. harzianum*

P. pulmonarius, *Agaricus bisporus Bjerkandera adusta*, *Trametes versicolor*, and *Lentinula edodes* were tested for the degradation of persistent organic substances, (Deshmukh et al. 2016; Mori et al. 2017; Purnomo et al. 2017).

9.7 Role in Reducing Stress in Plant

The application of plant growth promoting microbes (PGPM) and arbuscular mycorrhizal (AM) fungi can enhance plant growth and survival under adverse conditions (Nadeem et al. 2014). The microbes use different biochemical and molecular mechanisms to promote plant growth and development during stress conditions, e.g., nutrient mobilization, production of exopolysaccharide, by regulating hormonal and nutritional balance, and inducing resistance against phytopathogens

Table 9.3 Exogenously applied and native microbes helping in biodegradation of toxic and nontoxic substances

Nature of component	Name of the component	Microbes degrading the component
Toxic	DDT in soil	*Escherichia coli*, *Enterobacter aerogenes*, *Enterobacter cloacae*, *Klebsiella pneumoniae*, *Pseudomonas putida*, *Bacillus* species, and *Hydrogenomonas*
	DDT (activated sludge)	*Pseudomonas* sp., *Pseudomonas aeruginosa*, *Micrococcus*, *Bacillus pumilus*, *Bacillus circulans*, *Bacillus* sp., and *Flavobacterium* sp.
	Endosulfan	*Pseudomonas* sp., *Bacillus* sp., and *Flavobacterium* sp.
	Lindane	*B. thiooxidans*, *Basea thiooxidans*, *Sphingomonas paucimobilis*, *Streptomyces* sp., and *Pleurotus ostreatus*
	Atrazine, monocrotophos,	*Arthrobacter*, *Clavibacter*, *Nocardia*, *Rhodococcus*
	alachlor, and 4-	*Nocardioides, and Streptomyces*
	Chlorophenol	
	Pentachloronitrobenzene	*Rhizoctonia solani*, *Botrytis* spp., *Aspergillus* spp., *Penicillium* spp., *Fusarium* spp., *Sclerotinia* spp., *and Tilletia caries*
	Nitrobenzene	*Pseudomonas pseudoalcaligenes* JS45 and *Pseudomonas putida* HS12, *Comamonas* sp. JS765, *Pseudomonas mendocina* KR-1, *Pseudomonas pickettii* PKO1, *cyanobacterium*, *Microcystis aeruginosa*, *Arthrobacter* sp. NB1, *Serratia* sp. NB2, and *Stenotrophomonas* sp. NB3, Comamonadaceae, and *Clostridium* spp., *and Rhodotorula mucilaginosa* strain Z1
	Cypermethrin	*E. coli*, *S. aureus*, *P. aeruginosa*, *B. subtilis*, *Enterobacter asuburiae, and Pseudomonas stutzeri*
	Carbofuran	*Pseudomonas*, *Flavobacterium*, *Achromobacterium*, *Sphingomonas, and Arthrobacter*
	Carbendazin	*Pseudomoans*
	Lambda-cyhalothrin	*Klebsiella* sp., and *Pseudomonas oleovorans*
	Fenvalerate	*Bacillus cereus*, *Pseudomonas viridiflava*, *Arthrobacter* sp., *Flavobacterium* sp., *Pseudomonas* sp., *Phanerochaete chrysosporium*, *Phlebia* sp., and *Bjerkandera* sp.
	Endrin and Aldrin	*Trichoderma viridae*, *Pseudomonas* sp., *Micrococcus* sp., and *Bacillus* sp.
	Dieldrin	*Pseudomonas* sp.

(Vardharajula et al. 2011; Spence and Bais 2015). Besides these, microbes have sigma factors to change gene expression and produces various key enzymes such as chitinase, ACC-deaminase, and glucanase under stress conditions (Farooq et al. 2009;

Gupta et al. 2013). AM is the most commonly found mycorrhizae in agricultural field and it plays an important role in absorption, translocation, and cycling of translocated nutrients. These mechanisms of microbes help in mitigating the negative impact of stress and maintaining actual growth under stressed environment.

9.7.1 Drought Stress

Drought is a major agriculture problem from which almost all of major agriculture land world are affected. It creates food security and economy problem on human society (Disante et al. 2011; Mishra and Singh 2010). Drought stress promotes senescence by producing ethylene and reactive oxygen species, which result in decreased crop productivity (Tiwari et al. 2016). The drought tolerant microbes have ability to promote growth under water-deficient condition. The microbes use different mechanisms to manage the negative impact of drought on plant under drought stress conditions. The mechanism used by microbes is the production of Bacterial exopolysaccharides (EPS), growth promoting phytohormones, ACC deaminase and induced systemic tolerance. The ability of PGPR to synthesize plant growth-stimulating hormones such as IAA that regulates cell division, vascular tissue differentiation, adventitious, lateral root differentiation, and shoot growth during stress conditions can play a crucial role to cope with drought stress (Goswami et al. 2015). Another important growth regulator produced by inoculated PGPR is ABA, which minimizes the effect of drought stress by regulating root hydraulic conductivity and transcription of drought-related genes (Jiang et al. 2013). For example, in the study Cohen et al. (2014) *Arabidopsis thalianato* showed growth under drought condition mainly due to enhancement of ABA level when inoculated with *Azospirillum brasilenseameliorates*. When a plant goes through stress conditions, a succession hormone ethylene is produced, which enhances the process of wilting and necrosis. The intermediate precursor of ethylene, 1-Aminocyclopropane-1-carboxylate (ACC), which can be hydrolyzed by bacterial ACC deaminase into ammonia and alpha-ketobutyrate (Bal et al. 2012). These drought stress-tolerant and PGPR inoculants reduce the effect of drought by reducing oxidative stress, decreasing water loss, and increasing water potential via enhancing production of proline, free amino acid, and sugar in plants (Vardharajula et al. 2011).

9.7.2 Salinity Stress

The uncontrolled application of agrochemicals and some natural causes have increased the deposition of salt in agricultural fields. Mitigating soil salinity is a challenging task for worldwide agricultural scientists. The negative impact of accumulating imbalance in nutrients, Na, and Cl ions can be seen on plant growth and microbial diversity and activities. It is reported that the inoculation of endophytic and PGP microbes can mitigate the negative salt effect on plant by various direct and indirect mechanisms. The direct mechanisms include nitrogen fixation, nutrient mobilization, phytohormones production (e.g., auxin, cytokinin, gibberellins, and ethylene), and siderophore production (Hayat et al. 2010) whereas, indirect mechanisms include reduction in the frequency of disease-causing plant pathogens, produce ACC deaminase, and formation of biofilm by microbes under salinity stress (Kasim et al. 2016). The enzyme Rhizobitoxine inhibit ethylene production and enhance nodulation under stress condition. The direct and indirect mechanisms lead to increase in root growth (length, surface area, and root number) and accumulate osmolytes in their cytoplasm that help in improving uptake of nutrients and water (Egamberdieva et al. 2015). The microbe *Bacillus pumilus* and *B. subtilis* isolated from saline soil showed PGPR traits such as phosphate solubilization, production of ammonia, IAA, and hydrogen cyanide (HCN) along with these they are showing salt stress tolerance (Damodaran et al. 2013). Bano and Fatima (2009) demonstrated that co-inoculation of PGPR strains of *Rhizobium* and *Pseudomonas* can cope up and facilitate plant growth under saline soil conditions.

9.7.3 Heavy Metal Stress

The uncontrolled anthropogenic activities such as industrialization, intensive agricultural practices, mining, and dumping of waste lead to heavy metals contamination in soil. As the name implies heavy metals are metallic elements having a density above 4 g cm^{-3}. It is non-degradable and poisonous even at low concentrations due to which a severe impact is caused on plants and animals health (Ma et al. 2016). It is reported that a wide range of metal tolerant microorganisms are present which can promote plant growth and development under metal stress conditions, e.g., *Rhizobacteria, Mycorrhiza, and Firmicutes*. Metal stress-tolerant microbes use various mechanisms such as efflux, enzymatic detoxification, EPS sequestration, impermeability to metals, volatilization, metal complexation formation production of phytohormones, and enhance nutrient availability to plant (Glick 2010; Ahmad et al. 2011; Verma et al. 2013). The bioaccumulation of heavy metals by microbes is a potent removal method from contaminated soil due to microbial cell wall properties (Vijayaraghavan and Yun 2008). Zhang et al. (2015) reported that *Proteobacteria*, *Firmicutes*, and *Actinobacteria* have the potential to remove higher concentrations of Mn, Pb, and As from heavy metal contaminated soils. It is reported that the concentration of copper (Cu) above 1 mM can damage the growth of *Vicia faba* plant, but inoculation with rhizobia and PGPR has reduced the negative impact of copper (Fatnassi et al. 2015). AM fungi help in reducing harmful effects of cadmium stress by reducing production of hydrogen peroxide and malonaldehyde (Hashem et al. 2016). The bacteria *Enterobacter* sp. and *Klebsiella* sp. showed efficient tolerance and removal of Cd, Pb, and Zn metals from contaminated soils along with producing plant growth substances (Jing et al. 2013).

9.7.4 Temperature Stress

The temperature stress becoming significant abiotic stress which impacts worldwide crop productivity and food security. Increase in the frequency and intensity of temperature stress can be seen that is mainly caused by climate change. The temperature stress shows a major effect on reduced photosynthesis activity, enzyme functioning, rate of transpiration, cell division, and plant growth. The tropical and subtropical regions, including India, are greatly impacted by climate change (Rodell et al. 2009; Alam et al. 2017). It is the most serious abiotic stress, plant hormone concentration and responses are mostly influenced by it, e.g., increase in concentration of jasmonic acid (JA) by many folds during stress conditions. Plants have a complex regulatory mechanism and induce many physiological changes that enable them to acclimatize and survive in variable temperature conditions. Plants' various mechanisms include production of heat shock proteins (HSP20, HSP60, HSP70, HSP90, HSP100), reactive oxygen species (ROS)-scavenging enzymes (ascorbate peroxidase and catalase), and osmolytes to overcome heat stress (Qu et al. 2013; Kotak et al. 2007). But there is an urgent need of mechanism for tolerating extreme temperatures as most of the crops are unable to tolerate extreme temperatures. Microbes are able to successfully colonize majority of earth biosphere with the ability to tolerate extreme temperature (high and low). At low temperatures, the ability of plant growth promotion was shown by *Brevundimonas terrace*, *Pseudomonas cedrina*, and *Arthrobacter nicotianae* (Yadav et al. 2014). Meena et al. (2015) demonstrated that the PGPR isolated from low temperature growing pea plant root nodule can act as an efficient biofertilizer in low temperature. Similarly, isolation of thermo tolerant phosphate solubilizing microbes can be used in agricultural field which can act as multifunctional bio fertilizer as it helps in bio-geochemical phosphorus cycling in agriculture field and transform insoluble phosphorus to soluble forms through acidification (Chang and Yang 2009).

9.8 Plants, Microbes, and Animals Cross-Talk

Biofertilizers have been identified as an alternative to chemical fertilizers to increase soil fertility and crop production in sustainable farming. These are the product containing living cells of different types of microorganisms, which have the ability to convert nutritionally important elements from its unavailable to available forms through biological process (Kundu et al. 2009; Kumar et al. 2016). In recent years, biofertilizers have emerged as an important component of the integrated nutrient supply system and hold a great promise to improve crop yield through environmentally better nutrient supplies (Molla et al. 2012; Meena et al. 2017). Strains of *Azotobacter*, *Rhizobium*, *Bradyrhizobium*, *Azospirillum*, *Pseudomonas*, *Bacillus*, and *Acetobacter*, etc. have been used as biofertilizers for cereals including wheat, pulses, vegetables, oilseeds, cotton, and sugarcane (Adesemoye et al. 2009; Ghormade et al. 2011). Bioorganic fertilizers (BOFs) combine functional microbes with a suitable substrate and have been shown to effectively suppress soil-borne diseases and promote plant growth (Zhao et al. 2018).

Despite the increase, the interest in organic farming is not much known at the molecular level and rhizospheric changes. Tenea et al. (2011) reported differential expression of many genes in organically and chemically cultivated wheat. In a similar way, microbes interacting with plants can also upregulate or downregulate plant transcription machinery. It has been shown that change in regulation is modulated by sRNAs that are delivered by various microbes (Weiberg et al. 2013; Qiao et al. 2015). However, in response to microbes plants show Host-Induced Gene Silencing (HIGS) in which invading microbes colonization are counteracted by delivering sRNAs to microbes (Wang et al. 2016). Still, it is unknown whether long RNA or a siRNA gets silenced by the transferred molecule from the plant or microbes (Baulcombe 2015). For past decades, it was considered that protein biosynthesis was completely regulated by long RNA molecule but Mizuno et al. (1984) showed small RNAs first time in *Escherichia coli* which regulate gene expression by a unique mechanism. Thereafter, Genome-wide searches revealed that small RNAs exist as noncoding RNA with diverse functions in all kingdoms of life and most of them serve as regulators of gene expression (Weiberg et al. 2015). The first time RNA silencing was reported by Jorgensen and coworkers while working on the enrichment of flower pigmentation they were aiming to overexpress chalcone synthase (CHS) in petunia but they observed unexpected white flowers (Napoli 1990). For modulating gene expression in any kingdom of life transfer of sRNA or siRNA is important but little is known about the transfer mechanism, although recent evidence suggests that extracellular vesicles or exosomes could act as a vehicle for delivery (Cai et al. 2018). Exosomes are nanovesicles of endocytic origin that are about 30–100 nm in diameter, surrounded by a lipid bilayer membrane, and contain proteins, nucleic acids, and other molecules (Liao et al. 2018). Currently, great consent is given toward the study of plant drives Exosome-like particles due to its participation in interspecies communication (Mu et al. 2014). The vesicles are also known as plant-derived edible nanoparticles (PDENs) are similar in structure and function to their mammalian counterparts and located in the paramural space of plants (An et al. 2007). The exosome vesicles consist of different factors that determine its delivery efficiency, such as long-distance communication factors. Plant-derived exosome also consist of plant source unique compounds with physiologically relevant bioactivities that also delivered to their targets via exosome-like vesicles. Zhang et al. (2016) reported presence of high amount of phosphatidic acid in exosomes derived from ginger. Furthermore, Rutter and Innes (2017) reported the role of exosome in plant immunity during biotic stress based on its protein profile and high expression in the apoplasts of leaves. As the mammalian host cell homeostasis is greatly influenced by the plant-derived diet. The differential expression of genes in organic and chemical cultivated plants whose all gene functions are not known, may also influence the mammalian host. Currently, great consent is given toward the communication between plant–microbe and plant–animal, and the reported literature has shown that exosome is the key player of the respective signaling (Mu et al. 2014). Various studies have shown the role of exosomes in animal systems, i.e., they communicate and regulate the expression along with its applications. In 2014, Mu et al have shown in mice model that orally administrated dose of plant-derived exosome induces nuclear translocation of macrophage Nrf2 and activation of intestinal Wnt/TCF4 pathway. Both nuclear translocations of Nrf2 and Wnt/TCF4 pathway activation play a critical role in anti-inflammatory responses. This information may provide the molecular basis of using multiple plant-derived agents for a better therapeutic effect.

Since, very few studies are present on communication and exosomes constituent in different interactions between microbe to plant and plant to animal. The exosome can act as a natural nanodelivery system with increased bioavailability as some nutrients and phytochemicals can pass through the digestive tract without being catabolized (Akuma et al. 2019). More research is needed on the offloading process, biogenesis of plant-derived exosome and response to various environmental conditions (biotic and abiotic stresses) along with a role in delivery vesicles within the plant and animal systems.

9.9 Soil Carbon Sequestration

Soil act as a major reservoir of carbon than vegetation and atmosphere collectively (Tan et al. 2014). Soil stored about 2344–2500 Gt of carbon, which includes 1550 Gt of carbon in organic forms (Decomposed plants, animals and microorganism) and 950 Gt stored in inorganic forms (lime and carbonates) (Chan 2008; Stockmann et al. 2013). The soil ecology and properties are dependent on types of soil organic matter and its composition. There are two types of organic matter, active and passive, in which active contain easily digestible sugars and proteins whereas passive is hard to decompose by microbes (Hoorman et al. 2009). The organic carbon in soil comes from atmospheric CO_2 which is fixed by plants and biomass which is consumed by animals and microbes as food. Some portions pass through respiration where carbon returns to the atmosphere in the form of CO_2 and the rest is converted into organic matter when an organism dies and stored in soil. Nowadays, the rate of carbon emissions has increased due to deforestation, industrialization, and some natural activity. There is a need to reduce carbon emission, which causes a negative impact on our climate, environmental quality and water resources, and soil fertility and properties (Qafoku 2015). Due to these negative effects, Soil Carbon Sequestration is the possible solution for reducing atmospheric carbon concentration. Olson (2013) defined the term carbon sequestration as transferring atmospheric CO_2 into the land soil through plants, plant residues, and other organic solids in order to be retained in soil organic matter. Carbon could be kept steady long-lived forms in the soil as organic matter and increasing concentration of hard to decompose organic substance will lead to reduction of carbon emission (Powlson et al. 2011). Carbon sequestration depends on several biological processes (photosynthesis, respiration, and decomposition) and anthropogenic management practices (such as overgrazing, cultivation, and stubble burning) (Chan 2008; Ontl and Schulte 2012). The rate of carbon sequestration depends on carbon fixation processes and carbon emission processes. The microbial inoculants can help in achieving an increased rate of carbon sequestration as photosynthetic microorganisms and some microbial communities can increase carbon inputs and decrease soil carbon emission (Macreadie et al. 2014; Fang et al. 2014). The different microbes behave differently in different ecosystem (Fig. 9.2), selection of right microbial communities for specific land play vital role in increasing carbon sequestration. Different microbial communities use different pathways and mechanisms in sequestrating carbon, such as metabolic activities that help in atmospheric CO_2 capturing, depositing as carbonates, forming recalcitrants, and forming soil aggregates (Smith et al. 2014). The soil's physical and chemical characteristics are contributed by microbes in their living, dead, and dormant states. the soil consists of 56% dead microorganisms, 42% dormant, and 0.1–2% active of its total microbial biomass (Blagodatskaya and Kuzyakov 2013).

Soil consists of a large number of bacterial genera compared to fungal genera but some reports show that fungi contribute more in carbon sequestration than bacteria due to its large biomass, having ability to form recalcitrant biomass tissues and by-products, form soil aggregates, which enhance carbon protection and stabilization (Li et al. 2015). The fungal carbon sequestration depends on the fungal biomass, by-products and necromass degradation production which shows a discrepancy species to species (Fernandez and Koide 2011; Ekblad et al. 2013). Fungal mycelium can grow up to hundreds of kilograms per ha per year which is vegetative tissues provide water and nutrients to the fungi (Wallander et al. 2013). Fungal mycelium could be considered as a vital carbon sink as it is made up of carbon, so the fast and large growth ability of mycelia helps in capturing more amount of carbon in the soil. Fungal necromass has the ability to resist decomposition, which varies according to the composition of fungal tissues and fungal species. The fungal necromass can be considered as stable organic matter due to the ability of resistance from degradation for a long time (Peng et al. 2013; Koide et al. 2013). The plant growth promoting microbes (PGPMs) can be used as useful tool in carbon sequestration, as PGPMs produces different type of plant hormones, which help in increasing plant above and below ground biomass (Nie et al. 2015; Kumar et al. 2016). The plant above ground biomass help in terrestrial carbon sequestration in the form of timber whereas belowground biomass helps in soil carbon sequestration in the form of roots and roots exudates. The PGPMs have the potential to mitigate climate change by sequestration of the atmospheric CO_2. Zhang et al. (2011) reported that *Bacillus mucilaginosus* produces carbonic anhydrase that captures atmospheric CO_2 and then through bacterial metabolism it fixes atmospheric CO_2. Some bacterial strains showed the ability to reduce CO_2 and some showed carbonate induction, it indicates that bacteria have the potential to help in carbon sequestration by various pathways (Peng et al. 2013; Nie et al. 2015). These pathways induce carbonate, organic substance, and carbonated minerals formation by fixing CO_2. In Xiao et al. (2015) study, an engineered *Escherichia coli* was formed by using carbonic anhydrase genes from *Bacillus mucilaginosus* and results showed production of carbonic anhydrase which form $CaCO_3$ in a relatively short time. The bacterial strains such as *Bacillus pumilus*, *Bacillus cereus*, *Bacillus mucilaginosus*, *Bacillus pasteurii* were reported as having the ability to sediment carbonates (Stocks-Fischer

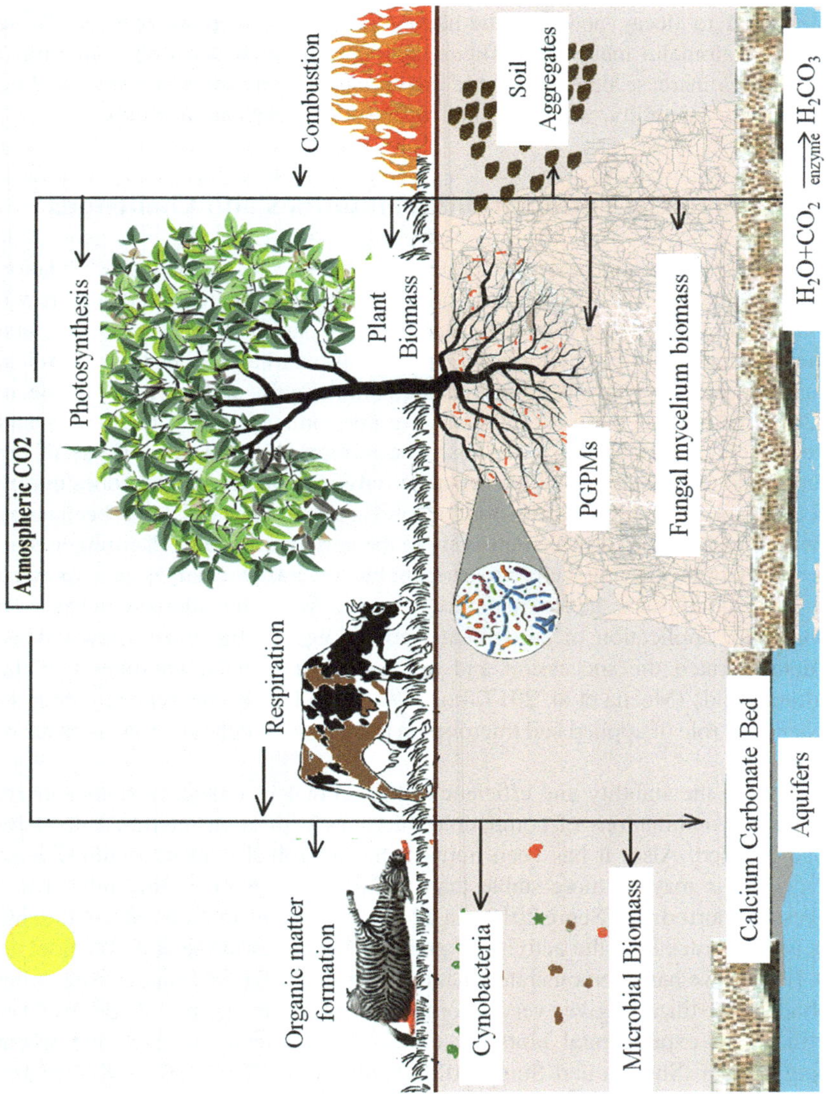

Fig. 9.2 Schematic diagram of carbon sequestration with the help of microbial community

et al. 1999; Zhang et al. 2011; Han et al. 2013; Komala and Khun 2014). The fungal genera *Entrophospora*, *Gigaspora*, *Scutellospora*, *Glomus*, and *Acaulospora* produces a glue-like proteinaceous substance Glomalin, which is produced from hyphae and spores in order to protect them from drying out (Singh 2012; Peng et al. 2013; Pal and Pandey 2014). Glomalin is recalcitrant in nature due to which it becomes hard to decay and its glue-like property form soil aggregates. These properties of glomalin make it an efficient soil carbon sequester at the same time, soil aggregates increase other favorable conditions such as soil structure stability, nutrient supply capability, water-holding capacity, porosity, and aeration.

9.10 Commercial Formulations; Limitations and Challenges

Although various microbial species facilitate atmospheric N_2-fixation, solubilization of nutrients (phosphorus, potassium, and zinc), and production of plant growth promoting hormones, siderophores, antagonistic substances, antibiotics, and enzymes for degradation of biological (lignin and protein) as well as toxic chemical substances (pesticides) have been formulated commercially and marketed (Nadeem et al. 2015). However, these formulations are not commonly used by farmers and other commercial agricultural producers. The rationale for not general use of the agrochemicals was instability of microbial activity and viability of microbes during production, storage, marketing, distribution, and application in soil. It has been noted that their efficiency goes down extensively in the agricultural fields (Berninger et al. 2018). Furthermore, living cells are susceptible to abrupt changes and extreme agroclimatic conditions, hence may not compete with the adopted native soil microbes after application of agroclimatic conditions. So far, only a few studies have demonstrated the succession and survival of inoculated microbe(s) in the agricultural fields (Meena et al. 2017) however, these studies are very important to understand the role of applied soil microbes in plant productivity and maintenance of soil fertility.

To increase the stability and efficiency of microbial inoculants various carriers and different combinations of compatible microbes with large microbial diversity can also be used. Also, it has been noticed that microbial consortia with diverse species richness may be more stable than a single species or closely related few microbes (Zomorrodi and Segrè 2018). In addition to that, availability of soil organic matter is very crucial for the activity of modified microbes in the soil (Wang et al. 2017). Earlier, we have demonstrated that organic matrix-based granules containing microbial biofertilizers give very promising results in terms of efficacy of biofertilizers in experimental plots for multiple crops including rice and wheat cropping system (Sharma and Singh 2011; Kumar et al. 2015, 2016; Kumar and Singh 2017).

In addition to the technical and technological shortage of production and marketing of authentic and effective microbial formulations for control of diseases and pests through non-pesticidal management for biocontrol, bionutrition of plants by microbial modification, and biodegradation of toxic and nontoxic organic substances. Also, lack of quality control mechanisms, end user education on toxicity of agrochemicals, and easily available alternatives are the major limitations for its non-acceptability by the farmers at a large scale. The economic size of centralized agrochemical production industries and their multinational nature influence the market and the entire extension and distribution channels. Contrary to this, the production of microbial inoculants for various uses can be done on small scale even, if market tycoons are not entering into it. The localized production by small-scale or medium-scale cottage industries can save transportation costs and marketing-related expenditures (Bashan et al. 2014). The local industrial units based on the production of these ecological agro-inputs can also provide a significant quantum of jobs to the skilled, semiskilled, and unskilled youth of rural, peri-urban, and urban regions.

9.11 Conclusion and Future Prospects

Although, rhizospheric engineering with inoculated microbial consortia for multiple agronomic benefits can enrich and even modify the natural microbiome of the soil, it can also invent new species diversity and species supremacy by the ecological succession. The dominant microbial inoculants that may be helpful to the specific agroecosystems being climate tolerant and more suitable for the requirement of new crops. Biodegradation of nontoxic animal and plant waste increases the organic matter of soil, whereas biodegradation of toxic agrochemicals, e.g., pesticides, as well as absorption of heavy metals can lead to contamination of soil and water and this can be fatal for the population of native and inoculated microflora and fauna of the soil. The major concern with the inoculation of non-native microbes may be related to dominance of certain exotic species that may suppress the growth and activity of the native rhizospheric microbes. This aspect is also needed to be addressed and further studies are warranted in this direction, especially in rhizosphere engineering projects. It should be taken into consideration that the native microbial population in soil as well as in rhizosphere may not be the same all the time due to frequent agroclimatic changes and emerging threats of global warming and climate change. The inoculated microbes, however, may modify the composition and function of rhizosphere, which can be designed carefully as per the emerging needs of the new agroclimatic situations and new cropping systems. The intervention of technology and innovations should not be always seen as a factor causing the negative impact on the conservation of soil biodiversity and practices of the ecological agriculture. The new soil microbial diversity can retain by various conditions, more suitable to the new requirements of the changing agroclimatic conditions and emerging needs of the new combination of crops. We are more focused toward enhancing total soil microbial diversity, carbon sequestration capacity, and multiple ecosystems and economic services of the ground and above ground agroecosystems.

Acknowledgments RPS is thankful to UGC New Delhi for financial support in form of the MRP project (MRP-MAJOR-ENVI-2013-11912/ F.No.43-320/2014(SR)). PK is thankful to DBT for financial support in form of JRF.

References

Adesemoye AO, Torbert HA, Kloepper JW (2009) Plant growth-promoting rhizobacteria allow reduced application rates of chemical fertilizers. Microb Ecol 58:921–929

Adisa IO, Pullagurala VLR, Peralta-Videa JR, Dimkpa CO, Elmer WH, Gardea-Torresdey J, White J (2019) Recent advances in nano-enabled fertilizers and pesticides: a critical review of mechanisms of action. Environ Sci Nano 6(7):2002–2030. https://doi.org/10.1039/c9en00265k

Ahmad M, Zahir ZA, Asghar HN, Asghar M (2011) Inducing salt tolerance in mung bean through coinoculation with rhizobia and plant-growth-promoting rhizobacteria containing 1-aminocyclopropane-1-carboxylate deaminase. Can J Microbiol 57(7):578–589. https://doi.org/10.1139/w11-044

Ajar NY, Priyanka V, Divjot K, Kusam LR (2017) Plant microbiomes and its beneficial multifunctional plant growth promoting attributes. Int J Environ Sci Nat Res 3(1):555601. https://doi.org/10.19080/IJESNR.2017.03.555601

Akuma P, Okagu OD, Udenigwe CC (2019) Naturally occurring exosome vesicles as potential delivery vehicle for bioactive compounds. Front Sustain Food Syst 3:23. https://doi.org/10.3389/fsufs.2019.00023

Alam MA, Seetharam K, Zaidi PH, Dinesh A, Vinayan MT, Nath UK (2017) Dissecting heat stress tolerance in tropical maize (*Zea mays* L.). Field Crop Res 204:110–119. https://doi.org/10.1016/j.fcr.2017.01.006

Ali Q, Ashraf S, Kamran M, Ijaz M (2019) Affirmative plant-microbe interfaces toward agroecosystem sustainability. In: Microbiome in plant health and disease. Springer, Cham, pp 145–170. https://doi.org/10.1007/978-981-13-8495-0_7

Alori ET, Dare MO, Babalola OO (2017) Microbial inoculants for soil quality and plant health. In: Lichtfouse E (ed) Sustainable agriculture reviews. Springer, Cham, pp 281–307. https://doi.org/10.1007/978-3-319-48006-0_9

Alvarenga N, Birolli WG, Meira EB, Lucas SC, de Matos IL, Nitschke M et al (2018) Biotransformation and biodegradation of methyl parathion by Brazilian bacterial strains isolated from mangrove peat. Biocatal Agric Biotechnol 13:319–326

Amer GA, Utkhede RS (2000) Development of formulation of biological agents for the management of root rot lettuce and cucumber. Can J Microbiol 46:809–816

An Q, Van Bel AJE, Hückelhoven R (2007) Do plant cells secrete exosomes derived from multivesicular bodieș? Plant Signal Behav 2:1. https://doi.org/10.4161/psb.2.1.3596

Arahal DR (2016) Beijerinckia. Bergey's manual of systematics of archaea and bacteria, pp 1–18. https://doi.org/10.1002/9781118960608.gbm00795.pub2

Attermeyer K, Premke K, Hornick T, Hilt S, Grossart HP (2013) Ecosystem-level studies of terrestrial carbon reveal contrasting bacterial metabolism in different aquatic habitats. Ecology 94:2754–2766

Azadi D, Shojaei H, Mobasherizadeh S, Naser AD (2017) Screening, isolation and molecular identification of biodegrading mycobacteria from Iranian ecosystems and analysis of their biodegradation activity. AMB Express 7(1):180. https://doi.org/10.1186/s13568-017-0472-4.

Bal HB, Nayak L, Das S, Adhya TK (2012) Isolation of ACC deaminase producing PGPR from rice rhizosphere and evaluating their plant growth promoting activity under salt stress. Plant Soil 366 (1–2):93–105. https://doi.org/10.1007/s11104-012-1402-5

Baldani VLD, Baldani JI, Döbereiner J (2000) Inoculation of rice plants with the endophytic diazotrophs *Herbaspirillum seropedicae* and *Burkholderia* spp. Biol Fertil Soils 30:485–489

Baldrian P (2019) The known and the unknown in soil microbial ecology. FEMS Microbiol Ecol 95 (2):fiz005. https://doi.org/10.1093/femsec/fiz005

Bano A, Fatima M (2009) Salt tolerance in *Zea mays* (L.) following inoculation with *Rhizobium* and *Pseudomonas*. Biol Fertil Soils 45:405–413

Bashan Y (1998) *Azospirillum* plant growth-promoting strains are nonpathogenic on tomato, pepper, cotton, and wheat. Can J Microbiol 44:168–174

Bashan Y, de-Bashan LE, Prabhu SR, Hernandez JP (2014) Advances in plant growth-promoting bacterial inoculant technology: formulations and practical perspectives. Plant Soil 378:1–33. https://doi.org/10.1007/s11104-013-1956-x

Baulcombe DC (2015) VIGS, HIGS and FIGS: small RNA silencing in the interactions of viruses or filamentous organisms with their plant hosts. Curr Opin Plant Biol 26:141–146. https://doi.org/10.1016/j.pbi.2015.06.007

Bellenger JP, Darnajoux R, Zhang X, Kraepiel AML (2020) Biological nitrogen fixation by alternative nitrogenases in terrestrial ecosystems: a review. Biogeochemistry 149:53–73. https://doi.org/10.1007/s10533-020-00666-7

Berendsen RL, van Verk MC, Stringlis IA, Zamioudis C, Tommassen J, Pieterse CMJ, Bakker PAHM (2015) Unearthing the genomes of plant-beneficial *Pseudomonas* model strains WCS358, WCS374 and WCS417. BMC Genomics 16(1):539. https://doi.org/10.1186/s12864-015-1632-z

Berninger T, López ÓG, Preininger C, Sessitsch A (2018) Maintenance and assessment of cell viability in formulation of non-sporulating bacterial inoculants. Microb Biotechnol 11 (2):277–301. https://doi.org/10.1111/1751-7915.12880

Bertrand H, Plassard C, Pinochet X, Touraine B, Normand P, Cleyet-Marel JC (2000) Stimulation of the ionic transport system in *Brassica napus* by a plant growth-promoting rhizobacterium (*Achromobacter* sp.). Can J Microbiol 46:229–236. https://doi.org/10.1139/w99-137

Bhardwaj D, Ansari M, Sahoo R, Tuteja N (2014) Biofertilizers function as key player in sustainable agriculture by improving soil fertility, plant tolerance and crop productivity. Microb Cell Fact 13:66. https://doi.org/10.1186/1475-2859-13-66

Bhattacharjee RB, Jourand P, Chaintreuil C, Dreyfus B, Singh A, Mukhopadhyay SN (2011) Indole acetic acid and ACC deaminase-producing *Rhizobium leguminosarum* bv. *trifolii* SN10 promote rice growth, and in the process undergo colonization and chemotaxis. Biol Fertil Soils 48 (2):173–182. https://doi.org/10.1007/s00374-011-0614-9

Blagodatskaya E, Kuzyakov Y (2013) Active microorganisms in soil: critical review of estimation criteria and approaches. Soil Biol Biochem 67:192–211

Bothe H, Schmitz O, Yates MG, Newton WE (2010) Nitrogen fixation and hydrogen metabolism in cyanobacteria. Microbiol Mol Biol Rev 74:529–551

Brück WM, Brück TB, Self WT, Reed JK, Nitecki SS, McCarthy PJ (2010) Comparison of the anaerobic microbiota of deep-water *Geodia* spp. and sandy sediments in the Straits of Florida. ISME J 4(5):686–699. https://doi.org/10.1038/ismej.2009.149

Cai Q, Qiao L, Wang M, He B, Lin F-M, Palmquist J et al (2018) Plants send small RNAs in extracellular vesicles to fungal pathogen to silence virulence genes. Science 360 (6393):1126–1129. https://doi.org/10.1126/science.aar4142

Canarini A, Wanek W, Merchant A, Richter A, Kaiser C (2018) Root exudation of primary metabolites: mechanisms and their roles in plant responses to environmental stimuli. Front Plant Sci 10:157. https://doi.org/10.3389/fpls.2019.00157

Cardinale M, Grube M, Erlacher A, Quehenberger J, Berg G (2015) Bacterial networks and co-occurrence relationships in the lettuce root microbiota. Environ Microbiol 17(1):239–252

Chagas FO, Pessotti R d C, Caraballo-Rodríguez AM, Pupo MT (2018) Chemical signaling involved in plant–microbe interactions. Chem Soc Rev 47(5):1652–1704. https://doi.org/10.1039/c7cs00343a

Chan Y (2008) Increasing soil organic carbon of agricultural land. NSW Dep Prim Ind 735:1–5

Chang CH, Yang SS (2009) Thermo-tolerant phosphate-solubilizing microbes for multifunctional biofertilizer preparation. Bioresour Technol 100:1648–1658

Chattopadhyay P, Banerjee G, Mukherjee S (2017) Recent trends of modern bacterial insecticides for pest control practice in integrated crop management system. 3 Biotech 7(1):60

Chauhan N, Sankhyan NK, Sharma RP, Singh J, Gourav (2020) Effect of long-term application of inorganic fertilizers, farm yard manure and lime on wheat (*Triticum aestivum* L.) productivity, quality and nutrient content in an acid alfisol. J Plant Nutr 43:1–10. https://doi.org/10.1080/01904167.2020.1783298

Chen L, Liao H (2017) Engineering crop nutrient efficiency for sustainable agriculture. J Integr Plant Biol 59(10):710–735. https://doi.org/10.1111/jipb.12559

Chen YP, Rekha PD, Arun AB, Shen FT, Lai WA, Young CC (2006) Phosphate solubilizing bacteria from subtropical soil and theirtricalcium phosphate solubilizing abilities. Appl Soil Ecol 34:33–41

Chen S-L, Tsai M-K, Huang Y-M, Huang C-H (2017) Diversity and characterization of *Azotobacter* isolates obtained from rice rhizosphere soils in Taiwan. Ann Microbiol 68(1):17–26. https://doi.org/10.1007/s13213-017-1312-0

Cheng M, Cao L, Ning K (2019) Microbiome big-data mining and applications using single-cell technologies and metagenomics approaches toward precision medicine. Front Genet 10:972. https://doi.org/10.3389/fgene.2019.00972

Chinakwe EC, Nwogwugwu NU, Ibekwe VI, Chinakwe PO, Egbadon EO, Adeleye SA (2019) Isolation and evaluation of bacteria exhibiting multiple plant growth traits in the Rhizosphere of yellow bell pepper (*Capsicum chinense*). J Adv Microbiol 14(4):1–6. https://doi.org/10.9734/jamb/2019/v14i430070

Cohen AC, Bottini R, Pontin M, Berli FJ, Moreno D, Boccanlandro H et al (2014) *Azospirillum brasilenseameliorates* the response of *Arabidopsis thalianato* drought mainly via enhancement of ABA levels. Physiol Plant 153(1):79–90. https://doi.org/10.1111/ppl.12221

Collins DP, Jacobsen B (2003) Optimizing a *Bacillus subtilis* isolate for biological control of sugar beet cercospora leaf spot. Biol Control J 26:153–161

Contreras-Cornejo HA, Macias-Rodriguez L, Cortes-Penagos C, Lopez-Bucio J (2009) Trichoderma virens, a plant beneficial fungus, enhances biomass production and promotes lateral root growth through an auxin-dependent mechanism in arabidopsis. Plant Physiol 149 (3):1579–1592. https://doi.org/10.1104/pp.108.130369

Contreras-Cornejo HA, Macías-Rodríguez L, Del-Val E, Larsen J (2016) Ecological functions of *Trichoderma* spp. and their secondary metabolites in the rhizosphere: interactions with plants. FEMS Microbiol Ecol 92(4):fiw036. https://doi.org/10.1093/femsec/fiw036

Dal Cortivo C, Barion G, Visioli G, Mattarozzi M, Mosca G, Vamerali T (2017) Increased root growth and nitrogen accumulation in common wheat following PGPR inoculation: assessment of plant-microbe interactions by ESEM. Agric Ecosyst Environ 247:396–408. https://doi.org/10.1016/j.agee.2017.07.006

Damodaran T et al (2013) Isolation of salt tolerant endophytic and rhizospheric bacteria by natural selection and screening for promising plant growth-promoting rhizobacteria (PGPR) and growth vigour in tomato under sodic environment. Afr J Microbiol Res 7:5082–5089

Danger M, Gessner MO, Bärlocher F (2016) Ecological stoichiometry of aquatic fungi: current knowledge and perspectives. Fungal Ecol 19:100–111

Datta A, Singh RK, Kumar S, Kumar S (2015) An effective and beneficial plant growth promoting soil bacterium "*Rhizobium*": a review. Ann Plant Sci 4(1):933–942

Dave BP, Dube HC (2000) Chemical characterization of fungal siderophores. Indian J Exp Biol 38:56–62

Dave BP, Anshuman K, Hajela P (2006) Siderophores of halophilic archaea and their chemical characterization. Indian J Exp Biol 44:340–344

Debnath S, Pachauri SP, Srivastava PC (2015) Improving use efficiency of applied phosphorus fertilizer by zinc fertilization in basmati rice-wheat cropping system. Indian J Agric Res 49:414–420

Deng Z-S, Kong Z-Y, Zhang B-C, Zhao L-F (2019) Insights into non-symbiotic plant growth promotion bacteria associated with nodules of *Sphaerophysa salsula* growing in northwestern China. Arch Microbiol 202(2):399–409. https://doi.org/10.1007/s00203-019-01752-7

Deshmukh R, Khardenavis AA, Purohit HJ (2016) Diverse metabolic capacities of fungi for bioremediation. Indian J Microbiol 56(3):247–264. https://doi.org/10.1007/s12088-016-0584-6

Desmond E (2016) The legitimation of development and GM crops: the case of Bt Cotton and Indebtedness in Telangana, India. World Dev Perspect 1:23–25. https://doi.org/10.1016/j.wdp.2016.05.008

Disante KB, Fuentes D, Cortina J (2011) Response to the drought of Zn-stressed *Quercus suber* L. seedlings. Environ Exp Bot 70:96–103

Dos Santos PC, Addo MA (2020) Distribution of nitrogen fixation genes in prokaryotes containing alternative nitrogenases. ChemBioChem 21(12):1749–1759. https://doi.org/10.1002/cbic.202000022

Durán J, Delgado-Baquerizo M, Dougill AJ, Guuroh RT, Linstädter A, Thomas AD, Maestre FT (2018) Temperature and aridity regulate spatial variability of soil multifunctionality in drylands across the globe. Ecology 99(5):1184–1193. https://doi.org/10.1002/ecy.2199

Eberl L, Vandamme P (2016) Members of the genus *Burkholderia*: good and bad guys. F1000Research 5:1007. https://doi.org/10.12688/f1000research.8221.1

Ebrahimi M, Safari Sinegani AA, Sarikhani MR, Mohammadi SA (2017) Comparison of artificial neural network and multivariate regression models for prediction of *Azotobacteria* population in soil under different land uses. Comput Electron Agric 140:409–421. https://doi.org/10.1016/j.compag.2017.06.019

Eckert B, Weber OB, Stoffels M, Hartmann A, Kirchhof G, Halbritter A (2001) *Azospirillum doebereinerae* sp. nov., a nitrogen-fixing bacterium associated with the C4-grass Miscanthus. Int J Syst Evol Microbiol 51(1):17–26. https://doi.org/10.1099/00207713-51-1-17

Egamberdieva D, Jabborova D, Hashem A (2015) Pseudomonas induces salinity tolerance in cotton (*Gossypium hirsutum*) and resistance to *Fusarium* root rot through the modulation of indole-3-acetic acid. Saudi J Boil Sci 22:773–779

Ekblad A, Wallander H, Godbold DL, Cruz C, Johnson D, Baldrian P et al (2013) The production and turnover of extramatrical mycelium of ectomycorrhizal fungi in forest soils: role in carbon cycling. Plant Soil 366(1–2):1–27. https://doi.org/10.1007/s11104-013-1630-3

Essen SA, Johnsson A, Bylund D, Pedersen K, Lundstrom US (2007) Siderophore production by Pseudomonas *stutzeri* under aerobic and anaerobic conditions. Appl Environ Microbiol 73 (18):5857–5864

Estrada-delos Station P, Bustitio-Cristales R, Caballero-Mallado J (2001) Burkholderia, a genus rich in plant-associated nitrogen fixers with wide environmental and geographic distribution. Appl Environ Microbiol 67:279–2798

Etesami H, Adl SM (2020) Plant growth-promoting rhizobacteria (PGPR) and their action mechanisms in availability of nutrients to plants. In: Kumar M, Kumar V, Prasad R (eds) Phyto-microbiome in stress regulation. Environmental and microbial biotechnology. Springer, Singapore. https://doi.org/10.1007/978-981-15-2576-6_9

Fang H, Cheng S, Wang Y, Yu G, Xu M, Dang X et al (2014) Changes in soil heterotrophic respiration, carbon availability, and microbial function in seven forests along a climate gradient. Ecol Res 29(6):1077–1086. https://doi.org/10.1007/s11284-014-1194-6

Farooq M, Wahid A, Kobayashi N, Fujita D, Basra SMA (2009) Plant drought stress: effects, mechanisms and management. In: Sustainable agriculture. Springer, Cham, pp 153–188. https://doi.org/10.1007/978-90-481-2666-8_12

Fatnassi IC, Chiboub M, Saadani O, Jebara M, Jebara SH (2015) Impact of dual inoculation with *Rhizobium* and PGPR on growth and antioxidant status of *Vicia faba* L. under copper stress. C R Biol 338(4):241–254. https://doi.org/10.1016/j.crvi.2015.02.001

Felten J, Legué V, Ditengou FA (2010) Lateral root stimulation in the early interaction between *Arabidopsis thaliana* and the ectomycorrhizal fungus *Laccaria bicolor*: is fungal auxin the trigger? Plant Signal Behav 5(7):864–867. https://doi.org/10.4161/psb.5.7.11896

Fernandez CW, Koide RT (2011) The role of chitin in the decomposition of ectomycorrhizal fungal litter. Ecology 93(1):24–28

Fierer N (2017) Embracing the unknown: disentangling the complexities of the soil microbiome. Nat Rev Microbiol 15(10):579–590. https://doi.org/10.1038/nrmicro.2017.87

Fuentes-Ramirez LE, Caballero-Mellado J (2005) Bacterial biofertilizers. In: PGPR: biocontrol and biofertilization. Springer, Cham, pp 143–172. https://doi.org/10.1007/1-4020-4152-7_5

Ganz T (2018) Iron and infection. Int J Hematol 107(1):7–15. https://doi.org/10.1007/s12185-017-2366-2

Garcia J, Kao-Kniffin J (2018) Microbial group dynamics in plant rhizospheres and their implications on nutrient cycling. Front Microbiol 9:1516. https://doi.org/10.3389/fmicb.2018.01516

Garnica-Vergara A, Barrera-Ortiz S, Muñoz-Parra E, Raya-González J, Méndez-Bravo A, Macías-Rodríguez L et al (2015) The volatile 6-pentyl-2H-pyran-2-one from *Trichoderma atroviride* regulates *Arabidopsis thaliana* root morphogenesis via auxin signaling and ETHYLENE INSENSITIVE 2 functioning. New Phytol 209(4):1496–1512. https://doi.org/10.1111/nph.13725

Germida JJ, Siciliano SD (2001) Taxonomic diversity of bacteria associated with the roots of modern, recent and ancient wheat cultivars. Biol Fertil Soils 33:410–415

Gessner MO, Gulis V, Kuehn KA, Chauvet E, Suberkropp K (2007) Fungal decomposers of plant litter in aquatic ecosystems. In: Environmental and microbial relationships, vol 4. Springer, Cham, pp 301–324

Ghormade V, Mukund V, Kishore D, Paknikar M (2011) Perspectives for nano-biotechnology enabled protection and nutrition of plants. Biotechnol Adv 29(6):792–803

Glick BR (2010) Using soil bacteria to facilitate phytoremediation. Biotechnol Adv 28:367–374

Gomare KS, Mese M, Shetkar Y (2013) Isolation of *Azotobacter* and cost effective production of biofertilizer. Indian J Appl Res 3:54–56

Goswami D, Thakker JN, Dhandhukia PC (2015) Simultaneous detection and quantification of indole-3-acetic acid (IAA) and indole-3-butyric acid (IBA) produced by rhizobacteria from l-tryptophan (Trp) using HPTLC. J Microbiol Methods 110:7–14

Gregory PJ, Ingram JSI, Andersson R, Betts RA, Brovkin V, Chase TN et al (2002) Environmental consequences of alternative practices for intensifying crop production. Agric Ecosyst Environ 88(3):279–290. https://doi.org/10.1016/s0167-8809(01)00263-8

Gupta G, Panwar J, Jha PN (2013) Natural occurrence of *Pseudomonas aeruginosa*: adominant cultivable diazotrophic endophytic bacterium colonizing *Pennisetum glaucum* (L.) R. Br. Appl Soil Ecol 64:252–261

Güsewell S, Gessner MO (2009) N:P ratios influence litter decomposition and colonization by fungi and bacteria in microcosms. Funct Ecol 23:211–219

Han J, Lian B, Ling H (2013) Induction of calcium carbonate by *Bacillus cereus*. Geomicrobiol J 30 (8):682–689. https://doi.org/10.1080/01490451.2012.758194

Hashem A, Abd-Allah EF, Alqarawi AA, Al Huqail AA, Egamberdieva D, Wirth S (2016) Alleviation of cadmium stress in *Solanum lycopersicum* L. by arbuscular mycorrhizal fungi via induction of acquired systemic tolerance. Saudi J Biol Sci 23(2):272–281. https://doi.org/10.1016/j.sjbs.2015.11.002

Hassani MA, Durán P, Hacquard S (2018) Microbial interactions within the plant holobiont. Microbiome 6(1):58. https://doi.org/10.1186/s40168-018-0445-0

Hati KM, Swarup A, Singh D, Misra AK, Ghosh PK (2006) Long-term continuous cropping, fertilisation, and manuring effects on physical properties and organic carbon content of a sandy loam soil. Aust J Soil Res 44(5):487. https://doi.org/10.1071/sr05156

Hayat R, Ali S, Amara U, Khalid R, Ahmed I (2010) Soil beneficial bacteria and their role in plant growth promotion: a review. Ann Microbiol 60(4):579–598. https://doi.org/10.1007/s13213-010-0117-1

Hayat R, Ahmed I, Sheirdil RA (2012) An overview of plant growth promoting rhizobacteria (PGPR) for sustainable agriculture. In: Crop production for agricultural improvement. Springer, Cham, pp 557–579. https://doi.org/10.1007/978-94-007-4116-4_22

Hellmann C, Große-Stoltenberg A, Thiele J, Oldeland J, Werner C (2017) Heterogeneous environments shape invader impacts: integrating environmental, structural and functional effects by isoscapes and remote sensing. Sci Rep 7(1):4118. https://doi.org/10.1038/s41598-017-04480-4

Hieber M, Gessner MO (2002) Contribution of stream detrivores, fungi, and bacteria to leaf breakdown based on biomass estimates. Ecology 83:1026–1038

Hoorman JJ et al (2009) Using cover crops to convert to no-till. Crops Soil 42:9–13

Horino H, Ito M, Tonouchi A (2015) *Clostridium oryzae* sp. nov., from soil of a Japanese rice field. Int J Syst Evol Microbiol 65(Pt 3):943–951. https://doi.org/10.1099/ijs.0.000042

Huang Y, Wang J (2013) Degradation and mineralization of DDT by the ectomycorrhizal fungi, *Xerocomus chrysenteron*. Chemosphere 92(7):760–764. https://doi.org/10.1016/j.chemosphere.2013.04.002

Jacoby R, Peukert M, Succurro A, Koprivova A, Kopriva S (2017) The role of soil microorganisms in plant mineral nutrition-current knowledge and future directions. Front Plant Sci 8:1617. https://doi.org/10.3389/fpls.2017.01617

Jacquet P, Daude D, Bzdrenga J, Masson P, Elias M, Chabriere E (2016) Current and emerging strategies for organophosphate decontamination: special focus on hyperstable enzymes. Environ Sci Pollut Res Int 23(9):8200–8218. https://doi.org/10.1007/s11356-016-6143-1

Jataraf J, Radhakrim NV, Hannk P, Sakoof R (2005) Biocontrol of tomato damping-off caused by *Pythium aphanidermatum*. BioControl 15:55–65

Jiang S, Zhang D, Wang L, Pan J, Liu Y, Kong X et al (2013) A maize calcium-dependent protein kinase gene, *ZmCPK4*, positively regulated abscisic acid signaling and enhanced drought stress tolerance in transgenic *Arabidopsis*. Plant Physiol Biochem 71:112–120. https://doi.org/10.1016/j.plaphy.2013.07.004

Jiang Y, Deng A, Bloszies S, Huang S, Zhang W (2017) Nonlinear response of soil ammonia emissions to fertilizer nitrogen. Biol Fertil Soils 53:269–274

Jing YX, Yan JL, He HD, Yang DJ, Xiao L, Zhong T et al (2013) Characterization of bacteria in the rhizosphere soils of polygonum pubescensand their potential in promoting growth and Cd, Pb, Zn uptake by *Brassica napus*. Int J Phytoremediation 16(4):321–333. https://doi.org/10.1080/15226514.2013.773283

Jorjani M, Heydari A, Zamanizadeh HR, Rezaee S, Naraghi L (2011) Controlling sugarbeet mortality disease by application of new bioformulations. J Plant Prot Res 52:303–307

Kalayu G (2019) Phosphate solubilizing microorganisms: promising approach as biofertilizers. Int J Agron 2019:1–7. https://doi.org/10.1155/2019/4917256

Kasim WA, Gaafar RM, Abou-Ali RM, Omar MN, Hewait HM (2016) Effect of biofilm forming plant growth promoting rhizobacteria on salinity tolerance in barley. Ann Agric Sci 61 (2):217–227. https://doi.org/10.1016/j.aoas.2016.07.003

Kaymak H (2010) Potential of PGPR in agricultural innovations. In: Plant growth and health promoting bacteria, vol 18. Springer, Berlin, pp 45–79

Kennedy IR, Islam N (2001) The current and potential contribution of asymbiotic nitrogen fixation to nitrogen requirements on farms: a review. Aust J Exp Agric 41(3):447. https://doi.org/10.1071/ea00081

Khadse A, Rosset PM, Morales H, Ferguson BG (2017) Taking agroecology to scale: the Zero Budget Natural Farming peasant movement in Karnataka, India. J Peasant Stud 45(1):192–219. https://doi.org/10.1080/03066150.2016.1276450

Khan A, Hossain MT, Park HC, Yun D-J, Shim SH, Chung YR (2016) Development of root system architecture of *Arabidopsis thaliana* in response to colonization by *Martelella endophytica* YC6887 depends on auxin signaling. Plant Soil 405(1–2):81–96. https://doi.org/10.1007/s11104-015-2775-z

Koide RT, Fernandez C, Malcolm G (2013) Determining place and process: functional traits of ectomycorrhizal fungi that affect both community structure and ecosystem function. New Phytol 201(2):433–439. https://doi.org/10.1111/nph.12538

Komala T, Khun T (2014) Biological carbon dioxide sequestration potential of *Bacillus pumilus*. Sains Malays 43(8):1149–1156

Kotak S, Larkindale J, Lee U, von Koskull-Döring P, Vierling E, Scharf K-D (2007) Complexity of the heat stress response in plants. Curr Opin Plant Biol 10(3):310–316. https://doi.org/10.1016/j.pbi.2007.04.011

Krauss G-J, Solé M, Krauss G, Schlosser D, Wesenberg D, Bärlocher F (2011) Fungi in freshwaters: ecology, physiology and biochemical potential. FEMS Microbiol Rev 35:620–651

Kuehn KA, Francoeur SN, Findlay RH, Neely RK (2014) Priming in the microbial landscape: periphytic algal stimulation of litter-associated microbial decomposers. Ecology 95:749–762

Kumar M, Singh RP (2017) Enhancement in growth promotion and production of wheat (*Triticum aestivum* L.) by application of a native strain of *Trichoderma virens* (T2) in pot condition. Int J Sci Technol Soc 3(2):62–67

Kumar M, Bauddh K, Sainger M, Sainger PA, Singh RP (2015) Enhancing efficacy of azotobactor and bacillus by entrapping in organic matrix for rice cultivation. Agroecol Sustain Food Syst 39 (8):907–923. https://doi.org/10.1080/21683565.2015.1050146

Kumar V, Kumar M, Shrivastava N, Bisht S, Sharma S, Varma A (2016) Interaction among rhizospheric microbes. In: Soil, and plant roots: influence on micronutrient uptake and bioavailability. Springer, Cham, pp 169–185

Kumawat N, Kumar R, Kumar S, Meena VS (2017) Nutrient solubilizing microbes (NSMs): its role in sustainable crop production. In: Agriculturally important microbes for sustainable agriculture. Springer, Singapore, pp 25–61. https://doi.org/10.1007/978-981-10-5343-6_2

Kundu BS, Nehra K, Yadav R, Tomar M (2009) Biodiversity of phosphate solubilizing bacteria in rhizosphere of chickpea, mustard and wheat grown in different region of Haryana. Ind. J Microbiol 49:120–127

Kundu DK, Mazumdar SP, Ghosh D, Saha AR, Majumdar B, Ghorai AK, Behera MS (2016) Long-term effects of fertilizer and manure application on soil quality and sustainability of jute-rice-wheat production system in Indo-Gangetic plain. J Appl Nat Sci 8(4):1793–1800. https://doi.org/10.31018/jans.v8i4.1042

Kuzyakov Y, Razavi BS (2019) Rhizosphere size and shape: temporal dynamics and spatial stationarity. Soil Biol Biochem 135:343–360. https://doi.org/10.1016/j.soilbio.2019.05.011

Lareen A, Burton F, Schäfer P (2016) Plant root-microbe communication in shaping root microbiomes. Plant Mol Biol 90(6):575–587

Le XH, Franco CMM, Ballard RA, Drew EA (2015) Isolation and characterisation of endophytic actinobacteria and their effect on the early growth and nodulation of lucerne (*Medicago sativa* L.). Plant Soil 405(1–2):13–24. https://doi.org/10.1007/s11104-015-2652-9

Lea PJ, Miflin BJ (2018) Nitrogen Assimilation and its Relevance to Crop Improvement. Annu Plant Rev 42:1–40. https://doi.org/10.1002/9781119312994.apr0448

Lee SA, Kim Y, Kim JM, Chu B, Joa J-H, Sang MK et al (2019) A preliminary examination of bacterial, archaeal, and fungal communities inhabiting different rhizocompartments of tomato plants under real-world environments. Sci Rep 9(1):9300. https://doi.org/10.1038/s41598-019-45660-8

Li N, Xu Y-Z, Han X-Z, He H-B, Zhang X, Zhang B (2015) Fungi contribute more than bacteria to soil organic matter through necromass accumulation under different agricultural practices during the early pedogenesis of a Mollisol. Eur J Soil Biol 67:51–58. https://doi.org/10.1016/j.ejsobi.2015.02.002

Li H-B, Singh RK, Singh P, Song Q-Q, Xing Y-X, Yang L-T, Li Y-R (2017) Genetic diversity of nitrogen-fixing and plant growth promoting pseudomonas species isolated from sugarcane rhizosphere. Front Microbiol 8:1268. https://doi.org/10.3389/fmicb.2017.01268

Li M, Yang Y, He Y, Mathieu J, Yu C, Li Q, Alvarez PJJ (2018) Detection and cell sorting of *Pseudonocardia* species by fluorescence in situ hybridization and flow cytometry using 16S rRNA-targeted oligonucleotide probes. Appl Microbiol Biotechnol 102(7):3375–3386. https://doi.org/10.1007/s00253-018-8801-3

Li L, Wilson CB, He H, Zhang X, Zhou F, Schaeffer SM (2019) Physical, biochemical, and microbial controls on amino sugar accumulation in soils under long-term cover cropping and no-tillage farming. Soil Biol Biochem 135:369–378. https://doi.org/10.1016/j.soilbio.2019.05.017

Liao W, Du Y, Zhang C, Pan F, Yao Y, Zhang T, Peng Q (2018) Exosomes: the next generation of endogenous nanomaterials for advanced drug delivery and therapy. Acta Biomater 86:1–14. https://doi.org/10.1016/j.actbio.2018.12.045

Liu X, Zhang Y, Han W, Tang A, Shen J, Cui Z, Vitousek P, Erisman JW, Goulding K, Christie P, Fangmeier A, Zhang F (2013) Enhanced nitrogen deposition over China. Nature 494:459–462

Liu J, Tan L, Wang J, Wang Z, Ni H, Li L (2016) Complete biodegradation of chlorpyrifos by engineered *Pseudomonas putida* cells expressing surface-immobilized laccases. Chemosphere 157:200–207. https://doi.org/10.1016/j.chemosphere.2016.05.031

López-Bucio J, Campos-Cuevas JC, Hernández-Calderón E, Velásquez-Becerra C, Farías-Rodríguez R, Macías-Rodríguez LI, Valencia-Cantero E (2007) *Bacillus megaterium* rhizobacteria promote growth and alter root-system architecture through an auxin- and ethylene-independent signaling mechanism in *Arabidopsis thaliana*. Mol Plant-Microbe Interact 20(2):207–217. https://doi.org/10.1094/mpmi-20-2-0207

Ma Y et al (2016) Inoculation of Brassica oxyrrhina with plant growth promoting bacteria for the improvement of heavy metal phytoremediation under drought conditions. J Hazard Mater 320:36–44

Maathuis FJM, Diatloff E (2012) Roles and functions of plant mineral nutrients. In: Plant mineral nutrients. Humana Press, Totowa, NJ, pp 1–21. https://doi.org/10.1007/978-1-62703-152-3_1

Macreadie PI et al (2014) Quantifying and modelling the carbon sequestration capacity of seagrass meadows—a critical assessment. Mar Pollut Bull 83(2):430–439

Makarov MI (2019) The role of mycorrhiza in transformation of nitrogen compounds in soil and nitrogen nutrition of plants: a review. Eurasian Soil Sci 52(2):193–205. https://doi.org/10.1134/s1064229319020108

Malla MA, Dubey A, Yadav S, Kumar A, Hashem A, Abd Allah EF (2018) Understanding and designing the strategies for the microbe-mediated remediation of environmental contaminants using omics approaches. Front Microbiol 9:1132. https://doi.org/10.3389/fmicb.2018.01132

Mansoori M, Heydari A, Hassanzadeh N, Rezaee S, Naraghi L (2013) Evaluation of *Pseudomonas* and *Bacillus* bacterial antagonists for biological control of cotton Verticillium Wilt disease. J Plant Prot Res 53:11–14

Martínez-Hidalgo P, Hirsch AM (2017) The nodule microbiome: N2-fixing rhizobia do not live alone. Phytobiomes 1(2):70–82. https://doi.org/10.1094/pbiomes-12-16-0019-rvw

Mazid M, Khan TA (2014) Future of bio-fertilizers in Indian agriculture: an overview. Int J Agric Food Res 3:10–23

Meena RK et al (2015) Isolation of low temperature surviving plant growth–promoting rhizobacteria (PGPR) from pea (*Pisum sativum* L.) and documentation of their plant growth promoting traits. Biocatal Agric Biotechnol 4:806–811

Meena VS, Mishra PK, Bisht JK, Pattanayak A (2017) Agriculturally important microbes for sustainable agriculture. Springer, Singapore

Mehnaz S (2011) Plant growth-promoting bacteria associated with sugarcane. In: Bacteria in agrobiology: crop ecosystems. Springer, Berlin, pp 165–187. https://doi.org/10.1007/978-3-642-18357-7_7

Mercado-Blanco J, Rodríguez-Jurado D, Hervás A, Jiménez-Diaz RM (2004) Suppression of Verticillium wilt in olive planting stocks by root-associated fluorescent *Pseudomonas* spp. Biol Control 30:474–486

Mille-Lindblom C, Tranvik LJ (2003) Antagonism between bacteria and fungi on decomposing aquatic plant litter. Microb Ecol 45:173–182

Mirza MS, Rasul G, Mehnaz S, Ladha JK, So RB, Ali S, Malik KA (2000) Beneficial effects of inoculated nitrogen-fixing bacteria on rice. In: Ladha JK, Reddy PM (eds) The quest for nitrogen fixation in rice. International Rice Research Institute, Los Banõs, pp 191–204

Mishra AK, Singh VP (2010) A review of drought concepts. J Hydrol 39:202–216

Mizuno T, Chou MY, Inouye M (1984) A unique mechanism regulating gene expression: translational inhibition by a complementary RNA transcript (micRNA). Proc Natl Acad Sci 81 (7):1966–1970. https://doi.org/10.1073/pnas.81.7.1966

Mohanram S, Kumar P (2019) Rhizosphere microbiome: revisiting the synergy of plant-microbe interactions. Ann Microbiol 69:307–320. https://doi.org/10.1007/s13213-019-01448-9

Mohanty S, Swain CK (2018) Role of microbes in climate smart agriculture. In: Microorganisms for green revolution. Springer, Singapore, pp 129–140. https://doi.org/10.1007/978-981-10-7146-1_7

Molla AH, Haque M, Haque A, Ilias GNM (2012) Trichoderma-enriched biofertilizer enhances production and nutritional quality of tomato (*Lycopersicon esculentum* Mill) and minimizes NPK fertilizer use. Agric Res 1:265–272

Mondal R, Mukherjee A, Biswas S, Kole RK (2018) GC-MS/MS determination and ecological risk assessment of pesticides in aquatic system: a case study in Hooghly River basin in West Bengal, India. Chemosphere 206:217–230. https://doi.org/10.1016/j.chemosphere.2018.04.168

Mori T, Wang J, Tanaka Y, Nagai K, Kawagishi H, Hirai H (2017) Bioremediation of the neonicotinoid insecticide clothianidin by the white-rot fungus *Phanerochaete sordida*. J Hazard Mater 321:586–590. https://doi.org/10.1016/j.jhazmat.2016.09.049

Mu J, Zhuang X, Wang Q, Jiang H, Deng Z-B, Wang B et al (2014) Interspecies communication between plant and mouse gut host cells through edible plant derived exosome-like nanoparticles. Mol Nutr Food Res 58(7):1561–1573. https://doi.org/10.1002/mnfr.201300729

Nadeem SM et al (2014) The role of mycorrhizae and plant growth promoting rhizobacteria (PGPR) in improving crop productivity under stressful environments. Biotechnol Adv 32:429–448

Nadeem SM, Naveed M, Ahmad M, Zahir ZA (2015) Rhizosphere bacteria for crop production and improvement of stress tolerance: mechanisms of action, applications, and future prospects. In: Arora N (ed) Plant microbes symbiosis: applied facets. Springer, New Delhi., Isbn:978-81-322-2067-1. https://doi.org/10.1007/978-81-322-2068-8_1

Naik K, Mishra S, Srichandan H, Singh PK, Choudhary A (2020) Microbial formulation and growth of cereals, pulses, oilseeds and vegetable crops. Sustain Environ Res 30(1):Article number: 10. https://doi.org/10.1186/s42834-020-00051-x

Napoli C (1990) Introduction of a chimeric chalcone synthase gene into petunia results in reversible co-suppression of homologous genes in trans. Plant Cell 2(4):279–289. https://doi.org/10.1105/tpc.2.4.279

Narayanasamy G, Arora BR, Biswas DR, Khanna SS (2012) Fertilizers, manures and biofertilizers. In: Goswami NN et al (eds) Fundamentals of soil science. Indian Society of Soil Science, New Delhi, pp 579–621

Naylor D, Coleman-Derr D (2018) Drought stress and root-associated bacterial communities. Front Plant Sci 8:1–16. https://doi.org/10.3389/fpls.2017.02223

Nie M, Bell C, Wallenstein MD, Pendall E (2015) Increased plant productivity and decreased microbial respiratory C loss by plant growth-promoting rhizobacteria under elevated CO2. Sci Rep 5(1):9212. https://doi.org/10.1038/srep09212

Ning P et al (2014) New maize hybrids had larger and deeper post-silking root than old ones. Field Crop Res 166:66–71

Nuccio EE, Starr E, Karaoz U, Brodie EL, Zhou J, Tringe SG et al (2020) Niche differentiation is spatially and temporally regulated in the rhizosphere. ISME J 14:999–1014. https://doi.org/10.1038/s41396-019-0582-x

Ogawa S, Valencia MO, Ishitani M, Selvaraj MG (2014) Root system architecture variation in response to different NH4 + concentrations and its association with nitrogen-deficient tolerance traits in rice. Acta Physiol Plant 36(9):2361–2372. https://doi.org/10.1007/s11738-014-1609-6

Oggerin M, Arahal DR, Rubio V, Marin I (2009) Identification of *Beijerinckia fluminensis* strains CIP 106281T and UQM 1685T as *Rhizobium radiobacter* strains, and proposal of *Beijerinckia doebereinerae* sp. nov. to accommodate *Beijerinckia fluminensis* LMG 2819. Int J Syst Evol Microbiol 59(9):2323–2328. https://doi.org/10.1099/ijs.0.006593-0

Okon Y, Itzigsohn R (1995) The development of *Azospirillum* as a commercial inoculant for improving crop yields. Biotechnol Adv 13(3):415–424. https://doi.org/10.1016/0734-9750(95)02004-m

Okon Y, Labandera-Gonzalez CA (1994) Agronomic applications of azospirillum: an evaluation of 20 years worldwide field inoculation. Soil Biol Biochem 26(12):1591–1601. https://doi.org/10.1016/0038-0717(94)90311-5

Olson KR (2013) Soil organic carbon sequestration, storage, retention and loss in U.S. croplands: issues paper for protocol development. Geoderma 195–196:201–206

Ontl TA, Schulte LA (2012) Soil carbon storage. Nat Educ Knowl 3(10):35

Pahari A, Pradhan A, Nayak SK, Mishra BB (2017) Bacterial siderophore as a plant growth promoter. In: Microbial biotechnology. Springer, Singapore, pp 163–180. https://doi.org/10.1007/978-981-10-6847-8_7

Pal A, Pandey S (2014) Role of glomalin in improving soil fertility: a review. Int J Plant Soil Sci 3(9):1112–1129

Pal S, Singh HB, Farooqui A, Rakshit A (2015) Fungal biofertilizers in Indian agriculture: perception, demand and promotion. J Eco-friendly Agric 10:101–113

Pan B, Lam SK, Mosier A, Luo Y, Chen D (2016) Ammonia volatilization from synthetic fertilizers and its mitigation strategies: a global synthesis. Agric Ecosyst Environ 232:283–289

Peng S, Guo T, Liu G (2013) The effects of arbuscular mycorrhizal hyphal networks on soil aggregations of purple soil in southwest China. Soil Biol Biochem 57:411–417. https://doi.org/10.1016/j.soilbio.2012.10.026

Péret B, Li G, Zhao J, Band LR, Voß U, Postaire O et al (2012) Auxin regulates aquaporin function to facilitate lateral root emergence. Nat Cell Biol 14(10):991–998. https://doi.org/10.1038/ncb2573

Pierik R, Tholen D, Poorter H, Visser EJW, Voesenek LACJ (2006) The Janus face of ethylene: growth inhibition and stimulation. Trends Plant Sci 11(4):176–183. https://doi.org/10.1016/j.tplants.2006.02.006

Polese V, de Paula Soares C, da Silva PRA, Simões-Araújo JL, Baldani JI, Vidal MS (2017) Selection and validation of reference genes for RT-qPCR indicates that juice of sugarcane varieties modulate the expression of C metabolism genes in the endophytic diazotrophic *Herbaspirillum rubrisubalbicans* strain HCC103. Antonie Van Leeuwenhoek 110(12):1555–1568. https://doi.org/10.1007/s10482-017-0906-7

Powlson DS, Whitmore AP, Goulding KWT (2011) Soil carbon sequestration to mitigate climate change: a critical re-examination to identify the true and the false. Eur J Soil Sci 62(1):42–55. https://doi.org/10.1111/j.1365-2389.2010.01342.x

Pozzer A, Tsimpidi AP, Karydis VA, de Meij A, Lelieveld J (2017) Impact of agricultural emission reductions on fine-particulate matter and public health. Atmos Chem Phys 17:12813–12826

Prasad R, Kumar K, Varma A (2015) Role of PGPR in soil fertility and plant health. In: Plant-growth-promoting rhizobacteria (PGPR) and medicinal plants, vol 42. Springer, Cham, pp 247–260

Purnomo AS, Nawfa R, Martak F, Shimizu K, Kamei I (2017) Biodegradation of aldrin and dieldrin by the white-rot fungus *Pleurotus ostreatus*. Curr Microbiol 74(3):320–324. https://doi.org/10.1007/s00284-016-1184-8

Qafoku NP (2015) Climate change effects on soils: accelerated weathering, soil carbon, and elemental cycling. Academic, Advan Agro

Qiao Y, Shi J, Zhai Y, Hou Y, Ma W (2015) Phytophthora effector targets a novel component of small RNA pathway in plants to promote infection. Proc Natl Acad Sci 112(18):5850–5855. https://doi.org/10.1073/pnas.1421475112

Qu A-L, Ding Y-F, Jiang Q, Zhu C (2013) Molecular mechanisms of the plant heat stress response. Biochem Biophys Res Commun 432(2):203–207. https://doi.org/10.1016/j.bbrc.2013.01.104

Quintero JC, Moreira MT, Feijoo G, Lema JM (2008) Screening of white rot fungal species for their capacity to degrade lindane and other isomers of hexachlorocyclohexane (HCH). Cien Invest Agraria 35(2):159–167

Raghavendra MP, Nayaka SC, Nuthan BR (2016) Role of rhizosphere microflora in potassium solubilization. In: Meena VS, Maurya BR, Verma JP, Meena RS (eds) Potassium solubilizing microorganisms for sustainable agriculture. Springer, New Delhi, pp 43–59
Raj SA (2007) Bio-fertilizers for micronutrients. Biofertil Newslett 15:8–10
Rajkumar M, Prasad MNV, Swaminathan S, Freitas H (2013) Climate change driven plant–metal–microbe interactions. Environ Int 53:74–86. https://doi.org/10.1016/j.envint.2012.12.009
Rana A, Joshi M, Prasanna R, Shivay YS, Nain L (2012) Biofortification of wheat through inoculation of plant growth promoting rhizobacteria and cyanobacteria. Eur J Soil Biol 50:118–126. https://doi.org/10.1016/j.ejsobi.2012.01.005
Rangel M, Moniz T, Silva AMN, Leite A (2018) Tuning the anti(myco)bacterial activity of 3 hydroxy-4-pyridinone chelators through fluorophores. Pharmaceuticals 11:110
Rebelein JG, Stiebritz MT, Lee CC, Hu Y (2016) Activation and reduction of carbon dioxide by nitrogenase iron proteins. Nat Chem Biol 13(2):147–149. https://doi.org/10.1038/nchembio.2245
Reinhold-Hurek B, Egener T, Hurek T, Martin D, Sarkar A, Zhang L, Miche L (2002) Regulation of nitrogen fixation and assimilation of *Azoarcus* sp. BH72 new approaches to study biodiversity of grass endophytes. In: Vanderleyden J (ed) Proceedings of the nineth international symposium on nitrogen fixation with non-legumes. Katholique Universideit Leuven, Belgium, p 48
Ribeiro-Barros AI, Catarino S, Moura I, Ramalho JC, Romeiras MM, Ghodhbane-Gtari F (2018) Actinorhizal trees and shrubs from Africa: distribution, conservation and uses. Antonie Van Leeuwenhoek 112:31–46. https://doi.org/10.1007/s10482-018-1174-x
Richardson AE (2001) Prospects for using soil microorganisms to improve the acquisition of phosphorus by plants. Aust J Plant Physiol 28:897–906
Rodell M, Velicogna I, Famiglietti JS (2009) Satellite-based estimates of groundwater depletion in India. Nature 460:999–1002
Rodriguez PA, Rothballer M, Chowdhury SP, Nussbaumer T, Gutjahr C, Falter-Braun P (2019) Systems biology of plant microbiome interactions. Mol Plant 12(6):804–821. https://doi.org/10.1016/j.molp.2019.05.006
Roley SS, Duncan DS, Liang D, Garoutte A, Jackson RD, Tiedje JM, Robertson GP (2018) Associative nitrogen fixation (ANF) in switchgrass (*Panicum virgatum*) across a nitrogen input gradient. PLoS One 13(6):e0197320. https://doi.org/10.1371/journal.pone.0197320
Rutter BD, Innes R (2017) Extracellular vesicles isolated from the leaf apoplast carry stress-response proteins. Plant Physiol 173(1):728–741. https://doi.org/10.1104/pp.16.01253
Ryu C-M, Hu C-H, Locy RD, Kloepper JW (2005) Study of mechanisms for plant growth promotion elicited by rhizobacteria in *Arabidopsis thaliana*. Plant Soil 268(1):285–292. https://doi.org/10.1007/s11104-004-0301-9
Sadler JC, Currin A, Kell DB (2018) Ultra-high throughput functional enrichment of large monoamine oxidase (MAO-N) libraries by fluorescence activated cell sorting. Analyst 143:4747–4755. https://doi.org/10.1039/c8an00851e
Saha R, Saha N, Donofrio RS, Bestervelt LL (2012) Microbial siderophores: a mini review. J Basic Microbiol 52:1–15. https://doi.org/10.1002/jobm.201100552
Saraf CS (1988) Sulphur fertilization to pulses in yield and quality. In: Proceedings of TSI-FAI symposium. Sulphur in Indian agriculture. New Delhi, p S II/2-1
Saravanan VS, Kumar MR, Sa TM (2011) Microbial zinc solubilization and their role on plants. In: Maheshwari D (ed) Bacteria in agrobiology: plant nutrient management. Springer, Berlin, Heidelberg. https://doi.org/10.1007/978-3-642-21061-7_3
Sati D, Pandey S, Pande V, Upreti S, Gouri V, Joshi T, Gangola S, Debbarma P, Pandey A, Samant M (2020) Toward an enhanced understanding of plant growth promoting microbes for sustainable agriculture. https://doi.org/10.1016/B978-0-12-821265-3.00005-0
Schalk IJ, Hannauer M, Braud A (2011) New roles for bacterial siderophores in metal transport and tolerance. Environ Microbiol 13:2844–2854. https://doi.org/10.1111/j.1462-2920.2011.02556.x

Schwencke J, Caru M (2001) Advances in actinorhizal symbiosis: host plant- Frankia interaction, biology, and applications in arid land reclamation: a review. Arid Land Res Manag 15:285–327

Sharma V, Singh RP (2011) Organic matrix based slow release fertilizer enhances plant growth, nitrate assimilation and seed yield of Indian mustard (*Brassica juncea* L. cv Pusa Bold). J Environ Biol 32:619–624

Shen H, He X, Liu Y, Chen Y, Tang J, Guo T (2016) A complex inoculant of N2-fixing, P- and K-solubilizing bacteria from a purple soil improves the growth of Kiwifruit (*Actinidia chinensis*) plantlets. Front Microbiol 7:841

Sindhu SS, Parmar P, Phour M, Sehrawat A (2016) Potassium-solubilizing microorganisms (KSMs) and its effect on plant growth improvement. In: Potassium solubilizing microorganisms for sustainable agriculture. Springer, New Delhi, pp 171–185. https://doi.org/10.1007/978-81-322-2776-2_13

Singh PK (2012) Role of glomalin related soil protein produced by arbuscular mycorrhizal fungi: a review. Agric Sci Res J 2(3):119–125

Singh JS, Gupta VK (2018) Soil microbial biomass: a key soil driver in management of ecosystem functioning. Sci Total Environ 634:497–500. https://doi.org/10.1016/j.scitotenv.2018.03.373

Singh M, Dotaniya ML, Mishra A, Dotaniya CK, Regar KL, Lata M (2016) Role of biofertilizers in conservation agriculture. In: Bisht JK et al (eds) Conservation agriculture. Springer, Singapore, pp 113–134

Singh YP, Tomar SS, Singh AK, Yadav RP (2018) Nutrient management and irrigation scheduling effect on blackgram (*Vigna mungo*)-frenchbean (*Phaseolus vulgaris*) yield, economics, water productivity and soil properties. J Soil Water Conserv 17(1):58–64. https://doi.org/10.5958/2455-7145.2018.00009.7

Slatko BE, Gardner AF, Ausubel FM (2018) Overview of next-generation sequencing technologies. Curr Protoc Mol Biol 122(1):e59. https://doi.org/10.1002/cpmb.59

Small CC, Degenhardt D (2018) Plant growth regulators for enhancing revegetation success in reclamation: a review. Ecol Eng 118:43–51. https://doi.org/10.1016/j.ecoleng.2018.04.010

Smith AP, Marín-Spiotta E, de Graaff MA, Balser TC (2014) Microbial community structure varies across soil organic matter aggregate pools during tropical land cover change. Soil Biol Biochem 77:292–303. https://doi.org/10.1016/j.soilbio.2014.05.030

Smith-Moore CM, Grunden AM (2018) Bacteria and archaea as the sources of traits for enhanced plant phenotypes. Biotechnol Adv 36:1900–1916. https://doi.org/10.1016/j.biotechadv.2018.07.007

Spence C, Bais H (2015) The role of plant growth regulators as chemical signals in plant–microbe interactions: a double edged sword. Curr Opin Plant Biol 27:52–58

Stockmann U, Adams MA, Crawford JW, Field DJ, Henakaarchchi N, Jenkins M et al (2013) The knowns, known unknowns and unknowns of sequestration of soil organic carbon. Agric Ecosyst Environ 164:80–99. https://doi.org/10.1016/j.agee.2012.10.001

Stocks-Fischer S, Galinat JK, Bang SS (1999) Microbiological precipitation of CaCO3. Soil Biol Biochem 31(11):1563–1571. https://doi.org/10.1016/s0038-0717(99)00082-6

Sundara B, Natarajan V, Hari K (2002) Influence of phosphorus solubilizing bacteria on the changes in soil available phosphorus and sugar cane and sugar yields. Field Crop Res 77:43–49

Swartjes FA, Van der Aa M (2019) Measures to reduce pesticides leaching into groundwater-based drinking water resources: an appeal to national and local governments, water boards and farmers. Sci Total Environ 699:134186. https://doi.org/10.1016/j.scitotenv.2019.134186

Takahashi M, Morikawa H (2019) Nitrogen dioxide at ambient concentrations induces nitration and degradation of PYR/PYL/RCAR receptors to stimulate plant growth: a hypothetical model. Plants 8(7):198. https://doi.org/10.3390/plants8070198

Takahashi M, Matsubara T, Sakamoto A, Morikawa H (2007) Uptake, assimilation, and novel metabolism of nitrogen dioxide in plants. Methods in biotechnology. Humana Press, Totowa, pp 109–118. https://doi.org/10.1007/978-1-59745-098-0_9

Tan W-F, Zhang R, Cao H, Huang C-Q, Yang Q-K, Wang M, Koopal LK (2014) Soil inorganic carbon stock under different soil types and land uses on the Loess Plateau region of China. Catena 121:22–30. https://doi.org/10.1016/j.catena.2014.04.014

Tenea GN, Peres Bota A, Cordeiro Raposo F, Maquet A (2011) Reference genes for gene expression studies in wheat flag leaves grown under different farming conditions. BMC Res Notes 4(1):373. https://doi.org/10.1186/1756-0500-4-373

Thiem D, Złoch M, Gadzała-Kopciuch R, Szymańska S, Baum C, Hrynkiewicz K (2018) Cadmium-induced changes in the production of siderophores by a plant growth promoting strain of *Pseudomonas fulva*. J Basic Microbiol 58(7):623–632. https://doi.org/10.1002/jobm.201800034

Tiwari S, Lata C, Chauhan PS, Nautiyal CS (2016) *Pseudomonas putida* attunes morphophysiological, biochemical and molecular responses in *Cicer arietinum* L. during drought stress and recovery. Plant Physiol Biochem 99:108–117. https://doi.org/10.1016/j.plaphy.2015.11.001

Totsche KU, Amelung W, Gerzabek MH, Guggenberger G, Klumpp E, Knief C et al (2017) Microaggregates in soils. J Plant Nutr Soil Sci 181(1):104–136. https://doi.org/10.1002/jpln.201600451

Tran Vân V, Berge O, Ke SN, Balandreau J, Heulin T (2000) Repeated beneficial effects of rice inoculation with a strain of *Burkholderia vietnamiensis* on early and late yield components in low fertility sulphate acid soils of Vietnam. Plant Soil 218:273–284

Trimurtulu N, Rao DLN (2014) Liquid microbial inoculants and their efficacy on field crops. ANGRAU, Agricultural Research Station, Amaravathi, p 54

Twomlow S, Rohrbach D, Dimes J, Rusike J, Mupangwa W, Ncube B, Hove L, Moyo M, Mashingaidze N, Mahposa P (2008) Micro-dosing as a pathway to Africa's Green Revolution: evidence from broad-scale on-farm trials. Nutr Cycl Agroecosyst 88:3–15

Umesha S, Singh KPP, Singh R (2018) Microbial biotechnology and sustainable agriculture. In: Biotechnology for sustainable agriculture. Elsevier, Amsterdam, pp 185–205. https://doi.org/10.1016/b978-0-12-812160-3.00006-4

Van Dyk JS, Pletschke B (2011) Review on the use of enzymes for the detection of organochlorine, organophosphate and carbamate pesticides in the environment. Chemosphere 82(3):291–307. https://doi.org/10.1016/j.chemosphere.2010.10.033.

Van Nguyen T, Pawlowski K (2017) Frankia and actinorhizal plants: symbiotic nitrogen fixation. In: Rhizotrophs: plant growth promotion to bioremediation. Springer, Cham, pp 237–261. https://doi.org/10.1007/978-981-10-4862-3_12

Vandamme P, Goris J, Chen WM, de Vos P, Willems A (2002) *Burkholderia tuberum* sp. nov and *Burkholderia phymatum* sp. nov., nodulate the roots of tropical legumes. Syst Appl Microbiol 25:507–512

Vardharajula S et al (2011) Drought-tolerant plant growth promoting *Bacillus* spp.: effect on growth osmolytes, and antioxidant status of maize under drought stress. J Plant Interact 6:1–14

Verbon EH, Liberman LM (2016) Beneficial microbes affect endogenous mechanisms controlling root development. Trends Plant Sci 21(3):218–229. https://doi.org/10.1016/j.tplants.2016.01.013

Verma JP, Yadav J, Tiwari KN, Kumar A (2013) Effect of indigenous *Mesorhizobium* spp. and plant growth promoting rhizobacteria on yields and nutrients uptake of chickpea (*Cicer arietinum* L.) under sustainable agriculture. Ecol Eng 51:282–286. https://doi.org/10.1016/j.ecoleng.2012.12.022

Verma P, Yadav AN, Kumar V, Pratap SD, Saxena A (2017) Beneficial plant-microbes interactions: biodiversity of microbes from diverse extreme environments and its impact for crop improvement. In: Plant-microbe interactions in agro-ecological perspectives, vol 2. Springer, Cham, pp 543–580

Vijayaraghavan K, Yun YS (2008) Bacterial biosorbents and biosorption. Biotechnol Adv 26:266–291

Wallander H, Ekblad A, Godbold DL, Johnson D, Bahr A, Baldrian P et al (2013) Evaluation of methods to estimate production, biomass and turnover of ectomycorrhizal mycelium in forests soils—a review. Soil Biol Biochem 57:1034–1047. https://doi.org/10.1016/j.soilbio.2012.08.027

Wallner A, Blatzer M, Schrettl M, Sarg B, Lindner H, Haas H (2009) Ferricrocin, a siderophore involved in intra- and transcellular iron distribution in *Aspergillus fumigatus*. Appl Environ Microbiol 75:4194–4196

Wang J, Zhu L, Wang Q, Wang J, Xie H (2014) Isolation and characterization of atrazine mineralizing *Bacillus subtilis* strain HB-6. PLoS One 9(9):e107270. https://doi.org/10.1371/journal.pone.0107270

Wang M, Weiberg A, Lin F-M, Thomma BPHJ, Huang H-D, Jin H (2016) Bidirectional cross-kingdom RNAi and fungal uptake of external RNAs confer plant protection. Nat Plants 2 (10):16151. https://doi.org/10.1038/nplants.2016.151

Wang R, Zhang H, Sun L, Qi G, Chen S, Zhao X (2017) Microbial community composition is related to soil biological and chemical properties and bacterial wilt outbreak. Sci Rep 7:Article number: 343

Weiberg A, Wang M, Lin F-M, Zhao H, Zhang Z, Kaloshian I et al (2013) Fungal small RNAs suppress plant immunity by hijacking host RNA interference pathways. Science 342 (6154):118–123. https://doi.org/10.1126/science.1239705

Weiberg A, Bellinger M, Jin H (2015) Conversations between kingdoms: small RNAs. Curr Opin Biotechnol 32:207–215. https://doi.org/10.1016/j.copbio.2014.12.025

Weindling R (1932) *Trichoderma lignorum* as a parasite of other soil fungi. Phytopathology 22:837–845

Weindling R (1934) Studies on lethal principle effective in the parasitic action of *Trichoderma lignorum* on *Rhizoctonia solani* and other soil fungi. Phytopathology 24:1153–1179

Wilpiszeski RL, Aufrecht JA, Retterer ST, Sullivan MB, Graham DE, Pierce EM et al (2019) Soil aggregate microbial communities: towards understanding microbiome interactions at biologically relevant scales. Appl Environ Microbiol 85:e0324-19. https://doi.org/10.1128/aem.00324-19

Woo SL, Ruocco M, Vinale F, Nigro M, Marra R, Lombardi N, Pascale A, Lanzuise S, Manganiello G, Lorito M (2014) Trichoderma-based products and their widespread use in agriculture. Open Mycol J 8:71–126

Wood JL, Green PT, Vido JJ, Celestina C, Harms KE, Franks AE (2019) Microbial communities associated with distance- and density-dependent seedling mortality in a tropical rainforest. Plant Ecol 221:41–54. https://doi.org/10.1007/s11258-019-00989-y

Wu C, Cano M, Gao X, Lo J, Maness P, Xiong W (2020) A quantitative lens on anaerobic life: leveraging the state-of-the-art fluxomics approach to explore clostridial metabolism. Curr Opin Biotechnol 64:47–54. https://doi.org/10.1016/j.copbio.2019.09.012

Xiao L, Lian B, Hao J, Liu C, Wang S (2015) Effect of carbonic anhydrase on silicate weathering and carbonate formation at present day CO2 concentrations compared to primordial values. Sci Rep 5(1):7733. https://doi.org/10.1038/srep07733

Yadav L, Verma JP, Jaiswal DK, Kumar A (2014) Evaluation of PGPR and different concentration of phosphorus level on plant growth, yield and nutrient content of rice (*Oryza sativa*). Ecol Eng 62:123–128. https://doi.org/10.1016/j.ecoleng.2013.10.013

Yadeta KA, Thomma J (2013) The xylem as battleground for plant hosts and vascular wilt pathogens. Front Plant Sci 4:1–12

Yang J, Kloepper JW, Ryu C-M (2009) Rhizosphere bacteria help plants tolerate abiotic stress. Trends Plant Sci 14(1):1–4. https://doi.org/10.1016/j.tplants.2008.10.004

Yoneyama T, Terakado-Tonooka J, Minamisawa K (2017) Exploration of bacterial N2-fixation systems in association with soil-grown sugarcane, sweet potato, and paddy rice: a review and synthesis. Soil Sci Plant Nutr 63(6):578–590. https://doi.org/10.1080/00380768.2017.1407625

Zaidi A, Khan MS, Saif S, Rizvi A, Ahmed B, Shahid M (2017) Role of nitrogen-fixing plant growth-promoting rhizobacteria in sustainable production of vegetables: current perspective. In:

Microbial strategies for vegetable production. Springer, Cham, pp 49–79. https://doi.org/10.1007/978-3-319-54401-4_3

Zamioudis C, Mastranesti P, Dhonukshe P, Blilou I, Pieterse CMJ (2013) Unraveling root developmental programs initiated by beneficial *Pseudomonas* spp. bacteria. Plant Physiol 162 (1):304–318. https://doi.org/10.1104/pp.112.212597

Zhang Z, Lian B, Hou W, Chen M, Li X, Li Y (2011) *Bacillus mucilaginosus* can capture atmospheric CO2 by carbonic anhydrase. Afr J Microbiol Res 5:106–112

Zhang J, Wang L-H, Yang J-C, Liu H, Dai J-L (2015) Health risk to residents and stimulation to inherent bacteria of various heavy metals in soil. Sci Total Environ 508:29–36. https://doi.org/10.1016/j.scitotenv.2014.11.064

Zhang M, Viennois E, Prasad M, Zhang Y, Wang L, Zhang Z et al (2016) Edible ginger-derived nanoparticles: a novel therapeutic approach for the prevention and treatment of inflammatory bowel disease and colitis-associated cancer. Biomaterials 101:321–340. https://doi.org/10.1016/j.biomaterials.2016.06.018

Zhang Y, Kang X, Liu H, Liu Y, Li Y, Yu X et al (2018) Endophytes isolated from ginger rhizome exhibit growth promoting potential for *Zea mays*. Arch Agron Soil Sci 64(9):1302–1314. https://doi.org/10.1080/03650340.2018.1430892

Zhao J, Liu J, Liang H, Huang J, Chen Z, Nie Y et al (2018) Manipulation of the rhizosphere microbial community through application of a new bio-organic fertilizer improves watermelon quality and health. PLoS One 13(2):e0192967. https://doi.org/10.1371/journal.pone.0192967

Zhu Q, De Vries W, Liu X, Zeng M, Hao T, Du E et al (2016) The contribution of atmospheric deposition and forest harvesting to forest soil acidification in China since 1980. Atmos Environ 146:215–222. https://doi.org/10.1016/j.atmosenv.2016.04.023

Ziarati P, Tamaskani MZ, Shirkhan F, Mostafidi M, Hochwimmer B (2018) Potential health risks and concerns of high levels of nitrite and nitrate in food sources. Sci Fed Pharm J 1:1–13

Zomorrodi AR, Segrè D (2018) Synthetic ecology of microbes: mathematical models and applications. J Mol Biol 428(5):837–861

Zúñiga A, Poupin MJ, Donoso R, Ledger T, Guiliani N, Gutiérrez RA, González B (2013) Quorum sensing and indole-3-acetic acid degradation play a role in colonization and plant growth promotion of *Arabidopsis thaliana* by *Burkholderia phytofirmans* PsJN. Mol Plant-Microbe Interact 26(5):546–553. https://doi.org/10.1094/mpmi-10-12-0241-r

10 Adoption of Vertical Farming Technique for Sustainable Agriculture

Saloni Saraswat and Manjula Jain

Abstract

The extensive growth of population across the globe has elevated the need for growing crops efficiently and effectively. It has been observed that the vertical farming technique is the most dependable and innovative technique to grow crops throughout the year with the use of hydroponic technology. Vertical farming technique produces crops with the help of artificial and natural lighting in plant trays that help them in gaining maximum exposure to light by revolving inside a building premise. Furthermore, this technique is considered to be most valuable for preventing the problem of the shortage of food in an environment-friendly manner. Across the globe, it has been estimated that regions such as North America and Asia-Pacific are creating ample opportunities for the development of shipping container vertical farms and building based vertical farms that are further driving the production of pesticide-free crops along with saving ample utilization of water in these regions. Moreover, it has been observed that the countries in Asia-Pacific such as China, Japan, and India are searching for sustainable techniques for growing crops that help them in meeting the problem of food shortages in an eco-friendly method.

This chapter is an attempt to estimate and forecast the potential of vertical farming techniques, which is considered to be a sustainable agriculture alternative. Additionally, the chapter aims at providing detailed insights into the segments and market dynamics of vertical farming techniques across the globe. These classifications will further help in predicting and understanding the concept of vertical farming technique and how it is helping to attain sustainability worldwide, this will further help the key players to understand the prospective opportunities and benefits for investing in this technique.

S. Saraswat (✉) · M. Jain
SBM, IFTM University, Moradabad, India
e-mail: salonisaraswat01@gmail.com; jainmanjula76@gmail.com

A. Kaushik et al. (eds.), *Climate Resilience and Environmental Sustainability Approaches*, https://doi.org/10.1007/978-981-16-0902-2_10

Keywords

Vertical farming technique · Sustainability · Environment-friendly production · Hydroponics · Building-based vertical farms

10.1 Introduction

Extensive demand for food owing to the growing population across the globe has attributed to the need for growing crops efficiently and effectively. Several countries across the globe are constantly revamping their techniques and methodology for growing crops and vertical farming technique is considered to be the most dependable and innovative technique for growing crops throughout the year (Travis 2018).

Rapid industrialization and urbanization have encroached upon arable lands, thus decreasing the area of cultivable land, and due to the adoption of conventional agricultural practices, there are extensive negative impacts on the environment. To meet the sustainable goal of feeding the world population, alternative methods for growing crops need to get evolved. To conserve available water and energy resources along with conserving fast depleting land, an alternative growth strategy is required to be implemented by the commercial growers to accelerate the sustainable production of products. To achieve soil less production of plants, it is suggested by several research organizations to adopt a vertical farming technique that helps in growing healthy foods and considered to be the most efficient alternative in sustainable agriculture.

Vertical farming is a form of indoor farming technique that helps the commercial growers to grow crops efficiently in multiple levels on a vertical axis. This technique utilizes aquaponic, hydropic, and aeroponic growth mechanisms and further helps in generating a greenhouse skyscraper that is considered to be a proficient option in growing crops and meeting the needs of the ever-growing population. Additionally, the vertical farming technique utilizes controlled environment agriculture and indoor farming methods to control the external environmental factors and further helps in meeting food demands and producing crops in an environment-friendly manner (Brook n.d.-a, b; Bromfield 2012).

Furthermore, it has been observed that Asia-Pacific is creating ample opportunities for the vertical farming technique to flourish owing to favorable geographical conditions. This technique utilizes artificial and natural lighting in three-story plant trays that help them in gaining maximum exposure to light by revolving inside a building premise. In the present scenario, the region is employing both shipping container vertical farms and building-based vertical farms and is saving ample water resources due to extensive utilization of vertical farming techniques as compared to that of traditional farming techniques to grow crops. This study covers various growth mechanisms techniques of vertical farms along with structures that are used in vertical farms, and various innovative technologies that are used for the production of crops throughout the year (Markets and Markets 2017; Smith 2012).

The chapter aims at exploring the concept of vertical farms and how vertical farming techniques can enable the commercial crop growers to attain sustainability. The chapter further identifies various driving forces, trends for building vertical farms, and technologies that are utilized to grow crops in vertical farming techniques. Furthermore, the study classifies various techniques utilized by commercial growers to grow plants among several technological segmentation. Additionally, this chapter will further create new vistas for the investors to make judicious decisions before entering and investing in this market.

10.2 Vertical Farming an Emerging Approach

This section provides details regarding vertical farming, it is a technique that produces food on vertically inclined surfaces instead of farming plants on a field or at a single level and utilizes various structures for producing crops such as shipping containers or skyscraper. This section covers the information inscribed in various studies about the progressions that occurred over the years in vertical farming technology. This section further discusses the structures utilized in vertical farming techniques and whether it is an efficient option that can progressively alter conventional farming techniques.

Vertical farming techniques offer various advantages over conventional farming methods such as it includes resiliency, efficiency, and environmental benefits. Vertical farming technique is established in multistory buildings that facilitate land conservation and enhancing the sustainable agriculture process globally (Bromfield 2012). It is an advanced technology that is generally adopted in case of unavailability of land and this method deals with harvesting technique, crop cultivation, water management, and yielding process. Vertical farming structure adopts advanced technology such as hydroponic systems, artificial lighting systems, and efficient farming management in urban zones (Jegadeesh and Verapandi 2014).

This technique is widely being recognized as a technique that helps in diminishing the use of land in the agriculture field, is pesticide-free, and is an environment-friendly technique to grow crops. While growing crop farmers are required to monitor farming conditions that they had been practicing traditionally with naked eyes as access to the latest techniques and data management was comparatively low. In the present scenario, to gain precise data use of vertical farming monitoring system with the Internet of Things (IoT) had been developed that plays a significant role in monitoring systems that help the commercial growers in observing and controlling the environmental conditions to grow crops. Besides this system helps in reducing the burden of the user, ensures precise information and statistics to be delivered to the user, and helps in detecting significant changes in the output. It has been further concluded that use of IoT devices thus ensures conventional farmers to adopt vertical farming practices that are environmentally friendly and delivers improved farming practices to the users (Yap Shien Chin 2017).

Vertical farming technique is gaining popularity in the cities at a rapid pace as this technique is used for growing crops that it is a combination of both farms and a building developed inside the cities in a high-rise building that was not possible with traditional method of farming in the past. Various techniques offered in vertical farming formulate innovative farming methods in developing the vertical farming system in cities that support the country in reducing poverty, enhancing sustainability, food security, and enhance the well-being of the human. These include the general structure of vertical farming that differs from one city to another based on the demand of the crop and size of the vertical farm. To develop a vertical farm skin façade of the building is required to be made up of ETFE (Ethylene Tetra Fluoro Ethylene) that enables transmission of almost 95% of light. Furthermore, the use of LED (Light Emitting Diode) or HPS (High-Pressure Sodium) lights could be utilized that plays a crucial role in uplifting the growth of vegetation. Additionally, ventilation, temperature, and air conditioning also perform a significant role in developing vertical farms. The adoption of HVAC (Heating, ventilation, and air conditioning) system helps in maintaining consistency of moisture, indoor quality of air, energy savings, and heat in vertical farms provided by the shades of plants. Moreover, an amalgamation of various techniques is required to meet the requirements of the ever-growing food production system (Fatemeh Kalantari 2017).

Development of smart cities, rapid urbanization, aim to attain sustainability, and renewability is thus making urban agriculture more substantial. Vertical farming is considered to be capable of responding to the demographic challenges witnessed by smart cities as an alternative to traditional farming techniques and extensive growth in cities is attributing to the increasing demand for urban agriculture. Vertical farming is one such alternative that is significantly contributing and can further help in fulfilling the growing demand for food worldwide and provides organic, herbicides, and pesticides free food (Wigier and Kowalski 2018).

Agricultural leaders believe that cities would need to produce internally to avoid harmful pollution, immobilizing congestion, rising food prices, and ever-increasing population. Technological advancements such as aeroponics, aquaponics, and hydroponics have accelerated the demand for vertical farming techniques globally. On the other hand, the vertical farm is proficient in encountering issues related to expansive growth in population, associated greenhouse gas emissions, food security, and shortages of farmland. An effective vertical farm is one that will depend upon various local factors such as urban populations and their density, culture and eating habits, demand and supply of food, water and energy supply, weather conditions, and technological development (Al-Kodmany 2018).

The use of traditional farming methods to provide a clean and fresh food supply to the ever-expanding population will be a difficult task in the future. Commercial growers have been developing various alternative techniques to grow plants and soil less plantation is one such technique that could be achieved using hydroponic and aeroponics systems. Aeroponics system is an innovative, modern, and informative technology that is used for the production of plants without the use of soil. This system conserves water and energy and is considered to be a safe and environmentally friendly method of cultivation. Aeroponics systems utilize several components and atomizers such as droplet size and spray misters, pressure atomization nozzles, ultrasonic atomization fogger, pH, and EC (electrical conductivity) meter, computer intelligent control techniques, nutrient reservoir, misting frequency, dissolved oxygen concentration, and humidity along with light and temperature (Imran Ali Lakhiar 2018).

There is another technique used for growing plants in nutrient-enriched water without soil known as hydroponics or soil-less growth. Hydroponic technique helps commercial growers in significantly reducing the cost, controlling the environmental factors by compressed technology, and efficient utilization of water resources. Furthermore, hydroponics is a versatile technique efficient to be utilized in backyard production systems and village to the high tech space stations. This technique is further helpful in reducing air pollution, sustainable climate change, extending innovation in agriculture, and meeting the ever-increasing demands of the growing population (Reddy 2018).

Conventional farming technique utilizes approximately 70% of freshwater for soil-based farming and 80% of soil along with this constant change in weather and environment leads to intolerable utilization of resources that could ultimately result in disturbance of supply chain. Increasing implementation of IoT (Internet of Things) based monitoring in vertical farming techniques is resolving the issue related to land depletion across the globe. This technique is further enabling the commercial growers and world population to acquire crops throughout the year since these devices support vertical farming techniques to grow crops without any seasonal restrictions. Besides, IoT sensors are enabling commercial growers to increase the production of yield at an extensive rate with less intervention of manual labor in the vertical farming area. Implementation of IoT in vertical farms aims at eliminating the laborious efforts of humans and is rampantly changing the traditional methods of farming (Abhishek et al. 2018).

10.3 Market Dynamics

10.3.1 Trends

Increasing adoption of the latest technology such as IoT and large production of crops are the key trends observed in the growth of vertical farming techniques. IoT (Internet of Things) is the most prominent technology that is driving control and automation in the global market. The use of conventional farming techniques for growing crops is stimulating the Asia-Pacific region to produce a low yield of crops. With the implementation of IoT technique in the farming system, the crop production can be extensively increased throughout the year since this technique allows the commercial growers to monitor temperature, moisture, and humidity and make provisions, in order to protect the crops and avail desired nutrition. To reduce costs, gain maximum efficiency, and to expand crop yields agricultural industry is using IoT devices, which is another factor contributing toward the growth of the vertical farming market globally.

According to a report published by the Food and Agriculture Organization of the United Nations (2018), the production of crops is continuously growing to meet the ever-increasing demand of the population. In countries such as India and China, the trade has exponentially increased and commercial growers exporting and importing produces are increasing. Thus, the tremendous demands for growing crops from these countries are thus impelling the need to increase the adoption of vertical farming techniques in Asia-Pacific (Food and Agriculture Organization of the United Nations 2018).

10.3.2 Drivers

The extensive demand for pesticide-free food and environment-friendly production of fresh vegetables and fruits is thus driving the growth of the vertical farming market. The consumption of organic foods is considered to be more nutritious since it provides antioxidants as compared to food produced using traditional farming techniques. The consumption of organic food is thus helping the population of various countries to reduce their problems with allergies from chemicals and foods. The vertical farming technique expedites the rapid growth of organic foods as it eradicates the consumption of pesticides while producing crops. The employment of this technique in this region is thus helping them to conserve water, enhance soil fertility, reduce pollution and soil erosion, and utilize less amount of energy. Moreover, vertical farming technique uses LED (Light Emitting Diode) lights that imitate the sunlight and provide a complete spectrum of light to the crops and deliver water in an adequate and recycled manner. This technique is thereby significantly helping the region to conserve natural resources while producing crops and in a sustainable manner. Furthermore, the use of progressive technology in vertical farming techniques such as sensors, lighting, climate control, and hydroponics components is thus enabling the plant growers to produce organic foods in a manner they can ensure that the extensive demand for crops is met efficiently (Markets and Markets 2017).

The sensor technology mounted in the vertical farming system is used for tracking optimum levels of nutrients and light for every crop that can be accustomed in accordance to the taste and preference of the consumers. Moreover, the vertical farming technique is considered to be a cost-effective technique and once installed provides long-term benefits to the commercial growers. These factors are thereby driving various companies operating in Asia-Pacific such as Letcetra Agritech-India, Illumitex Inc. (U.S.), Aerofarms (U.S.) to gain advantages provided by vertical farming technique at an extensive rate in near future. Apart from these factors, independent farming techniques with no obstruction from the climatic conditions and ability of these farms to grow crops in limited spaces are other factors significantly aggravating toward the growth of the global vertical farming market (Allied Market Research 2019).

10.4 Types of Growth Mechanism Used in Vertical Farms

10.4.1 Aeroponics

Aeroponics is a type of growth mechanism technique particularly used in indoor gardening practice where the root structures of plants are hanged in air and are continuously being sprayed with water and nutrient solution that further enables the plants to nourish and grow (Maximum Yield 2018).

Aeroponics is extensively being utilized for the cultivation of plants without the utilization of substrate and soil culture (Imran Ali Lakhiar 2018). The use of soil is eliminated in this technique since the exposure of suspended roots with nutrient-rich mist makes them eligible to thrive and grow. Moreover, aeroponics is considered to be an efficient option to grow fruits, plants, and vegetables without the need to constantly potting and repotting the plants to make them accessible to soil rich with nutrients. Furthermore, it has been observed that this technique is an air–water cultivation system in which the roots of plants are suspended in a wrapped container that is exposed in the air under darkness and gains nutrients through atomizers. The leaves and crown of the plants are prolonged overhead the wet zone and the canopy and root of the plant remain separated with the help of an artificially developed arrangement.

Additionally, this whole process is implemented under controlled environmental conditions and nutrients are being sprayed with the help of nozzles and foggers to enable the plants to observe extensive growth (Nir 1982; Osvald et al. 2005). However, the controlled conditions in this system comprise uniform nutrients absorption, humidity, pH values and EC values, atomization spray time, interval time and frequency, light intensity, temperature, and disposal of oxygen. The plants in this system grow generously due to dispersion and availability of ample oxygen in the growth chamber along with a sterilized environment.

It has been observed that shortly due to extensive demand for food across the globe would develop several challenges for the human as well as animals. It has been projected to overcome such circumstances the aeroponics system could be implemented and considered as a secure and environmentally friendly method of plant cultivation Since, this system is considered to be a very convincing alternative for conserving energy and water resources efficiently, employs method of nutrient-rich water recirculation and compromises of lower energy and water inputs per unit of growing range. Besides, this system is considered to be extremely valuable, economical, and favorable for local farmers to execute efficient agricultural activities. It is further considered to be one of the best alternatives to provide clean, fresh vegetables, and deliver nutrition to promptly flourishing population (Imran Ali Lakhiar 2018).

10.4.2 Advantages of Aeroponics System

According to various studies it has been emulated that in the present scenario aeroponics technique is considered to be an appropriate and innovative system for growing plants. It is an indoor horticulture exercise that has extensive prospective of growing plants in huge quantities and is considered to be a viable option to grow plants in the area where the soil is not appropriate for growing plants. Moreover, the system supports a vertical farming technique by providing more space to the plants for expanding yield produce. This system is considered to be a sustainable alternative since it helps the commercial growers to reduce labor cost, utilize less energy and water consumptions by approximately 98%, reduce the usage of herbicides and pesticides by 100%, reduces utilization of fertilizer by approximately 60% and further supports the extensive production of plants by 45–75%.

The nutrient solution utilized in this technique can be further reutilized by recycling it and after the complete growth of a plant, it can be plucked out without concerning another plant. The system is not restricted to grow crops under various weather conditions and thereby the production of crops could be executed throughout the year in this system without the intrusion of pesticides, residues, and soil. Additionally, this system is a scalable system and can be produced from commercial to apartment size of gardens by producing crops with lower struggles. The plants that are majorly being produced across the globe by employing aeroponics techniques include leafy greens, vine plants, and herbs (Singleton, n.d.; PowerHouse Hydroponics 2018; Smith 2012).

10.5 Hydroponics

Hydroponic system in vertical farming technique is another mechanism of growing plants by employing nutrient solutions that eliminate the use of an inert medium such as rock wool, gravel, vermiculite, sawdust, peat moss, coconut fiber, and coir dust that extends mechanical assistance to the plants (Chow et al. 2017).

These systems can be modified and customized in accordance with the reusability and recycling ability of the supporting media and nutrient solution. The majorly utilized systems that provide support in vertical farms to extensively contribute toward the growth of plants include deep water culture, wrick, nutrient film technique (NFT), and ebb and flow. These techniques are depicted in Fig. 10.1.

1. *Deep Water Culture*—It is considered to be the most effective technique for growing large plants that comprised big roots or for growing fruits. This air is dispersed to the roots of the plants with the help of a diffuser directly and the roots of the plants are suspended in the nutrient solution. Additionally, the bubble bucket is the most pervasively used DWC hydroponic system across the globe.
2. *Wick*—This system is considered to be the most simplistic and energy-efficient system that does not utilize aerators, pumps, and electricity. The plants in this system are placed in the absorbent medium such as perlite, coco coir, and

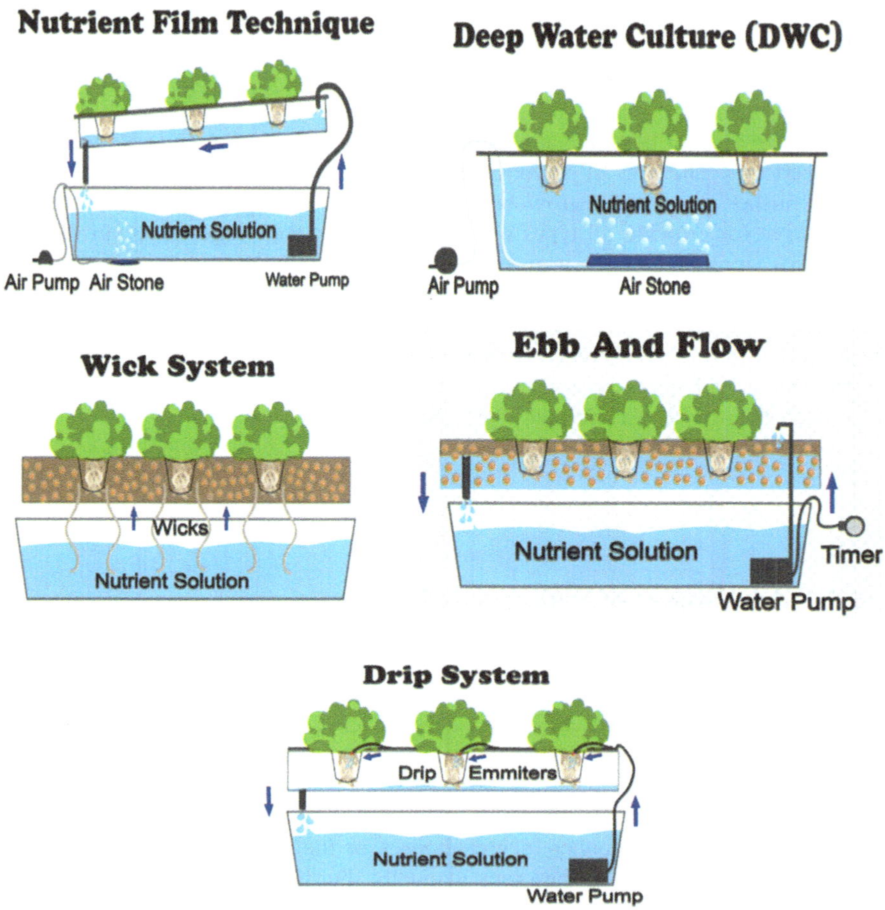

Fig. 10.1 Types of hydroponic systems [Adopted from NoSoilSolutions (n.d.)]

vermiculite with a nylon wick running from the plants to the tank comprising nutrient solution. The system is considered effective for small herbs and houseplants since this system is not capable of supplying ample nutrients to the plants.

3. *Nutrient Film Technique*—The channels in this system are slightly to maintain an effective flow of nutrients throughout the channel then ultimately to the dangling roots and simultaneously going back in the hydroponic reservoir. The nutrient solutions are thereby pumped into the channels that have the capability of holding varying amounts of plants. Furthermore, this system does not comprise the medium that helps plants grow although helps the plants to become secure. It is considered to be the most scalable technique in comparison to the other

techniques since it is simple to establish and works effectively for the plants that have small root systems for instance leafy greens.

4. *Ebb and Flow*—This system is usually employed in residential hydroponic gardeners and is commonly known as flood and drain. The plants are placed in large size of grow beds that are continuously flooded with nutrient solutions until it overflows. The flow of power is controlled by a timer that shuts off the pump that further enables the grow bed to drain away completely. This system is effective for growing all types of plants including root vegetables and is quite popular among the hobby growers.
5. *Drip*—These systems are considered to be an easy alternative of growing crops and the nutrient solutions are infused directly to the base of the plants with the help of a tube. This system works effectively for various types of plants and is thereby considered to be a very convenient option to grow any variety of plants (Nisha Sharma 2018; Daryadar 2015; NoSoilSolutions, n.d.).

10.5.1 Advantages of Hydroponic System

In the present scenario, it has been observed that the hydroponic technique is gaining ample popularity due to its user friendliness, hygienic, and no probability of development of soil-borne diseases. The various advantages of implementing a hydroponics system are as follows:

- The uncertainty occurring due to the weather is eliminated in this technique since this system is grown either in greenhouses or the indoors and the water and nutrients are provided to the crops manually. Furthermore, sunlight is also not considered to be a drawback of growing crops since artificial grow lighting is being utilized for replacing or either supplementing the sunlight.
- Plants grown in soil require 20% more space than the crops grown using the hydroponics technique and enable commercial growers to grow plants in very small spaces where plants cannot grow in the soil.
- The countries with limited arable land in several locations across the globe that have less access to the soil can now produce any type of crops using hydroponics techniques. The probabilities of cultivation can be enhanced in various geographies along with the conservation of water and the import of fresh produce can be reduced by implementing this system.
- The growth capacity of plants in this system is very high since this system customizes and controls the ideal environmental conditions for the plants and delivers them directly to the roots of the plants.
- In this system, an optimal pH level is maintained constantly and essential minerals for growing crops are effectively being maintained that enables and facilitates the growth of plants in this system.

- The running costs and labor costs involved in the setup of the hydroponics system are comparatively low (Max 2019).

10.6 Aquaponics

Aquaponics is the amalgamation of aquaculture (i.e., aquatic animal farming) and hydroponics (i.e., soil less technique for production of crops) to grow animal and plant products. In an aquaponic system, bacteria convert the waste excreted from aquatic animals into nutrients and further the plants improve the water quality for the aquatic animals and remove the nutrients of the plants. It is considered to be one of the alternatives for sustainable agriculture since it imitates the natural system, has less environmental influences, and is a water-efficient option. Aquaponics systems can be established for a variety of uses such as for community, for personal use, or for economic development and are considered as a medium for extensively enhancing the production of food in urban settings where production of plants is confined due to environmental constraints and reduced space (Rigers Bakiu 2017; Lennard 2004).

There are various forms of aquaponics systems although the common components are classified as follows below:

- *Media Filled Beds*—This technique comprises a media that is filled with a clay rock material and encompasses continuous flow of water. This is one of the simplest and most basic techniques of aquaponics.
- *Deep Water Culture*—This technique is also utilized in the hydroponics technique and is largely being used for commercially viable methods.
- *Nutrient Film Technique (NFT)*—Although this technique is also utilized in the hydroponics system and is not a common technique to be used in aquaponics systems. Additionally, in this method nutrient-rich water is pumped to the area where plants are being placed to absorb the nutrients and water. This technique is exclusively being adopted for growing larger plants and green leafy vegetables.

10.6.1 Advantages of Aquaponics System

Aquaponics is considered to be a sustainable production method and provides various benefits to the environment such as (McCarthy 2011; Travis 2018; Brook, n.d.-a, b).

- *Efficient Utilization of Nutrients and Water*—Water is a well-utilized resource in the aquaponics technique and is considered to be a crucial option for the roots of the plants and fish to gain life. The plants are cleared with the help of fish waste and the salinity of soil is reduced.
- *Elimination of Floor Use*—It has been observed that the quality of soil is poor at many places that further hamper the production of superior quality products. The

Aquaponics system is considered to be a great alternative for growing plants in areas where the quality of soil is poor and for desert areas as well.

- *Enhanced accessibility to the market*—An aquaponic system provides the benefit of establishing it at any place that is considered to be close to the market areas for instance food distribution centers and markets for selling it. This enables the users to gain accessibility of enhanced and fresh quality products and save plants from damage of post-harvest.
- *Land Conservation*—This technique enables commercial growers to increase the production rate of crops by utilizing multiple platforms on top of each other to grow plants. This enables the commercial growers to utilize less space and extensively contribute to the growth of the production of crops.
- *Organic Fertilization*—Fish water is the nutrient efficient fertilizer for plants that assist the grower to reduce the cost of fertilizer salts.

10.7 Analysis of Various Growth Mechanism Techniques

After studying and analyzing various reports, it has been observed that in 2016 among all the various growth mechanism systems hydroponics system is considered to be the most effective technique for growing plants in vertical farming. According to a study published by Markets and Markets in 2017, suggested that the hydroponics system is majorly contributing toward the swelling revenue of vertical farming technique globally (Markets and Markets 2017; P&S Market Research 2017).

Hydroponics is considered to be the most viable option to produce fresh vegetables in the countries across the globe that have very little arable land and composed of smaller areas for the production of plants and have to ultimately satisfy the needs of the ever-expanding population. Additionally, this technique is considered to be extremely beneficial to the countries that are leaders in tourism by facilitating the hotel to produce their plants instead of importing them thereby significantly reducing their cost and time (Anju 2018; Storey 2017; Brook, n.d.-a, b).

Furthermore, according to a report published by P&S Market Research aeroponics technique is estimated to witness the fastest growth during the forecast period (i.e., 2018–2023) since this technique is helping the commercial grower to gain sustainability by conserving water since this technique utilizes 95% less amount of water than the traditional method of farming thereby significantly attributing toward the growth of the vertical farming market globally (P&S Market Research 2017).

10.8 Types of Structure

10.8.1 Building-Based Vertical Farms

It has been observed that the building-based vertical farms are established in an abandoned building in various countries along with new building constructions for instance new multistory vertical farms are being developed near parking lot structure in the downtown area for instance in Wyoming. Additionally, building based structures ensure commercial growers to gain approximately 8 times more production area as compared to that of conventional farming area and helps in attaining cost efficiencies and food security for the city for an enhanced period of time. For instance, in July 2018, a project for developing the world's largest building based vertical farms in Dubai is anticipated to feed high fliers of the Middle East (CNN 2018; Despommier 2010).

10.8.2 Shipping Container Vertical Farms

This structure in vertical farms is gaining huge momentum as they acquire 40-ft shipping containers that carry service offerings worldwide. These types of structures are considered to be very inexpensive alternative and simple at handling for the commercial growers, Moreover, due to the low cost of acquisition involved in shipping based containers, various value-based companies that are entering into this market to gain cost efficiencies. These containers are inscribed with drip irrigation systems, LED lights, and vertically stacked shelves for producing various types of plants. Moreover, these shipping based containers comprise a self-contained computerized growth controlled systems that enable the commercial growers to view all systems from mobile devices (Smart Garden Guide, n.d.; Jegadeesh and Verapandi 2014).

10.8.3 Analysis of Structures of Vertical Farms

According to a report published by Mordor Intelligence in 2017, among the two structures of vertical farming technique, the shipping container-based vertical farms are anticipated to witness the fastest growth in CAGR (Compound Annual Growth Rate) during the forecast period (2019–2024) (Mordor Intelligence 2018). Shipping based containers are anticipated to gain increasing popularity among various countries across the globe shortly since it contains innovative techniques for growing plants and climate technology along with comprising a container with requisite environmental conditions throughout the year. Several countries are extensively investing in these types of vertical farming structures to enhance crop production and meet the demand for food shortages such as Japan, China, and other Asian countries (Markets and Markets 2017; P&S Market Research 2017).

10.9 Types of Support Systems Used for Growing Plants in Vertical Farms

The various types of devices that are used to produce plants in vertical farms include lighting, climate control, sensors, and hydroponic components. The constant technological upgradations and research being undertaken in vertical farming techniques about the automation and robotics are thus enhancing the adoption of this technique across the globe and is thereby considered as an alternative for traditional farming techniques and is the most promising method for achieving the objective of sustainability.

Vertical farming is the method of producing food in the form of used warehouses, skyscrapers, shipping containers, and vertically stacked layers. Vertical farming uses the method devised by indoor farming techniques and controlled-environment agriculture (CEA) technology. This technology possesses external environment factors and completely controls the environment. Also, this system controls the environmental conditions in an enclosed covering and protects the food from diseases, pests, or predator attacks. This ensures that the farms in which crops are grown are independent and do not require fertile arable land for the growth of crops. It also helps in growing the farms in any climatic region across the globe, even in extreme temperatures.

10.10 Challenges

The vertical farming technique requires high initial investments to set up the system and lack of technical expertise is the major factor that is hindering the adoption of this technique globally. Furthermore, the variety of crops grown using this technique is limited as a result of which this technique is not rapidly being adopted in the present scenario. This chapter discusses the objective of various countries to develop sustainable agriculture and is searching for various alternative techniques of growing crops to meet the shortages of food in an eco-friendly manner. The chapter tries to analyze whether the vertical farming technique serves as a leading alternative for growing crops globally. Furthermore, the existing studies do not emphasize the present and future market implications of vertical farming techniques and majorly focuses upon discussing the technical information of the same.

10.11 Conclusion

This chapter demonstrated and recognized vertical farming techniques as the most desirable technique that plays a vital role in resolving issues related to the unsustainability of agriculture across the globe. This technique is considerably prolonging the demand for organic food that is gaining huge popularity among the global consumers as compared to that of domestic supply since organic foods are considered to be more nutritious and healthy alternatives as compared to that of traditionally grown products. The vertical farming technique is rapidly helping the commercial growers to gain cost efficiencies and energy savings by the adoption of exceedingly proficient LED lighting technology that helps in the production of maximum and optimized plants by acquiring limited power. In addition to this, vertical farming technique require manual labor for packaging and planting of crops and for on-site harvesting which is considerably low as compared to that of labor required in traditional farming methods this reduces labor cost in vertical farming technique and further supports in uplifting the productivity.

Vertical farming accomplishes the objective of sustainability such as economic, environmental, and social aims. The implementation of new high-tech technologies and growth mechanisms in this farming method as discussed in this study such as hydroponics, aquaponics, and aeroponics that are thereby replacing the requirement of soil-based plantation system and helps in conservation of water and land at a rapid pace. Moreover, due to the features of mobility in shipping container based vertical farms, provide enhanced market access to the investors or the commercial growers for their produces and supports in significantly reducing transportation cost and storage cost arising due to the refrigeration of produces.

This chapter further concludes that the aquaponics growth mechanism system is anticipated to witness the fastest growth shortly due to its easy handling and simple setup along with its exclusive features such as the adoption of informative technology, modern system, and innovative technique. Furthermore, progressions in supporting technologies such as LED lighting, sensors, IoT (Internet of Things), and CEA (Controlled Environment Agriculture) technology are expected to extensively amalgamate the efficient production system of vertical farming techniques soon at an extensive rate. Additionally, various countries need to adopt a vertical farming technique that will enable them in developing fully automated vertical farms that will further help in resolving the problems of food shortages along with increasing the focus of countries towards environment-friendly production of fresh fruits and vegetables. Moreover, after the discussions of various restraints in this paper, it is thereby suggested that more developments are required in this technique for scaling up the pace of projects from various countries that enables the investors with a proficient return on investment and economic feasibility.

References

Abhishek AM et al (2018) Smart vertical farm using IOT. Int J Innov Res Comput Commun Eng 6 (4)

Al-Kodmany K (2018) The vertical farm: a review of developments and implications for the vertical city. MDPI

Chow YN, Lee LK, Zakaria NA, Foo KY (2017) New emerging hydroponic system. International Malaysia-Indonesia-Thailand Symposium on Innovation and Creativity (iMIT-SIC), vol 2, pp 353–358

Daryadar M (2015) Water stream hydroponics as a new technology for soilless production of valuable essential oil and medicinal plant peppermint. Acad J Agric Res 3(10):259–263

Despommier D (2010) The vertical farm: controlled environment agriculture carried out in tall buildings would create greater food safety and security for large urban populations. J Consum Protect Food Saf 6:233–236

Fatemeh Kalantari OM (2017) A review of vertical farming technology: a guide for implementation of building integrated agriculture in cities. Adv Eng Forum 24:76–91

Imran Ali Lakhiar JG (2018) Modern plant cultivation technologies in agriculture under controlled environment: a review on aeroponics. J Plant Interact 13(1):338–352

Jegadeesh M, Verapandi J (2014) An innovative approach on vertical farming techniques. SSRG Int J Agric Environ Sci 1(1):1–5

Lennard WA (2004) Aquaponics research at RMIT University, Melbourne Australia. Aquaponics J 35:18–24

Nisha Sharma SA (2018) Hydroponics as an advanced technique for vegetable production: an overview. J Soil Water Conserv 17(4):364–337

Osvald J, Petrovic N, Demsar J (2005) Sugar and organic acid content of tomato fruits (lycopersicon lycopersicum mill.) grown on aeroponics at different plant density. Acta Aliment 30(1):53–61

Rigers Bakiu CT (2017) First study about aquaponic systems in Albania. J Mar Biol Aquac Res 1 (1):1–7

Key Style Points: Other References

Paper Presented at a Conference

Reddy ED (2018) A study on water use efficiency and pollution control by hydroponic technology. International Conference on New Horizons in Green Chemistry & Technology (ICGCT) 2018

Yap Shien Chin LA (2017) Vertical farming monitoring system using the internet of things (IoT). Advances in electrical and electronic engineering: from theory to applications. Proceedings of the International Conference on Electrical and Electronic Engineering (IC3E 2017). American Institute of Physics

Online Publications Online Document

Allied Market Research (2019, Aug) Retrieved Aug 2019, from Allied Market Research: https://www.alliedmarketresearch.com/vertical-farming-market

Anju V (2018, Jan 11) Retrieved May 2019, from AgriTechTomorrow: https://www.agritechtomorrow.com/article/2018/05/growing-with-hydroponics-aeroponics-and-aquaponics/10733

Brook R (n.d.-a) Retrieved Jun 2019, from Home Aquaponics System: https://homeaquaponicssystem.com/basics/hydroponics-vs-aquaponics-which-is-better/

Brook R (n.d.-b) Retrieved Jun 2019, from Home Aquaponics System: https://homeaquaponicssystem.com/basics/top-10-benefits-of-having-aquaponics-at-home/

CNN (2018, Jul 10) Retrieved Jun 2019, from CNN travel: https://edition.cnn.com/travel/article/dubai-vertical-farm-emirates-catering/index.html

Food and Agriculture Organization of the United Nations (2018) Retrieved Jun 2019, from Food and Agriculture Organization: http://www.fao.org/3/I9542EN/i9542en.pdf

Markets and Markets (2017) Retrieved Jun 2019, from Markets and Markets: https://www.marketsandmarkets.com/Market-Reports/vertical-farming-market-221795343.html

Max (2019, Aug 6) Retrieved Aug 2019, from Green and Vibrant: https://www.greenandvibrant.com/advantages-disadvantages-of-hydroponics

Maximum Yield (2018, Apr 25) Retrieved May 2019, from Maximum Yield: https://www.maximumyield.com/definition/137/aeroponics

McCarthy M (2011, Feb 6) Retrieved May 2019, from aqua-pana-ponics: https://sites.google.com/site/aquapanaponics/4-project-updates/advantagesanddisadvantagesofaquaponics

Mordor Intelligence (2018) Retrieved Jun 2019, from Mordor Intelligence: https://www.mordorintelligence.com/industry-reports/vertical-farming-market

NoSoilSolutions (n.d.) Retrieved Jun 2019, from NoSoilSolutions: https://www.nosoilsolutions.com/6-different-types-hydroponic-systems/

P&S Market Research (2017) Retrieved 2019, from P&S Market Research: https://www.psmarketresearch.com/market-analysis/vertical-farming-market

PowerHouse Hydroponics (2018, Dec 30) Retrieved Jun 2019, from PowerHouse Hydroponics: https://www.powerhousehydroponics.com/15-benefits-of-aeroponic-growing/

Singleton B (n.d.) Retrieved Jun 2019, from SFGATE: https://homeguides.sfgate.com/plants-aeroponics-systems-52438.htm

Smart Garden Guide (n.d.) Retrieved 2019, from Smart Garden Guide: https://smartgardenguide.com/what-are-the-advantages-of-hydroponics/

Smith J (2012, Apr 18) Retrieved May 2020, from Hydroponics Equipment Co.: http://www.hydroponicsequipment.co/advantages-of-aeroponics-systems/

Storey A (2017, Jul 18) Retrieved Jun 2019, from Powered by Plenty: https://university.upstartfarmers.com/blog/aquaponics-vs-hydroponics-which-is-better

Travis A (2018, Apr 14) Retrieved May 2019, from Farming Method: https://farmingmethod.com/aquaponics-description-advantages-disadvantages/

Wigier M, Kowalski A (2018) The common agricultural policy of the European Union—the present and the future Non-EU Member States point of view. Institute of agricultural and food economics national research

Organization Site

Bromfield E (2012, 09) Retrieved July 2019, from http://www.fdnearth.org/files/2012/09/Why-vertical-farms_Final.pdf

Nir I (1982) Retrieved 2019, from International Society for Horticultural Science: https://www.ishs.org/ishs-article/126_49

Part II

Environmental Sustainability

This section discusses various aspects of alternate energy, eco-friendly waste management, and some sustainable approaches under the sub-theme "Environmental Sustainability."

The need for switching over to some renewable energy alternatives has been emphasized in the Sustainable Development Goals (SDGs) to mitigate carbon emissions and contain climate change. A study of bioenergy use in 25 OECD countries indicates that it is a promising energy alternative that ensures environmental sustainability and climate security in modern energy systems. Policy makers worldwide should use suitable assessment practices to influence the deployment of sustainable bioenergy.

Other sources of renewable energy (RE) such as hydropower and solar energy also have tremendous potential that need to be harnessed properly to meet the wide gap in demand and supply in many areas, particularly remote or hilly areas. A case study of Uttarakhand, India, shows how despite the RE alternative possibilities, the state faces geographical, constructional, administrative, financial, and environmental challenges to develop these, which need to be addressed by preparing a comprehensive road map for the next 10 years.

Microbial fuel cell (MFC) technology is another surging technology that has the unique potential of integrating two SDGs pertaining to clean water and clean energy. MFC performs wastewater treatment and simultaneously produces electricity by exploiting the biodegradative and bio-electrogenic characteristics of microbes. Technological innovations and the use of appropriate microbial inoculum show distinct potential for enhanced energy production with increased techno-economic feasibility of this technique in field.

Industrial growth has been seriously impacting the climate change phenomenon. While carbon footprinting of large industries has attracted great attention in the recent past, medium and small industries have not generally been studied. A case study of carbon footprints of medium-scale plywood industry in Yamunanagar, Haryana, India, shows significant impacts and suggests various measures and policy interventions to reduce carbon footprints of this industry.

Industrial growth has led to serious pollution problems, and wastewaters with different recalcitrant toxicants have been found to not only contaminate the water bodies but also upset the aquatic ecosystems and human health. Bioremediation of wastewaters containing toxic reactive dyes using specific extremophilic or indigenous microbial species or consortium of species has been found to be an eco-friendly, economical, and sustainable approach, in which the complex aromatic dyes are degraded, and ultimately mineralized to nontoxic forms.

Organochlorine pesticides are a group of recalcitrant and toxic chemicals, which have large environmental consequences. Degradation of pentachlorophenol (PCP) by bioremediation methods involving mixed microbial communities in soil and water using biochar and bioelectrochemical systems shows great promise as these methods are less costly, safer, ecologically sound, and sustainable when compared with conventional physicochemical methods.

Solid waste management is emerging as a massive challenge in urban areas in the present times due to increasing population, consumerism, and commercialization. A critical appraisal of solid waste generation, management practices, and policy requires to be done to understand the scenario of different countries. The solid waste should be converted to value-added products rather than letting it cause degradation of environment and health problems. Adopting innovative technologies, decentralized processing, involvement of workforce from the nonformal sector, planned outreach programs, and public–private partnership model have led to success stories in many cities of India indicating new opportunities of resource recovery.

Apart from solid waste management regulation at the government level, small social enterprises have shown how they can work successfully toward creating effective waste collection, recycling, and value addition and introducing it into a business model. Case studies of some social enterprises in India reveal how they have revolutionized the waste management practices by diffusing innovations and successfully developing mechanisms that leverage the capabilities of the unorganized and informal sector while providing them an occupational identity.

As urbanization and industrialization have thrown open several environmental challenges in modern times, the need for evolving or adopting strategies that lead to sustainable living is increasingly being understood. In this context, the importance of traditional ethnoecological knowledge (TEK) of indigenous communities has been immense and it has been found to have a link with all the 17 SDGs. A case study of Paudi Bhuyan tribe of Odisha, India, shows the rich TEK of the tribe and also how various drivers forced this tribe to move from hills to plains triggering a slow erosion of their ethnicity, which shows how important it is to preserve the framework of indigenous knowledge system.

While the SDGs come as a universal call from the United Nations under global partnership to end poverty and protect the planet, the policies and institutional mechanisms to achieve the same differ with different nations. Bangladesh, which has recently shown an upgradation in status from least developed countries (LDCs) to developing country, shows some degree of preparedness in achieving the goals, but there are enormous challenges. There is a need for integrating the goals into the country's strategic planning process and transforming the current education system in line with the needs of the modern market.

Health care and sanitation are key elements for sustainable development and accessibility to universal health care to all the citizens of a nation. An analysis of the "Ayushman Bharat" healthcare scheme of the Government of India shows that there is a significant requirement of funds and infrastructure for its successful implementation. A tangible model to achieve it is proposed through multiple stakeholder participation, particularly the corporates through their corporate social responsibility (CSR) funds and schemes.

Sustainability of Biofuels: The Dynamic Nexus Between CO_2 Emissions and Bioenergy Consumption in OECD Countries

11

Mohammad Younus Bhat, Arfat Ahmad Sofi, and R. S. Aswani

Abstract

Currently, the world is grappling with two actual scenarios: increasing demand for energy and the increasing climate change issue. It necessitates finding solutions by taking appropriate steps to reduce the steep trajectory of carbon dioxide emissions. It is believed that the environment and economy complement each other, i.e., both are dependent. The energy is the *sine qua non* for the growth of the economy. Therefore, this study seeks to investigate the interplay among carbon emissions, bioenergy consumption, nonrenewable energy consumption, and urbanization in a panel of 25 OECD countries. Results suggest a long-run relationship between the growth of the economy, energy consumption, urbanization, and carbon emission. Further, it is observed that economic growth, urbanization, and nonrenewable energy consumption generate carbon emissions in the OECD countries. However, results reveal that bioenergy consumption in the member countries significantly reduce carbon emission.

Keywords

Carbon emissions · Bioenergy consumption · Gross domestic production · Panel cointegration

M. Y. Bhat (✉)
Department of Economics and International Business, School of Business, University of Petroleum and Energy Studies, Dehradun, Uttarakhand, India
e-mail: younus.bhat@ddn.upes.ac.in

A. A. Sofi
Department of Economics, Birla Institute of Technology and Sciences (BITS) Pilani, K.K. Birla Goa Campus, Goa, India

R. S. Aswani
School for Life, University of Petroleum and Energy Studies, dehradun, Uttarakhand, India

A. Kaushik et al. (eds.), *Climate Resilience and Environmental Sustainability Approaches*, https://doi.org/10.1007/978-981-16-0902-2_11

11.1 Introduction

The development of modern economies is much dependent upon energy sources like coal, petroleum, natural gas, solar, wind, nuclear, and biofuel (Adewuyi and Awodumi 2017). The society needs all services related to energy to ensure that their basic needs are fulfilled. The invention of the steam engine gave coal dominance over the organic sources of energy (Owusu and Asumadu-Sarkodie 2016). Fossil fuels will have an advantage over renewable if energy supply is concerned for some time simply as fossils fuels possess immense energy density, evolution has been experienced by them. Most importantly, it is due to the time-constrained on technologies suitable for clean energy (Burke and Stephens 2018). Nevertheless, new sources for energy must be deployed in such ways that scale regarding the industrial revolution is equivalent (Duffy et al. 2019). Without proper planning, carbon emissions from the use of fossils will be double of what it is soon. For a very long time, inventions and technologies are being done in the field of fossil fuels-based power generation and the increasing population has furthermore increased the demand of these unconventional sources (Erias et al. 2016).

Globally, there are problems and challenges regarding the fulfilment of the demand for energy but at the same time keeping in mind the increasing climate change. Two major concerns are the environmental challenge and energy issues being faced by the world today (Tal 2019). There is a logical complementarity between environment and economy. Swelling of wealth as a function of mass consumerism has inversely affected the levels of the environment both in quality and number. To meet the consumer's demand, environmentally hazardous technologies are utilized to extract natural resources, in its wake spawning a disruptive ecosystem and degraded environment. Anthropogenic greenhouse gases most importantly carbon dioxide has a huge contribution toward the global climatic reverses (Nunez 2020). Several studies depict that increasing population specifically in the urban spaces has a key role in greenhouse gases (GHG) emissions. The dominance of fossil fuels and exponential rise in population impacts a growing requirement for energy resulting in the rapid growth of carbon dioxide emissions and this has become one of the greatest challenges today (Ritchie and Roser 2017). The population is projected to grow and so does the need for energy. Looking at the increasing challenge of climate change, the reduction of greenhouse gases is identified as necessary for the future and sustainability (Raj and Singh 2012). As policymakers cope with the issue of global climate change emanating from excessive use of fossil fuels, it is imperative to shift toward renewable sources of energy particularly hydro, solar, wind, and bioenergy.

Although fossils fuels remain basic to the production of electrical power, renewable energy has become widespread. New policies need to be formulated to promote renewables (Gielen et al. 2019). Bioenergy has become the best alternative to contain greenhouse gas (GHG) emissions. Understanding the need for bioenergy is the solution of today. The first comprehensive guideline that estimated greenhouse gas emissions, estimates that the net impact of GHG emission from bioenergy is zero such that CO_2 which is released when any biofuel is burned automatically returns to the already existing carbon dioxide in the atmosphere that was used by the plants for the process called photosynthesis (Khanna et al. 2011; Haberl et al. 2012). As renewables are being promoted throughout the world stakeholders attempt to measure the sustainability. Sustainability is another criterion that should be analyzed while being sure about shifting to the alternatives. Though it is believed that clean energy sources like solar, wind, and bioenergy are a great step toward a sustainable future but they should not affect food security, provision of ecosystem services, and threaten biodiversity (Owusu and Asumadu-Sarkodie 2016). Against this backdrop, exploring the dynamic relationship between bioenergy consumption, nonrenewable energy consumption, urbanization, economic advancement, and carbon emissions is of particular importance from a policy perspective toward a sustainable future. Therefore, understanding this context especially for OECD countries that are shifting their efforts toward modern energy sources particularly bioenergy is the overall aim of this chapter. Empirical findings from modern econometric techniques document that the variables under investigation revealed long-run cointegration. Long-run estimations highlight the inverse and significant relationship between bioenergy consumption and carbon emissions. In other words, bioenergy consumption facilitates a reduction in greenhouse gasses particularly carbon emissions.

This chapter is organized as Sect. 11.2 describing the context of OECD member countries. Section 11.3 presents the state of current literature, Sect. 11.4 presents the data and describes the empirical methodology, Sect. 11.5 is designated to results and discussion which explains the estimations derived from econometric analysis. Finally, Sect. 11.6 accomplishes the chapter and summarizes the findings.

11.2 The Organization of Economic Cooperation and Development Context

This section deals about the targets and importance that the Organization of Economic Cooperation and Development (OECD) countries play toward the Paris agreement. Paris agreement focuses on strengthening the response of the world toward the increasing climate change. Figure 11.1 presents graphically the history of carbon emissions, GDP, and energy use in per-capita of OECD member countries from 1960 to 2014. Almost every OECD nation has signed the Paris agreement and their efforts toward the aims of the Paris agreement are well-known. OECD focuses to improve the global economy and initiate trades. Talking about the mission, OECD nations promote policies regarding the social and economic welfare of the people. The main motive of the countries involved is to make sure that innovation, technological development that favor the environment and working toward a sustainable future (OECD 2012). OECD has great experience practices through collaborative

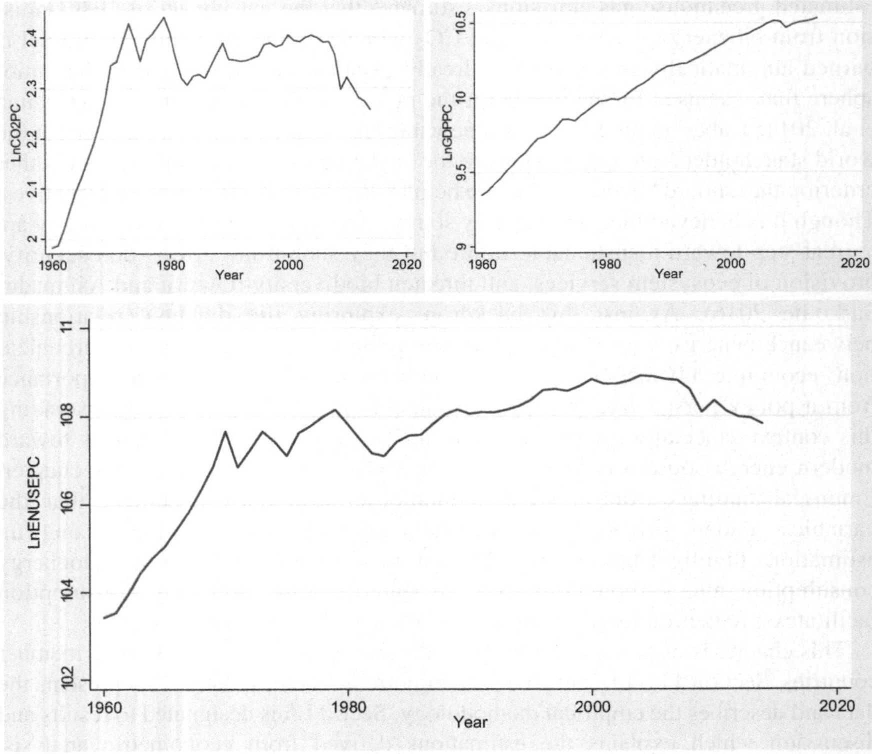

Fig. 11.1 Carbon emissions, GDP per-capita and energy consumption in OECD (1960–2014)

initiatives and policy recommendations. It helps other countries with policy analysis and guides them toward the NDC's and helps in planning and preparing their actions for mitigation and adaptation. Also looks toward the emission pathways, policies that can effectively be tailored to the national conditions of the countries, aligning policies as they focus on long-term planning to achieve their Paris agreement targets. OECD has taken an initiative to strengthen the resilience and focus on contributing to the global goal of climate change adaptation. While making other countries achieve and grow on their Paris agreement policies, OECD countries have actively contributed to supporting the climate change transparency and have developed one of the groups, i.e., OECD/IEA Climate Change Expert, the development committee and has also had set up targets for the countries in OECD itself.

Not only is OECD working on achieving their part but at the same time taking initiatives to support the government and advance the domestic and international actions and engaging directly with parties regarding the technical problems in the United Nations Framework Convention on Climate Change (UNFCCC) process and supporting the methodological work that may arise from Paris Agreement. OECD has also contributed to the development of green investment banks around its boundaries. The focus of such banks is to look after policies and finance regarding the mobilization of climate change. They also show that if price-based schemes are used for the support then many private investors will work toward such projects.

11.3 State of Current Literature

This section summarizes the review of previous literature on the topic which is under consideration in this study. Various factors influence carbon emission, which includes economic growth, increase in population, industrialization, urbanization, and continuous increase in energy consumption. For instance, Bilgili et al. (2016) analyze the influence that biomass has on the emission of carbon dioxide covering the US monthly observations ranging from 1984 to 2015. Also, the consumption of coal and other energy sources is studied. Wavelet coherence and wavelet partial coherence are used to study the frequency and time series of the variables to know about the usage on the biomass in the United States and its impact on the increasing carbon emissions. It is revealed that in the long-run biomass consumption mitigates carbon emissions. McKechnie (2011) explains that the use of bioenergy will not only decrease the dependence of the country on the nonrenewable sources but will also employ the investment in the forestry sector. He with his research tries to develop and integrate life cycle assessment with that of the carbon analysis derived from forests to know about the total greenhouse gas emissions linked with bioenergy from forests. The outcome is that a new technique for understanding the role of carbon dioxide emissions in knowing about the cost-effectiveness of changing climate. Birdsey et al. (2018) tell about the increase in combustion due to biomass from wood for energy. The study explores the impact of increasing bioenergy production from forests on the climate and greenhouse gases and develops a science-based idea about such incidences and identifies economic and environmental impacts of the production of bioenergy. Souza et al. (2017) describes challenges and issues in the current policies and also recommend new policy promotion of sustainable bioenergy in the world. Findings indicate that benefits can be generated through bioenergy cropping, the negative impacts will be solved by using correct crop yield and managerial skills should be the priority. Wu et al. (2018) describe the advantages that bioenergy has over other fossil fuels. The renewability and its abundant quantity provide great opportunities for this source of energy. The paper deals with the impact that bioenergy has on the environment. The findings show that bioenergy has negative effects on the quality and quantity of soil erosion and water. They also analyze China's energy market and find that China has a huge potential of bioenergy and that Asian country has huge opportunities for producing crops that will be used for the generation of energy. Booth (2018) assesses the emission from the use of biomass in the United States and also the wood pallet manufacturing sectors. Data is collected for a period from 2001 to 2016. The author evaluates the net emission impact of manufacturing and transport emission and defines the cumulative net emission to combustion ratio for 10 years. Findings indicate that the United States mostly burn black liquor with an NEI of 25% while the NEI ranges from 41 to 95% from plant residues. She also talks about the emissions that wood pallet has which is somewhat 14–20 tons by 40 years. Facilities have higher HPT emissions if they use fossil fuels to dry pallets. Efroymson et al. (2013) study the distribution and production process of the bioenergy chain. They assess that to have a sustainable future shifting toward renewable energy is the best decision one can make. Indicators are used to best analyze the impacts on the environment. They used panel data, time-series data to study the Biofuels supply and production, and also understand the trends in the demand for bioenergy. Their findings indicate that more sustainability indicators must be developed to understand the fossil fuels nature and bioenergy nature on the environment and toward having a sustainable future. The indicators must be tested before forming any policies and it should also be taken care that they provide useful and reliable measurements. While measuring the indicators small factors like the content of nitrate in water, soil erosion level must also be considered. The study calculates the level of energy produced by different plants; the amount of water required for the plantation and then forms their indicators regarding Biofuels sustainability.

The objective behind the study done by Seabra et al. (2011) was the assessment of life cycle energy use and greenhouse gas (GHG) emissions related to cane sugar and ethanol. GREET 1.8c.0 model was used to understand the parameters. The authors studied the data from 1990 to 2012 regarding the production and the emissions in the country. Monte Carlo analysis was used to get the life cycle emissions of ethanol. Eleven parameters were studied and their probability curve was analyzed from the statistical data collected from all industries present in the activity. The findings lead to knowing that fossil energy use and greenhouse gas emissions regarding the sugar production which was 721 kj/kg and 235 g CO_2eq/MJ and for ethanol, it was calculated to be 80 kj/MJ and 21.3g CO_2eq/MJ. Dauber et al. (2010) study and summarize the knowledge of the impact of bio-crops on biodiversity and identify the gaps in research and above all recommends policies for a sustainable future through bioenergy. The research concludes that abandonment landscapes for agriculture stimulate the rural economy. The research also shows the negative, neutral, and positive effects on biodiversity by biomass. ARDL and time series method are used to study the statistical data of 11 different countries. This study promotes conservation research program and biodiversity-friendly production of crops keeping in mind the food security of the nations. Rasool and Hemalatha (2016) deal with the need for bioenergy in today's world. Shifting from fossils to renewable and what impact would be faced by the industries in doing so. This article also estimates the conversion of oils to biodiesel, using transesterification or conversion of crops. Data regarding biodiesel and ethanol are analyzed from 2000 to 2012. Different sources of plants are studied in this research and the amount of energy each generates. Disadvantages and advantages of bioenergy are highlighted by the author. The conclusion of the research provides that the production of Biofuels can also be done through fungi and algae, engineering methods are to be used or should be used for the enhancement of the yields. In short, shifting to Biofuels will be best for reducing nonrenewable energy. Dufey (2006) studies about the challenges and opportunities of Biofuels which can also help to tackle the increasing climate change. Time series data was used regarding the production in different countries and also their policies involving sustainability goals. The study concludes that understanding the impact of sustainability is a complicated fact as the expected growth of Biofuels depends on the amount that is being traded. Challenge is to deal with domestic productions keeping in view the imports. Increase in the energy used globally due to the growing population and their basic need creates a matter of concern regarding the use of nonrenewable as a source of energy.

To understand the impact on GHG by bioenergy in Canada, Liu et al. (2017) identified the marginal land by intersecting maps and got land suitability and land cover. Analyses were done regarding the cost of the marginal lands and GHG emission was calculated through emissions models. The results indicate about 9.46 million ha of unused land is available. If the available land is completely used the production of Biofuels will be about 5.69–11.26 billion L. Canada being an oil-exporting nation it is calculated that the amount of biofuel production would be somewhat 14% of gasoline consumption done by Canada in the year 2014. The study examines the alternative uses of the marginal land available and also the amount of impact on soil quality, water resources, and biodiversity through analyzing the past data and using ADR and panel root test. It is calculated that about 29 mt CO_2/year of GHG emission can be reduced by the use of Biofuels from the marginal land. Zhang (2016) studies the advantage that bioenergy can have to tackle climate change. Findings indicate that biomass is a solution toward the decrease of GHG, as biomass will not only emit less of the carbon but it can be transformed into liquid, gas, or solids and will lead toward a positive influence on emissions. They calculate that throughout the world production of biomass by photosynthesis is 200 Gt, 99 Gtce/year. Al-Mulali (2015) investigates the impact of Biofuels on pollution, level of agriculture, and total production in 16 major countries, which have huge biofuel consumption. Panel model is used from 2000 to 2010 and it is concluded that bioenergy increases agricultural production and the prices of crops, leading to increase in GDP and reduces pollution. Therefore, it is recommended that steps toward the promotion of bioenergy should be considered but also keeping in mind that the agriculture price does not increase. Fritsche et al. (2010) summarize the major issues regarding the environmental assessment of bioenergy and aims to bring methodologies that are best tested and evaluated. Many a time's decisions are taken or policies are made without even considering the environment. This framework takes practical decision-making tools and has too served as a benchmark for future researches. Protection of biodiversity should be the priority before any policy formulation and the study also recommends policies like mapping must be done regarding the area where bioenergy projects must be conducted. Certain guidelines which can help in improving the environment must be set. A sound systematic plan must be considered. The choices of a plant in arid areas must be at low water wastage hence cropping system must be optimized through the efficiency of water. No drainage of wetland. The production system must be known about the risk involved and if that risk could not be eliminated project should be stopped for the optimization and yield cultivation system must be improved.

Reviewed studies present the relationship between bioenergy consumption and climate change, biodiversity, ecosystem services, etc., however, it is noticed that no proper study has been conducted wherein the relation between bioenergy consumption, economic growth, nonrenewable energy consumption, with carbon emissions can be noticed. The literature provides a relation between indicators separately and not altogether. Therefore, research on the topic will be the first wherein all different indicators that are urbanization, GDP per capita, and bioenergy consumption will be analyzed to study the linkage between them and CO_2 emissions.

11.4 Carbon Emission and Bioenergy Consumption Nexus in OECD Countries: An Overview of Econometric Methods

This section discusses variables, data, and econometric models used to obtain the estimates. Theoretically, we follow the same model specification where carbon emissions (CE) are affected by economic growth (EG), bioenergy consumption (BIOEC), nonrenewable energy consumption (NREC), and urbanization (URB). We considered a panel of 25 OECD countries that have blending targets to enhance the consumption of renewables to decrease carbon emissions. It can be examined by specifying a model while considering CO_2 emissions (CE) as the dependent variable, which is a function of multiple independent variables like EG, BIOEC, NREC,and URB. To investigate their influence on carbon emissions, Eq. (11.1) was established as below:

$$\mathrm{CE}_t = \beta_0 + \beta_1 \mathrm{EG}_t + \beta_2 \mathrm{BIOEC}_t + \beta_3 \mathrm{NREC}_{it} + \beta_4 \mathrm{URB}_t + \varepsilon_t \tag{11.1}$$

Given that panel data is used in this study to explore the dynamic relationship between carbon emissions and other explanatory variables, Eq. (11.1) can take the panel data form as follows:

$$\mathrm{CE}_{it} = \beta_0 + \beta_1 \mathrm{EG}_{it} + \beta_2 \mathrm{BIOEC}_{it} + \beta_3 \mathrm{NREC}_{it} + \beta_4 \mathrm{URB}_{it} + \varepsilon_{it} \tag{11.2}$$

In Eq. (11.2), $i = 1, \ldots, I$ signifies countries; $t = 1, \ldots, T$ denotes the time; β_0 is the intercept; $\beta_1 \ldots \beta_4$ are the model coefficients and ε_{it} represents error term. CE is per capita carbon emissions, EG is economic growth, BIOEC is per capita biofuel consumption, NREC is the per capita consumption of nonrenewable energy, respectively, and URB is urbanization. Data on all the variables like CO_2 emissions, EG, BIOEC, NREC, and URB is sourced from different sources including development indicators prepared by World Bank (WB) and the United States Energy Information Administration (EIA). Variables included in the model were transformed into natural logarithms (ln) so that efficient and systematic results can be generated and heteroscedasticity problems can be avoided (Lau et al. 2014).

To detect the level of stationarity, tests like Im, Pesaran, and Shin (IPS); Augmented-Dicky Fuller (ADF), and Phillips-Perron (PP) were performed (Maddala and Wu 1999; Levin et al. 2002; Im et al. 2003). This is required to determine the unit root test as well as to avoid spurious estimations. Further, the Pedroni test was used to determine cointegration among all the variables used in this study (Pedroni 1999, 2004). Cointegration test is conducted for the existence of a long-run association among the variables under consideration. The main characteristic of this test is that it allows for considerable heterogeneity across countries. Given the presence of cointegration, panel regression estimation was conducted using Long-run FMOLS (i.e., fully modified ordinary least squares) and DOLS (i.e., dynamic ordinary least squares) estimators (Chen et al. 2019).

11.5 Carbon Emissions, Economic Growth, Bioenergy Consumption, Nonrenewable Energy Consumption and Urbanization Nexus in OECD Countries

In this section, the results obtained by using the econometric methods mentioned in Sect. 11.3 are presented and explained. Specifically, discussing the result traced through using the test of Panel unit root, the estimates from panel cointegration estimate, and the long-run relationship among the variables.

11.5.1 Summary Statistics of Variables

Table 11.1 gives the descriptive statistics of all the variables that are being discussed in the study.

Mean CO_2 emissions turned out to be 11.413 kt per capita equivalent of oil with a maximum equal to 13.743 kt per capita. Mean of nonrenewable energy consumption is 8.247 kteo in the panel of countries considered in this study. At the same time mean of economic growth observed is 10.274 and mean of bioenergy consumption is 3.396 with maximum and minimum consumption being 4.406 and 1.477, respectively.

11.5.2 Unit Root Tests

Starting, the process of cointegration, the first step is to know about the stationary present is the chosen variables, i.e., CO_2 emission, economic growth, bioenergy consumption, nonrenewable energy consumption, and urbanization. Series needs to

Table 11.1 Descriptive statistics

Variables	CE	EG	NREC	BIOEC	URB
Mean	11.413	10.274	8.247	3.396	16.056
Median	11.109	10.464	8.233	3.408	15.853
Maximum	13.743	11.626	9.152	4.406	19.374
Minimum	8.903	8.614	6.854	1.477	12.641

Table 11.2 Panel unit root test results

Variables	Im, Pesaran, and Shin (IPS)	Augmented Dickey-Fuller (ADF)	Phillips-Perron (PP)
CE	3.092	30.454	33.998
ΔCE	−10.291***	210.314***	444.151***
EG	0.438	39.094	35.0748
ΔEG	−7.655***	151.217***	227.679***
NREC	1.607	42.240	54.404
ΔNREC	−31.7186***	650.881***	4510.1***
BIOEC	3.809	27.836	23.024
ΔBIOEC	−11.003***	218.553***	422.167***
URB	2.836	77.576	83.843
ΔURB	−10.594***	214.150***	6324.830***

Note: *** represents significance at 1% level

Table 11.3 Pedroni's cointegration test result: carbon emission equation

Within dimension (panel statistics) Pedroni (1999)			Weighted statistic Pedroni (2004)		
Test	**Statistic**	**Prob.**	**Test**	**Statistic**	**Prob.**
v-Statistic	−2.021	0.978	v-Statistic	−2.168	0.985
ρ-Statistic	1.757	0.961	ρ-Statistic	2.412	0.992
PP-Statistic	−1.840**	0.033	PP-Statistic	−2.325***	0.010
ADF-Statistic	−2.830***	0.002	ADF-Statistic	−5.315***	0.000
Between dimension (individual statistics)					
Test		**Statistic**	**Prob.**		
Group ρ-Statistics		3.381	0.999		
Group PP-Statistics		−2.627***	0.004		
Group ADF-Statistics		−2.979***	0.001		

Note: *** and ** represents significance at 1% and 5% level, respectively

be integrated to proceed to cointegration. Im, Persaran, and Shin (IPS); Augmented Dickey-Fuller (ADF), and Phillips-Perron (PP) tests are used to determine whether variables present in the data series are stationary. It can be seen from Table 11.2 that the null hypothesis of non-stationarity of all variables can be rejected in their first difference but not at levels for the panel. It is concluded or anticipated, variables appear to be in I (1) series.

11.5.3 Panel Cointegration Estimates

As per the order in which the values and series are stationary is defined, the step that should be followed next is the panel cointegration test. Pedroni cointegration is used, this was developed in (1999, 2004) by Pedroni who proposed seven statistics for the panel cointegration. These seven tests based are on the absence of cointegration. The results from the Table 11.3 show that cointegration exists among the variables revealed through the cointegration test conducted. In other words, there exists a long-run equilibrium relationship between the variables. Therefore, economic growth, bioenergy consumption, nonrenewable energy consumption, urbanization, and carbon emissions are moving toward the long-term relation. Hence, the null hypothesis stating the existence of no cointegration is rejected as per the values of maximum statistic and their robust probability values.

11.5.4 Long-Run Estimates

Finally, we assessed FMOLS and DOLS estimates after cointegration was determined between the variables. The estimated coefficients are presented in Table 11.4 demonstrating the impact of EG, BIOEC, NREC, and URB on CE. Figure 11.2 summarizes the long-run relationships between variables graphically. Estimation of long-run FMOLS and DOLS coefficient appraisals to a board of OECD nations shows that nonsustainable power source, i.e., NRENC, utilization has a positive and noteworthy relationship with carbon outflows. From Table 11.4, it is clear that a 1% surge in nonrenewable energy consumption will prompt 0.159% (resp. 0.527%) expansion in carbon emanations. Likewise, FMOLS and DOLS coefficient uncovers that BIOEC coefficient has a negative association with carbon outflows. It implies that an increase in its consumption will reduce carbon emanations, for example, a 1% increase in the consumption of bioenergy will reduce carbon emissions by 0.103% (resp. 0.106%). Moreover, the long-run elasticity coefficients of economic growth and urbanization reveal a positive correlation with carbon emissions.

Table 11.4 FMOLS and DOLS estimates

Dep. Var. : carbon emissions	FMOLS estimates	DOLS estimates
EG	0.013	0.003
BIOEC	−0.103**	−0.106***
NREC	0.159	0.527***
URB	0.790***	1.512***

Notes: *** and ** represents 1% and 5% significance, respectively

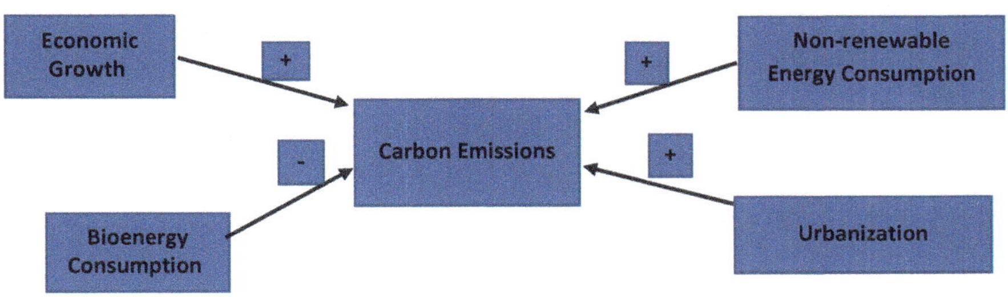

Fig. 11.2 Long-run relationships

11.6 Conclusion and Policy Recommendations

The main aim of this study is to explore the dynamic linkages among carbon emissions bioenergy consumption, nonrenewable energy consumption, urbanization, and economic growth in OECD countries and to develop policies that could lead to a sustainable future. A panel of 25 OECD countries is selected considering CO_2 emissions per capita, GDP per capita, and bioenergy consumption per capita and urbanization from 1990–1991 to 2014–2015. On employing robust econometric methods, we seek to understand the long-run integration across the chosen countries. The results of panel cointegration suggest a long-term association among the chosen variables. The panel results advocate that economic growth, nonrenewable energy consumption, and urbanization are associated with increased carbon emissions whereas bioenergy consumption is significantly associated with a reduction in the CO_2 emissions. Therefore, bioenergy is promising in ensuring environmental sustainability and climate security in modern energy systems. Policymakers worldwide should use suitable assessment practices to influence the deployment of sustainable bioenergy. Proper assessment of environmental impacts should take the cognizance of different characteristics of various bioenergy cropping systems, positive as well as negative. It is proposed that bioenergy production can be sustainable only if it is based and supported by good governance, reliable scientific evidence, strong institutions, voluntary certification, and proper management policies supporting the sustainable use of resources and benefits biological diversity. It is through these strategies' bioenergy can help in realizing its potential in mitigating the extraordinary environmental and climate change that threatens the future of humankind.

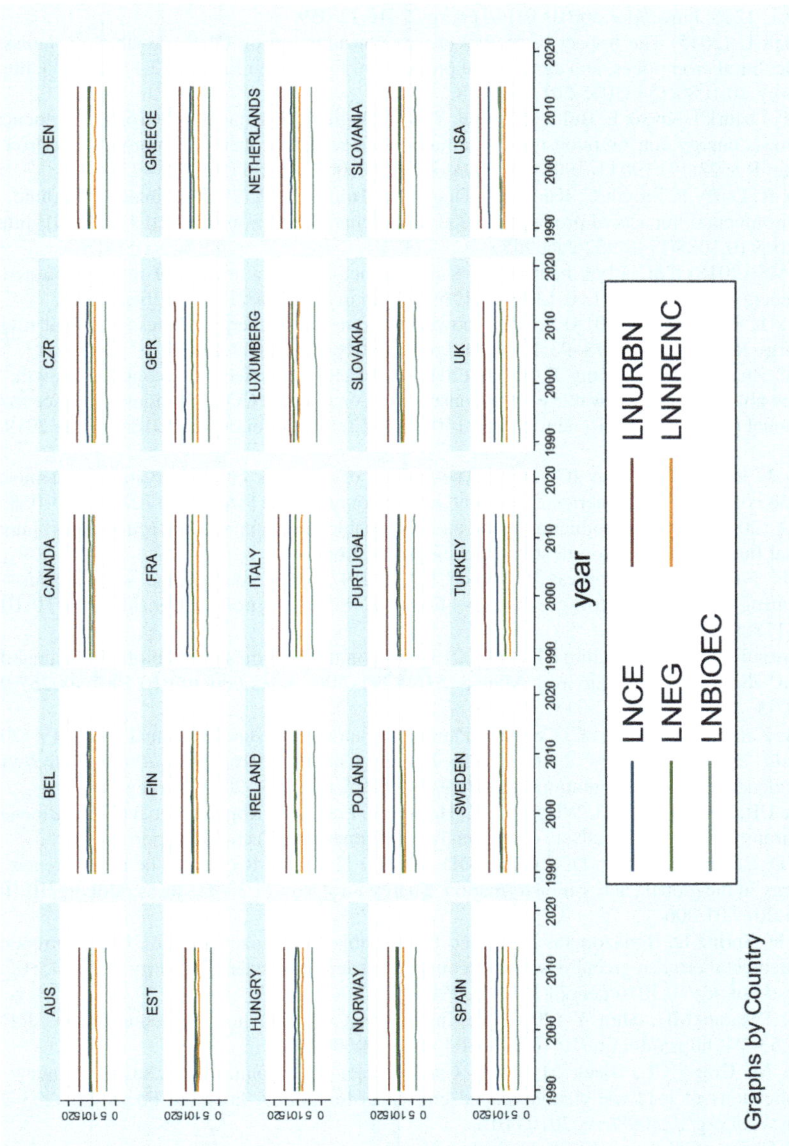
AUS
BEL
CANADA
CZR
DEN
EST
FIN
FRA
GER
GREECE
HUNGRY
IRELAND
ITALY
LUXUMBERG
NETHERLANDS
NORWAY
POLAND
PURTUGAL
SLOVAKIA
SLOVANIA
SPAIN
SWEDEN
TURKEY
UK
USA
0 5 10 15 20
1990 2000 2010 2020
year
LNCE
LNURBN
LNEG
LNNRENC
LNBIOEC
Graphs by Country

References

Adewuyi AO, Awodumi OB (2017) Biomass energy consumption, economic growth and carbon emissions: fresh evidence from West Africa using a simultaneous equation model. Energy 119:1–1246. https://doi.org/10.1016/j.energy.2016.12.059

Al-Mulali U (2015) The impact of biofuel energy consumption on GDP growth, CO_2 emission, agricultural crop prices, and agricultural production. Int J Green Energy 12:1100–1106. https://doi.org/10.1080/15435075.2014.892878

Bilgili F, Öztürk İ, Koçak E, Bulut Ü, Pamuk Y, Muğaloğlu E, Bağlıtaş HH (2016) The influence of biomass energy consumption on CO_2 emissions: a wavelet coherence approach. Environ Sci Pollut Res 23(19):19043–19061. https://doi.org/10.1007/s11356-016-7094-2

Birdsey R, Duffy P, Smyth C, Kurz WA, Dugan AJ, Houghton R (2018) Climate, economic, and environmental impacts of producing wood for bioenergy. Environ Res Lett 13:050201. https://doi.org/10.1088/1748-9326/aab9d5

Booth MS (2018) Not carbon neutral: assessing the net emissions impact of residues burned for bioenergy. Environ Res Lett 13:035001. https://doi.org/10.1088/1748-9326/aaac88

Burke MJ, Stephens JC (2018) Political power and renewable energy futures: a critical review. Energy Res Soc Sci 35:78–93. https://doi.org/10.1016/j.erss.2017.10.018

Chen Y, Zhao J, Lai Z, Wang Z, Xia H (2019) Exploring the effects of economic growth, and renewable and non-renewable energy consumption on China's CO_2 emissions: evidence from a regional panel analysis. Renew Energy 140:341–353. https://doi.org/10.1016/j.renene.2019.03.058

Dauber J, Jones MB, Stout JC (2010) The impact of biomass crop cultivation on temperate biodiversity. GCB Bioenergy 2:289–309. https://doi.org/10.1111/j.1757-1707.2010.01058.x

Dufey A (2006) Biofuels production, trade and sustainable development: emerging issues. International Institute for Environment and Development, London

Duffy P, Fitzpatrick C, Conway T, Lynch RP (2019) Energy sources and supply grids—the growing need for storage. Issues Environ Sci Technol. https://doi.org/10.1039/9781788015530-00001

Efroymson RA, Dale VH, Kline KL et al (2013) Environmental indicators of biofuel sustainability: what about context? Environ Manag 51(2):291–306. https://doi.org/10.1007/s00267-012-9907-5

Erias A, Karaka C, Grajetzki C, Carton J, Paulos M, Jantunen P, Baral P, Samal Bex JMV (2016) World energy resources 2016. World Energy Council 2016. https://doi.org/http://www.worldenergy.org/wp-content/uploads/2013/09/Complete_WER_2013_Survey.pdf

Fritsche UR, Hennenberg KJ, Wiegeman K, Herrera R, Franke B, Köppen S et al (2010) Bioenergy environmental impact analysis (bias): analytical framework. Technical report

Gielen D, Boshell F, Saygin D, Bazilian MD, Wagner N, Gorini R (2019) The role of renewable energy in the global energy transformation. Energy Strat Rev 24:38–50. https://doi.org/10.1016/j.esr.2019.01.006

Haberl H, Sprinz D, Bonazountas M, Cocco P, Desaubies Y, Henze M et al (2012) Correcting a fundamental error in greenhouse gas accounting related to bioenergy. Energy Policy 45:18–23. https://doi.org/10.1016/j.enpol.2012.02.051

Im KS, Pesaran MH, Shin Y (2003) Testing for unit roots in heterogeneous panels. J Econ 115:53–74. https://doi.org/10.1016/S0304-4076(03)00092-7

Khanna M, Crago CL, Black M (2011) Can biofuels be a solution to climate change? The implications of land use change-related emissions for policy. Interface Focus 1(2):233–247. https://doi.org/10.1098/rsfs.2010.0016

Lau LS, Choong CK, Eng YK (2014) Carbon dioxide emission, institutional quality, and economic growth: empirical evidence in Malaysia. Renew Energy 68:276–281. https://doi.org/10.1016/j.renene.2014.02.013

Levin A, Lin CF, Chu CSJ (2002) Unit root tests in panel data: asymptotic and finite-sample properties. J Econ 108:1–24. https://doi.org/10.1016/S0304-4076(01)00098-7

Liu T, Huffman T, Kulshreshtha S, McConkey B, Du Y, Green M et al (2017) Bioenergy production on marginal land in Canada: potential, economic feasibility, and greenhouse gas emissions impacts. Appl Energy 205:477–485. https://doi.org/10.1016/j.apenergy.2017.07.126

Maddala GS, Wu S (1999) A comparative study of unit root tests with panel data and a new simple test. Oxf Bull Econ Stat 61:631–652. https://doi.org/10.1111/1468-0084.61.s1.13

McKechnie J, Colombo S, Chen J, Mabee W, MacLean HL (2011) Forest bioenergy or forest carbon? Assessing trade-offs in greenhouse gas mitigation with wood-based fuels. Environ Sci Technol 45:789–795. https://doi.org/10.1021/es1024004

Nunez C (2020) Carbon dioxide levels are at a record high. Here's what you need to know. National geographic

OECD (2012) Better policies for better lives : the OECD at 50 and beyond. https://www.oecd.org/about/47747755.pdf

Owusu PA, Asumadu-Sarkodie S (2016) A review of renewable energy sources, sustainability issues and climate change mitigation. Cogent Eng 3. https://doi.org/10.1080/23311916.2016.1167990

Pedroni P (1999) Critical values for cointegration tests in heterogeneous panels with multiple regressors. Oxf Bull Econ Stat 61:653–670. https://doi.org/10.1111/1468-0084.0610s1653

Pedroni P (2004) Panel cointegration: asymptotic and finite sample properties of pooled time series tests with an application to the PPP hypothesis. Economet Theory 20:597–325. https://doi.org/10.1017/S0266466604203073

Raj B, Singh O (2012) Study of impacts of global warming on climate change: rise in sea level and disaster frequency. In Global warming—impacts and future perspectives. https://doi.org/10.5772/50464

Rasool U, Hemalatha S (2016) A review on bioenergy and biofuels: sources and their production. Braz J Biol Sci 3:3. https://doi.org/10.21472/bjbs.030501

Ritchie H, Roser M (2017) Fossil fuels. Published online at OurWorldInData.org. Retrieved from https://ourworldindata.org/fossil-fuels

Seabra JEA, Macedo IC, Chum HL, Faroni CE, Sarto CA (2011) Life cycle assessment of Brazilian sugarcane products: GHG emissions and energy use. Biofuels Bioprod Biorefin 5:519–532. https://doi.org/10.1002/bbb.289

Souza GM, Ballester MVR, de Brito Cruz CH, Chum H, Dale B, Dale VH et al (2017) The role of bioenergy in a climate-changing world. Environ Dev 23:57–64. https://doi.org/10.1016/j.envdev.2017.02.008

Tal H (2019) The future is now: science for achieving sustainable development. Alpha Omegan on Global Sustainable Report. https://doi.org/10.1016/j.aodf.2009.10.015

Wu Y, Zhao F, Liu S, Wang L, Qiu L, Alexandrov G, Jothiprakash V (2018) Bioenergy production and environmental impacts. Geosci Lett 5. https://doi.org/10.1186/s40562-018-0114-y

Zhang XY (2016) Developing bioenergy to tackle climate change: bioenergy path and practice of Tianguan group. Adv Clim Chang Res 7:17–25. https://doi.org/10.1016/j.accre.2016.06.001

Challenges and Solution for Renewable Energy (RE) Development in Uttarakhand, India

12

Hiranmoy Roy

Abstract

In Uttarakhand's perspective, there is huge scope for Renewable Energy (RE) such as hydro, small, mini, and micro-hydro, solar, biomass, geothermal, and waste to energy. At present demand for power in Uttarakhand is around 1500–1600 MW. Around 50% of the demand is met through imports and the rest is generated in the State. Since there is huge gap between demand and supply of RE in the state, it is suggested to deficit of power is 52% in the state. The main objectives of this chapter are (1) To identify and analyze the issues of RE sector in the state of Uttarakhand, (2) To suggest a suitable framework for RE development in the state. In this study, I have used descriptive and exploratory research methodology to through light on current issues of RE development in the state. I have also used the Delphi method to take expert opinions for this study. I have taken interviews with power sector experts and RE experts to get in-depth information on the issues and come out with solutions. The experts were from Central Electricity Regulatory Commission (CERC) and Uttarakhand Renewable Energy Development Association (UREDA). Interview with the experts gave me meaningful insights that a generalized public source will not be able to provide. Interviews were carried out with experts, which had open-ended questions to get meaningful insights about the issues. There is a huge deficit of power generation in the State compared to its potential. In Uttarakhand, there is a RE potential of 8115 MW (UREDA) that is much more than its demand of 1600 MW. Thus, for Uttarakhand's Future Electricity Road Map 2030, the emphasis is to take full advantage of the state's RE potential over the next few years, especially the solar,

H. Roy (✉)
Department of Economics and International Business, School of Business, UPES, Dehradun, Uttarakhand, India
e-mail: hiranmoy@ddn.upes.ac.in

A. Kaushik et al. (eds.), *Climate Resilience and Environmental Sustainability Approaches*, https://doi.org/10.1007/978-981-16-0902-2_12

small, mini, and micro hydropower projects. There are several challenges faced by the power sector in the state. In the vision 2030 for the state, there is a focus on simultaneous efforts to develop renewable energy in the form of small hydropower and solar. As there is a huge gap between the demand and supply of RE in Uttarakhand till 2030, it is suggested that a framework for an integrated policy strategy for rapid RE implementation be developed that complements both the existing and the planned power projects. This framework is based on the national and international best practices and expert opinions.

Keywords

Renewable · Hydro · Solar

12.1 Introduction

The ongoing debate on the mix of renewable and nonrenewable energy in India is that by 2022 we should be able to achieve 272 GW of Renewable energy, i.e., 40% RE (NITI Aayog 2015) since RE is environment friendly and having zero input cost. In Uttarakhand's perspective, there is huge scope for RE such as hydro, small, mini, and micro hydro, solar, biomass, geothermal, and waste to energy. In the State of Uttarakhand, the mix of power is 400 MW from coal-based thermal sources, 64 MW from gas-based thermal sources, and rest 1922 MW is from RE sources (GoUK). In the electricity mix, there are mainly RE sources of electricity with very little scope for thermal power, as transportation of coal is not possible in the hilly terrain of the State. The main objectives of this chapter are (1) To identify and analyze the issues of RE sector in the state of Uttarakhand, (2) To suggest a suitable framework for RE development in the state.

12.2 Methodology

In this study, descriptive and exploratory research methodology has been adopted to throw light on current issues of RE development in the state. Delphi method has been used to take expert opinions for this study. I have taken interviews with power sector experts and RE experts to get in-depth information on the issues and come out with solutions. The experts were from Central Electricity Regulatory Commission (CERC), Uttarakhand Renewable Energy Development Association (UREDA), and Uttarakhand Power Corporation Limited (UPCL). Interview with the experts gave meaningful insights that a generalized public source will not be able to provide. Interviews were carried out with experts, which had open-ended questions to get meaningful insights about the issues.

12.3 Demand Supply Gap

At present, demand for power in Uttarakhand is around 1500–1600 MW. Around 50% of the demand is met through imports and rest is generated in the State. It may be mentioned here that State has a solar power potential of 4077 MW and only 275.58 MW is realized which shows 94% potential remains untapped (Tyagi 2018). In small, mini and micro hydel projects, the State has the potential of 3000 MW and the achievement is only 210 MW, i.e., 7% of the potential (Tyagi 2018). Total Consumption in the year 2015–2016 was 10,298.14 MU (Directorate of Economics and Statistics 2015–2016) and total production was 4942.33 MU showing a demand–supply deficit of 52%. There is a huge deficit of power generation in the State compared to its potential. There was a marginal increase in electricity generation of 12% in the year 2015–2016 compared to 2014–2015 (Directorate of Economics and Statistics 2015–2016). Thus, the State needs to explore all the possibilities of hydro, small, mini and micro hydel, solar, biomass, geothermal, pine needles, waste to energy, etc. (Table 12.1).

In Uttarakhand, though RE remains the most significant source of power, there lies a huge gap between the potential and realization of RE power as shown in Tables 12.2, 12.3, and 12.4. Thus, there is a huge gap between the potential and realization of RE in Uttarakhand.

Apart from the above sources, there is also an installed capacity of 982 MW that can be supported with pine needles as feedstock for supplying electricity in rural

Table 12.1 Availability mix from firm sources in FY 15 (in MW)

Source	Latest firm entitlement (in MW)	% age of total firm availability own
Own Generating Stations (OGS)		
Hydro	1134.96	48.02
Small hydro	32.7	1.38
Subtotal—OGS	1167.66	49.40
Private Generating Stations (PGS)		
Non-solar (R.E.)	213.3	9.02
Solar	6.6	28.00
Vishnugarh (J.P.)	48	2.03
Subtotal—PGS	267.9	11.33
Central Generating Stations (CGS)		
Coal based	400.1	16.93
Gas based	69.64	2.95
Subtotal—thermal	469.75	19.87
Nuclear	31.24	1.32
Hydro based	427.13	18.07
Subtotal—CGS	928.12	39.27
Grand total	2361.08	

Source: 24 × 7 Power for all Uttarakhand, A Joint Initiative of Government of India and Government of Uttarakhand, CRISIL

Table 12.2 Potential of nonconventional energy in Uttarakhand

S. No.	Source	Potential (MW)
1	Small, mini and micro hydel segment (up to 25 MW)	3000
2	Biomass (forestry, crop residues, rice husk, bagasse)	262
3	Biogas (Cattle dung)	28.80
4	Wind	Not having much windy areas
5	Solar power project	4077
6	Geothermal	20
7	Waste to energy	100

Source: Uttarakhand Renewable Energy Development Association (UREDA)

Table 12.3 Renewable energy achievements: grid connected and off-grid

Grid connected projects		
1	Cogeneration-based power plant	72.60 MW
2	Solar PV plant (Ground mount and rooftop)	275.58 MW
3	Small hydro power plants (UJVNL)/UREDA/IPP)	210 MW
Off-grid projects		
4	Rice husk and W2E biomass power projects	59 MW
5	Solar decentralized power projects	3.66 MW
6	Micro and Mini Hydro power plants including watermill	5.90 MW

Source: UREDA

Table 12.4 Gap between potential and realization of RE in Uttarakhand

	Approx. potential (MW)	Realization (MW)	Gap between potential and realization (MW)
Small, mini and micro hydel segment (up to 25 MW)	3000	210	2790
Biomass (forestry, crop residues, rice husk, bagasse)	262	59	203
Biogas (Cattle dung)	28.80	0	28.8
Solar power project	4077	275.58	3801.42
Geothermal	20	0	20
Waste to energy	100	0	100

Source: UREDA

areas for 5 h a day. In terms of round-the-clock generation, an installed capacity of 205 MW can be supported by the pine needles as energy feedstocks (Kala and Subbarao 2017).

It may be mentioned here that we need to focus more on national best practices that can be implemented in Uttarakhand aligning with State conditions as we are bound by strong regulatory mechanisms in the Indian context.

12.4 Electricity Production and Consumption

The trend of growth of Electricity Consumption and production in Uttarakhand based on DES Data up to 2015 and we have projected up to 2030 based on linear trend as shown in Fig. 12.1. The graph shows in 2030 there will be a demand–supply gap of 10,670.13, which is widening continuously due to surge in consumption. Thus, GoUK must take urgent steps to fully realize the potential of renewables to bridge this gap of potential and realization.

However, there are huge challenges of RE in the state which are analyzed in the following section.

12.5 Challenges of Hydropower Sector in Uttarakhand

12.5.1 Administrative Challenges

- Hydropower projects require land acquisition over a vast area.
- The management has to spend a lot of time on identification of land, verification, preparation of papers for land acquisition.
- This further delays the start of the project. Even after obtaining all the clearances and implementing suggested measures, political interference creates problems during construction, and thus cost overrun adds up.
- Most of the projects are built in hilly regions of the State. In these areas, many a time, local people show resistance due to their own customs and traditions.
- Besides this, many rivers flow through more than one State. There is a delay in projects due to disputes over the sharing of river water.

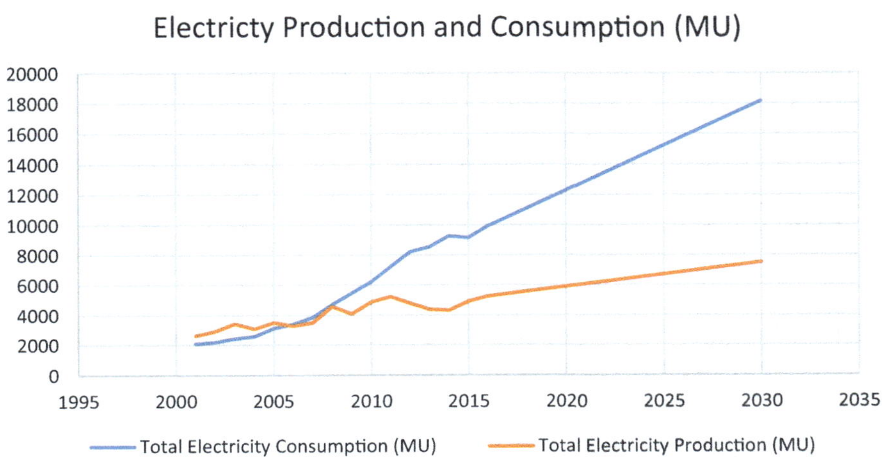

Fig. 12.1 Energy production and consumption pattern projections. Source: Directorate of Economics and Statistics (2015–2016)

12.5.2 Construction Challenges

There is a dearth of competent contracting agencies having technical, managerial construction, and financial resources required for constructing hydroelectric projects.

12.5.3 Geological Challenges

- The Himalayan ranges are geologically quite active and Uttarakhand is in the high seismic zone. Therefore, geological surprises cannot be avoided with the result that investigation works and planning of construction is time and money consuming.

12.5.4 Environmental Challenges

- Almost all hydro projects are responsible for deforestation.
- Due to deforestation, the possibilities of landslides, erosions, and floods are likely to occur more frequently.

In Uttarakhand, about 65% of land is under forests out of which about 15% of forest areas are either under National Parks or Sanctuaries/Biospheres. Some of the hydropower project sites are situated in these protected areas, thereby getting environmental clearances is one of the major issues in addition to the issues raised by local people with regard to employment, rights with regard to the use of water, etc. Some of the projects are multipurpose dam-based projects, which attract R&R (Rehabilitation and Resettlement) issues. Some projects, especially the Pacheshwar Dam (5000 MW), are to be situated in Indo-Nepal Border. Some time ago, the Govt. of Uttarakhand has decided to give preference to run-of-the-river projects by even changing the nature of dam-based projects to run-of-the-river type by having a cascade of projects rather than one dam based project. One good example in this regard is the Kotli Bhel Project (1000 MW) which was an earlier dam based and now three cascade-based run-of-the-river projects with a total estimated capacity of 1045 MW have been prepared through NHPC.

12.5.5 Unique Challenges and Reasons for Unsuccessful Small Hydropower Projects in Uttarakhand

- The power stations are located in remote hilly areas where even road linkages are not available.
- The small hydropower stations are prone to natural calamities such as flash floods due to cloud bursting, landslides, avalanches causing heavy damage, and long

shut downs. Road blockages and severe climatic conditions cause difficulty in construction, operation, and maintenance.

- Small Hydro Power Stations arc normally connected through service lines or weak grid connections, therefore the incidence of disruptions cause low generation.
- Due to their locations in far off and inaccessible locations of the State, it is difficult to provide necessary technical skills and spares in case of breakdown, necessitating long shut down requiring heavy expenditure and loss of revenue.
- Some of the commissioned plants have highly sophisticated machines and without the availability of skilled labor in remote areas, there is a lot of difficulty in operating and maintaining them. As a result, the machines are deteriorating and their efficiencies are declining.
- Many a time telephone connection is disrupted during monsoon.
- Long transmission lines are prone to frequent damage.
- The government chose to focus on small hydropower projects as well as gas and renewable power projects in the wake of the cloudburst in June 2013. However, it did not give details on the plans it has for the development of hydropower projects after the devastating floods that crippled the economy of most of the areas in the State.
- Uttarakhand requires a new State power policy with fresh strategies from planning to approvals, construction practices, and regulation. The State must also develop a strategy for energy conservation. Implementation of adaptive resilience and mitigation measures proposed in the draft State action plan for climate change (SAPCC) will also create sustainable developmental opportunities. The State Government will levy no entry tax on power generation, transmission equipment, and building material for projects. Free transfer of shares will be permitted in the companies' allotted projects as per the procedure laid down in this document. The Government of Uttarakhand has to earmark selected project sites for allocation to private developers. The Government should advertise and invite participation for developing these projects in lots.

To overcome the above challenges, GoUK as mentioned in the table may refer to the success factors of the Bhutan model.

The challenges for pine needle development are the patterns of fire disturbance and postfire regeneration in high elevation pine needle ecosystems are highly variable over space and time. While stand-replacing fires predominate in some regions, mixed-severity fire regimes that include low-intensity surface fires and crown fire components, appear most commonly. Fire disturbance provides critical regeneration opportunities for most high-elevation five-needle pines but fire exclusion over the last century is having some impacts—such as major declines of whitebark pine (*Pinus albicaulis* Englem). High-elevation pines are subject to damaging agents other than fire—such as localized insect infestations of bark beetles or the pine leaf adelgid and dwarf mistletoe infections.

Thus, it is suggested to conduct a study on ecological disturbances of pine needles immediately and by examining, the present policy of GoUK to implement revised new policy in the state for realizing the potential of pine needles in the state. It may be mentioned here that 16.36% (about 4 lakh ha) forest is covered by Chir Pine forest and 15 lakh MT pine litter is generated annually in Reserve and Van Panchyat forests (excluding wildlife area).

12.6 Policy Recommendations for RE Development in Uttarakhand

As there is a huge gap between the potential and realization of RE in Uttarakhand, it is suggested that a framework for an integrated policy strategy for rapid RE implementation be developed that complements both the existing and the planned conventional power projects. This framework is based on the national and international best practices adopted.

1. A new comprehensive national RE law and/or policy and its components.
2. Implementation of support mechanisms.
3. Reforms to ensure smooth grid integration of RE.
4. Energy access and off-grid RE considerations.
5. How?

Thus, the RE targets should integrate a suitable and measurable metric, e.g., generation, capacity, and share of consumption to monitor the progress in achieving the targets.

Therefore, a future road map is suggested that GoUK to follow success factors of the Bhutan model to augment the small and micro hydropower projects in Uttarakhand along with mitigating environmental—legal barriers. In addition, to pursue the realization of solar power potential by the speedy implementation of the solar power and rooftop projects.

12.7 Financial Support Required

The policy should evidently classify the source, level, and distribution mechanism for financial support for reducing the incremental cost of RE, which includes both generation and integration costs to the final buyers. This is needed due to already existing subsidized fossil fuel-based generation.

Most importantly, Integrated Energy Resources Planning needed with Inclusive and analytically sophisticated planning exercises should be conducted regularly in order to evaluate the benefits and costs of various aspects of the electricity sector, including supply-side resources, e.g., coal, hydro, gas, nuclear, RE, the transmission and distribution networks and their operation. The demand-side resources should also be evaluated (e.g., energy efficiency, and demand response). These planning exercises should explicitly and systematically account for various risk factors such as fuel availability, fuel costs, and other possible benefits and costs.

At the national and state level, the best practices prescribed for the RE sector are Support Mechanisms for Compliance and Timely Implementation with strong policy and legislation in place. Moreover, the emphasis on execution support will also be more desirable. The government, at both the Central and the State levels, will need to support compliance with mandatory requirements regarding RE on the power system. To ensure this, The following funtions need to be implemented.

Support from GoI is critical to promote RE in the State. At present, there are several promotional policies of GoI as noted below. GoUK should take advantage of such policies where Central funding available.

The Government of India has undertaken a number of policy measures for increasing the share of renewable energy in India's energy mix. These, inter-alia, include:

(a) Provision of Renewable Purchase Obligation (RPO) under the National Tariff Policy.
(b) Notification of the long-term growth trajectory of RPO for solar and nonsolar energy for next 3 years from 2016 to 2017, 2017 to 2018, and 2018 to 2019.
(c) Development of Solar Parks and Ultra Mega Solar Power Projects.
(d) Development of power transmission network through Green Energy Corridor project.
(e) Making rooftop solar as a part of housing loan provided by banks.
(f) Waiver of Inter-State Transmission Charges and losses.
(g) Repowering of Wind Power Projects for optimal utilization of wind resources.
(h) Offshore wind energy policy for the development of offshore wind energy in the Indian Exclusive Economic Zone.
(i) Supporting research and development on various aspects of renewable energy including industry participation.
(j) Financial incentives for off-grid and decentralized renewable energy systems and devices for meeting energy needs for cooking, lighting, and productive purposes.
(k) Permitting 100% Foreign Direct Investment in the sector through automatic route.

The above measures can as well be implemented in Uttarakhand aligning these policies with state conditions. In addition, few important specific measures as mentioned below will be able to foster RE development in the state.

12.7.1 Financial Support and Disbursal

A transparently designed uniform, simple financial support and disbursal mechanism directed at buyers and which offers certainty over a reasonable period could considerably support acceleration of RE growth. The financial support could be distributed through the new Intermediary Organization that ensures the bulk buyers are indifferent between RE and fossil fuel-based generation.

12.7.1.1 Streamlined Project Development

The most important impediment on speedy RE development is the lengthy and costly project development procedure that includes investment to grade RE resource assessments, access to land either acquisition or leasing, complementing infrastructure development such as roads, water and transmission interconnections. States themselves should lead the facilitation process to reduce soft costs in project development, e.g., siting, permitting, and supporting infrastructure with technical and logistical support from the Intermediary Institutions. This is essentially intended to minimizing risk in the sector and expediting RE growth.

12.7.1.2 Low-Cost Financing

If we compare the cost of RE and Non-Renewable technologies, RE technologies, unlike fossil-based energy technologies, have high capital costs but very low operating costs spread over 25–30 years. Thus, the cost of finance that is at present ranging from 12 to 14% in India forms a significant component of the power tariff from these sources. Bringing down the rate of interest for RE projects would reduce tariffs and hence scale-up demand for RE.

12.8 RE Grid Integration and Efficient Grid Operation

Finally, in addition to a strong policy and legislative framework, there is also a requirement for supportive implementation environment; grid interconnection and integration of RE are equally critical.

Intermittency is another very important aspects of RE, which denotes "Predictability or Uncertainty" and refers to the lack of accurate information about future RE generation, e.g., a sudden drop in wind power, which is not very different from fossil fuel-based generation/transmission systems, e.g., an unforeseen failure of a fossil-based generator or a transmission line. There is a strong need to forecast the future. "Variability" is the known natural variation in RE generation, e.g., wind peaking during monsoon and reduced availability in other seasons; just as exists on the demand side currently, e.g., low demand at midnight and high demand during the late afternoon.

Internationally where RE accounts for increasingly large shares of power system generation various changes to grid design, technology, and its operation have been implemented, which allow successful grid integration, i.e., minimizing or managing the variability and uncertainty aspects of RE.

These strategies can be classified into the following subcategories: (1) Upgrade grid technology, (2) Centralized RE forecasting mechanisms need to be tightly integrated with system operations, (3) Upgrade grid operation protocols that are already in place, (4) Grid Codes, (5) strengthening Scheduling and Dispatch, (6) Expand balancing areas, and (7) Promote flexible demand and supply resources.

Uttarakhand government currently following several best practices or initiatives (all these are available in the public domain by UREDA) to promote RE in the State.

Table 12.5 Energy saving target (up to 15% of energy demand) from renewable sources

BY FY	Demand in MU	Renewable energy share in MU				% share
		SHP	Solar	Others (Co-gen, Biomass, W2E)	Total	
2019–2020	15,027	746	663	322	1731	11.5
2023–2024	18,266	883	1109	647	2639	14.5
2029–2030	24,478	1121	1556	1405	4082	16.7

Source: UREDA

Table 12.6 Energy saving target (up to 25% of energy consumption)

Demand in MU	Sources of energy savings in MU					% Share
	LED bulbs	EE street lights	Solar water heaters	Other DSM	Total	
15,027	910	29	120	164	1223	8.13
18,266	As per that time applicable scheme of GoI and available new technologies					15
24,478	As per that time applicable scheme of GoI and available new technologies					25

Source: UREDA

These are Uttarakhand Solar Power Policy 2013, Policy for Development of Micro & Mini Hydro Power (Up to 2 MW), Policy for Development of Small Hydro Power (above 2 MW), Policy for Energy Generation from Pine Litter and Other Biomass—2018, Uttarakhand Green Energy Cess, etc. These best practices of the government to be continued. Moreover, important energy conservation programmes of the State such as Government Orders, UJALA Yojana, ECBC, Gram LED Light Training Programme, MuDSM, PAT, SECF, and School Awareness Programme should be continued as these conservation programs were found to be effective as reported by UREDA officials. The projected energy demand and to conserve or save up to 15–25% from RE sources as predicted by UREDA are given in Tables 12.5 and 12.6. The continuation of these best practices and initiatives will help the State to achieve the projected targets.

12.9 Conclusion

Thus, for Uttarakhand's Future Electricity Road Map 2030, the emphasis is to take full advantage of the state's RE potential over the next few years. Uttarakhand's renewable resources are abundant. To seize the benefits, the State would need to raise necessary capital, and adapt to managing the variability and uncertainty of renewable energy generation. There are several challenges faced by the power sector in the state. The most important challenges are the challenges faced by Discom, challenges of Hydropower Sector in Uttarakhand such as administrative challenges, construction challenges, geological challenges, and environmental challenges. In this chapter, best practices in the national and international contexts are highlighted, which may be followed by the state to expedite its RE, sector, as these will definitely serve as enablers. The current scenario, status, trends, best practices, challenges, and issues provide insights into the possible strategies to address the challenges and to draw a future road map for the state.

References

24 × 7 Power for all Uttarakhand, A Joint Initiative of Government of India and Government of Uttarakhand, CRISIL

Directorate of Economics and Statistics (2015–2016) Statistical Abstract Uttarakhand, Government of Uttarakhand

Directorate of Economics and Statistics (2015–2016) Uttarakhand Vision 2030

Kala LD, Subbarao PMV (2017) Pine needles as potential energy feedstock: availability in the Central Himalayan State of Uttarakhand, India", 2nd Euro Global Summit and Expo on BIOMASS AND BIOENERGY October 12-13, 2017 London, Uk, Posters & Accepted Abstracts: J Fundam Renewable Energy Appl. https://doi.org/10.4172/2090-4541-C1-045

NITI Aayog (2015) Report on India's renewable road map 2030

Tyagi AK (2018) Uttarakhand Renewable Energy Development. Government of Uttarakhand Association (UREDA)

Integrated Wastewater Treatment and Energy Production Using Microbial Fuel Cell Technology: A Sustainable Environment Management Approach

13

Singh Aradhana and Anubha Kaushik

Abstract

Microbial Fuel cell (MFC) has come up as a promising technology for the treatment of wastewater along with simultaneous energy generation. In the initial stages of development of MFC technology, it was mainly used for organic matter removal, specially COD, but nowadays, it is also explored for treatments that include heavy metal removal/recovery, dye degradation and colour removal, nutrient toxicity removal (phosphate, nitrate) and treatment of other xenobiotic compounds like nitrobenzene, chlorobenzene and others. While conventional wastewater treatment technologies are costly due to their dependence on energy and chemicals, the MFC technology recovers energy from the wastewater during treatment. Further, there is no generation of toxic sludge in this bioelectrochemical method in contrast to that of traditional physico-chemical treatment methods. The present chapter critically discusses various applications of MFC technology in wastewater treatment along with its power generation scenario. Challenges associated with MFC technology to finally upscale it to the field level can be addressed by further research insights, particularly in the areas including electrode material, separators, biofilm communities, design and configuration briefly discussed in this study. The chapter critically assesses the technical and economic feasibility of MFC technology that can play an important role towards achieving the sustainability goals of clean water and alternate energy through energy-supported wastewater treatment.

S. Aradhana · A. Kaushik (✉)
University School of Environment Management, Guru Gobind Singh Indraprastha University, Delhi, India
e-mail: akaushik@ipu.ac.in

A. Kaushik et al. (eds.), *Climate Resilience and Environmental Sustainability Approaches*, https://doi.org/10.1007/978-981-16-0902-2_13

Keywords

MFC · Wastewater treatment · Energy production · Challenges · Economic feasibility

13.1 Introduction

The ever increasing energy demand of the fast growing and economically developing human population in the twenty-first century has emerged as a big challenge across the world. At present, fossil fuels like coal, oil and petroleum are being used primarily for energy production, accounting for 86% of total global energy consumption. However, these reserves of non-renewable fossil fuels are finite and therefore, once depleted will not be replenished. Besides, burning of the fossil fuels leads to carbon emissions that are known to be one of the biggest contributors to the global climate change problem in near future. Therefore, in recent times, the focus is being shifted to renewable energy sources like solar energy, wind energy, hydropower energy, geothermal energy and biomass energy that are environmentally safer and have the potential of being regenerated. Energy from waste is one of the newer concepts getting more attention because of its dual advantages of waste reduction and energy production. The energy present in the wastewater is estimated to contain nine times higher energy as it is required to treat it. In this regard, Microbial fuel cell (MFC) technology is a unique invention which offers means of utilizing the wastewater energy while treating it in an eco-friendly manner and producing electricity. A major breakthrough in this direction was brought about by Habermann and Pommer (1991), who demonstrated simultaneous wastewater treatment and electricity generation by using Microbial fuel cell (MFC) technology. The organic substrate in the waste is degraded by the microorganisms present in waste or added to it as inoculum, and in the process electrons and protons released are passed on to the terminal electron acceptor and this is how electricity is generated. The schematic of a typical double chambered MFC (DCMFC) is represented in Fig. 13.1.

This technology was furthered by Allen and Bennetto (1993) who used the pure culture of bacteria to catalyze the oxidation of organics using artificial mediators for easy electron transfer from bacteria to the anode. Further advancement in MFC studies was brought about by Kim et al. (1999) who introduced the concept of electrochemically active bacteria that could facilitate oxidation of a variety of substances. Using electrogenic bacteria like *Shewanella putrefaciens*, *Geobacter sulferreducens*, *Geobacter metallireducens* and *Rhodoferax ferrireducens*, it has become possible to exclude the use of external chemical mediators.

The first MFC prototype was developed by Logan et al. (2006), which led to an improved power generation capacity by the MFC. While MFCs are being visualized for a wide range of potential applications, the technology is still in the stage of development and lot of research is in progress, and there is a need to enhance the performance of the MFCs. This chapter critically discusses the state of the art of MFC technology, its challenges, limitations and prospects for different applications

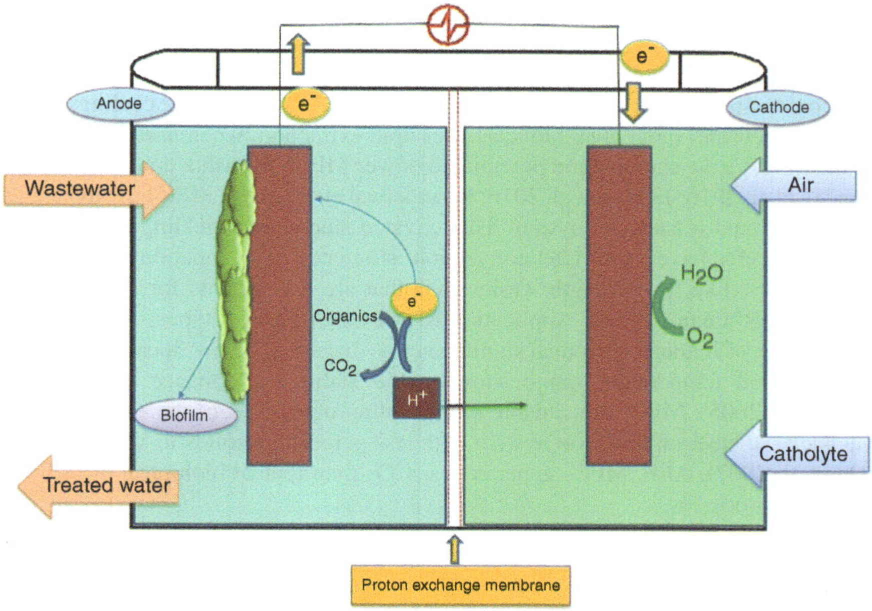

Fig. 13.1 Schematic of a typical double chambered MFC. Biofilm develops on the anode and its microbes degrade the organics present in the wastewater releasing protons (H^+) in the solution and electrons (e−) to the anode. Electrons move from anode to cathode through an outer circuit producing electricity, while protons move through the proton exchange membrane to the cathode. Sparging of air in cathode provides dissolved oxygen which facilitates acceptance of electrons and completes the reaction with H^+ to form water in the cathode, which carries some suitable catholyte with catalyst

as well as its economic feasibility for use at commercial or industry level for wastewater treatment in a relatively energy sustained manner.

13.2 Wastewater Treatment by MFCs

The wastewater treatment is the traditional and the commonly used application of MFC technology. In the beginning, it was mainly used for COD removal from wastewater. However, nowadays, it is explored for various other treatment that includes removal or recovery of heavy metals, dye degradation and color removal, nutrient toxicity removal (phosphates, nitrates) and treatment of other xenobiotic compounds like nitrobenzene, chlorobenzene, Polyaromatic hydrocarbons (PAHs) and many halogenated compounds. MFC technology uses the energy already present in the wastewater instead of using external energy. Thus, if the potential energy from wastewater can be captured efficiently, then instead of becoming a consumer, wastewater treatment can become an energy producer (Logan 2004). It has a great potential of saving energy as it requires only 10% of external energy when compared to the activated sludge process which is a conventional technique of treating wastewater (Gude 2016). Wastewater used as substrate in the anode chamber of the MFC is a key factor influencing the performance of MFC. Wastewaters from the food processing industry, beverage industry, dairy industry, paper recycling, distillery, human urine, spent algal biomass after lipid extraction, when used in the anode as feedstock, have shown good potential to power MFCs (Kaushik and Chetal 2013; Pandey et al. 2016; Jadhav et al. 2016; Khandelwal et al. 2018). MFC has advantages over traditional wastewater system like activated sludge and trickling filter. Microbial fuel cells have high rate conversion efficiency and capability of treating wastewaters (like low-strength wastewater) that are not suitable for the anaerobic digestion (Rittmann 2008). Application of MFCs in various types of wastewater treatment is of immense practical significance as discussed below. It can be operated under varied temperature range showing the stable performance (Rabaey and Verstraete 2005). MFCs are capable of harvesting up to 90% of the electrons from the bacterial electron transport system with less percentage solids to be disposed of (Du et al. 2007). Also, MFCs generate less CO_2 than typical biological wastewater treatment processes.

13.2.1 COD Removal

The most common application of MFC is the COD removal from wastewater. Almost all microorganisms use the chemically bound organic substrate as a feed material for their growth and metabolism. The microbes degrade and reduce the COD in an MFC and the conversion rate of chemically bound organic energy into electricity varies depending upon various factors. MFCs are found capable of COD removal from low COD strength wastewaters such as domestic wastewater, or vegetable and food processing wastewater (Mohan et al. 2010a, b; Behera et al. 2010a, b; Fogg et al. 2015) to very high COD strength wastewaters such as distillery effluent, dairy wastewater, livestock wastewater and many more (Singh and Kaushik 2019; Li et al. 2013; Mohan et al. 2010a, b). MFCs have been found to treat successfully the wastewaters with low COD concentration (20 mg/L) as well as high COD (127,500 mg/L) as reported by Xu et al. (2016). More than 99% of COD removal have been seen in MFCs in many studies (Parkash 2016; Sangeetha and Muthukumar 2011).

COD removal by MFC is a bioelectrochemical phenomenon affected by MFC components (proton exchange membrane, anode surface area, pH, temperature and inoculum). Inoculum selection is very important in deciding the fate of COD removal in MFC. Diverse and rich microbial source of inoculum generally results in high power generation. The COD doses are normally optimized to enhance the performance of MFC (Nam et al. 2010). Coupling of MFC with other wastewater modules has sometimes been tried, shown enhanced COD removal. Wang et al. (2014) found enhanced COD removal on coupling the MFC with a sequencing batch reactor. Coupling of the MFC with anaerobic digester showed more efficient COD removal from swine wastewater as compared to non-integrated system (Kim et al.

2015). Integration of constructed wetland with MFC has shown remarkable improvement in performance in terms of COD removal and thus attracted the attention of researchers (Araneda et al. 2018; Doherty et al. 2015).

13.2.2 Removal and Recovery of Heavy Metals

Heavy metals are common co-pollutants in most wastewaters along with other organic or inorganic pollutants. High toxicity of the heavy metals, even at very low concentration, causing serious health effects including kidney damage, bone deformation, neurological damage, skin and lung cancer and several others in humans is a major concern (Jaishankar et al. 2014), which is also a threat to the aquatic ecosystems and services (Kaushik et al. 2001). Conventional techniques of metal removal from wastewaters are highly energy dependent, carbon emitting, sludge generating, and have limited use when concentrations are high (Fu and Wang 2011). In recent years, MFCs are being used for the reduction of metal toxicity in addition to power generation (Mathuriya and Yakhmi 2014) making the metal removal process more cost efficient and less energy dependent. In general, the cathode chamber in MFC is used as destination of electrons and protons. Principally, the cathodic compound that is used as the final electron acceptor should have high redox potential. Oxygen, ferricyanide, permanganate, H_2O_2, nitrate and trichloroethene, and perchlorate have been demonstrated as effective catholytes that have been used as electron acceptors in the MFCs (Oh et al. 2004; Mathuriya and Sharma 2010; Tartakovsky and Guiot 2006; Clauwaert et al. 2007; Thrash et al. 2007; You et al. 2006).

Some of the metals in wastewaters have high redox potential, which could be gainfully utilized as electron acceptors in the cathode chamber to get reduced and or precipitated. Three types of reactions are involved in MFCs for wastewater treatment and simultaneous metal removal/recovery, i.e. the production of electrons, transfer of electrons and consumption. These reactions are chemical, electrochemical and bioelectrochemical.

In general, the bioelectrochemical reactions are involved in anode chamber, when the electrons are released from the microbial metabolism. These electrons gather at the anode and, subsequently, travel through an external load and arrive at the cathode chamber where the electrochemical reactions take over. The electrochemical reactions include the electrochemical reduction of dissolved oxygen to H_2O or H_2O_2 (Liu et al. 2011) and/or direct electrochemical reduction of metal ions. Researchers have demonstrated the removal of vanadium, chromium, copper, mercury, selenium and many others along with electricity generation using MFC (Zhang et al. 2013a, 2015; Wang et al. 2011; Li et al. 2008; Liu and Yin 2017), and also low redox potential metals like cadmium and nickel by using biocathode (Singh and Kaushik 2021). Recovery of precious metal such gold and silver was also made possible using MFC (Choi and Cui 2012; Choi and Hu 2013). Metal removal and recovery by MFC is influenced by several factors such as redox potential of the metal, metal concentration, pH, biofilm microbes, abiotic or biotic cathode, single culture or microbial consortium and presence of other co-metals are major determinants of treatment efficiency as well as electricity production by MFCs (Kaushik and Singh 2020).

13.2.3 Removal of Nitrates and Phosphates

Nitrates and phosphates are the nutrients that are removed from wastewaters using costly tertiary treatment methods. Recently the potential of nitrate as the terminal electron acceptor in MFC is being explored with a view to use this technology for the removal of nitrates that is toxic in high concentration. Due to high redox potential of nitrate for converting into nitrogen gas, its use as a terminal electron acceptor in the cathode of MFC is gaining attention. Along with organic matter removal in the anode chamber, there is possibility of nitrate removal in the cathode chamber without offsetting the electricity generation capacity of MFC. Gregory et al. (2004) observed reduction of nitrate to nitrite with the electrode serving as sole electron donor. Microbial cells attached to the electrode were responsible for the nitrate reduction. Carbon and nitrogen removal have also been demonstrated using biocathode MFC, taking combinations of different types of aerobic and anaerobic sludge and sediment as cathodic inoculum (Clauwaert et al. 2007). They have shown that the current production was in sync with the denitrification rate, which confirmed that electron balance corresponded with the nitrogen balance. Al-Mamun et al. (2016) analysed the efficiency of a sandwiched denitrifying biocathode for denitrification and power generation. Synthetic wastewater was used as substrate in the cathode chamber which was supplied with nitrate and domestic wastewater as substrate in anode chamber. Biocathode was sandwiched by the two chambers at the sides, which acted as bioanode. It was found that increase in nitrate loading resulted in increased current and power generation but decrease in denitrifying efficiency.

Phosphates can also be recovered from wastewater using MFC technology, as reported by some researchers (Fischer et al. 2011; Hirooka et al. 2011; Ichihashi et al. 2012). Ichihashi and Hirooka (2012) recovered phosphorus (27%) as struvite crystal precipitate from swine wastewater using air–cathode MFC. A 3-stage MFC/struvite extraction process system was developed by You et al. (2016) which facilitated struvite recovery along with COD removal and electricity generation using urine as substrate in the anode.

13.2.4 Removal of Dyes

Azo dyes are the most common synthetic dyes discharged from dye-manufacturing industries. The severe problem associated with the azo dyes is their intense color which when released to the water bodies affects light penetration and dissolved oxygen in the water bodies. Since azo dyes are very toxic in nature and affect the aquatic food chain, hence, their removal of great concern. In recent years, more emphasis is given to biological methods to remove toxic dyes from wastewaters over conventional treatment methods. MFC technology is also being explored for the decolorization of dyes. Sun et al. (2009) were the first to discover the potential of MFC in dye decolorization when they successfully decolorized the dye ABRX3 (up to 1500 mg/L) using MFC and found the process to be more efficient than the traditional anaerobic treatment.

Complex aromatic structure of the azo dyes is difficult to degrade, but mono azo dyes are more readily decolorized than poly azo dyes (Hsueh et al. 2009). In MFCs, the microorganisms in the anode chamber degrade the dye and release protons and electrons that finally reach the cathode, which help in the abiotic decolorization of the dye in the cathode. The transfer of electrons from microbes to the electrode and to the azo dye is the rate-limiting step for dye removal and electricity production. Sun et al. (2013) succeeded in enhancing the rate of this process by adding some redox mediators to the anolyte solution. The addition of redox mediator facilitates lowering the overpotential that is generated at anode and hence results in the enhanced reduction of the dye and electricity generation.

Concentration of the dye is important in determining the success of its removal in the MFC. Mu et al. (2009) observed that by increasing the concentration from 0.19 to 0.70 mM, decolorization of Acid orange 7 dye decreased from 78 to 35%. Integrating the MFC with some other conventional technique has been found to be particularly useful in dye decolorization. The performance of constructed wetland-MFC was evaluated by Fang et al. (2015) in azo dye degradation and electricity production under varying operating conditions. They found hydraulic retention time (HRT) and COD as determinant factors in color removal and electricity production by MFC. The decolorization of 93% and power density of 0.0619 W/m^3 occurred in 3 days HRT. While the COD of 300 mg/L led to enhanced decolorization of 96% and high power generation of 0.852 W/m^3.

13.2.5 Removal of Xenobiotics

MFCs also find application in the degradation of xenobiotic compounds (nitrobenzene, chlorobenzene, polyaromatic hydrocarbons and many halogenated compounds). In a recent study, 81% degradation of p-nitropehnol was achieved in 24 h operation in an MFC (Zhao and Kong 2018). They also found an enhancement in degradation when the operation time was increased. Reductive dechlorination of trichloroethane was demonstrated in MFC, which was found to be highly influenced by the cathode potential (Aulenta et al. 2011). Highly efficient biodegradation of diesel (82%) was found in MFC, which was much higher than that (31%) in anaerobic control cell (Morris et al. 2009). Moreover, the power production of 31 mW/m^2 was also achieved in the MFC. Likewise, toluene degradation has been studied by Lin et al. (2014). The degradation was achieved in 60 h; it was also observed that the addition of mediators led to an enhanced electricity generation in 11.09 mg/L toluene-fed MFC. Liu et al. (2018) studied the removal of benzene from groundwater in stacked tubular air-cathode MFC. They observed higher removal and power generation in the stacked MFC compared to single MFC. Intermittent aeration also improved the removal and power generation in the MFC. All these studies suggest that MFCs have tremendous potential of degrading the xenobiotic compounds, which pose one of the biggest challenges of treatment of contaminated waters.

13.3 Energy Production by MFC

13.3.1 Power Generation Scenario

Power generation by MFCs has improved over the years due to persistent efforts of various researchers from initial 0.001 to 0.01 mW/m^2 in 1999 to 2770 mW/m^2 in 2008 (Xing et al. 2008), 3.900 mW/m^2 in 2009 (Yi et al. 2009) and has now increased to 4200 mW/m^3 (Sharma and Li 2010; Zinadini et al. 2017). However, single MFC is insufficient to produce a considerable amount of electricity, and size increase of MFC generally leads to increase in internal resistance and reduced power density.

A 250 L module consisting of two MFC units achieved a relatively low power density of 0.47 W/m^3 (Feng et al. 2014) and hence miniature MFCs and stacking of MFCs are preferred to increase the power output. To increase the power output, series stacking of individual MFC units are needed to be done because the power generated from single MFC unit in insufficient for practical applications. Stacking is done by connecting many units/modules of MFCs that share a common feed passage in a system. A recent pilot-scale study has shown that an integrated 72-L stacked membrane-based MFC produced power output of 50.9 W/m^3 (Wu et al. 2016). A 90 L MFC system using brewery wastewater was able to prevent short circuit by using cloth separator between the anode and the cathode (Dong et al. 2015). A 200 L modular MFC treating primary effluent was able to give 75% of COD removal (Ge and He 2016) and generate electricity for the catholyte recirculation pump. A 1000 L modular MFC operated for 1 year led to 90% COD removal and 60 W/m^3 of power production treating real municipal wastewater (Liang et al. 2018). Thus, the use of stacked MFC modules is recommended to increase power output rather than using MFCs of large capacity.

13.3.2 Electrode Materials

Anode is one of the cardinal components of the MFC affecting its overall performance. The essential requisites of anode include properties like high biocompatibility with the microbial population, good electronic conductivity, large surface area, stability and the most importantly its cost effectiveness (Kumar et al. 2013). Carbon-based anodes have been a more common choice, which include carbon cloth, carbon fiber, graphite fibre brush, etc. (Rezaei et al. 2007; Feng et al. 2010; Logan et al. 2007) and a wide variation in their surface area (0.0045–7.11 m^2/g) is reported.

While various configurations of carbon have been used, the carbon brushes give the best result as they provide large surface area for optimal adhesion and growth of bacteria (Choi and Cui 2012). Jiang and Li (2009) used carbon cloth as anode taking domestic wastewater as a feed in double chambered MFC and found maximum power density (PD) of 5 W/m^3. Carbon brush used as an anode with sludge as anodic feed in a double chambered MFC showed improved PD_{max} of 4.25 W/m^2 (Choi and Cui 2012).

Metal-based anodes such as stainless steel (SS) are used as an alternative to carbon-based anode having excellent mechanical properties, electrical conductivity and resistance to corrosion. In one of the studies, SS bioanode showed higher current generation as compared to graphite electrode (Pocaznoi et al. 2012). However, SS may resist thick biofilm formation and hence hamper power and current production. For this reason, the advanced electrode is now being explored such as three-dimensional (3D) SS foam with enhanced performance due to its high porosity (Ketep et al. 2014). Thus, selection of electrode material should be done considering properties like mechanical strength, the large surface area suitable for microbial film development, good conductance and corrosion resistance.

13.3.3 Electrode Modification for Enhanced Energy

Stability and efficacy of the electrodes used in MFCs play a decisive role in electricity production. The traditional carbon electrodes are much versatile but have low electrical conductivity. Modification of these traditional electrodes is likely to improve the performance and stability of MFCs. The common methods of electrode modification include chemical synthesis, covalent modification, electrode-position and soaking/adsorption (Mehdinia et al. 2014; Tang et al. 2014; Guo et al. 2013; Singh and Kaushik 2019; Lowy and Tender 2008).

In MFCs, two important issues related to microbial biofilm are bacterial adhesion to the anode and availability of substrate to the bacteria. These two factors ultimately decide the electron transfer rate and reduce the mass transfer limitation (Liu et al. 2010). Though a large surface area of anode is recommended for good adhesion of biofilm on its surface, but instead of increasing the size of the anode, working on its structure and modification are considered more important.

Anode modification is aimed at increasing the transfer of electrons from the microbes to the surface of the anode. Another way of increasing electron transfer is to add redox mediators into the anolyte. However, these mediators add to the cost of the MFCs and with time tend to get washed out from the system and enter the environment. Anode modification is therefore, a better option for increasing the electron transfer in MFC. Table 13.1 shows various types of anode modifications and associated increase in power generation by the MFCs. While heat and acid treatment increased power production by 34% (Feng et al. 2010), various other treatments like Riboflavin and humic acid treatments increased it by 40–60% (Huang et al. 2017). In recent years up to 100–200% increase has been recorded by using

Table 13.1 Increase of power density by anode modification in MFC

Anode modification	Anode material	Power density (compared to unmodified electrode)	References
Combined heat-acid treatment	Carbon brush	34% higher	Feng et al. (2010)
H_2SO_4-doped polyaniline	Carbon cloth	2.7 times higher	Lai et al. (2011)
Tartaric acid-doped polyaniline	Carbon cloth	4.1 times higher	Liao et al. (2015)
Humic acid electro-deposition	Graphite felt	66% higher	Huang et al. (2017)
Riboflavin electrodeposition	Graphite felt	40% higher	Huang et al. (2017)
Graphene/riboflavin composite	Graphite paper	5.3 fold increase	Wang et al. (2017)
Niacin doped	Graphite brush	2 times increase	Singh and Kaushik (2019)
Powdered activated carbon (PAC) applied onto C-fibre	Carbon fibre	77% increase	Gajda et al. (2020)

anode modifications with Niacin (Singh and Kaushik 2019) and tartaric acid (Liao et al. 2015).

13.4 Role of Biofilm Communities

Both electricity production and wastewater treatment by MFCs are majorly dependent on the biofilm developed on the anodic surface (Liu et al. 2010). Microbial communities in the biofilm feed on the organic matter present in the wastewater and release electrons. The overall performance of MFC system is dependent on the interactions taking place between the microorganisms and the anode surface. Microbes, which release electrons from inner to outer membranes and are able to release it out are called exoelectrogens. The two most studied exoelectrogens are *Geobacter* (Bond and Lovley 2003) and *Shewanella* (Gorby et al. 2006). However, instead of pure cultures, mixed cultures are preferred in MFC, because, it is easier to take mixed culture from different types of environments (Liu et al. 2010; Kaushik and Chetal 2013). Table 13.2 shows the effect of different types of inoculum sources on MFC performance.

Different sources of inoculum studied for power generation by MFCs include garden compost (Parot et al. 2008), press mud (Singh and Kaushik 2019), heat-treated soils (Niessen et al. 2006) activated sludge from fly ash leachate (Varanasi et al. 2017), Cow dung (Kaushik and Chetal 2013), industrial sludge (Chetal and Kaushik 2014), Anaerobic sludge (Lv et al. 2020), anaerobic sludge from wastewater treatment plant (Kim et al. 2007) and many others. Inoculum containing high

Table 13.2 Various inoculum sources for enhanced power density, COD removal and Columbic efficiency

Source of inoculum	COD removal	Coulombic efficiency	Electricity generation	References
Heat-treated soils	N/A	30%	170–200 mA/L	Niessen et al. (2006)
Anaerobic sludge from wastewater treatment plant	N/A	10%	488 mW/m^2	Kim et al. (2007)
Manure	N/A	N/A	4.21 mW/m^2	Scott and Murano (2007)
Garden compost	N/A	N/A	247 mW/m^2	Parot et al. (2008)
Cow dung	N/A	N/A	450 mW/m^3	Kaushik and Chetal (2013)
Industrial sludge	38%	N/A	640 mW/m^3	Chetal and Kaushik (2014)
Activated sludge from fly ash leachate	75%	10.6%	5.43 W/m^3	Varanasi et al. (2017)
Anaerobic sludge	N/A	19.5%	628 mW/m^2	Lv et al. (2020)

exoelectrogens and biofilm-forming bacteria will be capable of accelerating the formation of biofilm on the anode.

In electron shuttling, microbes themselves produce some natural mediators like phycocyanin, as produced by *Pseudomonas aeruginosa* (Rabaey et al. 2004), which mediates the transfer of electrons to the final electron acceptor. Nanowires, on the other hand, are conductive appendages observed in *Geobacter* (Bond and Lovley 2003) and *Shewanella* (Gorby et al. 2006) species which are the modified pili/extension of the outer membrane through which electrons are transferred to electron acceptor. These two are basically direct electron transfer, which takes place via physical contact of the bacterial cell membrane or a membrane organelle with the surface of the anode. Another way of electron transfer in MFC is through mediated electron transfer (MET), wherein compounds like quinone, phenazines, cysteine, which have right redox potential and which can mimic the activity of metabolites or proteins produced by microorganisms can help in transferring electron to the final electron acceptor (Uria et al. 2017).

13.5 Technical Challenges and Prospects

Being a bioelectrochemical system, power generation by MFC is greatly influenced by bio-chemical characteristics and MFC architecture (Mohan et al. 2014). Figure 13.2 shows various challenges faced by the MFC technology along with some approaches which can be followed to make it a field level-technology. The design and configuration of MFC affect its performance to a large extent. In H-shaped double chamber MFC (DCMFC), a high resistance is generally developed due to the larger distance between anode and cathode. In single chamber MFC (SCMFC), the

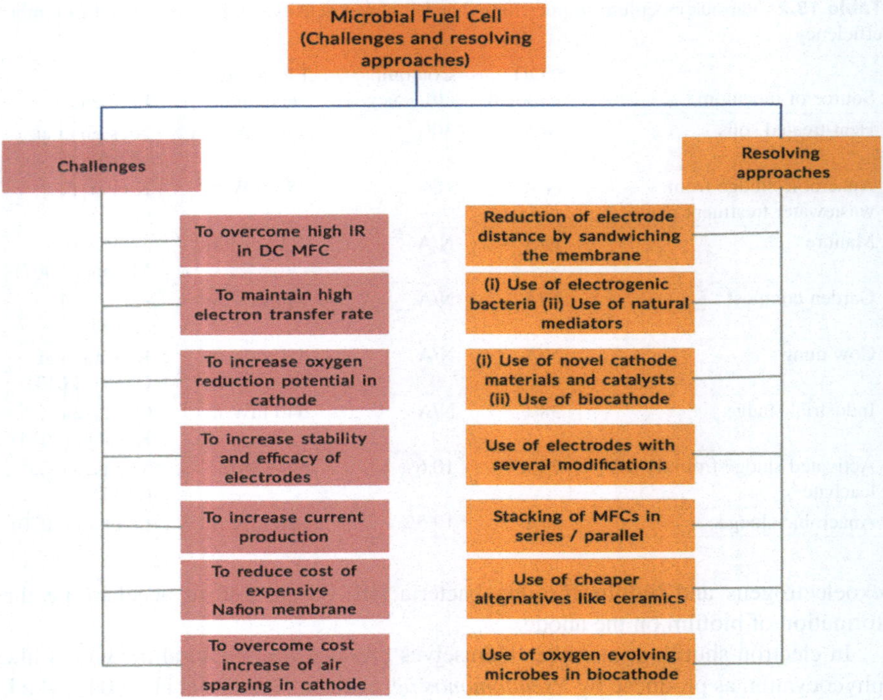

Fig. 13.2 Challenges and resolving approaches of MFC Technology

internal resistance (IR) is less, as the cathode is near the anode. Single MFCs produce a low voltage, hence, the MFCs are stacked in series or parallel to increase voltage generation. To reduce the internal resistance in H-shaped MFCs, a sandwiched model of MFC has come into the light, where the cation exchange membrane is sandwiched between anode and cathode chamber. A new design has come up as an up flow fixed-bed biofilm reactor, where there is continuous flow of feedstock through a porous anode (Jana et al. 2010).

Stability and efficacy of the electrodes used in MFCs are important issues. The traditional carbon electrodes are much versatile, but low electrical conductivity is their major limitation. Modification of traditional electrodes is therefore, required for developing high-performance MFCs. In DCMFCs, there is need for continuous oxygen supply in the cathode chamber for final reduction. Extraneous energy is therefore, provided to the cathode chamber, which offsets the net energy production by the MFC. A major parameter affecting the output of MFCs is the electron transfer rate. There are two ways by which electrons can get transferred from inside the body of the bacteria to the outside. One is by using electrogenic bacteria like *Geobacter* and *Shewanella* sp., which directly expel the electrons outside. Another way is by using some chemicals or natural mediators which help in releasing the electrons from inside the bacteria to outside. MFC technology is based on a biological process, and electricity production is naturally low as it is dependent on the activity of microorganisms. The use of mediators sometimes proves toxic to the microorganisms affecting their natural metabolism and sometimes, even destroy biofilm growth on the anode.

One of the recent approaches to increase the efficiency of MFCs is using novel cathode materials to improve its oxidation-reduction potential. Carbon paper with Pt catalyst is the most common cathode material used. However, Pt being expensive offsets the production cost of energy by the MFCs. Another concern is the separation of anode and cathode that requires a proton exchange membrane for efficient transfer of the protons. The membrane should have high mechanical strength and longevity to withstand various operational conditions. Nafion has been the most commonly used membrane, which meets all the criteria of a successful membrane (Oh and Logan 2006). However, its high cost necessitated looking for cheaper alternatives and some of the low-cost membranes, such as Ultrex CMI 7000, Hyflon and Zirfon (Harnisch et al. 2008; Ieropoulos et al. 2010; Pant et al. 2010) have been come up, but power generation capacity by MFCs has been found to be low when these membranes are used.

External resistance also plays an important role in delivering maximum power density in an MFC. According to Ohm's Law: $V = I\,R\text{ext}$. Thus, both voltage and current are affected by the external resistance. High and low external resistance can both lead to low power density. Li and Chen (2018) reported that due to the low redox potential created under low external resistance, the conditions for electron acceptance by microbes become unfavourable, whereas under high resistance, it is difficult for the microbes to transfer their electrons.

Another major challenge in MFC technology is the volume-surface ratio of MFC. To increase the performance, the surface area of electrodes is generally increased to enhance the growth of electrogenic microbes. However, increasing the surface area also leads to the increased volume of the MFC, which is required even in large-scale application. But when the compartment volume is increased, the performance of the MFC declines. Maximum power density of 4.3 W/m^2 using a reactor volume 30 mL was reported by Fan et al. (2012). Even a small increase in reactor volume reduces power density to lower than 1 W/m^2.

13.6 Economic Feasibility

Prospects of upscaling a new technology for the commercial application depend upon its economic feasibility. The cost of development of a typical MFC producing unit power density (1 W/m^2d) has been estimated to vary from $5 to 40 (Xu et al. 2018). However, the operating cost could be higher as it would include the costs of anolyte, catholyte, external energy and maintenance. The major expenditure in MFCs is due to the use of costly catalyst in the cathode and the proton exchange membrane. The commonly used cathode catalyst Pt/C increases the cost of MFC by more than 50% (Rozendal et al. 2008). Some metal-based catalysts have been found to be cheaper, but the use of microbial biocatalysts is now getting greater attention due to their ideal oxygen-reduction reaction catalysis at almost no external cost (Kaushik and Singh 2020). However, there are stability issues attributed to the complex microbial dynamics, which needs more intensive research to make the system resilient.

Ge and He (2016) studied the long-term operation of MFCs and found 60% of total material expenditure was incurred on the cation exchange membrane. The use of less expensive ceramic material for the proton transfer has come to light in the past years (Behera et al. 2010a, b; Behera and Ghangrekar 2011), but the type and thickness of the ceramic need to be more intensively explored for getting high output.

Electrode configuration and modification are the other major factors that increase the efficiency of MFCs along with their cost and may contribute 68–96% of the total MFC cost (Stoll et al. 2016). Singh and Kaushik (2019) have shown in their recent study, the effectiveness of electrode modification using novel niacin, in enhancing power production and COD removal by MFC. Developing cost-effective electrode modification techniques can go a long way in reducing the overall cost of MFCs.

The pH has a strong effect in maintaining the overall performance of MFC output (Kaushik and Chetal 2013) and regulation of pH in anode and cathode adds to recurring operational cost. Air sparging has been demonstrated to support high power generation, but costs for the external energy tend to increase the operation cost of MFCs. A new approach has come up in recent years, where a biological cathode (biocathode) is used in the MFCs and the microbes present in the biofilm of the cathode act as biocatalyst and help in the reduction process (Romo et al. 2019). These are some important approaches to cut down the cost of MFCs to make its electricity production capacity more sustainable. However, its variety of new applications as discussed in the present chapter gives it a strong competitive advantage over most of the conventional technologies.

13.7 Opportunities for Sustainable Environment Management

MFC technology is a new and sustainable approach that treats various types of wastewaters with a wide range of toxic pollutants and also converts the organic energy of the wastewater into electric energy directly. It has low carbon footprint and hence considered as environment friendly technique, based on biological and electrochemical principles. Two decades of intensive research has highly improved the power density generated by the MFCs that is now sustained for longer period of time. The MFC technology also makes possible the real-time monitoring of different parameters of wastewaters, and hence, offers opportunities for managing the quality of wastewater through prevention of toxicity and hazards.

Presently, the activated sludge process is the most commonly used treatment technique for wastewater, but it is energy intensive and involves high capital cost (Capodaglio and Olsson 2020). MFC technology, which is still in its developing stage, may not be so efficient for the treatment of wastewater at very large scale, but because it not only removes COD, but also treats other pollutants like toxic metals,

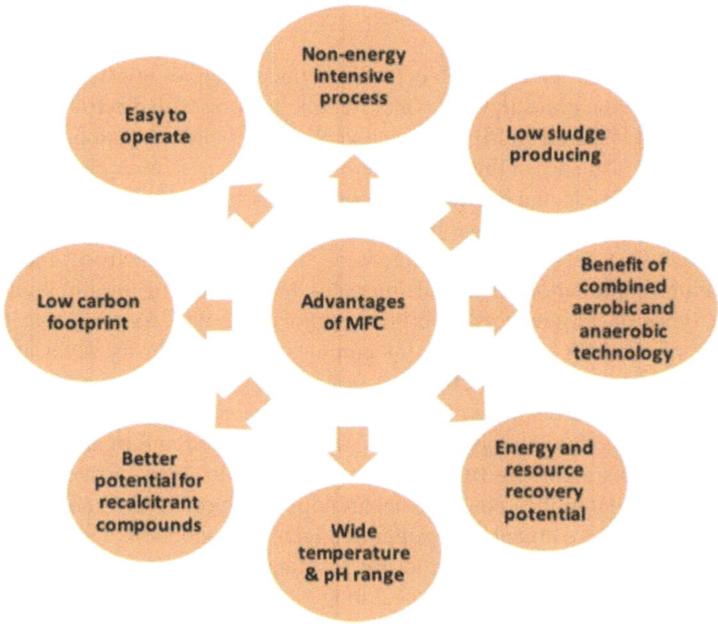

Fig. 13.3 Advantages of MFC over activated sludge process

dyes and xenobiotic compounds, for which very expensive tertiary treatment facilities are otherwise required, make it a promising technology of future. Figure 13.3 represents the advantages of MFC over the activated sludge process. Moreover, being a technology that recovers energy from the wastewaters, makes it a favoured choice. MFCs also have the advantage of producing very little sludge as compared to commonly used activated sludge system. One study reports production of around 0.4–0.8 g volatile suspended solids (VSS)/g COD in activated sludge system as compared to just 0.1 g-VSS/g COD by MFC (Zhang et al. 2013b). MFCs thus, reduce the extra cost input which is required for sludge treatment. MFCs have a better potential for the treatment of recalcitrant contaminants because of the presence of both anaerobic and aerobic conditions and hence, involve all those reactions in MFC that are not possible in strict anaerobic or aerobic treatment technologies (Li et al. 2014). MFCs have the added advantage of being capable of functioning in a wide temperature range, pH range and in varied climatic conditions, all of which support the development of this technology for industrial application in environmentally and economically sustained way.

13.8 Conclusion

MFCs have emerged as a novel bioelectrochemical technology with exceptionally wide range of applications in wastewater treatment along with electricity generation. Two decades of research on MFCs have revealed the challenges and prospects of this technology and also shown the areas where improvement is possible to enhance the efficiency of treatment and energy production at a low economic and environmental cost. The present study discussed how power density has been improved 1000-folds through intensive research involving MFC architecture, electrode material and modification, exploitation of useful microbes for developing biofilms and use of different types of wastewaters. A critical appraisal of the economic and technical feasibility of the MFC technology for commercial and industry level application brings forth the areas, where efficiency can be improved, and cost can be significantly reduced. Some distinctively superior features, such as low carbon footprints, little sludge production, energy recovery from wastewater and treatment of various types of pollutants including BOD, COD, heavy metals, dyes, nitrates, phosphates, antibiotics and xenobiotics put MFC technology as a preferred future option in the field of wastewater treatment with supported energy production. The MFC technology thus, shows tremendous potential for sustainable environment management by exploiting the biodegradative, bioremediating and bio-electrogenic characteristics of various microbes coupled with advanced electrochemical innovations for treatment of wastewaters so that the same after treatment could be reused for some designated purposes, and for the recovery of resources like metals or nutrients, and production of bioelectricity in an eco-friendly way.

References

Allen RM, Bennetto HP (1993) Microbial fuel-cells: electricity production from carbohydrates. Appl Biochem Biotechnol 39/40:27–40

Al-Mamun A, Lefebvre O, Baawain MS, Ng HY (2016) A sandwiched denitrifying biocathode in a microbial fuel cell for electricity generation and waste minimization. Int J Environ Sci Technol 13(4):1055–1064

Araneda I, Tapia NF, Lizama Allende K, Vargas IT (2018) Constructed wetland-microbial fuel cells for sustainable greywater treatment. Water 10(7):940

Aulenta F, Tocca L, Verdini R, Reale P, Majone M (2011) Dechlorination of trichloroethene in a continuous-flow bioelectrochemical reactor: effect of cathode potential on rate, selectivity, and electron transfer mechanisms. Environ Sci Technol 45(19):8444–8451

Behera M, Ghangrekar MM (2011) Electricity generation in low cost microbial fuel cell made up of earthenware of different thickness. Water Sci Technol 64(12):2468–2473

Behera M, Jana PS, Ghangrekar MM (2010a) Performance evaluation of low cost microbial fuel cell fabricated using earthen pot with biotic and abiotic cathode. Bioresour Technol 101 (4):1183–1189

Behera M, Jana PS, More TT, Ghangrekar MM (2010b) Rice mill wastewater treatment in microbial fuel cells fabricated using proton exchange membrane and earthen pot at different pH. Bioelectrochemistry 79(2):228–233

Bond DR, Lovley DR (2003) Electricity production by *Geobacter sulfurreducens* attached to electrodes. Appl Environ Microbiol 69(3):1548–1555

Capodaglio AG, Olsson G (2020) Energy issues in sustainable urban wastewater management: use, demand reduction and recovery in the urban water cycle. Sustainability 12(1):266
Chetal A, Kaushik A (2014) Influence of COD loading on electrochemical properties and COD removal by Microbial Fuel Cell. Indian J Environ Sci 18:45–49
Choi C, Cui Y (2012) Recovery of silver from wastewater coupled with power generation using a microbial fuel cell. Bioresour Technol 107:522–525
Choi C, Hu N (2013) The modeling of gold recovery from tetrachloroaurate wastewater using a microbial fuel cell. Bioresour Technol 133:589–598
Clauwaert P, Rabaey K, Aelterman P, De Schamphelaire L, Pham TH, Boeckx P, Boon N, Verstraete W (2007) Biological denitrification in microbial fuel cells. Environ Sci Technol 41 (9):3354–3360
Doherty L, Zhao Y, Zhao X, Hu Y, Hao X, Xu L, Liu R (2015) A review of a recently emerged technology: constructed wetland–microbial fuel cells. Water Res 85:38–45
Dong Y, Qu Y, He W, Du Y, Liu J, Han X, Feng Y (2015) A 90-liter stackable baffled microbial fuel cell for brewery wastewater treatment based on energy self-sufficient mode. Bioresour Technol 195:66–72
Du Z, Li H, Gu T (2007) A state of the art review on microbial fuel cells: a promising technology for wastewater treatment and bioenergy. Biotechnol Adv 25(5):464–482
Fan Y, Han SK, Liu H (2012) Improved performance of CEA microbial fuel cells with increased reactor size. Energy Environ Sci 5(8):8273–8280
Fang Z, Song HL, Cang N, Li XN (2015) Electricity production from Azo dye wastewater using a microbial fuel cell coupled constructed wetland operating under different operating conditions. Biosens Bioelectron 68:135–141
Feng Y, Yang Q, Wang X, Logan BE (2010) Treatment of carbon fiber brush anodes for improving power generation in air–cathode microbial fuel cells. J Power Sources 195(7):1841–1844
Feng Y, He W, Liu J, Wang X, Qu Y, Ren N (2014) A horizontal plug flow and stackable pilot microbial fuel cell for municipal wastewater treatment. Bioresour Technol 156:132–138
Fischer F, Bastian C, Happe M, Mabillard E, Schmidt N (2011) Microbial fuel cell enables phosphate recovery from digested sewage sludge as struvite. Bioresour Technol 102 (10):5824–5830
Fogg A, Gadhamshetty V, Franco D, Wilder J, Agapi S, Komisar S (2015) Can a microbial fuel cell resist the oxidation of tomato pomace? J Power Sources 279:781–790
Fu F, Wang Q (2011) Removal of heavy metal ions from wastewaters: a review. J Environ Manag 92(3):407–418
Gajda I, Greenman J, Ieropoulos I (2020) Microbial Fuel Cell stack performance enhancement through carbon veil anode modification with activated carbon powder. Appl Energy 262:114475
Ge Z, He Z (2016) Long-term performance of a 200 liter modularized microbial fuel cell system treating municipal wastewater: treatment, energy, and cost. Environ Sci Water Res Technol 2 (2):274–281
Gorby YA, Yanina S, McLean JS, Rosso KM, Moyles D, Dohnalkova A, Beveridge TJ, Chang IS, Kim BH, Kim KS, Culley DE (2006) Electrically conductive bacterial nanowires produced by *Shewanella oneidensis* strain MR-1 and other microorganisms. Proc Natl Acad Sci 103 (30):11358–11363
Gregory KB, Bond DR, Lovley DR (2004) Graphite electrodes as electron donors for anaerobic respiration. Environ Microbiol 6(6):596–604
Gude VG (2016) Wastewater treatment in microbial fuel cells—an overview. J Clean Prod 122:287–307
Guo K, Chen X, Freguia S, Donose BC (2013) Spontaneous modification of carbon surface with neutral red from its diazonium salts for bioelectrochemical systems. Biosens Bioelectron 47:184–189
Habermann W, Pommer EH (1991) Biological fuel cells with sulphide storage capacity. Appl Microbiol Biotechnol 35(1):128–133

Harnisch F, Schroder U, Scholz F (2008) The suitability of monopolar and bipolar ion exchange membranes as separators for biological fuel cells. Environ Sci Technol 42(5):1740–1746

Hirooka K, Yamamoto N, Li W, Nakai Y, Ichihashi O (2011) Power generation from animal wastewater by microbial fuel cell-microbial community analysis and behavior of nutrient. In: Proceedings of the 4th IWA-ASPIRE Conference & Exhibition, p 494

Hsueh CC, Chen BY, Yen CY (2009) Understanding effects of chemical structure on azo dye decolorization characteristics by *Aeromonas hydrophila*. J Hazard Mater 167(1–3):995–1001

Huang W, Chen J, Hu Y, Chen J, Sun J, Zhang L (2017) Enhanced simultaneous decolorization of azo dye and electricity generation in microbial fuel cell (MFC) with redox mediator modified anode. Int J Hydrog Energy 42(4):2349–2359

Ichihashi O, Hirooka K (2012) Removal and recovery of phosphorus as struvite from swine wastewater using microbial fuel cell. Bioresour Technol 114:303–307

Ichihashi O, Yamamoto N, Hirooka K (2012) Power generation by and microbial community structure in microbial fuel cell treating animal wastewater. J Jpn Soc Water Environ 35(1):19–26

Ieropoulos I, Greenman J, Melhuish C (2010) Improved energy output levels from small-scale microbial fuel cells. Bioelectrochemistry 78(1):44–50

Jadhav DA, Jain SC, Ghangrekar MM (2016) Cow's urine as a yellow gold for bioelectricity generation in low cost clayware microbial fuel cell. Energy 113:76–84

Jaishankar M, Tseten T, Anbalagan N, Mathew BB, Beeregowda KN (2014) Toxicity, mechanism and health effects of some heavy metals. Interdiscip Toxicol 7(2):60–72

Jana PS, Behera M, Ghangrekar MM (2010) Performance comparison of up-flow microbial fuel cells fabricated using proton exchange membrane and earthen cylinder. Int J Hydrog Energy 35 (11):5681–5686

Jiang D, Li B (2009) Novel electrode materials to enhance the bacterial adhesion and increase the power generation in microbial fuel cells (MFCs). Water Sci Technol 59:557–563

Kaushik A, Chetal A (2013) Power generation in microbial fuel cell fed with post methanation distillery effluent as a function of pH microenvironment. Bioresour Technol 147:77–83

Kaushik A, Singh A (2020) Metal removal and recovery using bioelectrochemical technology: the major determinants and opportunities for synchronic wastewater treatment and energy production. J Environ Manag 270:110826

Kaushik A, Jain S, Dawra J, Sahu R, Kaushik CP (2001) Heavy metal pollution of river Yamuna in the industrially developed state of Haryana. Indian J Environ Health 43:164–168

Ketep SF, Bergel A, Calmet A, Erable B (2014) Stainless steel foam increases the current produced by microbial bioanodes in bioelectrochemical systems. Energy Environ Sci 7(5):1633–1637

Khandelwal A, Vijay A, Dixit A, Chhabra M (2018) Microbial fuel cell powered by lipid extracted algae: a promising system for algal lipids and power generation. Bioresour Technol 247:520–527

Kim BH, Kim HJ, Hyun MS, Park DH (1999) Direct electrode reaction of Fe (III)-reducing bacterium, *Shewanella putrefaciens*. J Microbiol Biotechnol 9:127–131

Kim JR, Jung SH, Regan JM, Logan BE (2007) Electricity generation and microbial community analysis of alcohol powered microbial fuel cells. Bioresour Technol 98(13):2568–2577

Kim T, An J, Jang JK, Chang IS (2015) Coupling of anaerobic digester and microbial fuel cell for COD removal and ammonia recovery. Bioresour Technol 195:217–222

Kumar GG, Nahm KS, Sarathi VGS (2013) Recent advances and challenges in the anode architecture and their modifications for the applications of microbial fuel cells. Biosens Bioelectron 43:461–475

Lai B, Tang X, Li H, Du Z, Liu X, Zhang Q (2011) Power production enhancement with a polyaniline modified anode in microbial fuel cells. Biosens Bioelectron 28(1):373–377

Li S, Chen G (2018) Factors affecting the effectiveness of bioelectrochemical system applications: data synthesis and meta-analysis. Batteries 4(3):34

Li Z, Zhang X, Lei L (2008) Electricity production during the treatment of real electroplating wastewater containing Cr6+ using microbial fuel cell. Process Biochem 43(12):1352–1358

Li X, Zhu N, Wang Y, Li P, Wu P, Wu J (2013) Animal carcass wastewater treatment and bioelectricity generation in up-flow tubular microbial fuel cells: effects of HRT and non-precious metallic catalyst. Bioresour Technol 128:454–460
Li WW, Yu HQ, He Z (2014) Towards sustainable wastewater treatment by using microbial fuel cells-centered technologies. Energy Environ Sci 7(3):911–924
Liang P, Duan R, Jiang Y, Zhang X, Qiu Y, Huang X (2018) One-year operation of 1000-L modularized microbial fuel cell for municipal wastewater treatment. Water Res 141:1–8
Liao ZH, Sun JZ, Sun DZ, Si RW, Yong YC (2015) Enhancement of power production with tartaric acid doped polyaniline nanowire network modified anode in microbial fuel cells. Bioresour Technol 192:831–834
Lin CW, Wu CH, Chiu YH, Tsai SL (2014) Effects of different mediators on electricity generation and microbial structure of a toluene powered microbial fuel cell. Fuel 125:30–35
Liu WP, Yin XF (2017) Recovery of copper from copper slag using a microbial fuel cell and characterization of its electrogenesis. Int J Miner Metall Mater 24(6):621–626
Liu Y, Harnisch F, Fricke K, Schroder U, Climent V, Feliu JM (2010) The study of electrochemically active microbial biofilms on different carbon-based anode materials in microbial fuel cells. Biosens Bioelectron 25(9):2167–2171
Liu L, Yuan Y, Li FB, Feng CH (2011) In-situ Cr (VI) reduction with electrogenerated hydrogen peroxide driven by iron-reducing bacteria. Bioresour Technol 102(3):2468–2473
Liu SH, Lai CY, Ye JW, Lin CW (2018) Increasing removal of benzene from groundwater using stacked tubular air-cathode microbial fuel cells. J Clean Prod 194:78–84
Logan BE (2004) Peer reviewed: extracting hydrogen and electricity from renewable resources. Environ Sci Technol 38(9):160A–167A
Logan BE, Hamelers B, Rozendal R, Schroder U, Keller J, Freguia S, Aelterman P, Verstraete W, Rabaey K (2006) Microbial fuel cells: methodology and technology. Environ Sci Technol 40 (17):5181–5192
Logan B, Cheng S, Watson V, Estadt G (2007) Graphite fiber brush anodes for increased power production in air-cathode microbial fuel cells. Environ Sci Technol 41(9):3341–3346
Lowy DA, Tender LM (2008) Harvesting energy from the marine sediment-water interface III. Kinetic activity of quinone- and antimony-based anode materials. J Power Sources 185:70–75
Lv Y, Wang Y, Ren Y, Li X, Wang X, Li J (2020) Effect of anaerobic sludge on the bioelectricity generation enhancement of bufferless single-chamber microbial fuel cells. Bioelectrochemistry 131:107387
Mathuriya AS, Sharma VN (2010) Treatment of brewery wastewater and production of electricity through microbial fuel cell technology. Int J Biotechnol Biochem 6(1):71–80
Mathuriya AS, Yakhmi JV (2014) Microbial fuel cells to recover heavy metals. Environ Chem Lett 12(4):483–494
Mehdinia A, Ziaei E, Jabbari A (2014) Facile microwave-assisted synthesized reduced graphene oxide/tin oxide nanocomposite and using as anode material of microbial fuel cell to improve power generation. Int J Hydrog Energy 39(20):10724e30
Mohan SV, Mohanakrishna G, Sarma PN (2010a) Composite vegetable waste as renewable resource for bioelectricity generation through non-catalyzed open-air cathode microbial fuel cell. Bioresour Technol 101(3):970–976
Mohan SV, Mohanakrishna G, Velvizhi G, Babu VL, Sarma PN (2010b) Bio-catalyzed electrochemical treatment of real field dairy wastewater with simultaneous power generation. Biochem Eng J 51(1–2):32–39
Mohan SV, Velvizhi G, Modestra JA, Srikanth S (2014) Microbial fuel cell: critical factors regulating bio-catalyzed electrochemical process and recent advancements. Renew Sust Energ Rev 40:779–797
Morris JM, Jin S, Crimi B, Pruden A (2009) Microbial fuel cell in enhancing anaerobic biodegradation of diesel. Chem Eng J 146(2):161–167
Mu Y, Rabaey K, Rozendal RA, Yuan Z, Keller J (2009) Decolorization of azo dyes in bioelectrochemical systems. Environ Sci Technol 43(13):5137–5143

Nam JY, Kim HW, Lim KH, Shin HS (2010) Effects of organic loading rates on the continuous electricity generation from fermented wastewater using a single-chamber microbial fuel cell. Bioresour Technol 101(1):S33–S37

Niessen J, Harnisch F, Rosenbaum M, Schroder U, Scholz F (2006) Heat treated soil as convenient and versatile source of bacterial communities for microbial electricity generation. Electrochem Commun 8(5):869–873

Oh SE, Logan BE (2006) Proton exchange membrane and electrode surface areas as factors that affect power generation in microbial fuel cells. Appl Microbiol Biotechnol 70(2):162–169

Oh S, Min B, Logan BE (2004) Cathode performance as a factor in electricity generation in microbial fuel cells. Environ Sci Technol 38(18):4900–4904

Pandey P, Shinde VN, Deopurkar RL, Kale SP, Patil SA, Pant D (2016) Recent advances in the use of different substrates in microbial fuel cells toward wastewater treatment and simultaneous energy recovery. Appl Energy 168:706–723

Pant D, Van Bogaert G, De Smet M, Diels L, Vanbroekhoven K (2010) Use of novel permeable membrane and air cathodes in acetate microbial fuel cells. Electrochim Acta 55(26):7710–7716

Parkash A (2016) Utilization of distillery wastewater for electricity generation using microbial fuel cell. J Appl Emerg Sci 6(2):79–86

Parot S, Délia ML, Bergel A (2008) Acetate to enhance electrochemical activity of biofilms from garden compost. Electrochim Acta 53(6):2737–2742

Pocaznoi D, Calmet A, Etcheverry L, Erable B, Bergel A (2012) Stainless steel is a promising electrode material for anodes of microbial fuel cells. Energy Environ Sci 5(11):9645–9652

Rabaey K, Verstraete W (2005) Microbial fuel cells: novel biotechnology for energy generation. Trends Biotechnol 23(6):291–298

Rabaey K, Boon N, Siciliano SD, Verhaege M, Verstraete W (2004) Biofuel cells select for microbial consortia that self-mediate electron transfer. Appl Environ Microbiol 70 (9):5373–5382

Rezaei F, Richard TL, Brennan RA, Logan BE (2007) Substrate-enhanced microbial fuel cells for improved remote power generation from sediment-based systems. Environ Sci Technol 41 (11):4053–4058

Rittmann BE (2008) Opportunities for renewable bioenergy using microorganisms. Biotechnol Bioeng 100(2):203–212

Romo DMR, Gutierrez NHH, Pazos JOR, Figueroa LVP, Ordonez LAO (2019) Bacterial diversity in the Cr(VI) reducing biocathode of a Microbial Fuel Cell with salt bridge. Rev Argent Microbiol 51:110–118

Rozendal RA, Hamelers HV, Rabaey K, Keller J, Buisman CJ (2008) Towards practical implementation of bioelectrochemical wastewater treatment. Trends Biotechnol 26(8):450–459

Sangeetha T, Muthukumar M (2011) Catholyte performance as an influencing factor on electricity production in a dual-chambered microbial fuel cell employing food processing wastewater. Energy Sources Part A 33(16):1514–1522

Scott K, Murano C (2007) Microbial fuel cells utilising carbohydrates. J Chem Technol Biotechnol 82(1):92–100

Sharma Y, Li B (2010) Optimizing energy harvest in wastewater treatment by combining anaerobic hydrogen producing biofermentor (HPB) and microbial fuel cell (MFC). Int J Hydrog Energy 35 (8):3789–3797

Singh A, Kaushik A (2019) Anode modification for increased power generation and COD removal in Microbial Fuel Cell. Asian J Microbiol Biotechnol Environ Sci 21:181–186

Singh A, Kaushik A (2021) Removal of Cd and Ni with enhanced energy generation using biocathode microbial fuel cell: insights from molecular characterization of biofilm communities. J Clean Prod 315:127940

Stoll ZA, Ma Z, Trivedi CB, Spear JR, Xu P (2016) Sacrificing power for more cost-effective treatment: a techno-economic approach for engineering microbial fuel cells. Chemosphere 161:10–18

Sun J, Hu YY, Bi Z, Cao YQ (2009) Simultaneous decolorization of azo dye and bioelectricity generation using a microfiltration membrane air-cathode single-chamber microbial fuel cell. Bioresour Technol 100(13):3185–3192

Sun J, Li W, Li Y, Hu Y, Zhang Y (2013) Redox mediator enhanced simultaneous decolorization of azo dye and bioelectricity generation in air-cathode microbial fuel cell. Bioresour Technol 142:407–414

Tang X, Li H, Du Z, Ng HY (2014) Spontaneous modification of graphite anode by anthraquinone-2-sulfonic acid for microbial fuel cells. Bioresour Technol 164:184–188

Tartakovsky B, Guiot SR (2006) A comparison of air and hydrogen peroxide oxygenated microbial fuel cell reactors. Biotechnol Prog 22(1):241–246

Thrash JC, Van Trump JI, Weber KA, Miller E, Achenbach LA, Coates JD (2007) Electrochemical stimulation of microbial perchlorate reduction. Environ Sci Technol 41(5):1740–1746

Uria N, Ferrera I, Mas J (2017) Electrochemical performance and microbial community profiles in microbial fuel cells in relation to electron transfer mechanisms. BMC Microbiol 17(1):1–12

Varanasi JL, Sinha P, Das D (2017) Maximizing power generation from dark fermentation effluents in microbial fuel cell by selective enrichment of exoelectrogens and optimization of anodic operational parameters. Biotechnol Lett 39(5):721–730

Wang Z, Lim B, Choi C (2011) Removal of Hg2+ as an electron acceptor coupled with power generation using a microbial fuel cell. Bioresour Technol 102(10):6304–6307

Wang YP, Zhang HL, Li WW, Liu XW, Sheng GP, Yu HQ (2014) Improving electricity generation and substrate removal of a MFC–SBR system through optimization of COD loading distribution. Biochem Eng J 85:15–20

Wang QQ, Wu XY, Yu YY, Sun DZ, Jia HH, Yong YC (2017) Facile in-situ fabrication of graphene/riboflavin electrode for microbial fuel cells. Electrochim Acta 232:439–444

Wu S, Li H, Zhou X, Liang P, Zhang X, Jiang Y, Huang X (2016) A novel pilot-scale stacked microbial fuel cell for efficient electricity generation and wastewater treatment. Water Res 98:396–403

Xing D, Zuo Y, Cheng S, Regan JM, Logan BE (2008) Electricity generation by *Rhodopseudomonas palustris* DX-1. Environ Sci Technol 42(11):4146–4151

Xu L, Zhao Y, Doherty L, Hu Y, Hao X (2016) The integrated processes for wastewater treatment based on the principle of microbial fuel cells: a review. Crit Rev Environ Sci Technol 46 (1):60–91

Xu F, Yuan Q, Zhou LL, Zhu YJ, Li YM, Du YD, Wang Q, Kong Q (2018) Economic benefit analysis of typical microbial fuel cells based on a cost–benefit analysis model. Desalin Water Treat 135:59–93

Yi H, Nevin KP, Kim BC, Franks AE, Klimes A, Tender LM, Lovley DR (2009) Selection of a variant of *Geobacter sulfurreducens* with enhanced capacity for current production in microbial fuel cells. Biosens Bioelectron 24:3498–3503

You SJ, Zhao QL, Jiang JQ, Zhang JN, Zhao SQ (2006) Sustainable approach for leachate treatment: electricity generation in microbial fuel cell. J Environ Sci Health A 41 (12):2721–2734

You J, Greenman J, Melhuish C, Ieropoulos I (2016) Electricity generation and struvite recovery from human urine using microbial fuel cells. J Chem Technol Biotechnol 91(3):647–654

Zhang Y, Wu Z, Wang Q, Yang L, Li M, Lin ZQ, Banuelos G (2013a) Removal of selenite from wastewater using microbial fuel cells inoculated with *Shewanella oneidensis* MR-1. In: Bañuelos GS, Lin ZQ, Yin X (eds) Selenium in the environment and human health. Taylor & Francis, London, pp 210–211

Zhang F, Ge Z, Grimaud J, Hurst J, He Z (2013b) Long-term performance of liter-scale microbial fuel cells treating primary effluent installed in a municipal wastewater treatment facility. Environ Sci Technol 47(9):4941–4948

Zhang B, Tian C, Liu Y, Hao L, Liu Y, Feng C, Liu Y, Wang Z (2015) Simultaneous microbial and electrochemical reductions of vanadium (V) with bioelectricity generation in microbial fuel cells. Bioresour Technol 179:91–97

Zhao H, Kong CH (2018) Enhanced removal of p-nitrophenol in a microbial fuel cell after long-term operation and the catabolic versatility of its microbial community. Chem Eng J 339:424–431

Zinadini S, Zinatizadeh AA, Rahimi M, Vatanpour V, Rahimi Z (2017) High power generation and COD removal in a microbial fuel cell operated by a novel sulfonated PES/PES blend proton exchange membrane. Energy 125:427–438

Carbon Footprinting: A Study of Plywood Industry in District Yamunanagar (India) 14

Ritu Rani and Tejinder Sharma

Abstract

In wake of intensifying climate change, the issue of estimating carbon footprinting is taking the centre stage and the stakeholders are getting vigilant towards its estimation. While its being done for the large industries, there is less research on the MSME sector. There is multifaceted impact on the environment by the plywood industry.

This study identifies the sources/activities of carbon emission and estimates carbon footprint in the plywood cluster of Yamunanagar district in Haryana state of northern India. The study uses IPCC methodology and is based on estimating carbon footprints across various stages of plywood manufacturing by calculating the total carbon dioxide emitted using standard emission factors published by various government agencies, and intergovernmental agencies. Carbon emissions emerging from direct deforestation (scope 1), timber processing (scope 2) and the tertiary manufacturing and distribution support services used by the industry (scope 3) have been estimated. It is found that scope 2 activities contribute a large share in carbon footprinting.

The study recommends adequate data measurement mechanisms at the firm level so that the firms could estimate their emissions and hence work for their reduction.

Keywords

Carbon footprinting · Carbon emission · GHG · Carbon trading

R. Rani (✉)
Haryana Education Department, Government of Haryana, Panchkula, Haryana, India

T. Sharma
Department of Commerce, Kurukshetra University, Kurukshetra, Haryana, India
e-mail: tsharma@kuk.ac.in

A. Kaushik et al. (eds.), *Climate Resilience and Environmental Sustainability Approaches*, https://doi.org/10.1007/978-981-16-0902-2_14

14.1 Introduction

In wake of intensifying climate change, the issue of estimating carbon footprinting is taking the centre stage and the stakeholders are getting vigilant towards its estimation. While its being done for the large industries, there is a need for conducting research on the industries that use environmentally dependent inputs and are into the MSME sector primarily. In view of increasing economic prosperity and boost by the government of India to the housing sector, there is big surge in the demand for various inputs used for construction. Each of these inputs comes at the cost of the environment. In India, the use of timber is very popular and presently the plywood is emerging as a popular form of construction input. Yamunanagar district, located on the banks of river Yamuna in the north Indian state of Haryana has emerged as a major cluster of plywood manufacturing and caters to a big chunk of the demand for plywood in the country. Plywood industry has a two-pronged impact on the environment. Firstly, using timber as a manufacturing input, it is causing a direct depletion of the forests. Secondly, the manufacturing process is traditional and is not very efficient, leading to loss of output as well higher consumption of energy. The study aims at estimating the carbon emissions emerging form the plywood units located in Yamuna-nagar district of Haryana. The study has been done with the help of primary data and the standard methodologies used to estimate the carbon emissions of the industries.

14.1.1 About Yamunanagar

Yamunanagar is well known for its industries and lies on the western bank of river Yamuna which flow about 11 km to the eastern boundary of the district. There are a large number of industries in Yamunanagar district such as steel industry, plywood industry, sugar mills, paper mills, utensil manufacturing, etc. The Yamunanagar district has an abundance of forest cover (Table 14.1) and there are plenty of poplar and eucalyptus tree plantations, due to which the plywood industry is a booming

Table 14.1 Forest cover in district Yamunanagar

Forest type		Area (ha)
Reserve forest		6944.29
Protected forest	Compact	8420.80
		1693.80
	Rail	173.23
	Canal	2747.86
	Bund	57.39
Protected forest (total)		13,093.08
Unclassed		348.72
Others		1411.13
Total		21,797.22

Source: Yamunanagar.nic.in

industry here. A lot of revenue of the state of Haryana is contributed by these plywood manufacturers of Yamunanagar and the top-class quality of timber and wood manufactured here is exported to different parts of the country. There are a total of 3250 industrial units, out of which 1159 are registered. Together, these units provide employment to about 17,500 persons of different skill sets.

14.1.2 Carbon Emission in Plywood Industry

As per IPCC methodology, the total CO_2 emitted from an organization can be divided into three types of emissions: scope 1, scope 2, scope 3 (GHG-protocol 2011). Table 14.2 shows the constituents of the carbon dioxide emissions in different scopes (categories). Direct emissions include those activities which are controlled and owned by the company and indirect emissions include those activities which are controlled and owned by another company.

14.2 Review of Literature

Carbon footprinting is an emerging field of measuring the environmental impact of the industries. Barnett et al. (2013) observed that the predecessor to carbon footprinting is ecological footprinting. The phrase carbon footprinting did not appear in literature until later, as described by Wiedmann and Minx (2008). Life cycle analysis (LCA) is a predecessor to product carbon footprinting used to compare products. LCA has always been used as a method for comparing the set of products based on benchmarks, such as cost of manufacture, energy consumption, water consumption, etc. However, it was not until carbon footprinting first became popular that greenhouse gas emission was included in such analysis. In 1988, the Intergovernmental Panel on Climate Change (IPCC) was formed. Kyoto Protocol is another major milestone in carbon footprinting (United Nations 1998). There are several standards to define carbon footprinting. The first standard to define carbon footprinting was the Greenhouse Gas Protocol (WRI 2000; WBCSD 2000). This formed the basis for most other carbon footprinting methods. In the UK, legislation has been passed to reduce carbon emission, in the form of carbon reduction commitments (CRCs) the international standard organization (ISO 2009) based on the greenhouse gas protocol.

There are various methods of carbon footprinting used by the companies, (Andrews 2009a, b) a majority of them fall into two categories: corporate and

Table 14.2 Carbon dioxide emission categories

Emission	Scopes	Constituents
Direct	Scope 1	Emissions that are controlled and owned by the company
Indirect	Scope 2	Emission from the generation of electricity, heat and oil
	Scope 3	All other indirect emissions not included in scope 2

product and it depends upon the objectives of the firm and the purpose which they are supposed to serve e.g. the purpose of carbon footprinting exercise done by Sony Ericsson was to generate awareness internally to reduce the carbon footprints. Rigot (2013) analyzed two real-life industrial cases, on data from 2011 to 2012. The first cases concern inbound flows to the UK and the second case concerns outbound flows from the UK. The research presented in this paper has attempted to map the end-to-end carbon footprint for given supply chain by two cases companies in two different industrial sectors. The interesting aspect of this study is the variety of transport modes used in the supply chain including road transport, containership, Ro-Ro vessel and rail transport. The research covers only the emission from transport and warehousing operation. CO_2 equitable for the emission of carbon dioxide greenhouse gases are used for the expression of carbon foot printing provided by Wieldmann and Minx (2008). The result from both the cases shows that maritime leg remains the main part of the end-to-end supply chain's carbon emission, it is the most efficient transport mode, it is still the important flow for concerning the reduction of emission as nearly 90% of international trade uses maritime transport and result of case 2 shows that local sourcing does not always have a lower emission benefit when different transport modes are used. Data availability and reliability play an important role in modelling process and results may be considered with regard to the restriction and approximately used. Srivastav (2013) estimated carbon emissions for cities. The 5 year study showed rapidly increasing emissions rising from 1.07 metric tonnes per capita in 2007 to 1.43 metric tonne per capita in 2011. Transport and residential sectors are identified as the key drivers from the 5 year emission profile calculation for city, it was established that compared to cities in developed countries and emission from the city are very low. The aim of this thesis was to identify the opportunities and barriers to the use GHG inventories, Although it is possible to conduct a GHG inventory within the existing data and governance scenario in mid-scale Indian cities. The significant gap in emission profiles between developed and developing nations cannot just be attributed to the differences in the city infrastructure. The intent of this research was to increase awareness among Indian local authorities of carbon reduction in urban management.

In another study, carbon emissions from four ports located in Antwerp, Gothenburg, Los, Angeles and Valencia to determine different emission reduction action (Sarbring 2014). The study includes the 22 most extensive projects out of 160 construction and maintenance related projects. The total CO_2 emission for the 22 projects is approximately 1340 tonnes and in mass approximately 1060 tonnes (excluding fuels). Two strategies dematerialization (increase in service exchange for a product) and trans-materialization (substitution of the current product with other product) strategies to reduce the emission in construction and maintenance operations Hertwich and Peter (2009). In another study conducted on 72 countries and 15 aggregated region an analysis of carbon emissions from 8 categories such as construction, food, mobility shelter, manufactured products, clothing service and trade was made and through multiregional input and output (MRIO) modes The research finds that carbon footprinting of poor countries is higher than in rich countries. Lee and Cheong (2011) focused the study on the identification of carbon emission and measurement practices in SCM of the automobile industry. For the purpose, Hyundai motors and its first-level suppliers were explored using a case study approach. The study concludes that Hyundai Co. proactively conducted a self and suppliers evaluation along with regular on-site monitoring and investigation of 15 participating suppliers self-developed environmental criteria and achieved positive results by visualizing 16,187 CO_2 reduction in just 2 years period. The empirical study suggests that quantitative identification and measurement of CO_2 emission in the first to implement the carbon management practices in SCM .the study also puts forward a CF measurement framework as well as measurement criteria and scope within the supply chain. Muller et al. (2013) based their study on two real-life industrial cases, with activity data from 2011 to 1012. The first case concerned about inbound flows to the UK and the second case concerned outbound flows from the UK. The study focused mainly on the transport and warehousing operations these based on direct CO_2 emission factors. Results of the first case revealed that end-to-end logistic-related carbon emissions can be reduced by 16–21% through direct delivery to the UK as opposed to indirect channels. Results of the second case demonstrated that in the case of distant overseas countries as origin or destinations, the maritime leg remained the main part of the end-to-end supply chains' carbon emission and local sourcing does not always have lower emission benefit when different transport modes are used. Researchers have also examined various corporate approaches dealings with carbon management accounting which influence the measure of the total carbon footprint of a corporation and the representation of carbon performance (Gibassier and Schaltegger 2015). Through the examination of the carbon management accounting of a company Donone, researchers identify three major types of carbon management accounting for different purpose: organization carbon accounting, product carbon accounting and project carbon accounting. Donone's carbon management accounting system is composed of two different accounting approaches; product carbon accounting and the organization, site-based carbon accounting. An analysis of 50 available online carbon calculators was done and it was concluded generally carbon calculators provide no information on carbon footprint assessment approaches or provide limited information and use different methodological approaches for making the carbon footprint (Filimonau 2012). Padgett et al. (2008) studied carbon dioxide calculators, which are very important in the calculation of carbon footprint. There are a large number of carbon calculator are available on the internet. These calculators are provided by different organization and agencies such as govt. organizations, non-govt. organisations, and private companies. These calculators calculate carbon footprint of an individual based on different activities like electricity consumption, transportation, etc. carbon footprinting calculated by each calculator is based on two or three main categories such as household activities, transportation, etc. these calculators use similar almost techniques for calculating carbon emission and because of similar technique and similar data. McKinnon (2010) examined in detail the issues of product-level carbon labelling and auditing by discussing the various practical problems associated with both the issues through informal discussions with senior managers and available literature. Three basis findings emerged in the study, first the underestimation of the scale and cost of carbon auditing. Second the non-justified investment of time, efforts and cost associated. And the third finding is the low contribution of product level-carbon auditing to carbon management of the supply chain. The paper concluded that product-level carbon auditing and labelling as a wasteful distortion given the security of environmental crisis facing own planet (Finkbeiner 2009). But there are some challenges like scope of emission, life cycle stages, system boundaries, offsetting and availability of data, etc. are raised to deal with carbon footprinting. The need for carbon calculating online courses to reduce household consumption has also been professed Crompton et al. (2002) and Dawe et al. (2004) described the methodology and outcomes of a sustainability audit carried out at Holme lacy college, Herefordshire, the UK using "triple bottom line" assessment which comprised of number of key steps namely viz. A scoping review, an environmental impact assessment results analysis showed that the college has a number of unsustainable features revealed the weak ecological performance of home lacy college significantly arising on account of waste and good, along with the high transport and energy consumption. Li et al. (2019) described that national total emission includes a significant proportion of household carbon emission. This paper investigates household carbon emission using a unique nationwide survey data and study the impact of various subjective measures of social awareness on household carbon emission. Abeydeera et al. (2019) investigated that greenhouse gases such as carbon dioxide, sulphur dioxide and nitrogen dioxide were recognized as a prime cause of global warming. CO_2 is considered as a prominent gas among various green house gases which is prime cause of climate change and motivated researchers to explore carbon reduction strategies. A comprehensive study on carbon emission is not available. This study reviewed the global literature on carbon emission from 1981 to 2019 and the findings of this research revealed the increasing trend of carbon emission related publications mostly from China, the USA and England. Dunnan Liu (2019) this investigation is based on world input output database from 1995 to 2009. It investigated the GHG emission growth using structure decomposition analysis. Factors those contribute to GHG emission vary in different countries. China and India are the key contributors to CO_2 and CH_4.

14.3 The Research Problem

In wake of the background and the review of literature described above, the research problem is identified as—*estimating the carbon footprint of plywood industry in district Yamunanagar*. Since the plywood industry affects the environment very significantly, there is a need to estimate the carbon footprints so that the reactive strategies for environmental conservation can be designed and implemented by the firms in this industry.

Research Methodology

For calculating the carbon footprint, the total carbon dioxide emitted from the activities of the organization is calculated. For this purpose, data is collected with the help of a structured questionnaire from 15 firms engaged in plywood manufacturing. There are a total of 34 registered plywood manufacturing firms in district Yamunanagar and data was collected from the firms who were willing to participate in the survey and accounted for a relatively major portion of their market share. The questionnaire contained the questions eliciting data on each of the activities lying in scopes 1, 2 and 3 so that the carbon emission can be estimated from each of these activities. Since the data was collected personally, the relevant observations of the production processes, their inventories, etc. were also made to verify the information supplied by the respondents. Once data is collected for each activity then emission factors for each activity were traced from the published sources and these emission factors were used to convert activity data to emission values. Emission factors are published by various government agencies, and inter-governmental agencies. Desired mathematical calculations were done for estimating the carbon emissions.

14.4 Results and Calculations

14.4.1 Scope 1

One category is included in scope 1: purchased wood. There are two options of purchasing wood: purchase directly from the farmer or may be purchased from the broker. There are five categories of product are produced by respondent units and average production capacity is given. Total wood purchased is calculated as under (Table 14.3).

14.4.2 Scope 2

Transportation: Factory It includes the transportation used from forest to factory. Farmers grow poplar trees and supply it to wood factories. They use various vehicles for the supply of wood to plywood factories like tractor, truck, etc. tractors and trucks use diesel as fuel. Table 14.4 shows the carbon emissions emanating from various means of transport used in the plywood industry.

Transportation: Market It includes the transportation used between factory and market. The factory uses transportation for the supply of the finished product (plywood) to various markets. The data related to this category is calculated by using the average monthly supply of finished goods (plywood) to market.

Table 14.3 Calculation of wood purchased

1 kg wood of wood is holding about 1.83 kg of CO_2
And 35,216,928 kg wood is holding about
35,216,928 kg × 1.83 kg of CO_2 = 64,446,978.24 kg CO_2
64,446,978.24 kg CO_2 × 34 = 2,191,197,260.16 kg CO_2

Table 14.4 Carbon emission from transportation (a)

Transportation medium	Percentage	Distance travelled	Emission factor	Carbon footprint
Tractor trolley	75.66 %	22,507.2	233 g CO_2/km	5,942,056.87 g CO_2
Mini truck	41.67%	14,169.13	329.6 g CO_2/km	4,670,145.25 g CO_2
Large pick-up truck city	66.67%	22,669.93	373.5 g CO_2/km	8,467,218.88 g CO_2
Diesel truck heavy	41.67%	14,169.13	870 g CO_2/km	12,327,143.1 g CO_2
				31,406,564.10 g CO_2 Or 31,406.56 kg CO_2
Total carbon footprint = 31,406.56 × 34 = 1,067,823.04 kg CO_2				

Table 14.5 Carbon emission from transportation (b)

Transportation medium	Percentage	Distance travelled	Emission factors	Carbon footprint
Tractor trolley	58.33	5862.09	233 g CO_2/km	1,365,866.97 g CO_2
Mini truck	41.67	4187.78	329.6 g CO_2/km	1,380,292.28 g CO_2
Large pick-up truck city	83.33	8374.57	373.5 g CO_2/km	3,127,898.16 g CO_2
Diesel truck heavy	100	10,049.88	870 g CO_2/km	8,743,386.9 g CO_2
Total	283.33	28,474.32 km		14,617,444.31 g CO_2 Or 14,617.45 kg CO_2
Carbon footprint = 14,617.45 kg CO_2 × 34 = 496,993.3 kg CO_2				

Source: Researchers' computations

The average distance plywood manufacturing unit and market where they supply finished product is 139.58 km and they supply plywood to market approx. 17 times in month. Seventy-five percent of the selected plywood manufacturing units uses tractor trolley as their medium of transportation, 41.67% manufacturing unit use mini truck, 66.67% manufacturing units use large pick-up truck city and 41.6% manufacturing units use diesel heavy truck for the supply of raw material. For calculation carbon emission we have to estimate we have to use proportionate kilometres of distance travelled. These calculations are shown in Table 14.5 wherein the carbon footprint from different means of transport has been calculated.

Transportation: Warehouse Sometimes plywood factories use warehouse for the storage of plywood. And the transportation used between factory and warehouse is also a cause of carbon dioxide emission. The data is collected by questionnaire and according to the response given by plywood manufacturing units that there is not any single manufacturing unit that is using the warehouse for storing the finished product. So we allocate carbon emission from transportation in equal proportion for market and warehouse.

Transportation: Market = 496,993.3 kg $CO_2/2$ = 248,496.65 kg CO_2
Transportation: Warehouse = 496,993.3 kg $CO_2/2$ = 248,496.65 kg CO_2

Electricity: Factory Electricity also contributes to CO_2 emission. The electricity used in plywood factories for the production of plywood. Electricity is used in various stages of the production process like in cutting veneers, pressing plywood, plywood drying, etc. And electricity bills are used to calculate the amount of electricity used in the production process.

Electricity: Warehouse If warehouse is used by factories for storing plywood then electricity used in warehouse is also included in the calculation of carbon footprint. The electricity bills of the warehouse are used to calculate the units of electricity used in the warehouse.

Selected units of the plywood industry are not using any warehouse for storing their finished product (plywood). They store their manufactured plywood sheets at their own factory location until it is supplied to market or we can say that they are using their factory as warehouse for storing their finished goods therefore we allocate electricity consumption 50% to factory and 50% to warehouse.

The emission associated with the electricity consumption is as follows:
53,200 kW * 0.991 kg CO_2/kWh = 52,721.2 kg CO_2-eq
Total electricity consumption in a year by plywood industry = 52,721.2 * 34 = 1,792,520.8 kg CO_2
Factory: electricity = 896,260.4 kg CO_2
Warehouse: electricity = 896,260.4 kg CO_2

14.4.3 Scope 3

There are a large number of categories that can be included in scope 3 but according to our requirement, we include only three categories: staff commuting, waste treatment and water supply. Scope 3 data is not easily available; we have to estimate it on the basis of other data.

Staff Commuting For exact calculation of carbon footprint, we have to use the total amount of distance travelled by employees and workers by each transportation source. Some plywood factories provide residence to their workers near the factory and in the rest of factories the workers are from the local area or within 15 km.

Table 14.6 Carbon emission from the transportation of employees

Transportation medium	Percentage	Distance travelled (kms)	Emission factors	Carbon footprint
Bus	7.42	13,758.68	1035 g CO_2/km	14,240,233.8 g CO_2
Two wheeler	7.42	13,758.68	93 g CO_2/km	1,279,557.24 g CO_2
Auto rikshaw	10.04	18,616.87	233 g CO_2/km	4,337,730.71 g CO_2
				19,857,521.75 g CO_2 = 19,857.52 kg CO_2
Total carbon footprint = 19,857.52 kg CO_2 × 34 = 675,155.68 kg CO_2				

Source: Researchers' computations

There is no accurate data available about distance travelled by employees and workers, we had estimated that approximately. In this category, we include the total distance travelled by workers of the plywood manufacturing units. There are total 1873 workers are employed in selected plywood manufacturing units and the average distance between their residence and plywood unit in which they are employed is 8.25 km. There are many transportation categories used by workers out of which only two transportation categories are related to carbon dioxide emission. 14.84% workers use bus/two wheeler and 10.04% workers use auto rickshaw. Emission factors of bus and two wheeler are different so we assume that out of 14.84% workers 50% use bus and 50% use auto rickshaw. The calculations of carbon footprints by the means of transportation used by the employees for coming to the work are shown in the following Table 14.6.

Waste Treatment Gathering data about waste was not very tough job because there is no any type of waste in plywood process which is harmful and contribute in carbon footprinting. Only eco-friendly waste (wood) is left like the bark of wood, wood chips, etc. some factories burn wood residuals for energy. It includes wood residuals from wood trimming and wood dust. Carbon dioxide emitted from these wood residuals is included in GHG emission results carbon dioxide released by burning these residuals. The data related to wood residuals are collected from the manufacturer. Carbon released by burning wood residuals is absorbed by plants during photosynthesis, it is called carbon neutral and sometimes this is not included in GHG emission.

Water Supply There is no need for excess water in plywood-making process. Only a little amount of water is used for the soaking of wood which is rarely changed in a year.

Carbon Footprint

Table 14.7 above shows that 2,191,197.26 tonne CO_2 is stored in wood, and according to this calculation, 4032.49 tonnes CO_2 is emitted by plywood industry. 38.8% of total emission is emitted by transportation used in the supply of raw

Table 14.7 Carbon Footprint of plywood industry in Yamunanagar

Scope	Sources	Emission per kg CO_2eq/month	Emission per kg CO_2eq/year	Emission per tonne CO_2eq/year	Percentage of total
Scope 1	Direct emission				
	Purchased wood	(182,599,771.68)	(2,191,197,260.16)	(2,191,197.26)	
Scope 2	Indirect emission				
	Transportation: Factory	88,985.25	1,067,823.04	1067.82	26.48
	Transportation: Market	20,708.05	248,496.65	48.50	6.16
	Transportation: Warehouse	20,708.05	248,496.65	248.50	6.16
	Electricity: Factory	74,688.31	896,259.72	896.26	22.23
	Electricity: Warehouse	74,688.31	896,259.72	896.26	22.23
Scope 3	Other indirect emission				
	Staff commuting	56,262.97	675,155.68	675.15	16.74
	Waste treatment	0	0	0	0
	Water supply	0	0	0	0
Total		336,040.93	4,032,491.46	4032.49	100

Source: Researchers' computations

material and supply of finished product to market, 44.46% of total emission is emitted by electricity and remaining 16.74% emission is because of staff commuting.

Scope 2 emissions are the largest source of emission. 497 tonnes CO_2/year is emitted by transportation used between market and plywood manufacturing unit which is allocated equally to market and warehouse because the warehouse is not used by any plywood manufacturing unit. Electricity is also one of the factors which contribute the largest share in carbon emission. Carbon emission from electricity is 1792.52 tonnes CO_2/year out of which 896.26 tonnes CO_2 from electricity used in factory and 896.26 tonnes kg CO_2 from electricity consumed as warehouse. Hence total carbon emission from electricity is equally allocated to both factory and warehouse. The next category is staff commuting under scope 3. Emission from transportation used by staff employees and workers is 675.15 tonnes CO_2/year Water supply and waste treatment-related data is not available because all waste from the plywood manufacturing process is eco-friendly, and there is no need for excess water in plywood-making process. Only a little amount of water is used for the soaking of wood which is rarely changed in a year. According to the above table plywood industry emitted about 4032.49 tonnes CO_2/year transportation and electricity is the largest source of emission.

14.5 Major Findings and Discussion

On the basis of the study, one of the key things to be estimated is that how much wood is used in the production process of plywood, wood is the only construction material which is absorbed CO_2 from the atmosphere when produced, not emitted more during its production. Wood—absorbed 1.7 tonnes of CO_2 from atmosphere, over and above the energy growing, harvesting and processing. We exclude that the carbon generated during forest growth and wood quantity used in the production process. If these activities are included, then wood products have a negative carbon footprint (they store more carbon than they consume during manufacturing). Transportation used during supply from various locations generates carbon emission which is included in scope 2. Another source of carbon emission in scope 2 is electricity consumption. Table 14.8 shows the carbon footprint calculations of inbound and outbound logistics and the electricity used for manufacturing and warehousing.

Table 14.8 Components of carbon emission

Transportation medium	Carbon footprint	%
Transportation: factory	1067.82	26.48
Transportation: market	248.50	6.16
Transportation: warehouse	248.50	6.16
Electricity: factory	896.26	22.23
Electricity: warehouse	896.26	22.23
Total emission: scope 2	3357.34	83.26

Table 14.9 Scope-wise proportion of carbon emission

Scope	Sources	Emission per tonne CO_2eq/year	Percentage of total	Percentage
Scope 1	Direct emissions & Indirect emissions			
	Purchased wood	(2,191,197.26)		
Scope 2	Transportation: factory	1067.82	26.48	83.26%
	Transportation: market	248.50	6.16	
	Transportation: warehouse	248.50	6.16	
	Electricity: Factory	896.26	22.23	
	Electricity: Warehouse	896.26	22.23	
Scope 3	Other indirect emission			
	Staff commuting	675.15	16.74	16.74%
	Waste treatment	0	0	
	Water supply	0	0	
Total		4032.49	100	

Scope 3 includes indirect emission activities from other source which includes staff commuting, waste treatment, water supply, etc. staff commuting contribute 16% share in total carbon emission. 90% of the workers use cycle as their medium of transport, which do not use any fuel and not make any contribution to carbon footprint only 10% uses any transport medium like bus or auto rickshaw, which contribute a small share in carbon emission.

Another activity in scope 3 is waste treatment, and we above said that there is no any type of waste in wood-based industries which need any waste management devices. Last category included in scope 3 is water supply, during plywood manufacturing process only a little amount of water is used for the soaking of wood which is rarely changed in a year. This activity contributes a negligible amount of GHG emission. Table 14.9 shows the scope-wise proportion of carbon emission estimates in the plywood industry.

The study estimates the carbon emissions of the plywood industry, which come from various scopes, i.e. the types of sources of carbon emission. Interestingly, the quantum of carbon emission is seen highest in scope 2, which pertains to the operational performance of the plywood industry. Transportation is one of the biggest contributors to the carbon emissions in this industry. Scope 1 has the potential of some carbon compensation by way of afforestation, but the MSMES engaged into plywood manufacturing in district Yamunanagar of Haryana, a north Indian state, do not use modern technologies to reduce the carbon emissions emerging out of their processes.

14.6 Future Research Prospects

There is a need for more case studies. The main reason is, less research studies have been done about CO_2 emission or carbon footprint, so this type of research can be valuable addition to the knowledge base. There is no proper procedure developed in India for calculating carbon emission. There is no availability of adequate data. Data needs to be available for the correct calculation of carbon footprint. It is recommended that data must be stored or recorded to improve accessibility for future carbon footprint processes.

References

Andrews SL (2009a) A classification of carbon footprint methods used by companies. Master of Engineering in Logistics, pp 50–54

Andrews S (2009b) A classification of carbon footprint methods used by companies. https://ctl.mit.edu/sites/ctl.mit.edu/files/library/public/theses_2009_Andrews_Executive%20Summ.pdf. Accessed 20 Apr 2020

Barnett A, Barraclough RW, Becerra V, Nasuto S (2013) A history of product carbon footprinting. https://www.reading.ac.uk/web/files/tsbe/Barnett_TSBE_Conference_Paper_2013.pdf. Accessed Apr 20 2016

Crompton S, Roy R, Caird S (2002) Household ecological footprinting for active distance learning and challenge of personal lifestyles. Int J Sustain High Educ 3(4):313–323

Dawe GFM, Vetter A, Martin S (2004) An overview of ecological footprinting and other tools and their application to the development of sustainability process: audit and methodology at Holme Lacy College, UK. Int J Sustain High Educ 5(4):340–371

Filimonau V (2012) Carbon calculators as a tool for carbon impact appraisal of holiday travel: a critical review. Worldwide Hosp Tour Themes 4(4):302–331

Finkbeiner M (2009) Carbon footprinting—opportunities and threats. The International Journal of Life Cycle Assessment 14:91–94

Gibassier D, Schaltegger S (2015) Carbon management accounting and reporting in practice: a case study on converging emergent approaches. Sustain Account Manag Policy J 6(3):340–365. https://doi.org/10.1108/SAMPJ-02-2015-0014

Hertwich EG, Peters GP (2009) Carbon footprint of nations: a global, trade-linked analysis. Environ Sci Technol 43(16):6414–6420

Lee KH, Cheong IM (2011) Measuring a carbon footprint and environmental practice: the case of Hyundai Motors Co. (HMC). Ind Manag Data Syst 111(6):961–978

Li J, Zhang D, Su B (2019) The impact of social awareness and lifestyles on household carbon emissions in China. Ecol Econ 160:145–155

Liu D, Guo X, Xiao B (2019) What causes growth of global greenhouse gas emissions? Evidence from 40 countries. Sci Total Enviro 661:750–766. issn:0048-9697. https://doi.org/10.1016/j.scitotenv.2019.01.197

McKinnon AC (2010) Product-level carbon auditing of supply chains: environmental imperative or wasteful distraction? Int J Phys Distrib Logist Manag 40(1/2):42–60

Padgett, J. P., Steinemann, A. C., Clarke, J. H., & Vandenbergh, M. P. (2008). A comparison of carbon calculators. Environmental impact assessment review, 28(2-3), 106-115.

Rigot-Muller P, Lalwani C, Mangan J, Gregory O, Gibbs D (2013) Optimising end-to-end maritime supply chains: a carbon footprint perspective. Int J Logist Manag 24(3):407–425

Sarbring A (2014) A carbon footprint assessment on construction and maintenance operations for the port of Gothenburg with emphasis on emission reduction actions. Masters Thesis,

Department of Energy and Environment, Chalmers University of Technology. https://odr.chalmers.se/bitstream/20.500.12380/203855/1/203855.pdf. Accessed 21 Apr 2016
Srivastav S (2013) Carbon footprint as a policy tool in Indian cities: process, inventory, application. Doctoral dissertation, Cardiff University
The Greenhouse Gas Protocol (2000) World Resource Institute (Rev edn). https://ghgprotocol.org/sites/default/files/standards/ghg-protocol-revised.pdf
Udara Willhelm Abeydeera LH, Wadu Mesthrige J, Samarasinghalage TI (2019) Global research on carbon emissions: a scientometric review. Sustainability 11:3972
United Nations (1998) https://unfccc.int/resource/docs/convkp/kpeng
WBCSD (2000) https://ghgprotocol.org/about-wri-wbcsd
Wiedmann T, Minx J (2008) A definition of 'carbon footprint': chapter 1. In: Pertsova CC (ed) Ecological economics research trends. Nova Science Publishers, Hauppauge, NY, pp 1–11. https://www.novapublishers.com/catalog/product_info.php?products_id=5999

Websites

http://www.carbondiet.ca/going_green/carbonfootprint.html. Accessed 17 Mar 2016
http://timeforchange.org/what-is-a-carbon-footprint-definition. Accessed 17 Mar 2016
http://whatis.techtarget.com/definition/carbon-footprint. Accessed 17 Mar 2016
http://climatekids.nasa.gov/greenhouse-effect/. Accessed 19 Apr 2016
http://m.livescience.com/37743-greenhouse-effect.html. Accessed 19 Apr 2016
http://dcmsme.gov.in/dips/yamunana.pdf. Accessed 19 Apr 2016
https://www.epa.gov/ghgemissions/overview-greenhouse-gases. Accessed 20 Apr 2016
http://www.nmsea.org/Curriculum/Primer/Global_Warming/fossil_fuels_and_global_warming.htm. Accessed 21 May 2016
http://www.ucsusa.org/global_warming/science_and_impacts/science/each-countrys-share-of-co2.html#.V5s3UPmF7IU. Accessed 21 May 2016
http://time.com/4059051/india-indc-climate-change-carbon-emissions/. Accessed 21 Jun 2016
http://blogs.wsj.com/indiarealtime/2015/12/03/indias-rich-have-a-smaller-carbon-footprint-than-rich-countries-poor/. Accessed 25 Jun 2016
http://www.dwellsmart.com/News-Information/Tips-for-Green-and-Healthy-Living/Reducing-Carbon-Footprint. Accessed 30 Jun 2016
https://www.google.co.in/search?q=plywood+industries+in+ynr&rlz=1C1RLNS_enIN684IN684&oq=ply&aqs=chrome.1.69i57j69i59j69i60l2.4743j0j8&sourceid=chrome&ie=UTF-8. Accessed 2 Jul 2016
http://www.jyotiplywood.com/about.php. Accessed 2 Jul 2016
http://www.business-standard.com/article/markets/price-of-carbon-credits-may-rebound-this-year-112052500017_1.html. Accessed 8 Aug 2016
http://timesofindia.indiatimes.com/business/india-business/Companies-holding-carbon-credits-stare-at-real-loss/articleshow/31803387.cms. Accessed 8 Aug 2016
http://www.rediff.com/money/2008/feb/05inter1.htm. Accessed 8 Aug 2016
http://www.livestrong.com/article/183436-the-importance-of-reducing-a-carbon-footprint/. Accessed 8 Aug 2016
https://www.reading.ac.uk/web/files/tsbe/Barnett_TSBE_Conference_Paper_2012.pdf. Accessed 9 Sept 2016
www.indiancarbon.org/the-carbon-credit-market.html. Accessed 9 Sept 2016
http://www.biotecharticles.com/issues-article/carbon-trading-the-future-money-venture-for-india-3384.html. Accessed 24 Apr 2017
http://www.legalservicesindia.com/article/carbon-trading-in-india-264-1.html. Accessed 24 Apr 2017
http://www.investopedia.com/terms/c/carbon_credit.asp. Accessed 24 Apr 2017
http://www.sourcewatch.org/index.php/carbon_trading. Accessed 24 Apr 2017

https://doi.org/10.1108/IJLM-01-2013-0002. Accessed 16 May 2019
https://doi.org/10.1108/14676370410561063. Accessed 16 May 2019
https://doi.org/10.1108/17554211211255675. Accessed 16 May 2019
https://doi.org/10.1108/09600031011018037. Accessed 16 May 2019
https://www.greenbiz.com/sites/default/files/document/EIARVol28Issue2-3pgs106-115.pdf. Accessed 16 May 2019
https://odr.chalmers.se/bitstream/20.500.12380/203855/1/203855.pdf. Accessed 16 May 2019
https://res.mdpi.com/d_attachment/sustainability/sustainability-11-03972/article_deploy/sustainability-11-03972-v3.pdf. Accessed 16 May 2019
https://yamunanagar.nic.in/economy/ Accessed 24 Apr 2017

Eco-Friendly Bioremediation Approach for Dye Removal from Wastewaters: Challenges and Prospects

15

Yogita Prabhakar, Anshu Gupta, and Anubha Kaushik

Abstract

Discharge of huge quantities of toxic dyes along with textile wastewaters causes irreversible damage to the aquatic ecosystems on entering the rivers or streams. The conventional physico-chemical treatment methods being costly, energy intensive and less efficient in dye removal have necessitated looking for novel methods of treatment. The bioremediation approach involving biological systems has emerged as a more effective, less expensive, less energy demanding and eco-friendly treatment method for dyes. The present chapter focuses on the use of pure and mixed bacterial cultures isolated from different habitats for bioremediation of Reactive dyes, a widely used and toxic group of azo dyes. The effects of various process parameters including both nutritional and physical factors viz. carbon and nitrogen source, temperature, pH, agitation and presence or absence of oxygen on dye decolorization have been critically reviewed along with possibilities of process optimization for enhanced dye degradation using Response Surface Methodology (RSM) models. The chapter also highlights the biotransformation pathways followed by different types of microbes and the role of enzymes responsible for dye degradation. Biotoxicity assays that could be conducted to examine the dye detoxifying capability of the microbe have also been discussed. A critical appraisal of the bioremediation approach, its limitations, major challenges and future perspectives opens avenues for its application as an economically and environmentally sustainable treatment method for dye-laden wastewaters, so that the treated waters could be used safely

Y. Prabhakar · A. Gupta (✉) · A. Kaushik
University School of Environment Management, Guru Gobind Singh Indraprastha University, Delhi, India
e-mail: anshurcy@yahoo.com

A. Kaushik et al. (eds.), *Climate Resilience and Environmental Sustainability Approaches*, https://doi.org/10.1007/978-981-16-0902-2_15

for some designated purposes, thus helping in meeting the goals of sustainable water management.

Keywords

Reactive dyes · Bioremediation · Biodegradation · Biotoxicity assays

15.1 Introduction

15.1.1 Synthetic Dyes: Reactive Dyes

Textile industry is one of the prime industries in India as well as in the world. Its importance can be seen from the fact that about 20% of the total industrial production in India is due to textile industries. However, they are also identified to be responsible as one of the key contributors to increasing environmental pollution and deterioration with more than 10,000 dyes used commercially. Synthetic origin and complex structure make the degradation of these dyes very difficult (Saratale et al. 2010). Besides dyes, discharge of alkali, organic salts, surfactants and organic acids also contribute to pollution from textile industries (Khatri et al. 2015). Among different groups of synthetic dyes, reactive dyes are the choice of many industries owing to their brilliant shades, application flexibility and excellent colour fastness. Reactive dyes are mostly organic substances with stable structure. According to Ahmed (1995), the first commercially available reactive dyes consisted of dichloro-*s*-triazine group, which is reactive in nature. The most commonly used reactive groups, with increasing reactivity consist of trichloropyrimidine, aminochloro-*s*-triazine, sulphatoethylsulphone, dichloroquinoxaline, aminofluoro-*s*-triazine, difluoro-chloropyrimide and dichlorotriazin (Shore 2002; Khatri et al. 2015). The set of reactive dyes widely used today consists mainly of two groups and are known as bifunctional dyes (Taylor et al. 2001).

The most predominant dye-fibre combination in the apparel industry is reactive dyes and cotton fibre (King 2007). They are resistant to fading towards many agents including water, sweat, light and oxidizing agents (Nigam et al. 2000). Reactive groups bind to the cotton and cellulosic fibre by forming covalent bonds with the fibre's hydroxyl groups, in alkaline conditions (Khatri et al. 2015). Besides cotton, these dyes can be also be used for silk and wool (Lewis et al. 2008).

During binding, dye fixation is around 50–80% (Smith 2003), and the rest is released into the environment. Reactive dyes are soluble anionic dyes, so get repelled by the negative surface of cotton fibre in aqueous solution (Gordon and Hsieh 2007). Effluents discharged from cotton dyeing are highly polluted with reactive dyes rendering it coloured along with high salt load and oxygen demand (Khatri et al. 2015). These industries consume large amount of freshwater (Kalyani et al. 2009a, b). Around 280,000 tonnes of textile effluent containing dyes are generated annually worldwide (Jin et al. 2007), which needs to be treated. Untreated discharge of such coloured industrial wastewaters into the waterbodies results in the reduction of dissolved oxygen which becomes hazardous and fatal to living aquatic organisms.

It is therefore, extremely important to treat the dye-containing wastewaters so that the toxic dyes are either detoxified or removed from the system and water becomes safe for reuse. Several physico-chemical treatment methods available for dyes removal from wastewater are generally expensive and are high energy and chemical intensive in nature. Most of the time they are incapable of complete removal of these dyes and/or their organic metabolites from the contaminated environment. This results in substantial amount of sludge generation thus causing secondary pollutant issues, and hence not sustainable (Zhang et al. 2004). Efforts are now focused on developing wastewater treatment methods that are cost-effective, environmentally safe and sustainable (Wang and Ren 2014). Biological treatment methods are emerging as more economical and safer alternatives and in the past two decades, dye removal using several groups of microbes counting bacteria, cyanobacteria, fungi and algae has been explored extensively (Mona et al. 2011a, b). Bioremediation is thus, surging as an alternative solution to safe, sustainable and cost-effective treatment of different types of dye comprising wastewaters. The present chapter is aimed at critically examining the effectivity of biological degradation and detoxification of reactive dyes using various types of bacteria and gain insight into various important determinant parameters that can be controlled and modelled during the operation leading to enhanced response and understand the mechanism of biodegradation of the dyes. It further analyzes the techno-economic feasibility of the bioremediation technique for commercial application as a step towards sustainable wastewater management.

15.1.2 Bioremediation Approach for Dye Removal

Bioremediation involving microbial application to remove or transform toxic dyes present in wastewaters is emerging as an important approach. Application of different biosorbents for dye removal using inactive or dead biomass of microbial systems has been widely investigated (Aksu and Donmez 2005; Mona et al. 2011a; Vijayaraghavan and Yun 2008). A wide range of microbes has been explored for dye removal including bacteria (Saratale et al. 2009), fungi (Humnabadkar et al. 2008), yeasts (Saratale et al. 2009), cyanobacteria (Mona et al. 2011b) and algae (Preeti et al. 2015). Biosorption is a physico-chemical technique that is metabolism-independent (Asgher and Bhatti 2012). Though the biosorption process carries out the dye removal at faster rate, but disposal of the dye-laden adsorbent is a challenge. Along with microbes, microbial polysaccharides are also found to adsorb dyes from wastewater. Microbial exo-polysaccharides are biopolymers that are secreted by the cells into the extracellular medium (Rehm 2009). Exo-polysaccharides are produced by various microbial systems including fungi, algae, bacteria, yeast, cyanobacteria and microalgae (Ates 2015). But disposal of the generated sludge and exo-polysaccharide production expense are the major challenges to their application.

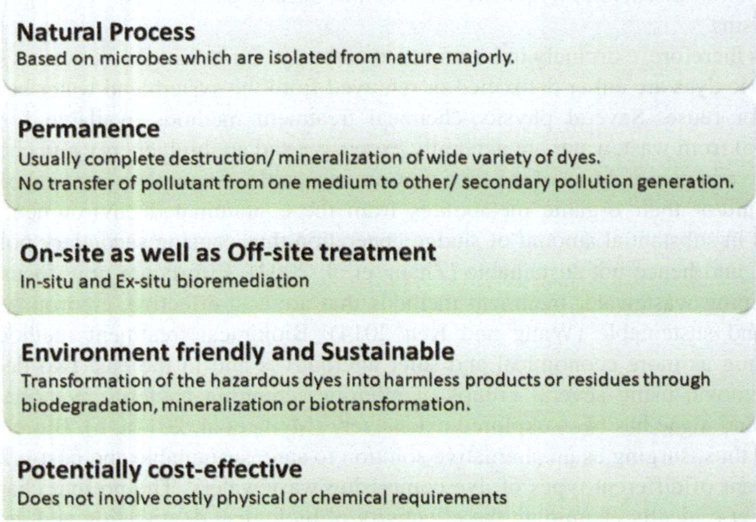

Fig. 15.1 Advantages associated with the use of bioremediation for dye removal

Biotransformation of dye is a more useful method as it converts the toxic dyes to non-toxic form, eliminating the problem of disposal of sludge and involves lower cost (Saratale et al. 2010). As reactive dyes are recalcitrant in nature, application of suitable microbial treatment method is important that may degrade the dye and transform the parent toxic compound to either non-toxic form or may lead to its complete mineralization (Banat et al. 1996). Effectivity of the microbial dye decolorization process depends on the tolerance and the competence of the microbe to degrade the dye and in recent years, a large number of microbial strains have been isolated, genetically modified and applied/checked for decolorization and degradation of many dyes (Pandey et al. 2007). The microbes used, the type of dye and its concentration and the prevailing physico-chemical and nutritional conditions that strongly influence the dye transformation or biodegradation are discussed here, as it is important to understand their influence, which ultimately may be modelled for optimization of the dye degradation process. Some of the advantages associated with the use of bioremediation as a promising approach for dye removal are presented in Fig. 15.1.

15.2 Pure Microbial Cultures and Consortia for Dye Decolorization

Application of pure bacterial cultures in dye degradation started in the 1970s using *Bacillus subtilis*, *B. cereus* and *Aeromonas hydrophila* (Wuhrmann et al. 1980). Later on, many bacterial species showed promising results for dye decolorization when used in single form. For instance, *Proteus mirabilis*, *Pseudomonas luteola*, *Pseudomonas* sp. and *Pseudomonas* sp. *SUK1* are reported for dye degradation under anaerobic conditions (Kalyani et al. 2009a, b; Saratale et al. 2011). In contrast to that, Nachiyar and Rajkumar (2003) have reported the decolorization of Navita Fast Blue S5R, a commercial tannery and textile dye using *Pseudomonas aeruginosa*, in the presence of glucose as a supplement under aerobic conditions. In the application of pure bacterial strains for dye decolorization, it is not only easy to maintain the culture requirements of single microbe but detailing out the dye removal mechanism is also convenient using various biochemistry and molecular biology tools.

Though the bacterial decolorization process is considered to be efficient with the use of pure bacterial strains, however, complete mineralization of dyes is many a time not achieved (Table 15.1). The generated intermediates may be potentially carcinogenic in nature, hence need to be further degraded (Joshi et al. 2008). Mixed microbial consortia are found to show better decolorization and complete mineralization, due to synergistic metabolic actions of the different microbial species (Saratale et al. 2010). During the process, individual strains may target different positions on the dye molecule or metabolites produced during the breakdown of the dye may further be utilized/decomposed by the coexisting strain(s) (Saratale et al. 2009). Thus, many a times, microbial consortia collectively have been found to be more successful as compared to pure strain (Banat et al. 1996).

15.3 Factors Impacting Microbial Dye Degradation and Process Optimization

In the microbiological treatment methods various operational conditions, such as temperature, agitation, dissolved oxygen, pH, initial dye concentration, electron donor and supplements of carbon and nitrogen sources are important that affect the bacterial decolorization efficiency (Fig. 15.2). Some of the major factors and their effects are discussed in detail here.

15.3.1 Temperature

Since all processes related to microbial vitality are influenced by temperature, hence it plays a significant role in dyes biodegradation. Studies associated with microbial decolorization have shown that a narrow temperature range may be necessary for decolorization (Dos Santos et al. 2007). Temperature changes may result in abrupt alteration of activation energy. In addition, it affects biomass yield, bacterial growth and reaction mechanism (Blaga et al. 2008). Decolorization rate has often been found to increase till the optimal temperature is reached, and then decreases marginally, which may be due to loss of cell viability and/or denaturation or reduced activity of enzymes involved in decolorization like azoreductase (Saratale et al. 2009). However, in some whole cell bacterial systems, there is development of

Table 15.1 Dye decolorization using bacteria from different habitats

Name of bacterial isolates	Habitat	Name of dye	pH	Treatment time (h)	Decolorization (%)	References
Bacillus sp. CH12	Alkaline lake	Reactive red 239	9.0–10.0	24–96	95–100	Guadie et al. (2017)
Bacillus sp. (N1–N6)	Non-contaminated soil	Mixture of seven dyes	7.0	336	40–47	Olukanni et al. (2006)
Bacillus cereus	Alkaline soda soil sample	Direct blue 151	9.5	120	93	Lalnunhlimi and Krishnaswamy (2016)
Shewanella haliotis DW01	Alkaline lake water sediment	Reactive blue 172	9.5	12	93	Birmole et al. (2014)
Moderately alkaliphilic bacterial consortium	Soil sample of saline environment	Direct blue 151; direct red 31	9.5	140	97.6; 95.3	Lalnunhlimi and Krishnaswamy (2016)

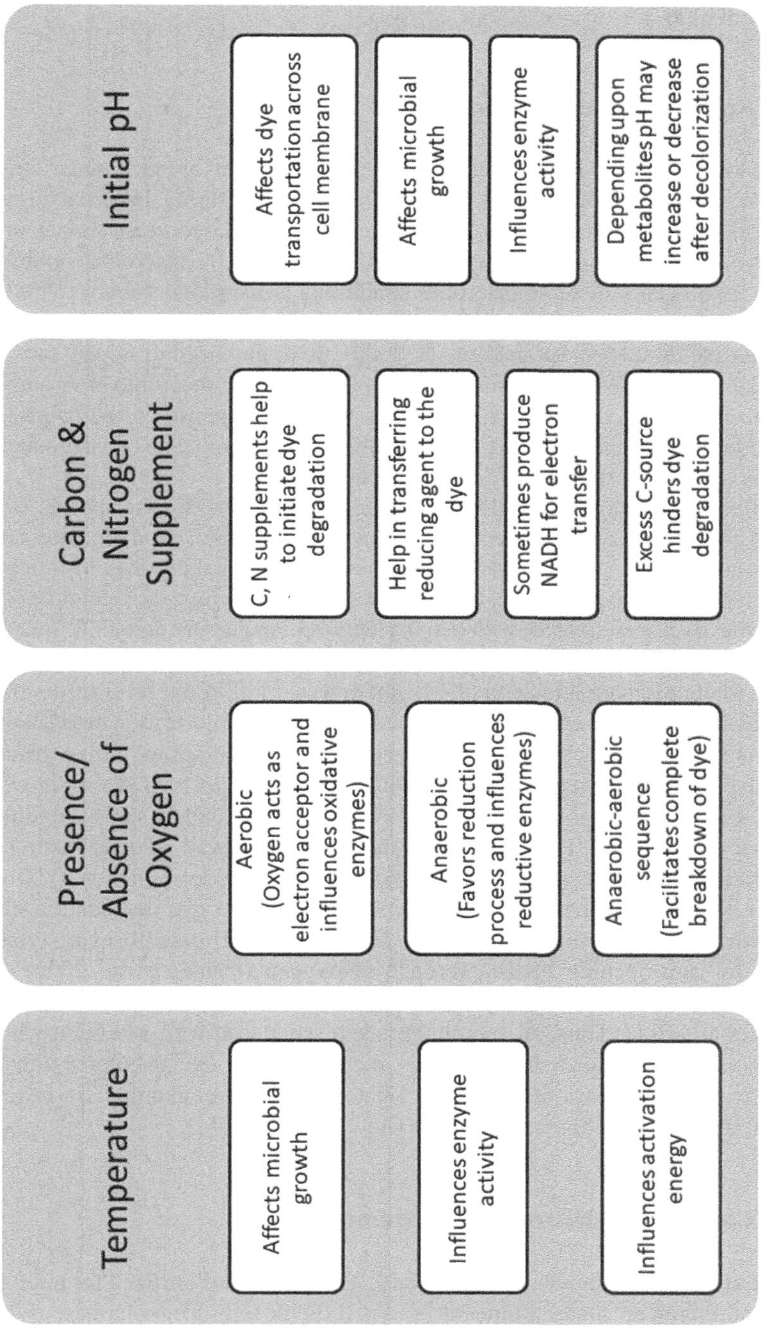

Fig. 15.2 Effects of different factors on dye removal efficiency

thermostable enzymes, which are active even at 60 °C, though for short durations (Pearce et al. 2003).

15.3.2 Agitation and Oxygenation

Dye decolorization may occur utilizing different groups of bacteria under strictly aerobic, anaerobic or, facultative anaerobic conditions. During bacterial growth, most of the dyes get reduced to amines under anoxic conditions (Banat et al. 1996). Dye decolorization is generally more efficient in strictly anaerobic conditions, though it also occurs in semi-anaerobic conditions (Knapp and Newby 1995), as shown for *Pseudomonas* sp. SUK1 (Kalyani et al. 2008) and *Micrococcus glutamicus* NCIM-2168 (Saratale et al. 2009). It might be due to the fact that reductive enzyme activities are considered to be higher in anaerobic environment. However, during dye degradation, sometimes a little oxygen might be helpful for some oxidative enzymes' activity, though, the effect of oxygen on azoreductase enzyme is found to be irreversible (Pearce et al. 2003).

The effect of agitation on dye decolorization varies from species to species. Some microbes prefer static conditions while some others show good decolorization efficiency at different shaking speeds. The preference of shaking may also depend upon the type and concentration of dye as well as incubation time. For example when reactive red dye was treated with *Escherichia* sp. under an agitation speed of 200 rpm, the dissolved oxygen level reduced to 0.5 mg/l with no significant colour removal, whereas under static conditions, there was a fall in the level of dissolved oxygen in the culture medium to near zero and fast decolorization was attained (Chang and Kuo 2000). We have observed similar observations during reactive violet 1 dye treatment by using *Nesterenkonia lacusekhoensis* EMLA3 strain where efficient results on dye decolorization were obtained under static conditions (Prabhakar et al. 2019a). However, when the same strain was used for methyl red decolorization, higher dye removal rate was obtained under agitation (100 and 200 rpm) compared to that of static condition. During azo dye degradation, there is the formation of intermediate aromatic compounds that undergo hydroxylation followed by ring opening in the presence of oxygen (Pandey et al. 2007). The aerobic condition is helpful in the complete mineralization of aromatic molecules (Saratale et al. 2011). Thus, an anerobic process sequenced with aerobic treatment might be a suitable choice for achieving more effective dye decolorization and effluent treatment (Khehra et al. 2006). Treatment of dye-containing wastewaters under different oxygenation conditions is shown in Table 15.2.

15.3.3 Carbon and Nitrogen Supplement

Azo dyes are generally unable to supply sufficient carbon and nitrogen for microbial growth and therefore, many a times it is very difficult to biodegrade these dyes in the absence of any external supplement. Complex organic sources like peptone,

Table 15.2 Dye decolorization as influenced by conditions of aeration (static, aerobic, anaerobic and microaerophilic)

Name of bacterium/ consortium	Name of dye	Maximum decolorization with different conditions	References
Bacterial consortium (six classes of α-, β-, and γ-proteobacteria, Bacilli, Cytophagia and Nitrospirale	Reactive red 3BS	90% (aerobic)	Tan et al. (2013)
Pseudomonas sp. ARa, *Bacillus* sp., *Bacillus* sp. and *Ochrobactrum* sp.	Reactive red 195	>70% (microaerophilic)	Khan et al. (2014)
Consortium of *Pseudomonas* sp. strain SUK1, LBC2 and LBC3	Reactive orange 16	100% (static)	Jadhav et al. (2010)
Clostridium bifermentans SL186	Reactive red 3B-A, reactive black-5, and reactive yellow 3G-P	90% (anaerobic)	Joe et al. (2008)
Brevibacterium sp. VN-15	Mix of reactive black 5, reactive red 198, reactive yellow 107, direct blue71	95% (microaerophilic)	Franciscon et al. (2012)

yeast and carbohydrates are often required for microbial dye degradation, which transfer reducing agents to the dye (Khehra et al. 2006). Under anaerobic conditions, the decolorization is found to be affected by the supplement of carbon sources in many cases. These may be simple substrates such as glucose, acetate, starch and ethanol, or complex supplements like whey and tapioca. During anaerobic conditions, acidogenic bacteria change soluble carbohydrates to simple volatile organic acids like acetic acid or alcohols such as methanol. Some studies, however, show that extraneous addition of peptone, beef extract, yeast extract and urea like organic nitrogen sources helps in dye degradation through NADH generation, which acts as electron donor during microbial dye decolorization (Chang and Lin 2000; Chang et al. 2001). In contrast to that, carbon sources are sometimes found less useful in dye degradation due to the cell preference towards the assimilation of carbon source prior to dye compound (Saratale et al. 2009). It has been found that 2% starch supplemented as carbon source promotes degradation of reactive dyes, while 3% starch supplement leads to decreased degradation (Prabhakar et al. 2019a). Thus, the supplement of carbon or other nutritional sources needs to be optimized for different microbial cultures and dyes for effective biodegradation or bioremoval.

15.3.4 Ligno-Cellulose Amendment

The presence of lignocellulosic substrates is known to enhance dye decolorization due to the enhanced production of lignolytic enzymes. Lignocelluloses are a major constituent of agricultural and forest residues or wastes, and also that of certain industrial effluents like sugar, textile, pulp and paper industries (Saratale et al. 2008). Dye decolorization of Scarlet R and Reactive Green 19A dye using bacterial consortium GR and *Micrococcus glutamicus* NCIM-2168, respectively in presence of various agricultural wastes, including extracts of sugarcane baggase powder, wood straw, rice straw and rice husk have been found to enhance decolorization process (Saratale et al. 2009). In another study, *Comamonas* sp. UVs were found to decolorize Direct blue GLL dye along with simultaneous production of lignolytic enzymes (Jadhav et al. 2009).

15.3.5 Initial pH

The pH of the medium or wastewater plays very important role in dye decolorization using microbes. The pH may affect dye transportation across the cell membrane, which is one of the major steps in the dye decolorization process (Chang and Lin 2000). During biological reduction of azo bond, pH increases due to the formation of aromatic amines since these products are more basic in nature compared to the original dye compound. While in some cases, the initial pH of the media may decrease after dye degradation probably due to some acidic metabolites produced during microbial growth and degradation of dyes (Bhattacharya et al. 2017).

The dye decolorization rate has been found to be maximum at optimum pH, which is species specific, and then decreases at strong acidic or alkaline pH. In most of the studies, the optimal pH for bacterial dye decolorization has been found majorly to be in the range of 6.0–10.0 (Kilic et al. 2007). It has also been noted that within the pH range of 7.0–9.5, there is very little variation in dye reduction efficiency, whereas, on increasing the pH (5.0–7.0), the reduction increased nearly 2.5-fold (Chang et al. 2001). However, in some studies, it has been observed that biodegradation of dyes is not pH dependent. In a study on colour removal of Brilliant G dye, Jadhav et al. (2008a) concluded that the dye decolorization by a microbial consortium was pH independent as they observed complete decolorization of dye within pH range of 5–9. Similarly, very good decolorization of Reactive red 190 dye was obtained under strongly acidic (pH 4.0) as well as strongly alkaline (pH 12.0) using *Citrobacter* sp. CK3 (Wang et al. 2009).

Varied range of pH tolerance is necessity for practical suitability of the treatment of dyeing mill wastewaters (Wang et al. 2009). The optimized pH for biodegradation/bioremoval of some of the commonly used dyes along with their percent decolorization is presented in Table 15.3.

Table 15.3 Decolorization of some reactive dyes under optimized pH

Name of reactive dye	pH (optimized)	Maximum decolorization (%)	References
Reactive orange-16	7.0	100	Jadhav et al. (2010)
Reactive black-5	8.0	80	Kılıç et al. (2008)
Reactive brilliant blue	7.0	93	Dong et al. (2003)
Reactive blue-19	7.0	95	Giwa et al. (2012)
Reactive orange	7.0	98	Shah et al. (2013)
Reactive red 239	9.0–10.0	95	Guadie et al. (2017)
Reactive orange	7.0	98.2	Shah et al. (2012)
Reactive blue 172	9.5	93	Birmole et al. (2014)
Reactive golden yellow HER	7.0	87	Gomare et al. (2009)
Methyl red	11.5	97	Bhattacharya et al. (2017)
Reactive violet 1	11.5	96	Prabhakar et al. (2019a)

15.3.6 Optimization of Dye Degradation Response Using Models

One of the most extensively used models for optimizing dye degradation is based on Response surface methodology (RSM), which is an assembly of mathematical and statistical methods. Dye decolorization is modelled in response to the interaction of three or more parameters under a set of designed experiments through which the process is optimized (Behbahani et al. 2011; Khedmati et al. 2017; Haghshenas et al. 2015). The core of RSM is based on mathematical model fitting, which includes linear, square polynomials and other functions (Witek-Krowiak et al. 2014). The objective of RSM is to determine the optimum operational conditions for acquiring the region in accordance with operating specifications (Biswas et al. 2017). The process of RSM application for simulation and optimization of process parameters follow six steps, (1) independent factor screening and selection of desired responses, (2) selection of experimental design, (3) conduction of experiments and result collection, (4) mathematical model fitting to the obtained results, (5) model confirmation using graphs and analysis of variance, (6) determination of optimized conditions (Karimifard and Alavi Moghaddam 2018).

RSM has proved to be an effective tool in knowing the effect of different parameters on dye decolorization or degradation. For instance, Jadhav et al. (2013), studied the effects of various parameters, viz. pH, temperature and cell mass concentration on bacterial decolorization of sulphonated Remazol orange dye using the RSM approach. RSM helps in determining the predicted dye degradation response under a set of interacting factors that can be tested statistically for its fitness to the employed model. In another study, Senthilkumar et al. (2013) worked on the degradation of reactive red 195 dye using *Pseudomonas* sp. and modelled dye degradation response to factors like initial dye concentration, nitrogen source and carbon source using central composite design model. The predicted results complemented well with the experimental results. Application of RSM model was also demonstrated for optimization of degradation of Reactive orange 4 dye using *Bacillus cereus* (Garg et al. 2015), which enhanced the decolorization upto 100%, which was not attained during the one factor at a time approach.

Thus, RSM comes with added benefits of optimizing, checking and enhancing the decolorization process, as the designs build up various combinations which help analyze their effect on the decolorization process. The use of RSM approach is particularly useful in scale-up of the degradation process where appropriate combinations of parameters are required to be maintained for maximized response in the reactors.

15.4 Dye Degradation and Biotransformation Pathways

15.4.1 Anaerobic Degradation

It is believed that dye decolorization undergoes through relatively simple and non-specific process in anaerobic environment (Wang et al. 2009). Low redox potential (<50 mV) which develops under anaerobic conditions helps in effective dyes decolorization (Bromley-Challenor et al. 2000). The azo bond reduction activity is usually inhibited in presence of oxygen as aerobic respiration dominates NADH utilization and thus hampers the electrons transfer from NADH to dye's azo bond (Chang et al. 2001; Stolz 2001).

Saratale et al. (2011) in their review on biodegradation of dyes have mentioned that anaerobic biomass generates various reduced compounds and dye decolorization happens because of their non-specific extracellular reactions (Saratale et al. 2011).

15.4.2 Aerobic Degradation

Several dye degrading aerobic bacterial strains have been isolated in the past few years. It is found that under aerobic conditions, oxygen is first incorporated into the aromatic structure with the help of mono and di oxygenase enzymes followed by ring fission (Shah 2014). However, very few bacterial species can degrade reactive dyes in completely aerobic conditions using reducing enzymes (Saratale et al. 2011). Aerobic azoreductase uses both NADPH and NADH as cofactors and cleaves carboxylated growth substrate, and can also degrade the sulphonated structural analogues. This kind of azoreductase has been reported in *Pseudomonas* sp. strains K22 and KF46 (Nachiyar and Rajkumar 2005; Zimmermann et al. 1984). Most of the bacteria can exploit azo dye as growth substrate under aerobic conditions only in presence of some additional carbon sources (Stolz 2001).

15.4.3 Role of Enzymes

Biodegradation of complex and recalcitrant dyes is mediated by the enzymatic action of different enzymes such as laccase (Hatvani and Mecs 2001), lignin peroxidase (Duran and Esposito 2000), tyrosinase (Zhang and Flurkey 1997), NADPH-DCIP reductase (Bhosale et al. 2006), hexane oxidase (Saratale et al. 2007) and aminopyrine *N*-demethylase (Salokhe and Govindwar 2003).

The oxi-reductive enzymes play a key role in generating reactive free radicals which undergo through a sequence of spontaneous reactions causing dye structure cleavage. These enzymes work more effectively when the concentration of the target pollutant is highest and the other contaminants are lowest. Other contaminants interfere with the treatment process of the enzymatic degradation. As mentioned earlier, a number of different types of oxi-reductive enzymes including lignin peroxidase, laccases, azoreductases and others are involved in bacterial dye degradation. Azoreductase enzyme which catalyzes azo bonds cleavage has been explored in bacterial cytoplasm, as cytoplasmic reductases and is being recommended for its application in environmental biotechnology (Saratale et al. 2011; Maier et al. 2004). On functional basis, these enzymes are characterized as flavin-dependent and flavin-independent azoreductases (Blumel and Stolz 2003; Zimmermann et al. 1982). The flavin-dependent azoreductases are further classified on the basis of cofactors i.e. NADH and NADPH as only one of them (Chen et al. 2005) or both of them (Wang et al. 2007). These coenzymes serve as electron donors. Though many bacterial species are reported to have intracellular azoreductase activity, dye degradation becomes difficult many times as the azo dyes due to their complex structure and high polarity make its diffusion through the cell membranes difficult and hence reduced dye cleavage is attained (Telke et al. 2008; Saratale et al. 2009).

The role of riboflavin reductase in dye degradation has also been demonstrated (Russ et al. 2000). When non-enzymatic reduction of free flavins by NADH and NADPH is rather slow, the microbes possess riboflavin reductase as the enzyme system to catalyze the reaction. This enzyme helps in the reduction of different types of flavins, including riboflavin, flavin mononucleotide (FMN) and flavin adenine dinucleotide (FAD) (Ingelman et al. 1999). The riboflavin is known to play role in the degradation of Mordant yellow10 and Brilliant blue G by anaerobic granular sludge and consortium, respectively (Field and Brady 2003; Jadhav et al. 2008a).

Many oxidative enzymes like laccase, lignin peroxidase and tyrosinase are reported to play important role in the treatment of a number of recalcitrant pollutants like synthetic dyes, PAHs (polyaromatic hydrocarbons) and PCBs (polychlorinated biphenyls) (Dawkar et al. 2009; Duran and Esposito 2000). For dye reduction, peroxidases like lignin peroxidase oxidize phenolic groups for the production of radical at the azo-linkages (Chivukula and Renganathan 1995). The significance of oxidative enzymes in the decolorization of azo dyes having sulphonated reactive groups has been extensively studied.

Laccase or phenol oxidase is a copper-containing enzyme, which catalyzes aromatic substances like phenols by reducing oxygen to water (Majcherczyk et al. 1998). This enzyme is known to decolorize some of the azo dyes without directly cleaving the azo bonds and thus this mechanism avoids the production of toxic

Table 15.4 Role of enzymes in reactive dye degradation

Substrate	Enzyme	Bacterium	References
Reactive yellow, reactive black, reactive red, direct blue	Azoreductase	*Staphylococcus arlettae* VN-11	Elisangela et al. (2009)
Reactive black 5	Azoreductase	*Pseudomonas entomophila*	Khan and Malik (2016)
Reactive red 2	Lignin peroxidase and azoreductase	*Pseudomonas* sp.	Kalyani et al. (2009a)
Reactive dye green HE4BD	Oxidative enzymes	Consortia	Saratale et al. (2010)
Reactive blue 172	Azoreductase	*Providencia rettgeri*	Lade et al. (2015)
Reactive black 5, Reactive orange 16	Azoreductase, NADH-DCIP reductase	*Providencia rettgeri* strain HSL1 and *Pseudomonas* sp. SUK1	Lade et al. (2015)
Reactive red 141	Oxidative and reductive enzyme	*Rhizobium radiobacter* MTCC 8161	Telke et al. (2008)

aromatic amines (Kalme et al. 2009). Laccase activity is most predominant in lignolytic fungi (Abdullah et al. 2000), though it is reported in some bacterial species also (McMahon et al. 2007). Purified extracellular laccase from the *Pseudomonas* sp. LBC1 was applied successfully for the removal of bisphenol by Telke et al. (2009) and Kalme et al. (2009).

Tyrosinase also designated as monophenol monooxygenase, is one of the oxidative enzymes used in dyes degradation. This enzyme catalyzes the oxidation of phenol (Wu et al. 2001). Kalme et al. (2007) and Jadhav et al. (2008b) have shown the involvement of tyrosinase in the degradation of Direct blue-6 and Disperse brown 3REL by *Pseudomonas desmolyticum* NCIM 2112 and a microbial consortium, respectively.

The reported studies show that azo dyes degradation is initiated by reductive enzyme followed by complete mineralization through oxidative enzymes. A list of different dyes degraded by different types of microbes with names of enzymes involved is presented in Table 15.4.

15.4.4 Degradation Pathways

Permeation of azo dyes into the cells through the cell membrane is known to be the chief rate-limiting factor for dye decolorization. Dye decolorization pathway generally involves azo bond (–N=N–) cleavage due to the action of azoreductase enzyme with the generation of colourless amines in the process (Khehra et al. 2006; Prabhakar et al. 2021).When microbial consortium is used for the treatment, the

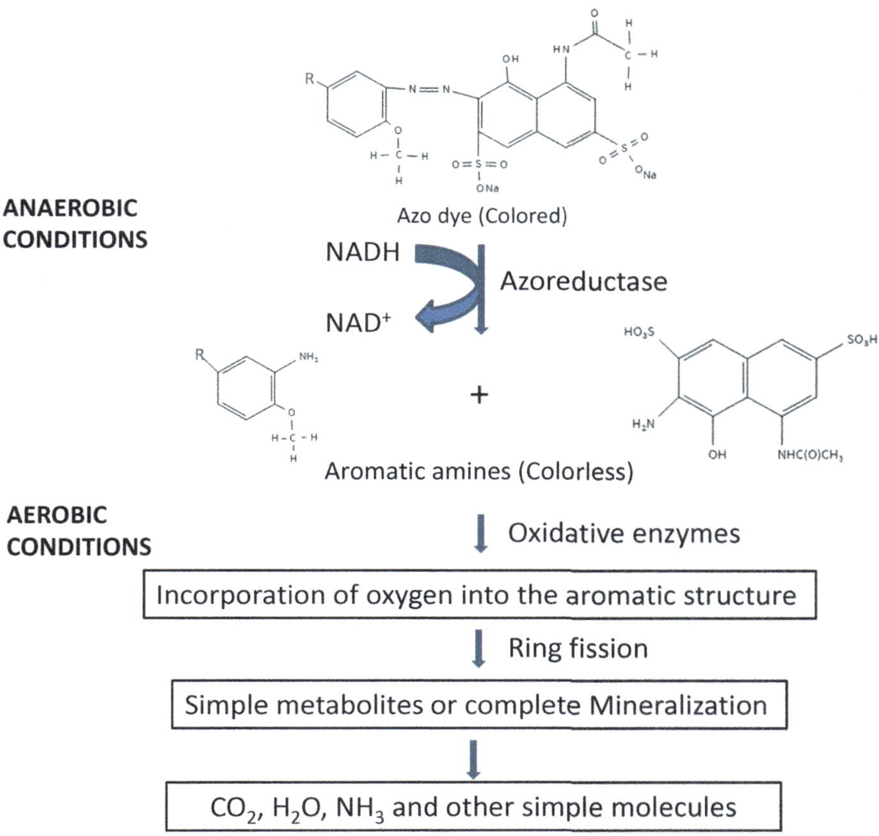

Fig. 15.3 Possible pathway of microbial degradation of reactive azo dyes

produced toxic amines are further degraded by coexisting organisms, thus, making the process more effective (Moosvi et al. 2007). The resultant intermediates/ metabolites get further cleaved aerobically or anaerobically (Joshi et al. 2008).

Degradation proceeds at the azo linkage through two stages consisting of the transfer of two electrons at each stage to the dye, the dye plays the role of final electron acceptor and thus resulting in dye decolorization (Chang and Lin 2000). Figure 15.3 presents possible pathway of microbial degradation of reactive azo dyes.

15.5 Bioassays to Assess Reusability of Wastewaters with Biodegraded Dyes

Reactive dyes are very toxic and are known to cause asthma, dermatitis (Nilsson et al. 1993), carcinogenicity (De Roos et al. 2005), genotoxicity (Dogan et al. 2005) and mutagenicity (Mathur et al. 2006). While the major concern is the removal of colour from dye-containing wastewater, it is equally important that the degraded metabolites should be much less toxic. As the dyes are degraded by microbes, complete mineralization to inorganic compounds may or may not be achieved depending upon the type of dye, the microbe, and the prevailing conditions. Since many aromatic metabolites may be formed by microbial degradation, which may be even more toxic than the original dye, therefore, toxicity assessment of the degraded dyes after complete treatment with the microbes is very important. Short-term screening methods to detect mutagenicity and carcinogenicity have played important role in screening suspected range of chemicals. Toxicity assessment is frequently done using bioassays like single-celled alga *Selenastrum capricornutum* (Hao et al. 2000), Ames test for assessing mutagenic potential or mitotic index reduction owing to DNA damage potential of toxic chemicals (Khan et al. 2019). Genotoxicity of sewage water and industrial effluents treated with microbes have also been assessed for biotoxicity using *Allium cepa* test based on the development of root tip cells (Sudhakar et al. 2001). Seed germination of some suitable crop species, such as *Vigna radiata* have also been used as bioassays for testing the toxicity of microbially treated dye-containing wastewaters (Prabhakar et al. 2021). Most of such bioassay studies have revealed that the dye-containing wastewaters, which exhibit high toxicity to various biological systems show much reduced or no toxicity effects after treatment with the dye degrading microbes, indicating detoxification of the harmful dyes either into non-toxic compounds or complete mineralization that is generally achieved when anaerobic treatment is followed by an aerobic sequence of treatment. This forms an essential component of the bioremediation protocol to ensure the safe reuse of the water for some designated purpose.

15.6 Major Challenges and Techno-Economic Feasibility

Dye accumulation and dye-laden wastewaters create aesthetic issues as well as severe environmental pollution. With regulations becoming more stringent and due to highly toxic nature of dye-containing effluents, there is an urgent need for technically viable, efficient, cost-effective and sustainable technology for the treatment of such wastewaters before disposal. Bacterial degradation can play an effective role in remediating high-strength dye containing wastewaters. Though lot of work has been done using pure or single and mixed bacterial cultures, still there is a need of generating related performance data on bacterial remediation behaviour for industrial effluents. With various physico-chemical parameters significantly affecting the dye degradation process, it becomes important to optimize these parameters for effective biotreatment. However, the optimization process is species and dye dependent, hence needs to be modelled for the specific dyes in relation to the species.

Anaerobic degradation leads to aromatic amines formation after the azo bonds' cleavage, thus acts as a limitation. This issue can however, be resolved by adopting sequential treatment method of first breaking down the dye molecule under anaerobic condition and then subjecting it to aerobic condition for further non-toxic mineralization. The possible mechanism for decolorization can be more elaborately explored, by studying the enzymes involved and analysis of intermediates and final products formed. Scale-up of the process in dynamic and continuous mode for treatment of real textile wastewaters is needed for commercial application of the bioremediation technique as an eco-friendly sustainable approach. In general, the usage of living microbes is yet not successful at commercial scale for the treatment of dye-containing real effluents. One reason for this is that industrial effluents contain many toxic organic or inorganic co- contaminants which may interfere with the dye degradation by influencing the microbe (Prabhakar et al. 2019b) Most of the microbes are tested for the dye alone at lab scale, hence, they are unable to work with real industrial effluents, which have a complex nature possessing other co-pollutants. If lab-scale and pilot-scale studies are conducted using real industrial effluent, there are more chances that microbes will be acclimatized to the toxicants and dye removal conditions could be optimized according to real conditions which would further help in increasing treatment efficiency during industrial/commercial applications.

The continuous biological mode of treatment for dye removal is also in infancy and there are still many problems in its applications. For instance, the biotreatment is generally hampered if the treatment process is operated for too long or once the toxicant becomes too high in concentration leading to increased toxicity of pollutants and reduced treatment efficiency (Kulandaivel et al. 2014). Use of immobilized preparation of microbes and/or enzyme in place of free counterpart might play a very significant role in this direction due to numerous merits of immobilization majorly being stability and reusability of immobilized product. Immobilized preparations are generally more tolerant to local agitations like fluctuations in pH, temperature and toxic compounds (Tallur et al. 2015).

Another major issue in the field of treatment of wastewater using microbes is poor settling of biomass after treatment which reduces the quality of effluent. The use of immobilized system comes with the additional benefit of easy separation and recovery of immobilized preparation and also prevents biomass washout (Suganya and Revathi 2016; Usha et al. 2010). Immobilization of cells increases cells density and thus facilitates biomass separation from treated effluent. Solid-liquid separation in settling tanks after microbial treatment is one of the major difficulties in wastewater treatment plants which can be solved by using immobilization (Usha et al. 2010). The cost of the bioremediation technique can be cut down by using tolerant microbes that can work at the prevailing pH and TDS of the wastewater, so that there is minimal requirement of the addition of chemicals to moderate the conditions. Some extremophiles like halo-alkaliphilic *Nesterenkonia lacusekhoensis* EMLA3, *Bacillus cohnii* MTCC 3616 have been found successful in degrading reactive dyes under prevailing high pH and salts (Bhattacharya et al. 2017; Prabhakar et al. 2019a; Prasad and Rao 2013). Exploitation of the anaerobic degradation pathway of microbes needs to be explored further as it is less energy demanding due to the absence of shaking or aeration requirements.

These research insights (Fig. 15.4) can be immensely helpful for researchers and policy makers for filling the lacuna in developing efficient bioremediation protocols for the removal of dyes from industrial wastewaters at large scale.

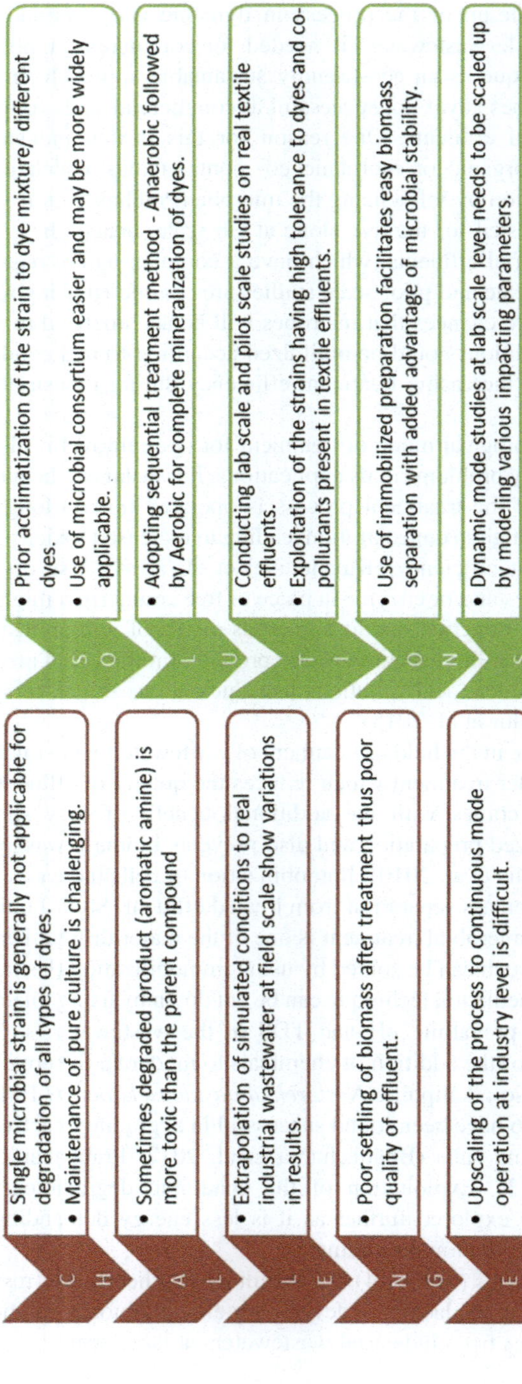

Fig. 15.4 Some challenges and solutions regarding the success of bioremediation for dye degradation

15.7 Conclusion

Sustainability of wastewater treatment and management practices have become extremely important in the present-day scenario, as clean water availability is one of the major Sustainable Development Goals. Polluted waters and wastewaters need to be treated in such a manner that they become reusable. Treatment of dye-containing wastewaters is a major problem worldwide due to the recalcitrant and toxic nature of many dyes. In view of various demerits of conventional physico-chemical treatment methods, majorly their expensive and non-ecofriendly nature, the focus is being shifted towards bioremediation technologies, which are efficient, environment-friendly and cost-effective. The great diversity of microbial species with their diverse range of tolerance, metabolic preferences and enzyme systems offer immense opportunities for their application in the degradation of different types of toxic dyes into non-toxic or less toxic forms. Intensive research on bioremediating potential of various individual species as well as consortia has clearly indicated the importance of certain specific factors that have major impact on dye decolorization and degradation. Interestingly, it has been found possible to model the degradation response to the maximum for the microbes by using specific sets of conditions through RSM models like BBD or CCD, which help in optimizing the operational conditions and also enhance the efficiency of the process. However, the biotreatment methods also have few limitations and challenges, particularly for efficient upscaling at commercial scale. Some of these limitations may be addressed by using innovative techniques, such as exploitation of strains with high tolerance to the pollutants and co-pollutants, optimizing the impacting factors for modelled response, and for greater stability of the microbial system to the polluted milieu. The lab-scale batch studies need to be upgraded for continuous mode column reactor studies using real industrial effluents. There is an urgent need to reduce the hydraulic retention time or contact time, so that the dye-containing effluents could be decolorized in the desired time range to make the bioremediation method practicable. Bioremediation is thus, a highly promising technique for safe, eco-friendly and low-cost treatment of dye-containing wastewaters employing suitable consortia of microbes under optimized set of operational conditions.

Bioremediation approach has the potential to gradually substitute the costly, energy demanding and toxic sludge generating conventional methods by upscaling the present lab-scale technique with further modelling and simulation to meet the industry scale application requirements in a sustainable manner.

References

Abdullah E, Tzanov T, Kosta S, Robra KH, Cavaco-Paulo A, Gubitz G (2000) Decolorization and detoxification of textile dyes with a laccase from *Trametes hirsuta*. Appl Environ Microbiol 66:3357–3362

Ahmed A (1995) Reactive dyes development: a review. Text Dye Print 28:19–24

Aksu Z, Donmez G (2005) Combined effects of molasses sucrose and reactive dye on the growth and dye bioaccumulation properties of *Candida tropicalis*. Process Biochem 40:2443–2454

Asgher M, Bhatti HN (2012) Evaluation of thermodynamics and effect of chemical treatments on sorption potential of citrus waste biomass for removal of anionic dyes from aqueous solutions. Ecol Eng 38:79–85

Ates O (2015) Systems biology of microbial exoploysaccharides production. Front Bioeng Biotechnol 3:200

Banat IM, Nigam P, Singh D, Marchant R (1996) Microbial decolorization of textile-dye-containing effluents: a review. Bioresour Technol 58:217–227

Behbahani M, Alavi M, Reza M (2011) Techno-economical evaluation of fluoride removal by electrocoagulation process: optimization through response surface methodology. Desalination 271:209–218

Bhattacharya A, Singhal N, Gupta A (2017) Degradation of azo dye methyl red by alkaliphilic, halotolerant *Nesterenkonia lacusekhoensis* EMLA3: application in alkaline and salt-rich dyeing effluent treatment. Extremophiles 21:479–490

Bhosale S, Saratale G, Govindwar S (2006) Mixed function oxidase in *Cunninghamella blakesleeana* (NCIM-687). J Basic Microbiol 46:444–448

Birmole R, Patade S, Sirwaiya V, Bargir F, Aruna K (2014) Biodegradation study of reactive blue 172 by *Shewanella haliotis* DW01 isolated from lake sediment. Indian J Sci Res 5:139–152

Biswas G, Kumari M, Adhikari K, Dutta S (2017) Application of response surface methodology for optimization of biosorption of fluoride from groundwater using Shorea robusta flower petal. Appl Water Sci 7:4673–4690

Blaga A, Tatyana A, Lilyana S, Sava M (2008) Temperature effect on bacterial azo bond reduction kinetics: an Arrhenius plot analysis. Biodegradation 19:387–393

Blumel S, Stolz A (2003) Cloning and characterization of the gene coding for the aerobic reductase from *Pigmentiphaga kullae* K24. Appl Microbiol Biotechnol 62:186–190

Bromley-Challenor K, Knapp JS, Zhang Z, Gray NCC, Hetheridge M, Malcol EMR (2000) Dye reduction of an azo dye by unacclimated activated sludge under anaerobic conditions. Water Res 34:4410–4418

Chang JS, Kuo TS (2000) Kinetics of bacterial decolorization of azo dye with *Escherichia coli* NO3. Bioresour Technol 75:107–111

Chang JS, Lin YC (2000) Fed-batch bioreactor strategies for microbial decolorization of azo dye using a *Pseudomonas luteola* strain. Biotechnol Prog 16:979–985

Chang JS, Chou C, Chen SY (2001) Decolorization of azo dyes with immobilized *Pseudomonas luteola*. Process Biochem 36:757–763

Chen H, Hopper SL, Cerniglia CE (2005) Biochemical and molecular characterization of an azoreductase from *Staphylococcus aureus*, a tetrameric NADPH-dependent flavoprotein. Microbiology 151:1433–1441

Chivukula M, Renganathan V (1995) Phenolic azo dye oxidation by laccase from *Pyricularia oryzae*. Appl Environ Microbiol 61:4374–4377

Dawkar VV, Jadhav UU, Telke AA, Govindwar SP (2009) Peroxidase from Bacillus sp. VUS and its role in the decolourization of textile dyes. Biotechnol Bioprocess Eng 14:361–368

De Roos AJ, Ray RM, Gao DL, Wernli KJ, Fitzgibbons ED, Zinding F, Astrakianakis G, Thoma DB, Checkoway H (2005) Colorectal cancer incidence among female textile workers in Shanghai, China: a case-Cohort analysis of occupational exposures. Cancer Causes Control 16:1177–1188

Dogan EE, Yesilada E, Ozata L, Yologlu S (2005) Genotoxicity testing of four textile dyes in two crosses of Drosophila using wing somatic mutation and recombination test. Drug Chem Toxicol 28:289–301

Dong X, Zhou J, Liu Y (2003) Peptone-induced biodecolorization of reactive brilliant blue (KN-R) by *Rhodocyclus gelatinosus* XL-1. Process Biochem 39:89–94

Dos Santos AB, Cervantes FJ, Van Lier JB (2007) Review paper on current technologies for decolorization of textile wastewaters: perspectives for anaerobic biotechnology. Bioresour Technol 98:2369–2385

Duran N, Esposito E (2000) Potential applications of oxidative enzymes and phenoloxidase-like compounds in wastewater and soil treatment: a review. Appl Catal B Environ 28:83–99

Elisangela F, Andrea Z, Dias F, Ragagnin C, Lucia D, Artur C (2009) Biodegradation of textile azo dyes by a facultative *Staphylococcus arlettae* strain VN-11 using a sequential microaerophilic/ aerobic process. Int Biodeterior Biodegradation 63:280–288

Field JA, Brady J (2003) Riboflavin as a redox mediator accelerating the reduction of the azo dye mordant yellow 10 by anaerobic granular sludge. Water Sci Technol 36:1497–1504

Franciscon E, Grossman MJ, Paschoal JAR, Reyes FGR, Durrant LR (2012) Decolorization and biodegradation of reactive sulfonated azo dyes by a newly isolated *Brevibacterium* sp. strain VN-15. Springplus 1:37

Garg S, Tripathi M, Lal N (2015) Statistical design for optimization of process parameters for biodecolorization of Reactive Orange 4 Azo Dye by Bacillus cereus isolate. Res J Microbiol 10:502–512

Giwa A, Giwa FJ, Ifu BJ (2012) Microbial decolourization of an anthra- quinone dye C.I. reactive blue 19 using *Bacillus cereus*. Am J Chem Sci 2:60–68

Gomare SS, Tamboli DP, Kagalkar AN, Govindwar SP (2009) Eco-friendly biodegradation of a reactive textile dye golden yellow HER by *Brevibacillus laterosporus* MTCC 2298. Int Biodeterior Biodegradation 63:582–586

Gordon S, Hsieh YL (2007) Cotton: science and technology. Ch. 13. Woodhead Publishing and CRC Press, Cambridge

Guadie A, Tizazu S, Melese M, Guo W, Ngo HH, Xia S (2017) Biodecolorization of textile azo dye using *Bacillus* sp. strain CH12 isolated from alkaline lake. Biotechnol Rep (Amst) 15:92–100

Haghshenas H, Khodaii A, Khedmati TS (2015) A mathematical model for predicting stripping potential of hot mix asphalt. Constr Build Mater 75:488–495

Hao OJ, Kim H, Chiang PC (2000) Decolorization of wastewater. Crit Rev Environ Sci Technol 30:449–505

Hatvani N, Mecs I (2001) Production of laccase and manganese peroxidase by *Lentinus edodes* on malt containing by-product of the brewing process. Process Biochem 37:91–496

Humnabadkar IP, Saratake GD, Govindwar SP (2008) Decolorization of purple 2R by *Aspergillus ochraceus* (NCIM-1146). Asian J Microbiol Biotechnol Environ Sci 10:693–697

Ingelman M, Ramaswamy S, Nivière V, Fontecave M, Eklund H (1999) Crystal structure of NAD (P)H:flavin oxidoreductase from *Escherichia coli*. Biochemi 38:7040–7049

Jadhav SU, Jadhav MU, Anuradha NK, Govindwar SP (2008a) Decolorization of brilliant blue G dye mediated by degradation of the microbial consortium of *Galactomyces geotrichum* and *Bacillus* sp. J Chin Inst Chem Eng 39:563–570

Jadhav SU, Jadhav MU, Anuradha NK, Govindwar SP (2008b) Biodegradation of disperse dye brown 3REL by microbial consortium of *Galactomyces geotrichum* MTCC 1360 and *Bacillus* sp. VUS. Biotechnol Bioprocess Eng 13:232–239

Jadhav SU, Dawkar VV, Telke AA, Govindwar SP (2009) Decolorization of direct blue GLL with enhanced lignin peroxidase enzyme production in *Comamonas* sp UVS. J Chem Technol Biotechnol 84:126–132

Jadhav JP, Kalyani DC, Telke AA, Phugare SS, Govindwar SP (2010) Evaluation of the efficacy of a bacterial consortium for the removal of color, reduction of heavy metals, and toxicity from textile dye effluent. Bioresour Technol 101:165–173

Jadhav SB, Surwase SN, Phugare SS, Jadhav JP (2013) Response surface methodology mediated optimization of Remazol Orange decolorization in plain distilled water by *Pseudomonas aeruginosa* BCH. Int J Environ Sci Technol 10:181–190

Jin X, Liu G, Xu Z, Tao W (2007) Decolourisation of a dye industry effluent by *Aspergillus fumigatus* XC6. Appl Microbiol Biotechnol 74:239–243

Joe MH, Lim SY, Kim DH, Lee IS (2008) Decolorization of reactive dyes by *Clostridium bifermentans* SL186 isolated from contaminated soil. World J Microbiol Biotechnol 24:2221–2226

Joshi T, Iyengar L, Singh K, Garg S (2008) Isolation, identification and application of novel bacterial consortium TJ-1 for the decolorization of structurally different azo dyes. Bioresour Technol 99:7115–7121

Kalme S, Parshetti GK, Jadhav S, Govindwar S (2007) Biodegradation of benzidine based dye direct blue-6 by *Pseudomonas desmolyticum* NCIM 2112. Bioresour Technol 98:1405–1410

Kalme S, Jadhav S, Jadhav M, Govindwar S (2009) Textile dye degrading laccase from *Pseudomonas desmolyticum* NCIM 2112. Enzym Microb Technol 44:65–71

Kalyani DC, Patil PS, Jadhav JP, Govindwar SP (2008) Biodegradation of reactive textile dye Red BLI by an isolated bacterium Pseudomonas sp. SUK1. Bioresource Technol 99(11):4635–4641

Kalyani DC, Telke AA, Dhanve RS, Jadhav JP (2009a) Eco-friendly biodegradation and detoxification of reactive red 2 textile dye by newly isolated *Pseudomonas* sp. SUK1. J Hazard Mater 163:735–742

Kalyani DC, Telke AA, Govindwar SP, Jadhav JP (2009b) Biodegradation and detoxification of reactive textile dye by isolated *Pseudomonas* sp. SUK1. Water Environ Res 81:298–307

Karimifard S, Alavi Moghaddam MR (2018) Application of response surface methodology in physicochemical removal of dyes from wastewater: a critical review. Sci Total Environ 640–641:772–797

Khan S, Malik A (2016) Degradation of reactive black 5 dye by newly isolated bacterium *Pseudomonas entomophila* BS1. Can J Microbiol 62:220–232

Khan Z, Jain K, Soni A, Madamwar D (2014) Microaerophilic degradation of sulphonated azo dye–reactive red 195 by bacterial consortium AR1 through co-metabolism. Int Biodeterior Biodegradation 94:167–175

Khan S, Anas M, Malik A (2019) Mutagenicity and genotoxicity evaluation of textile industry wastewater using bacterial and plant bioassays. Toxicol Rep 6:193–201

Khatri A, Peerzada MH, Mohsin M, White M (2015) A review on developments in dyeing cotton fabrics with reactive dyes for reducing effluent pollution. J Clean Prod 87:50–57

Khedmati M, Khodaii A, Haghshenas HF (2017) A study on moisture susceptibility of stone matrix warm mix asphalt. Constr Build Mater 144:42–49

Khehra MS, Saini HS, Sharma DK, Chadha BS, Chimni SS (2006) Biodegradation of azo dye C.I. Acid red 88 by anoxic-aerobic sequential bioreactor. Dyes Pigments 70:1–7

Kilic NK, Nielsen JL, Yuce M, Donmez G (2007) Characterization of a simple bacterial consortium for effective treatment of wastewaters with reactive dyes and Cr(VI). Chemosphere 67:826–831

Kılıç Z, Atakol O, Emregül E, Çelikkol P, Aras S, Cansaran-Duman D (2008) Biosorption characteristics of zinc(II) from aqueous solutions by *Pseudevernia furfuracea*. Presented at 6th international Conference of Chemical Societies of South Eastern European Countries, 10–14 Sept, Sofia

King D (2007) Dyeing of cotton and cotton products. In: Gordon S, Hsieh YL (eds) Cotton: science and technology. Woodhead Publishing, Cambridge, pp 353–377

Knapp JS, Newby PS (1995) The microbiological decolorization of an industrial effluent containing a diazo-linked chromophore. Water Res 29:1807–1809

Kulandaivel S, Kaleeswari P, Mohanapriya P (2014) Decolorization and adsorption of dyes by consortium of bacteria with agriculture waste. Int J Curr Microbiol App Sci 3:865–882

Lade H, Kadam A, Paul D, Govindwar S (2015) Biodegradation and detoxification of textile azo dyes by bacterial consortium under sequential microaerophilic/aerobic processes. EXCLI J 14:158–174

Lalnunhlimi S, Krishnaswamy V (2016) Decolorization of azo dyes (direct blue 151 and direct red 31) by moderately alkaliphilic bacterial consortium. Braz J Microbiol 47:39–46

Lewis DM, Broadbent PJ, Vo LTT (2008) Covalent fixation of reactive dyes on cotton under neutral conditions. AATCC Rev 8:35–41

Maier J, Kandelbauer A, Erlacher A, Cavaco-Paulo A, Guebitz G (2004) A new alkali-thermostable azoreductase from *Bacillus* sp. strain SF. Appl Environ Microbiol 70:837–844

Majcherczyk A, Johannes C, Huttermann A (1998) Oxidation of polycyclic aromatic hydrocarbons (PAH) by laccase of *Trametes versicolor*. Enzym Microb Technol 22:335–341

Mathur N, Bhatnagar P, Bakre P (2006) Assessing mutagenicity of textile dyes from Pali (Rajasthan) using Ames bioassay. J Appl Ecol Environ Res 4(1):111–118
McMahon AM, Doyle EM, Brooks S, O'Connor KE (2007) Biochemical characterisation of the coexisting tyrosinase and laccase in the soil bacterium Pseudomonas putida F6. Enzyme Microb Technol 40:1435–1441
Mona S, Kaushik A, Kaushik CP (2011a) Biosorption of reactive dye by waste biomass of *Nostoc linckia*. Ecol Eng 37:1589–1594
Mona S, Kaushik A, Kaushik CP (2011b) Waste biomass of *Nostoc linckia* as adsorbent of crystal violet dye: optimization based on statistical model. Int Biodeterior Biodegradation 65:513–521
Moosvi S, Kher X, Madamwar D (2007) Isolation, characterization and decolorization of textile dyes by a mixed bacterial consortium JW-2. Dyes Pigments 74:723–729
Nachiyar CV, Rajkumar GS (2003) Degradation of tannery and textile dye, Navitan Fast Blue S5R by *Pseudomonas aeruginosa*. World J Microbiol Biotechnol 19:609–614
Nachiyar CV, Rajkumar GS (2005) Purification and characterization of an oxygen insensitive azoreductase from *Pseudomonas aeruginosa*. Enzym Microb Technol 36:503–509
Nigam P, Armour G, Banat IM, Singh D, Marchant R (2000) Physical removal of textile dyes from effluents and solid-state fermentation of dye adsorbed agricultural residues. Bioresour Technol 72:219–226
Nilsson R, Nordlinder R, Wass U, Meding B, Belin L (1993) Asthma, rhinitis, and dermatitis in workers exposed to reactive dyes. Br J Ind Med 50:65–70
Olukanni OD, Osuntoki AA, Gbenle GO (2006) Textile effluent biodegradation potentials of textile effluent-adapted and non-adapted bacteria. Afr J Biotechnol 5:1980–1984
Pandey A, Singh P, Iyengar L (2007) Bacterial decolorization and degradation of azo dyes. Int Biodeterior Biodegr 59:73–84
Pearce CI, Lioyd JR, Guthrie JT (2003) The removal of colour from textile wastewater using whole bacterial cells: a review. Dyes Pigments 58:179–186
Prabhakar Y, Gupta A, Kaushik A (2019a) Enhanced decolorization of reactive violet dye 1 by halo-alkaliphilic *Nesterenkonia* strain: process optimization, short acclimatization and reusability analysis in batch cycles. Process Saf Environ Prot 131:116–126
Prabhakar Y, Gupta A, Kaushik A (2019b) Effect of some organic co-pollutants on decolorization of reactive violet 1 dye by an indigenous microbial strain from textile wastewater. Environ We Int J Sci Tech 14:159–168
Prabhakar Y, Gupta A, Kaushik A (2021) Microbial degradation of reactive red-35 dye: upgraded progression through Box–Behnken design modeling and cyclic acclimatization. J Water Process Eng 40:101782
Prasad ASA, Rao KVB (2013) Aerobic biodegradation of azo dye by *Bacillus cohnii* MTCC 3616, an obligately alkaliphilic bacterium and toxicity evaluation of metabolites by different bioassay systems. Appl Microbiol Biotechnol 97:7469–7481
Preeti R, Kaushik A, Kaushik CP (2015) Exploring biosorption potential of dry immobilized biomass of cyanobacteria for removal of triphenylmethane dye under thermal conditions. Int J Multidiscip Res Dev 2:748–753
Rehm B (2009) Microbial production of biopolymers and polymer precursors. Caister Academic, Norfolk
Russ R, Rau J, Stolz A (2000) The function of cytoplasmic flavin reductases in the reduction of azo dyes by bacteria. Appl Environ Microbiol 66:1429–1434
Salokhe MD, Govindwar S (2003) Inducibility of biotransformation enzymes in *Serratia marcescens*. World J Microbiol Biotechnol 19:199–200
Saratale G, Humnabadkar R, Govindwar S (2007) Study of mixed function oxidase system in Aspergillus ochraceus (NCIM 1146). Ind J Microbiol 47:304–309
Saratale GD, Chen SD, Lo YC, Saratale RG, Chang JS (2008) Outlook of biohydrogen production from lignocellulosic feedstock using dark fermentation-a review. J Sci Ind Res 67:162–179

Saratale RG, Saratale GD, Kalyani DC, Chang JS, Govindwar SP (2009) Enhanced decolorization and biodegradation of textile azo dye scarlet R by using developed microbial consortium-GR. Bioresour Technol 100:2493–2500

Saratale RG, Saratale GD, Chang J, Govindwar SP (2010) Decolorization and biodegradation of reactive dyes and dye wastewater by a developed bacterial consortium. Biodegradation 21:999–1015

Saratale RG, Saratale GD, Change JS, Govindwar SP (2011) Bacterial decolorization and degradation of azo dyes: a review. J Taiwan Inst Chem Eng 42:138–157

Senthilkumar S, Prabhu HJ, Perumalsamy M (2013) Response surface optimization for biodegradation of textile azo dyes using isolated bacterial strain *Pseudomonas* sp. Arab J Sci Eng 38:2279–2291

Shah K (2014) Biodegradation of azo dye compounds. Int Res J Biochem Biotechnol 1:5–13

Shah PD, Dave SR, Rao MS (2012) Enzymatic degradation of textile dye reactive orange 13 by newly isolated bacterial strain *Alcaligenes faecalis* PMS-1. Int J Biodeter Biodegr 69:41–50

Shah MP, Patel KA, Nair SS, Darji AM (2013) Microbial degradation and decolorization of reactive orange dye by strain of *Pseudomonas* spp. Int J Environ Bioremediat Biodegr 1:1–5

Shore J (2002) Chemistry of reactive dyes. In: Shore J (ed) Colorants and auxiliaries, vol 1. Society of Dyers and Colourists, Bradford, pp 356–443

Smith B (2003) Wastes from textile processing. In: Andrady AL (ed) Plastics and the environment. Wiley, Hoboken, pp 293–295

Stolz A (2001) Basic and applied aspects in the microbial degradation of azo dyes. Appl Microbiol Biotechnol 56:69–80

Sudhakar R, Gowda NKN, Venu G (2001) Mitotic abnormalities induced by silk dyeing industry effluents in the cells of Allium cepa. Cytologia 66:235–239

Suganya K, Revathi K (2016) Decolorization of reactive dyes by immobilized bacterial cells from textile effluents. Int J Curr Microbiol App Sci 5:528–532

Tallur PN, Mulla SI, Megadi VB, Talwar MP, Ninnekar HZ (2015) Biodegradation of cypermethrin by immobilized cells of Micrococcus sp. strain CPN 1. Braz J Microbiol 46:667–672

Tan L, Ning S, Xia H, Sun J (2013) Aerobic decolorization and mineralization of azo dyes by a microbial community in the absence of an external carbon source. Int Biodeterior Biodegradation 85:210–216

Taylor JA, Pasha K, Phillips DA (2001) The dyeing of cotton with hetero bi-functional reactive dyes containing both a monochlorotriazinyl and a chloroacetylamino reactive group. Dyes Pigments 51:145–152

Telke A, Kalyani D, Jadhav J, Govindwar S (2008) Kinetics and mechanism of reactive red 141 degradation by a bacterial isolate *Rhizobium radiobacter* MTCC 8161. Acta Chim Slov 55:320–329

Telke A, Kalyani D, Jadhav U, Parshetti G, Govindwar S (2009) Purification and characterization of an extracellular laccase from a *Pseudomonas* sp. LBC1 and its application for the removal of bisphenol A. J Mol Catal B Enzym 61:252–260

Usha MS, Sasirekha B, Bela RB, Devi S, Kamalini C, Manasa GA, Neha PM (2010) Batch, repeated batch and continuous degradation of reactive black 5 and reactive red 120 dye by immobilized bacteria. J Sci Ind Res 71:504–510

Vijayaraghavan K, Yun YS (2008) Bacterial biosorbents and biosorption. Biotechnol Adv 26:266–291

Wang H, Ren ZJ (2014) Bioelectrochemical metal recovery from wastewater: a review. Water Res 66:219–232

Wang CJ, Hagemeier C, Rahman N, Lowe E, Noble M, Coughtrie M, Sim E, Westwood I (2007) Molecular cloning, characterisation and ligand-bound structure of an azoreductase from *Pseudomonas aeruginosa*. J Mol Biol 373:1213–1228

Wang H, Su JQ, Zheng HW, Tian Y (2009) Bacterial decolorization and degradation of the reactive dye reactive red 180 by *Citrobacter* sp. CK3. Int Biodeterior Biodegradation 63:395–399

Witek-Krowiak A, Chojnacka K, Podstawczyk D, Dawiec A, Pokomeda K (2014) Application of response surface methodology and artificial neural network methods in modelling and optimization of biosorption process. Bioresour Technol 160:150–160
Wu C, Liu X, Wei D, Fan J, Wang L (2001) Photosonochemical degradation of phenol in water. Water Res 35:3927–3933
Wuhrmann K, Mechsner K, Kappeler T (1980) Investigation on rate-determining factors in the microbial reduction of azo dyes. Eur J Appl Microbiol Biotechnol 9:325–338
Zhang X, Flurkey WH (1997) Phenoloxidases in portabella mushrooms. J Food Sci 62:97–100
Zhang F, Yediler A, Liang X, Kettrup A (2004) Effects of dye additives on the ozonation process and oxidation by-products: a comparative study using hydrolyzed C1 reactive red 120. Dyes Pigments 60:1–7
Zimmermann T, Kulla HG, Leisinger T (1982) Properties of purified orange II azoreductase, the enzyme initiating azo dye degradation by *Pseudomonas* KF46. Eur J Biochem 129:197–203
Zimmermann T, Gasser F, Kulla H, Leisinger T (1984) Comparison of two bacterial azoreductases acquired during adaptation to growth on azo dyes. Arch Microbiol 138:37–43

Degradation and Biotransformation of Pentachlorophenol by Microorganisms

16

Madan Kumar, Asmita Gupta, and Shaili Srivastava

Abstract

Chlorinated phenols are a group of commercially produced substituted phenols and cresols, also referred to as chlorophenols and chlorocresols that are prepared by direct chlorination or hydrolysis of higher chlorinated derivatives of benzene. Pentachlorophenol (PCP) is an organochlorine pesticide that has found application as a wood treatment agent and biocide. Its recalcitrant and toxic nature leading to serious environmental consequences has resulted in the need for its remediation and also a ban in many countries. Although, in nature, PCP is mostly recalcitrant, yet numerous microorganisms such as certain bacteria, fungi, algae and their consortia have been shown promising PCP biodegradation and detoxification potential. This chapter presents a brief account of PCP degradation by various microorganisms along with their transformation pathways. The chapter discussed physicochemical removal of PCP and by mixed microbial communities in soil and water using biochar and bioelectrochemical systems.

Keywords

Chlorinated phenols · Pentachlorophenols (PCP) · Bacteria · Fungi · Biodegradation

M. Kumar
School of Environmental Sciences, Jawaharlal Nehru University, Delhi, India

A. Gupta
Department of Botany, Daulat Ram College, University of Delhi, Delhi, India

S. Srivastava (✉)
Amity School of Earth and Environmental Sciences, Amity University Haryana, Gurgaon, Haryana, India
e-mail: shailisrivastava05@gmail.com

A. Kaushik et al. (eds.), *Climate Resilience and Environmental Sustainability Approaches*, https://doi.org/10.1007/978-981-16-0902-2_16

16.1 Introduction

Chlorinated phenols (CPs) are industrial chemicals having wide applications, such as pesticides, herbicides, disinfectants, biocides and wood preservatives (Keane 2005). Chlorophenols (CPs) are aromatic ring structures containing at least one chlorine atom (–Cl) and one hydroxyl (–OH) group at the benzene rings. The group of CPs comprises 19 different chlorinated phenolic compounds, including pentachlorophenol (PCP), 3 tetrachlorophenols (TeCP), 6 trichlorophenols (TCP), 6 dichlorophenols (DCP), and 3 monochlorophenols (MCP) (Agency for Toxic Substances and Disease Registry 1999). CPs are industrially produced by direct chlorination of phenol with chlorine gas, as well as other reactions, such as hydrolysis and hydro dechlorination, or by chlorination of less chlorinated phenols in the presence of aluminium or iron trichloride.

These compounds are widely used as mothproofing agents, miticides, germicides, algicides, fungicides and wood preservatives as well as for the synthesis of dyes and drugs (Muller and Caillard 2011).

CPs have been introduced into the environment via anthropogenic activities. The major sources of contamination are industrial wastes, pesticides, herbicides and complex chlorinated hydrocarbons (Olaniran and Igbinosa 2011). CPs are industrially produced by direct chlorination of phenol with chlorine gas, as well as other reactions, such as hydrolysis and hydrodechlorination or by chlorination of less chlorinated phenols in the presence of aluminium or iron trichloride (Kim and Carraway 2003). CPs are also by products of industrial processes, such as the bleaching of pulp using chlorine or chlorine dioxide (Patel and Suresh 2008). CPs have been detected in groundwater, surface water, wastewater, air and soils as a result of their improper disposal, leaching from landfills and the incineration of chlorinated wastes (Agency for Toxic Substances and Disease Registry 1999). People may be exposed to CPs by eating or drinking substances that contain them or through skin contact. CPs and their derivatives are highly toxic to living beings due to their carcinogenic, mutagenic and cytotoxic properties (Igbinosa et al. 2013). The toxicity of CPs increases with the degree of chlorination, but decreases with the degree of their dissociation (Pera-Titus et al. 2004). As a result, at low pH, where the non-dissociated form of CP is dominant, their toxicity is greater. While PCP has been identified as the most toxic chlorophenol (Tanjore and Viraraghavan 1994), many CPs are of environmental concern due to their acute toxicity and resistance to degradation. Phenol and most of the chlorophenolic compounds can induce mutations in living being including bacteria and fungi.

Several conventional methods which include chemicals and physical methods such as adsorption, ion exchange, liquid–liquid extraction and chemical oxidation and advanced oxidation processes have been used for the removal of CPs from wastewater and soil. These methods are expensive; require a lot of energy and not ecofriendly due to the formation of hazardous compounds as by products (Olaniran and Igbinosa 2011). Therefore the requirements of energy-efficient, cost-effective and green technology treatments are needed. Conversely, bioremediation is an effective and ecofriendly method of removing CPs from the environment.

Biodegradation of CPs has gained attention due to the complete mineralization of CPs by microorganisms in the environment. The use of enzymatic activities is one of the latest technologies used in the treatment and monitoring of effluents with a high content of organochlorinated compounds (Arora and Bae 2014).

16.1.1 Properties and Types of Chlorinated Phenols

Chlorophenols are basically made by a benzene ring associated with one -OH group, and one or more chlorine atoms. Chlorophenol's group comprises several different compounds differing from each other for the number of OHs and/or chlorine atoms and their relative positions on the aromatic ring, and the presence of other substituents. Chlorophenols show different physical-chemical properties due to different in molecular structure. The Cluster Rule, introduced by the US Environmental Protection Agency (U.S. EPA 2014) has regulated the level of 12 chlorophenols: 2,4,5-trichlorophenol, 2,4,6-trichlorophenol, 2,3,4,6-tetrachlorophenol, pentachlorophenol, 3,4,5-trichlorophenol, 3,4,6-trichlorophenol, tetrachlorocatechol, 3,4,5-trichloroguaiacol, 3,4,6-trichloroguaiacol, 4,5,6-trichloroguaiacol, tetrachloroguaiacol and trichlorosyringol. Other chlorophenols that have frequently been found in the environment (Fig. 16.1) and have some level of toxicity are 2-chlorophenol, 3-chlorophenol, 2,4-dichlorophenol, and 2,6 dichlorophenol (Shiu et al. 1994).

16.2 Chlorinated Compounds and Their Significance

Organochlorine compounds have been a matter of great concern due to uses in mothproofing agents, miticides, germicides, algicides, fungicides and wood preservatives and synthesis in dyes and drugs for the last decade. These compounds are also used for the preservation of leather. The toxicity of chlorinated compounds increases with the increasing number of chlorine atoms on the organic compounds. It is generally believed that they are of precisely the right size to fit as the key in the lock in some vital molecules of the living cells and by finally attached to this site, they prevent the normal functioning of the cell.

16.2.1 Pentachlorophenol and Its Significance

Pentachlorophenol and its sodium salt had a variety of other applications, for instance as a biocide, pesticide, disinfectant, defoliant, anti-sapstain and antimicrobial agent and preservation of leather. PCP is also being used in the preservation of wood, food rubber, oil paints, cardboard, and unintentionally formed in pulp and paper mill (WHO 2015; Agency for Toxic Substances and Disease Registry (ATSDR) 2001).

Fig. 16.1 Chemical structure of different chlorinated phenols

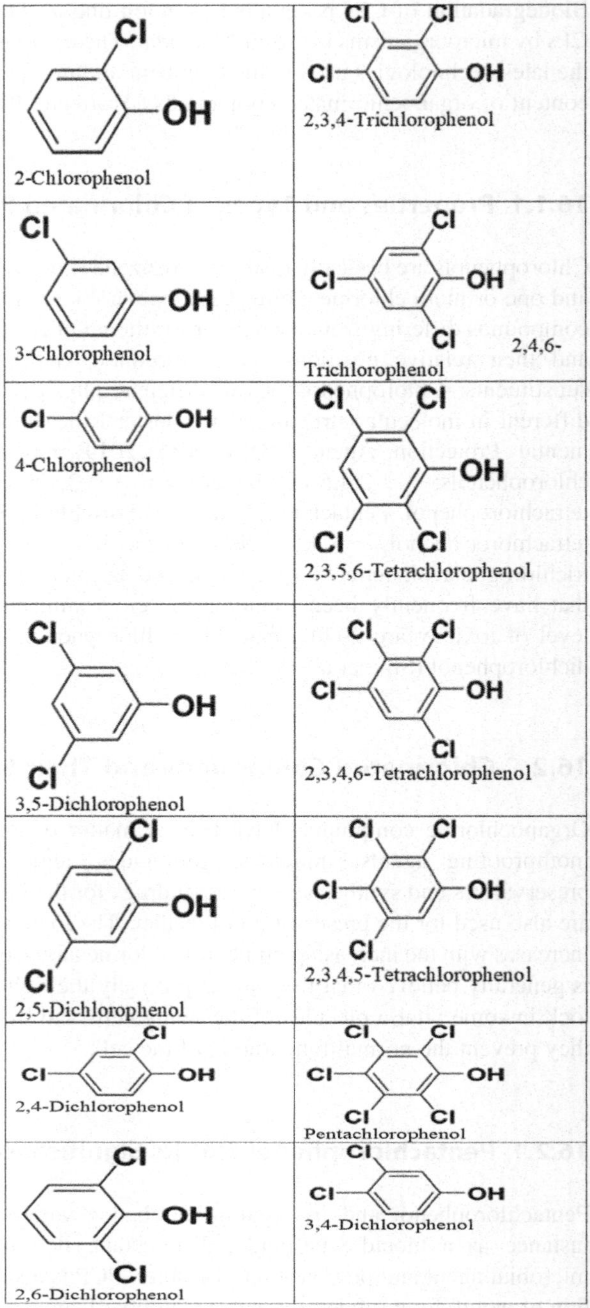

The worldwide production of PCP and Na-PCP is 150,000 tons/year. The US Environmental Protection Agency has registered 578 products containing Pentachlorophenol (Saber and Crawford 1985). It is estimated that 300–400 tonnes of PCP are used annually by UK industries over recent years. Over 100 tons are released directly via spillages, volatilization or associated with wastewater. The USEPA has regarded PCP as priority toxic chemical (Wild et al. 1992).

A United States patent granted the use of mixture containing 80% pentachlorophenol and 0.5–10% thymol for the protection of materials against biological attack. Pentachlorophenol has been extensively used in the construction and lumber industries and used to control mold and termite infestation and for the control of powder post beetles and wood-boring insects (Agency for Toxic Substances and Disease Registry (ATSDR) 2001). Chloranil (Tetrachloro-*P*-hydroquinone) an oxidation product of pentachlorophenol has been used as a fungicide under the trade name of "spergon" to protect seeds and bulbs. Sodium pentachlorophenol has been used as a molluscicide for the destruction of the snails, intermediate host of human schistosomes (Bevenue and Beckman 1967). A chemical with so wide and so varied usage presents a potential problem of environmental contamination.

16.2.2 Properties of Pentachlorophenol

Pentachlorophenol is a chlorinated phenol. All its valencies are substituted by chlorine. The widely used salt of PCP is the sodium Pentachlorophenol (Fig. 16.2).

16.2.2.1 Physical Properties of Pentachlorophenol

PCP forms colorless solid needle-like monoclinic crystals with a phenolic odor and insoluble in water. Therefore, the rapidly water-soluble salt (Na-PCP) is substituted in many industrial applications. The general properties of PCP are described in the following Table 16.1.

Pentachlorophenol is rapidly and extensively degraded by U.V. radiation and by sunlight (when it is in soluble form) to several identifiable products (Agency for Toxic Substances and Disease Registry (ATSDR) 2001). The degradation of pentachlorophenol by U.V. radiation or by sunlight is first-order reaction. The aqueous

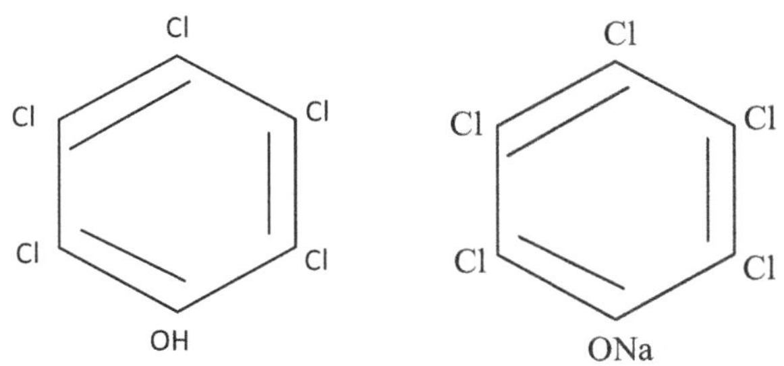

Fig. 16.2 Chemical structure of PCP (I) sodium PCP (II)

Table 16.1 Physical properties of pentachlorophenol

S.N.	Physical properties	Value
1.	Molecular weight	266.36
2.	Melting point	190 °C
3.	Boiling point	293 °C
4.	Density	1.85
5.	Vapour pressure (20–100 °C)	0.00011–0.12 mm Hg

Bevenue and Beckman (1967)

Table 16.2 Solubility of PCP in the different organic solvent

Solvent	Solubility (gm of PCP/100 gm solvent)
Methanol	57–65
Diethyl ether	53–60
Ethanol	47–52
Acetone	21–33
Xylene	14–17
Benzene	11–14
Carbon tetrachloride	2–3

solution of sodium pentachlorophenol expose to sunlight for 10 days and it decomposed in products chloranilic acid and chlorinated benzoquinones.

16.2.2.2 Chemical Properties of Pentachlorophenol

Pure pentachlorophenol is relatively inert and not subjected to coupling or substitution reactions, common to most phenols (Monsanto 1963). It is a weak acid ($Ka = 10^{-5}$) and reacts with strong bases to give corresponding water-soluble salt. The data on the solubility of Pentachlorophenol (gm of PCP/100 gm of solvent) in different solvents as recorded by Bevenue and Beckman (1967) are given in Table 16.2.

Strong oxidizing agents decompose pentachlorophenol. For example nitric acid treatment converts PCP to a mixture of tetrachloro-*o*- and *p*-quinones, which can be measured colorometrically (Monsanto 1963). Among its salt sodium and potassium salts are highly water soluble, ammonium salt 1% soluble in water and other metallic salt such as silver (golden yellow crystals) copper (purple crystals) and mercury (lemon yellow crystals) the solubility is of the same order of magnitude as pentachlorophenol (Monsanto 1963). Pentachlorophenol can be readily converted to either derivative, a property that is utilized for its analysis by gas chromatography.

16.2.2.3 Toxicology of Pentachlorophenol

Pentachlorophenol is highly toxic and recalcitrant compound. The recalcitrant nature of this compound is due to the carbon halogen bond and low electron density of the aromatic ring (Allison et al. 1983). Several illness and some deaths have resulted from the careless handling of this hazardous chemical. The contamination of water with toxic chemicals from run off waste, sewage and industrial wastes and its effect on the environment has created considerable interest.

Pentachlorophenol and its biotransformation products manifest adverse or toxic effects. The PCP gets access to the body system through inhalation and skin. It has been observed that lipoproteins have high affinity with PCP and the bonding between blood carrier is reversible (Gomez et al. 1991).

Pentachlorophenol may cause irritation of the skin, conjunctiva and upper respiratory tract and also cause demonstrable systemic absorption. Exfoliation of the epidermal layer of human hands occurred at 1–2 days after adjusting a weir dispensing 20% Na-PCP. Nine men died after dipping timber by hand and without any protection in a 1.5–2% solution for periods varying 3–30 days with average 13 days. Both hemolytic anemia and aplastic anemia have been reported to be associated with exposure to PCP (Hassan et al. 1985). The occupationally exposed cases had a significantly higher mean level of urinary PCP than the controls in 88 wood treaters in Hawaii (Gilbert et al. 1990). Chlorophenol derivatives catechol, chlorocatechols, guaiacol, chloroguaiacols and syringol exhibit toxic properties including cytotoxic, mutagenic and cancerogenic activity (Michałowicz and Majsterek 2010). Chlorophenol ozonation considers the degradation of single pollutants and/or the enhancement of the oxidative potential using catalysts or UV light (Aken et al. 2019).

The biochemical action of PCP is similar to that of the dinitro compounds, namely uncoupling of oxidative phosphorylation. The mechanism of uncoupling involves binding to the mitochondrial protein. A study of Mg^{++} ATPase and of Na^{+}, K^{+} ATPase from various tissues of rat revealed very complex reactions, suggesting that PCP uncouples oxidative phosphorylation at low concentrations and inhibit at higher concentration (De Saiah 1978). In vitro effect of PCP on mammalian or molluscan mitochondria included inhibition of mitochondrial and myosin adenosine triphosphatase, inhibition of glycolytic phosphorylation, inactivation of respiratory enzyme and damage to the mitochondrial structure.

16.3 Physicochemical Methods for Treatment of Pentachlorophenol

Several physicochemical processes have been developed for the removal of pollutants from industrial effluents.

16.3.1 Physical Method for the Degradation of PCP

Traditional clean up methods for removal of PCP from the effluent included the use of activated charcoal. Edgehill (1998) studied the potential of using carbonized pine bark as a substitute for activated carbon. The absorption capacity for phenol and PCP at pH 2 and pH 8 were evaluated. The adsorption capacity of carbonized bark was much lower than for phenol. The decomposition of PCP by sonication was investigated at two frequencies (20 and 500 MHz) and two concentrations (20 and 60 fm), observed by products of sonication include tetrachloro-*o*-benzoquinone, tetrachlorocatechol, oxalate and chloride (Weavers et al. 2000). Gunavardana et al. (2011) have been reported zero valent iron (ZVI) and iron-based bimetallic systems have the potential to efficiently dechlorinate CPs. The removal of dissolved CPs in iron-water systems occurs via dechlorination, sorption and co-precipitation. Although ZVI has been commonly used for the dechlorination of CPs, its long-term reactivity is limited due to surface passivation over time.

16.3.2 Chemical Removal of PCP

Various chemical techniques are used to remove PCP from effluent such as coagulation, precipitation, reverse osmosis, etc. A chemical method for oxidative degradation of PCP in soil under unsaturated conditions and neutral pH was developed. Reagents used were haeme (Fe++) as a catalyst and hydrogen peroxide as an oxidant. Chen et al. (1999) reported that haeme and peroxide (Fenton reagent) could degrade PCP efficiently in short period of time either in liquid or unsaturated soil systems. However, the problems as underlying the industrial non-acceptability of the physicochemical treatment technology are those associated with cost and reliability.

16.4 Bacterial Degradation of PCP

Complete mineralization of Cps and Pentachlorophenol is possible through biodegradation; It could effective, economic and sustainable approach. PCP is more recalcitrant to bacterial degradation than less chlorinated Cps due to the presence of five chlorine atom in the phenolic ring and structure. Several strains of bacteria and fungi have been reported to degrade CP and PCP by aerobic and anaerobic degradation processes. A variety of bacteria have been isolated and applied for the degradation of pentachlorophenol (Srivastava and Thakur 2007).

The degradation of pentachlorophenol by a *Flavobacterium* sp. was investigated using PCP as the sole carbon source, and the degradation actively increased with PCP concentration. However, a lag phase was observed which was more pronounced at higher concentration of PCP (Gonzalez and Hu 1991). Edgehill (1998) studied degradation of PCP by *Arthrobacter* strain with naturally immobilized cells on glass bead in a column. PCP was removed from mineral salt medium 4–5 days after inoculation of the column with PCP acclimatized cells. Four strains of PCP-degrading bacteria, (*Arthrobacter*, *Flavobacterium*, *Pseudomonas* and *Sphingomonas*) were isolated by Ederrer et al. (1997) from the different geographical regions of the USA. The PCP degradation was reported in tannery effluent by *Acinetobacter* sp. and *Aspergillus* sp. using sequential bioreactor (Srivastava et al. 2007).

P. aeruginosa completely degraded PCP up to 800 mg/L in 6 days. The PCP was degraded to CO_2 with the release of chloride (Premlatha and Rajkumar 1994). Mineralization of PCP by *Pseudomonas* sp. was demonstrated by Resnick and

Chapman (1994), by loss of detectable PCP from the growth medium, stoichiometric of chloride release and formation of biomass consistent with the concentration of PCP mineralized.

Saboo et al. (1998) isolated bacterium from a pentachlorophenol contaminated site which can grow in the presence of 50 mg PCP/mL but were not able to degrade it in either liquid medium or in the presence of 1% sterile potting solid support.

A Pentachlorophenol degrading bacterium was isolated from possible PCP-contaminated soil from Pusan, Korea and identified as a member of the genus *Pseudomonas*. This strain was able to degrade a much higher concentration of PCP (4000 mg/L) than any previously reported PCP-degrading bacteria and fungi. It was suggested that this strain could be used for the bioremediation of highly PCP-contaminated soils, water or wood products (Lee et al. 1998).

Crawford et al. (1999) isolated four pentachlorophenol degrading bacteria from geographically diverse areas which have been examined in detail as regards to their physiology and phylogeny. According to traditional biochemical methods, these strains had been classified as members of genera *Arthobacteria*, *Flavobacterium*, *Pseudomonas* and *Sphingomonas*.

16.4.1 Aerobic Degradation of Pentachlorophenols by Bacteria

In *Sphingomonas chlorophenolicum* (Takeuchi et al. 2001), PCP degradation is initiated at the para-position hydroxylation by either PCP-4-monooxygenase (Fetzner 1998; Xun and Orser 1991) or cytochrome P-450 type enzyme. This step converts PCP into the formation of tetrachlorohydroquinone (TeCHQ), encoded by PcpB gene (Orser et al. 1993a, b; Lange et al. 1996). Afterwards, TCHQ is sequentially dehalogenated into 2,6-dichloro-1,4-hydroquinone (2,6-DCHQ) by a TCHQ-reductive dehalogenase (RDase). This reaction is encoded by the pcpC gene (Orser et al. 1993a, b). In the next step, 2,6-DCHQ is degraded to 2-chloromaleylacetate by the 2,6-DCHQ-1,2 dioxygenase (Ohtsubo et al. 1999). This step is crucial due to the cleavage of the aromatic ring. PcpE further converts 2-chloromaleylacetate into 3-oxoadipate via maleylacetate. Finally, 2-chloromaleylacetate is degraded in the TCA cycle (Fig. 16.3).

In *Mycobacterium chlorophenolicum* PCP-1 and *Mycobacterium fortuitum* CG-2 (formerly *Rhodococcus* strains), PCP is hydroxylated to TeCHQ by a membrane-bound cytochrome P-450 type enzyme (Uotila et al. 1992). Subsequently, TeCHQ

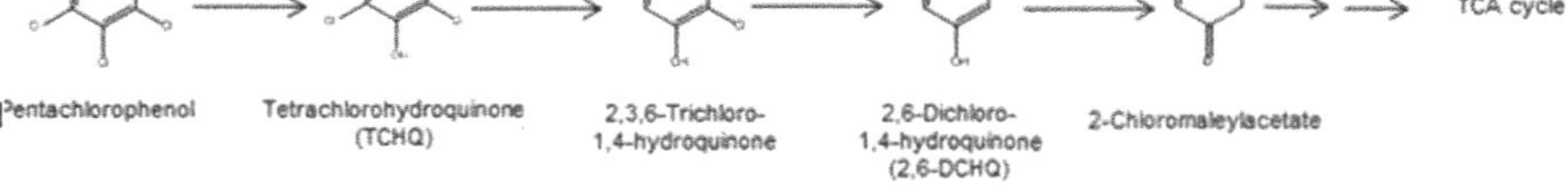

Fig. 16.3 Bacterial degradation pathway of pentachlorophenol (PCP) in *Sphingobium chlorophenolicum* [Adapted from Fetzner (1998) and Steiert and Crawford (1986)]

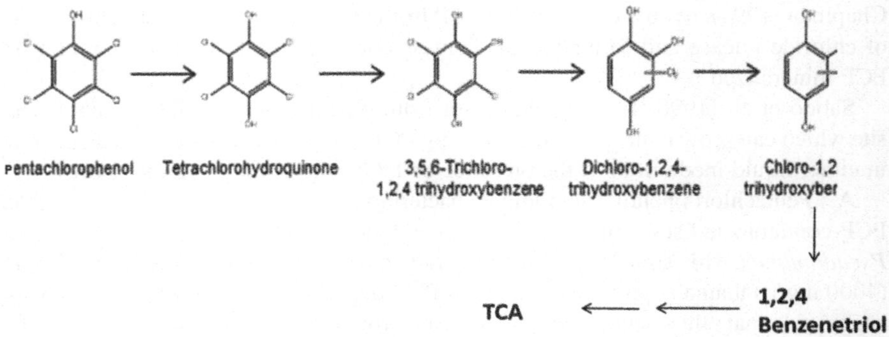

Fig. 16.4 Bacterial degradation pathway of PCP in *Mycobacterium* strains [Adapted from Uotila et al. (1992) and Apajalahti and Salkinoja-Salonen (1987)]

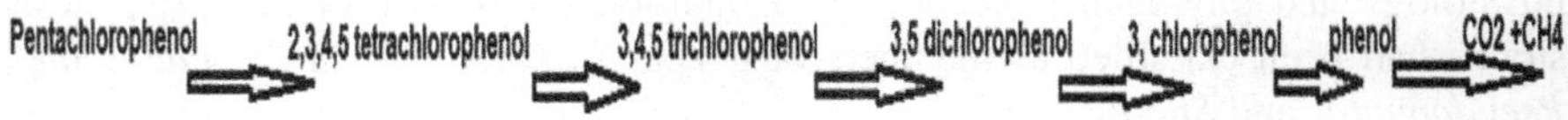

Fig. 16.5 Anaerobic dehalogenation of PCP in genus *Desulfitobacterium*

undergoes hydrolytic dehalogenation followed by reductive dehalogenation to form dichloro-1,2,4-trihydroxybenzene, which produces Benzene triol after two successive reductive dehalogenation (Fig. 16.4) (Apajalahti and Salkinoja-Salonen 1987).

16.4.2 Anaerobic Degradation of PCP

Anaerobic degradation of PCP are reported in many microorganisms, it has been reported that degradation occurs by series of reductive dehalogenase reaction, resulting in the formation of phenol which finally converted into CH4 and CO_2 by anaerobic prokaryotes such as *Desulfitobacterium* (Fig. 16.5) (Mikesell and Boyd 1986).

16.5 Biodegradation of Pentachlorophenol by Fungi

Various fungi have been isolated and applied and their potentialities for the bioremediation of PCP have been tested. Different fungi exhibited varying rates and extent of PCP degradation. Several studies have shown that members of Ascomycota are more efficient in degrading PCP (Hechmi et al. 2016; Vacondio et al. 2015; Carvalho et al. 2009).

In one study conducted by Bosso et al. (2016), two fungal strains *Byssochlamys nivea* and *Scopulariopsis brumptii* were tested PCP sensitivity and degradation. It was found that both while strains had low sensitivity to 25 mg PCP/L, *B. nivea* and *S. brumptii* showed complete depletion of 12.5 and 25 mg PCP/L in liquid culture, respectively, after 28 days of incubation and at 28 °C. Carvalho et al. (2009) assessed 17 fungal strains from cork-colonizing community for co- and direct metabolism of PCP and reported PCP degradation under co-metabolic conditions by 15 of the isolates, and complete degradation through direct metabolism by 5 isolates viz., *Trichoderma longibrachiatu*, *Chrysonilia sitophila*, *Mucor plumbeus*, *P. glandicola* and *Penicillium janczewskii*. In a recent study conducted by Coelho et al. (2019), a thermostable enzyme Laccase (Dc-Lac) isolated from fungus *Deconica castanella* was shown to biodegrade 23% of 100μg/L concentration of PCP, enzyme optima being at 55 °C temperature and 2.5 pH.

Hechmi et al. (2016) studied PCP bioremediation and detoxification potential of two fungal species *Byssochlamys nivea* and *Scopulariopsis brumptii* in soil microcosms and observed a 60% and 62% PCP removal by *B. nivea* and *S. brumptii*, respectively in bioaugmented microcosms. A synergetic effect on PCP removal was reported in the case of co-inoculation of *B. nivea* and *S. brumptii*, with PCP concentration decreasing to 95% and 80% from initial concentrations of 12.5 and 25 mg/kg, respectively. The fungal strains were also shown to detoxify bioaugmented PCP-contaminated soils as confirmed by toxicity assays.

16.6 Pathways and Enzymes Involved in Degradation

Xiao and Kondo (2020) investigated the degradation and transformation of PCP by *Phlebia* species in liquid culture and reported that *Phlebia acanthocystis* had higher tolerance to PCP in potato dextrose agar (PDA) medium than other *Phlebia* species. Also, after incubating for 10 days, *P. acanthocystis* was found to remove 76% and 100% of PCP (25μM) in potato dextrose broth and low-nitrogen media, respectively. Transformation of PCP via methylation and oxidation to pentachloroanisole and *p*-tetrachlorohydroquinone, with the latter further undergoing methylation to tetrachloro-4-methoxyphenol and tetrachloro-1,4-dimethoxybenzene, was also reported to be mediated by enzymes cytochrome P450 monooxygenase, lignin peroxidase and manganese peroxidase of *P. acanthocystis* (Fig. 16.6). The results emphasized the importance of *P. acanthocystis* to be promising tool for the remediation of PCP-contaminated sites. Earlier, Ning and Wang (2012) had confirmed the occurrence of cytochrome P450 and P450-mediated PCP oxidation in a white-rot fungus *Phanerochaete chrysosporium*.

In another study, the marine-derived fungus *Trichoderma harzianum* CBMAI 1677 isolated from marine ascidian *Didemnun ligulum* was shown to completely degrade PCP with an increasing presence of pentachloroanisole (PCA) and 2,3,4,6-tetrachloroanisole (2,3,4,6-TeCA) in a liquid medium (containing 3% malt and initial PCP concentration of 20 mg/L), after 1 week of incubation. Further, *T. harzianum* CBMAI 1677 was also observed to biodegrade the main metabolites of PCP, that is PCA and 2,3,4,6-TeCA (Vacondio et al. 2015). Another study to understand the mechanisms of PCP degradation was conducted by Ruiz-Lara et al. (2020), wherein secretome and the intracellular proteome analysis of *Rhizopus oryzae* ENHE, showed the presence of 14 and 37 proteins, respectively, the latter

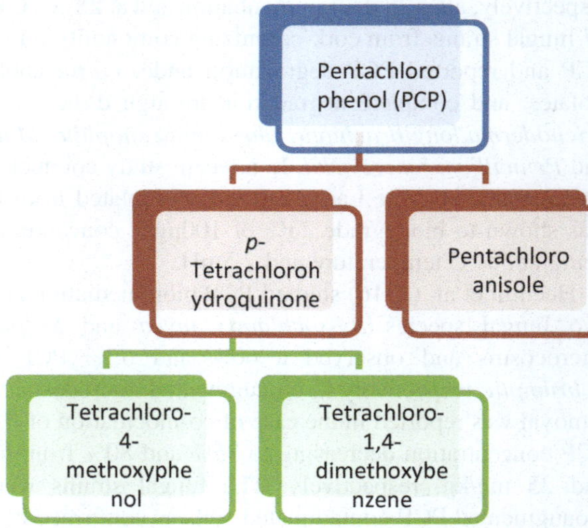

Fig. 16.6 Flowchart of proposed partial pathways for the degradation of PCP by *Phlebia acanthocystis* [Adapted from Xiao and Kondo (2020)]

also including the PCP-degrading peroxidases and methyltransferase. The kinetics of PCP sorption by non-viable biomass of *Rhizopus oryzae* ENHE had already been reported by León-Santiesteban et al. (2014).

16.7 Biodegradation of Pentachlorophenol by Algae

The degradation of PCP was mainly reported for bacteria and fungi and there very studies available on PCP degradation by algae. The idea of PCP degradation by algae was proposed by Oswald and Gotaas in 1957 but this process has recently gained interest due to several advantages such as the use of photosynthesis as energy source, uses inorganic nitrogen and phosphorus for growth, cost-effective production and maintenance and biomass to value-added products in form biofuel and biochemicals. A research group reported unicellular green algae *Scenedesmus obliquus* for the comparative degradation of chlorophenols and it was observed that with the increase in toxicity of chlorophenols the degradation of toxic compound increase with low biomass growth of microalgae (Fig. 16.7). It was observed that PCP initially inhibited the growth of algae but reported 89.1% degradation of PCP on the fifth day (Papazi et al. 2019). de Morais et al. reported PCP degradation study at environmental concentrations by a cyanobacterium *Microcystis aeruginosa* and microalga *Chlorella vulgaris*. It was observed that 1 μg/L PCP concentration promoted the growth of *M. aeruginosa* and above this concentration, growth inhibition was observed with removal of some PCP from the medium. But in the case of *C. vulgaris*, stabilization effect was observed with different concentration of PCP. No toxicity was detected at the lower concentration of PCP but at higher concentration of PCP, cell count decrease with no effect on chlorophyll a was observed (de Morais et al. 2014). A study reported degradation of phenolic effluent that

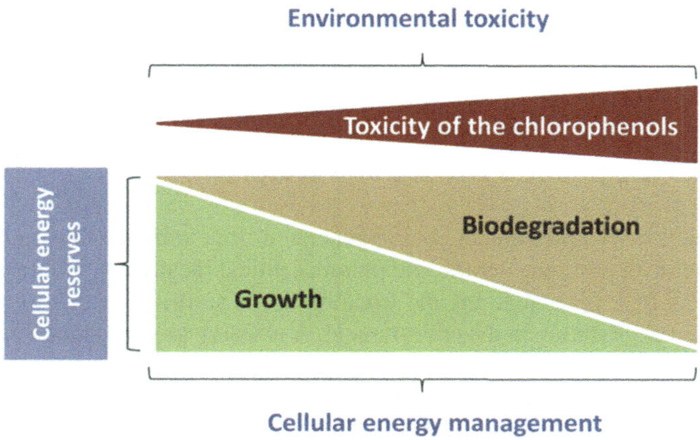

Fig. 16.7 Representation of bioenergetics strategy of microalgae [Adapted from Papazi et al. (2019)]

contains various phenolic derivatives including chloroderivatives from coal-based gasification plant by freshwater microalga *chlorella pyrenoidos*. It was observed that this alga at biomass concentration of 4 g/L in nutrient rich medium can degrade more than 95% of the initial total phenolic content, i.e. ~1500 mg/L with 7 days. Since oxygen is required for aromatic cleavage, it was observed that the addition of nutrients enhanced the rate of photosynthesis as result oxygen is released by the algae for degradation (Stephen and Ayalur 2017).

16.8 Pentachlorophenol Degradation by Mixed Microbial Communities

Mixed culture or consortia might overcome some of the problems degradative monocultures face in the environment such as nutrient limitation and unutilizable or toxic substrates. Consortia may be based on the exchange of specific nutrients, removal of growth inhibitory products and co-metabolism (Slater and Lovatt 1984). The association of microorganism to completely mineralize the toxic compound by sequential metabolism is referred to as microbial communities, consortia, syntrophic association or synergistic association. Simultaneous lignocellulose deconstruction and chlorophenols dehalogenation through two developed microbial consortia (CS-5 and BC-4) for enhanced methane production was performed in a ecofriendly process. The microbial consortia resulted in chlorophenol (Individual and mixed) degradation and its toxicity reduction to almost 90% after 15 days and their synergistic action further enhanced the biogas and methane content by 113.7%. Methanosataceae family was found to be the most dominant methanogenic archaea detected in the process of anaerobic digestion. High production of hemicellulolytic enzymes along with other hydrolytic enzymes played important role in the biomass deconstruction of Catalpa sawdust (Ali et al. 2020).

16.8.1 PCP Degradation in Soil and Water

PCP and their derivatives that are being generated from various industrial processes such as chemical, pesticide, textile and pharmaceutical mostly contaminate the water and soil. The PCP is recalcitrant and toxic lifeforms so it is essential to understand their fate and degradation in the environment. A research group reported the effect of carbon source addition on PCP dechlorination in soil under anaerobic conditions and it was observed that citrate and lactate addition enhanced the dechlorination compared to glucose. Microbial community dynamics indicated a shift in microbial population with abundance of Clostridiales in citrate and lactate amended soil with enhanced PCP degradation was observed (Li et al. 2019). The addition of citrate or lactate to the PCP-contaminated soil can improve the degradation of PCP. The effect of vermicomposting (sterile and non-sterile) on PCP degradation in the soil was studied and change in microbial (bacteria and fungi) community dynamics and its role in PCP degradation was evaluated. The earthworm and the microbial diversity of compost synergistically enhanced the degradation of PCP. The community dynamics analysis indicated the presence of bacteria from Pseudomonadaceae, Sphingobacteriaceae, Flavobacteriaceae and Xanthomonadaceae family and fungi from Trichocomaceae, Mucoraceaea and Tremellaceae that helped in enhanced degradation of PCP in non-sterile compost (Lin et al. 2016). Similar finding was observed in flooded soil where methanogenesis and PCP degradation were synergistically promoted in presence of methanogenesis cofactor (coenzyme M) and inhibited in presence of inhibitor 2-bromoethanesulfonate (Zhu et al. 2019). It was observed that PCP changes the microbial community responsible for acetate assimilation especially involved in soil redox and electron transfer in the anaerobic soil environment. Microbial community analysis indicated that Clostridiales and Burkholderiales are the dominant microbes for acetate accumulation and PCP degradation. The PCP has little effect on iron reduction but inhibited sulphate reduction by suppressing sulphur reducing microbes Desulfarculaceae, Desulfobulbaceae and Desulfobacteraceae (Xu et al. 2019). Vanadium (V) and PCP were anaerobically treated in synthetic groundwater and it was observed that both are competitively bioreduced both extracellularly (cytochrome c, other proteins) and intracellularly (nicotinamide adenine dinucleotide (NADH) mediated) by microbes. The Vanadium (V) was reduced to insoluble Vanadium (iV) and PCP was reductively dechlorine dated to intermediated with reduced toxicity by *Pseudomonas*, *Soehngenia* and *Anaerolinea* (Zhang et al. 2019a, b).

16.8.2 Biochar Based PCP Degradation

Biochar is one of the cost-effective and promising materials that can be used for the degradation of PCP. There are several recent reports that highlight the importance of biochar in the degradation of PCP due to its inherent properties such as redox functional group and electrical conductivity. But the application of biochar for PCP degradation is debatable due to its high sorption potential that decreases the bioavailability of PCP for microbial degradation and can also produce secondary pollution (Zhao et al. 2020; Xu et al. 2020). A study reported redox activity of humic acid prepared from protein, lignocelluloses and lignin-rich compost for microbial reductive PCP dechlorination. It was observed that lignin-rich compost outperformed in microbial reductive PCP dechlorination. Aromatic systems are the major electron-accepting groups present in humic acids and polyphenols and amino acids are major components to synthesize redox functional groups in ligni-rich compost. The microbial community analysis indicated the presence of microbial groups in order of their dominance Proteobacteria, Firmicutes, Bacteroidetes, Actinobacteria, Spirochaetae and Tenericutes were mainly responsible for the generation of precursor and redox intermediates for reductive dechlorination of PCP (Zhao et al. 2020). The biochar obtained from different materials (rice husk, bamboo, caragana and Garbage) were used to study their effect on reductive dechlorination of PCP by a PCP dechlorinating mixed culture of *Sulfurospirillum*, *Clostridium* and *Dehalobacter*, as inoculums. Among the above-mentioned biochar, only caragana showed consistent electron transfer for PCP dechlorination due to its highest electricity conductivity property (Zhang et al. 2019a, b). The effect of biochar (derived from rice husk) on biotic (aerobic and anaerobic) and abiotic transformation of PCP in soil was studied. It was observed that with the amendment of biochar, the dehalogenation/degradation of PCP decreased under both aerobic and anaerobic conditions due to the high sorption affinity of biochar for PCP resulting in reduced bioavailability. It was also inferred that inhibition of PCP degradation was more prominent in aerobic conditions compared to anaerobic conditions pointing towards the importance of facultative dechlorinators in PCP degradation. The aerobic culture stimulated the growth of PCP degraders *Bacillus* and *Sphingomonas* whereas under anaerobic condition *Desulfovibrio*, *Anaeromyxobacter*, *Geobacter* and *Desulfomonile* were mainly involved in PCP degradation. High sorption affinity of biochar for PCP can lead to the generation of secondary pollution in the soil (Xu et al. 2020).

16.8.3 Bioelectrochemical System for PCP Degradation

The degradation of PCP in the Bioelectrochemical system is an attractive technology because power is generated along with the decomposition of contaminants such as PCP. The research is mainly focused on enhancement in PCP degradation and power generation from MFC. PCP degradation has been tried in the different setup of MFC. This is a promising technology for remediation of soil and effluents but its scale-up is technically challenging that needs further research (Cai et al. 2020; Khan et al. 2018; Huang et al. 2013). A study reported the bioelectrochemical cell formation at optimal potential for bioanode and biocathode to improve the PCP degradation and power production from microbial fuel cells (MFC). Microbial consortia, their biocatalytic reaction and composition were also evaluated. Compared to biocathode the dechlorination of PCP was more efficient at bianode with detection of new degradation products. The different types of dominant electrotroph were identified on bianode (*Dechlorospirillum* sp., *Desulfovibrio carbinoliphilus*) and biocathode (*Commamonas* sp., *Desulfovibrio marrakechensis*) that are actively involved in PCP degradation and power generation (Huang et al. 2013). A study reported the combined effect of co-substrate (glucose and acetate) and enrichment procedure (simultaneously and sequentially) on microbial diversity and MFC performance in terms of PCP degradation, power generation, acclimatization time was evaluated. Simultaneous addition of glucose significantly improved the acclimatization time as well as PCP shock load in the MFC. The call gammaproteobacteria and phylum Lentisphaerae were the dominant microbial group observed in the MFC during the process (Wang et al. 2012). A research group reported that PCP could be degradation in single-chambered and dual-chambered MFC as co-metabolite along with power generation. The dual-chambered MFC operated under aerobic conditions showed effective PCP degradation compared to anaerobic treatment in single-chambered MFC (Khan et al. 2018). A soil bioelectrochemical system (SBES) was conducted with biochar prepared at different temperatures to remediate the PCP-contaminated soil. A conductive network was observed between biochar and microbiota in soil matrix for long-distance electron transfer for effective remediation of PCP contaminated soil. The introduction of biochar as low cost conductive material was increased the effective area of remediation by four times around the electrode. The microbial diversity analysis indicated the selective enrichment of Geobacter and Desulfitobacterium responsible for microbial-biochar conductive network establishment in SBES (Cai et al. 2020). The findings indicated that SBES can be a promising technology for the effective bioremediation of contaminated soil.

16.9 Conclusion

The recalcitrant and toxic nature of pentachlorophenol has resulted in soil, water and groundwater contamination. The urgent need for its remediation led to the development of various technologies out of which biological method has proved to be one of the most cost-effective and efficient ones. This chapter highlights recent trends in PCP bioremediation by various microorganisms including bacteria, algae, fungi and consortia and their associated pathways. Recent trends biochar and bioelectrical system mediated in PCP biodegradation along with enzymes have been reported. Advanced techniques such as biochar based and bioelectrical systems mediated biotransformation by mixed microbial consortia have shown promising results in PCP biodegradation.

References

Agency for Toxic Substances and Disease Registry (1999) Toxicological profile for chlorophenols. Agency for Toxic Substances and Disease Registry, U.S. Department of Health and Human Services, Atlanta

Agency for Toxic Substances and Disease Registry (ATSDR) (2001) Toxicological profile for pentachlorophenol; update. U.S. Department of Health and Human Services, Public Health Service, Atlanta, GA

Aken PV, Lambert N, Van den Broeck R, Degrève J, Dewil R et al (2019) Advances in ozonation and biodegradation processes to enhance chlorophenol abatement in multisubstrate wastewaters: a review. Environ Sci Water Res Technol 5:444–481. https://doi.org/10.1039/C8EW00562A

Ali SS, Kornaros M, Manni A, Sun J, El-Shanshoury AERR, Kenawy ER, Khalil MA (2020) Enhanced anaerobic digestion performance by two artificially constructed microbial consortia capable of woody biomass degradation and chlorophenols detoxification. J Hazard Mater 389:122076

Allison N, Skinner AJ, Cooper RA et al (1983) Dehalogenases of a 2, 2-dichloropropionate degrading bacterium. J Gen Microbiol 129:1293–1297

Apajalahti JHA, Salkinoja-Salonen MS (1987) Complete dechlorination of tetrachlorohydroquinone by cell-extracts of pentachlorophenol-induced *Rhodococcus chlorophenolicus*. J Bacteriol 169:5125–5130

Arora PK, Bae H (2014) Bacterial degradation of chlorophenols and their derivatives. Microb Cell Fact 13(31):1–17

Bevenue A, Beckman H (1967) Pentachlorophenol: a discussion of its properties and its occurrence as a residue in human and animal tissues. Res Rev 19:86–730

Bosso L, Scelza R, Varlese R, Meca G, Testa A, Rao MA, Cristinzio G et al (2016) Assessing the effectiveness of *Byssochlamys nivea* and *Scopulariopsis brumptii* in pentachlorophenol removal and biological control of two *Phytophthora* species. Fungal Biol 120(4):645–653

Cai X, Yuan Y, Yu L, Zhang B, Li J, Liu T, Yu Z, Zhou S et al (2020) Biochar enhances bioelectrochemical remediation of pentachlorophenol-contaminated soils via long-distance electron transfer. J Hazard Mater 391:122213

Carvalho MB, Martins I, Leitão MC, Garcia H, Rodrigues C, San Romão V, McLellan I, Hursthouse A, Pereira CS et al (2009) Screening pentachlorophenol degradation ability by environmental fungal strains belonging to the phyla Ascomycota and Zygomycota. J Ind Microbiol Biotechnol 36(10):1249–1256

Chen ST, Kang SY, Muller R et al (1999) Pentachlorophenol and crystal violet degradation in water and soils using haeme and hydrogen peroxides. Water Res 33:3657–3665

Coelho GD, Ballaminut N, Thomaz DV, Gomes Machado KM et al (2019) Characterization of a thermostable *Deconica castanella* Laccase and application toward pentachlorophenol degradation. Prep Biochem Biotechnol 49(9):908–915

Crawford RL, Ederer MM, Laskin AI, Whete DC et al (1999) Phylogeny of *Sphingomonas* species that degrade pentachlorophenol. J Ind Microbiol Biotechnol 23:320–325

de Morais P, Stoichev T, Basto MCP, Ramos V, Vasconcelos VM, Vasconcelos MTS (2014) Cyanobacterium *Microcystis aeruginosa* response to pentachlorophenol and comparison with that of the microalga *Chlorella vulgaris*. Water Res 52:63–72

De Saiah D (1978) Effects of PCP on the ATPase in rat tissue. In: Rao KR (ed) Pentachlorphenol: chemistry, pharmacology and environmental toxicology. Plenum Press, New York, pp 277–283

Ederrer MM, Crawford RL, Herwig RP, Orser CS et al (1997) PCP degradation mediated by closely related strains of the genus *Sphingomonas*. Mol Biol 6:39–49

Edgehill RU (1998) Effect of copper chrome arsenate (CCA) components on PCP degradation by *Arthrobacter* strain ATCC-33790. Bull Environ Contam Toxicol 57:258–263

Fetzner S (1998) Bacterial dehalogenation. Appl Microbiol Biotechnol 50:633–657

Gilbert FJ Jr, Minn CE, Dunean RG, Wekinson J et al (1990) Effect of pentachlorophenol and other preservatives on the health of wood treating workers in Hawaii. Arch Environ Contam Toxicol 19:603–609

Gomez Catalan J, To Tigueres J, Rodamilans M, Corbella J et al (1991) Transport of organochlorine residues in the rat and human blood. Arch Environ Cont Toxicol 20:61–66

Gonzalez JF, Hu WS (1991) Effect of glutamate on the degradation of pentachlorophenol by *Flavobacterium* sp. Appl Microbiol Biotechnol 35:100–104

Gunawardana B, Singhal N, Swedlund P et al (2011) Degradation of chlorinated phenols by zero valent iron and bimetals of iron: a review. Environ Eng Res 16(4):187–203

Hassan AB, Seligeman H, Bassan HM (1985) Intravascular haemolysis induced by PCP. Br Med J 291:21–22

Hechmi N, Bosso L, El-Bassi L, Scelza R, Testa A, Jedidi N, Rao MA et al (2016) Depletion of pentachlorophenol in soil microcosms with *Byssochlamys nivea* and *Scopulariopsis brumptii* as detoxification agents. Chemosphere 165:547–554

Huang L, Wang Q, Quan X, Liu Y, Chen G et al (2013) Bioanodes/biocathodes formed at optimal potentials enhance subsequent pentachlorophenol degradation and power generation from microbial fuel cells. Bioelectrochemistry 94:13–22

Igbinosa EO, Odjadjare EE, Chigor VN, Igbinosa IH, Emoghene AO, Ekhaise FO, Igiehon NO, Idemudia OG et al (2013) Toxicological profile of chlorophenols and their derivatives in the environment: the public health perspective. Sci World J 2013:11

Keane MA (2005) A review of catalytic approaches to waste minimization: case study—liquid-phase catalytic treatment of chlorophenols. J Chem Technol Biotechnol 80:1211–1222

Khan N, Khan MD, Nizami AS, Rehan M, Shaida A, Ahmad A, Khan MZ et al (2018) Energy generation through bioelectrochemical degradation of pentachlorophenol in microbial fuel cell. RSC Adv 8(37):20726–20736

Kim YH, Carraway ER (2003) Dechlorination of chlorinated phenols by zero valent zinc. Environ Technol 24:1455–1463

Lange CC, Schneider BJ, Orser CS et al (1996) Verification of the role of pcp 4-monooxygenase in chlorine elimination from pentachlorophenol by *Flavobacterium* sp. strain ATCC 39723. Biochem Biophys Res Commun 219:146–149

Lee SG, Yoon BD, Park YH, On HM et al (1998) Isolation of novel pentachlorophenol degrading bacterium *Pseudomonas* sp. Bu 34. J Appl Microbiol 85:1–8

León-Santiesteban HH, Wrobel K, García LA, Revah S, Tomasini A et al (2014) Pentachlorophenol sorption by *Rhizopus oryzae* ENHE: pH and temperature effects. Water Air Soil Pollut 225 (5):1947

Li H, Jiang Y, Chen L, Chen Y, Wen X, Tao L et al (2019) Carbon sources mediate microbial pentachlorophenol dechlorination in soils. J Hazard Mater 373:716–724

Lin Z, Bai J, Zhen Z, Lao S, Li W, Wu Z, Li Y, Spiro B, Zhang D (2016) Enhancing pentachlorophenol degradation by vermicomposting associated bioremediation. Ecol Eng 87:288–294

Michałowicz J, Majsterek I (2010) Chlorophenols, chlorocatechols and chloroguaiacols induce DNA base oxidation in human lymphocytes (in vitro). Toxicology 268:171–175

Mikesell MD, Boyd SA (1986) Complete reductive dechlorination and mineralization of pentachlorophenol by anaerobic microorganisms. Appl Environ Microbiol 52:861–865

Monsanto Chemical Co. (1963) Tech Bull SC-8, St. Louis, Mol

Muller F, Caillard L (2011) Chlorophenols. In: Ullmann's encyclopedia of industrial chemistry. Wiley, New York

Ning D, Wang H (2012) Involvement of cytochrome P450 in pentachlorophenol transformation in a white rot fungus *Phanerochaete chrysosporium*. PLoS One 7(9):e45887

Ohtsubo Y, Miyauchi K, Kanda K, Hatta T, Kiyohara H, Senda T, Nagata Y, Mitsui Y, Takagi M et al (1999) PcpA, which is involved in the degradation of pentachlorophenol in *Sphingomonas chlorophenolica* ATCC39723, is a novel type of ring-cleavage dioxygenase. FEBS Lett 459:395–398

Olaniran AO, Igbinosa EO (2011) Chlorophenols and other related derivatives of environmental concern: properties, distribution and microbial degradation processes. Chemosphere 83:1297–1306

Orser CS, Lange CC, Xun L, Zahrt TC, Schneider BJ et al (1993a) Cloning, sequence analysis, and expression of the *Flavobacterium* pentachlorophenol-4-monooxygenase gene in *Escherichia coli*. J Bacteriol 175:411–416

Orser CS, Dutton J, Lange C, Jablonski P, Xun L, Hargis M et al (1993b) Characterization of a *Flavobacterium glutathione* S-transferase gene involved reductive dechlorination. J Bacteriol 175:2640–2644

Oswald WJ, Gotaas HB (1957) Photosynthesis in sewage treatment. Trans Am Soc Civ Eng 122 (1):73–105

Papazi A, Karamanli M, Kotzabasis K et al (2019) Comparative biodegradation of all chlorinated phenols by the microalga *Scenedesmus obliquus*—the biodegradation strategy of microalgae. J Biotechnol 296:61–68

Patel UD, Suresh S (2008) Electrochemical treatment of pentachlorophenol in water and pulp bleaching effluent. Sep Purif Technol 61:115–122

Pera-Titus M, Garcia-Molina V, Banos MA, Gimenez J, Esplugas S et al (2004) Degradation of chlorophenols by means of advanced oxidation processes: a general review. Appl Catal B Environ 47:219–256

Premalatha A, Raj Kumar GS (1994) Pentachlorophenol degradation by *Pseudomonas aeruginosa*. World J Microbiol Biotechnol 10:334–337

Resnick S, Chapman PJ (1994) Physiological properties and substrate specificity of pentachlorophenol degrading *Pseudomonas* species. Biodegradation 50:47–54

Ruiz-Lara A, Fierro F, Carrasco U, Oria JA, Tomasini A et al (2020) Proteomic analysis of the response of *Rhizopus oryzae* ENHE to pentachlorophenol: understanding the mechanisms for tolerance and degradation of this toxic compound. Process Biochem 95:242–250

Saber DL, Crawford RL (1985) Isolation and characterization of *Flavobacterium* strains that degrade pentachlorophenol. Appl Environ Microbiol 50:1512–1518

Saboo VM, Gealt MA, Fenster TS et al (1998) Gene sequences of the pcpB of pentachlorophenol-degrading *Sphingomonas chlorophenolica* found in soil. Can J Microbiol 44:667–675

Shiu WY, Ma KC, Varhanickova D, Mackay D et al (1994) Chlorophenols and alkylphenols—a review and correlation of environmentally relevant properties and fate in an evaluative environment. Chemosphere 29:1155–1224

Slater JH, Lovatt D (1984) Biodegradation and the significance of microbial communities. In: Gibson DT (ed) Microbial degradation of organic compounds. Marcel Dekker, New York, pp 439–485

Srivastava S, Thakur IS (2007) Evaluation of biosorption potency of *Acinetobacter* sp. for removal of hexavalent chromium from tannery effluent. Biodegradation 18:637–646

Srivastava S, Ahmad AH, Thakur IS (2007) Bioremediation potentiality of microorganism for removal of chromium and pentachlorophenol in tannery effluent. Bioresour Technol 98:1128–1132

Steiert JG, Crawford RL (1986) Catabolism of pentachlorophenol by a *Flavobacterium* sp. Biochem Biophys Res Commun 141:825–830

Stephen DP, Ayalur BK (2017) Effect of nutrients on *Chlorella pyrenoidosa* for treatment of phenolic effluent of coal gasification plant. Environ Sci Pollut Res 24(15):13594–13603

Takeuchi M, Hamana K, Hiraishi A et al (2001) Proposal of the genus *Sphingomonas sensu stricto* and three new genera, *Sphingobium*, *Novosphingobium* and *Sphingopyxis*, on the basis of phylogenetic and chemotaxonomic analyses. Int J Syst Evol Microbiol 51:1405–1417

Tanjore S, Viraraghavan T (1994) Pentachlorophenol—water pollution impacts and removal technologies. Int J Environ Stud 45:155–164

United States Environmental Protection Agency (2014) Priority pollutant list. https://www.epa.gov/eg/toxic-andpriority-pollutants-under-clean-water-act#priority

Uotila JS, Kitunen VH, Saastamoinen T, Coote T, Haggblom MM, Salkinoja-Salonen MS et al (1992) Characterization of aromatic dehalogenases of *Mycobacterium fortuitum* CG-2. J Bacteriol 174:5669–5675

Vacondio B, Birolli WG, Ferreira IM, Seleghim MH, Gonçalves S, Vasconcellos SP, Porto AL et al (2015) Biodegradation of pentachlorophenol by marine-derived fungus *Trichoderma harzianum* CBMAI 1677 isolated from ascidian *Didemnun ligulum*. Biocatal Agric Biotechnol 4(2):266–275

Wang S, Huang L, Gan L, Quan X, Li N, Chen G, Lu L, Xing D, Yang F et al (2012) Combined effects of enrichment procedure and non-fermentable or fermentable co-substrate on performance and bacterial community for pentachlorophenol degradation in microbial fuel cells. Bioresour Technol 120:120–126

Weavers LK, Malmstadt N, Haffman NR et al (2000) Kinetics and mechanisms of pentachlorophenol degradation by sonication and ozonation. Environ Sci Technol 34:1280–1285

Wild SR, Harrad SJ, Jones KC et al (1992) Pentachlorophenol in the UK environment. I. A budget and source inventory. Chemosphere 24:833–845

World Health Organization (WHO) (2015) Pentachlorophenol. Available online: http://www.inchem.org/documents/ehc/ehc/ehc71.htm

Xiao P, Kondo R (2020) Biodegradation and biotransformation of pentachlorophenol by wood-decaying white rot fungus *Phlebia acanthocystis* TMIC34875. J Wood Sci 66(1):1–11

Xu Y, He Y, Egidi E, Franks AE, Tang C, Xu J et al (2019) Pentachlorophenol alters the acetate-assimilating microbial community and redox cycling in anoxic soils. Soil Biol Biochem 131:133–140

Xu Y, Liu J, Cai W, Feng J, Lu Z, Wang H, Franks AE, Tang C, He Y, Xu J (2020) Dynamic processes in conjunction with microbial response to disclose the biochar effect on pentachlorophenol degradation under both aerobic and anaerobic conditions. J Hazard Mater 384:121503

Xun L, Orser CS (1991) Purification and properties of pentachlorophenol hydroxylase, a flavoprotein from *Flavobacterium* sp. strain ATCC 39723. J Bacteriol 173:4447–4453

Zhang B, Cheng Y, Shi J, Xing X, Zhu Y, Xu N, Xia J, Borthwick AG (2019a) Insights into interactions between vanadium (V) bio-reduction and pentachlorophenol dechlorination in synthetic groundwater. Chem Eng J 375:121965

Zhang C, Zhang N, Xiao Z, Li Z, Zhang D (2019b) Characterization of biochars derived from different materials and their effects on microbial dechlorination of pentachlorophenol in a consortium. RSC Adv 9(2):917–923

Zhao X, Tan W, Peng J, Dang Q, Zhang H, Xi B (2020) Biowaste-source-dependent synthetic pathways of redox functional groups within humic acids favoring pentachlorophenol dechlorination in composting process. Environ Int 135:105380

Zhu M, Feng X, Qiu G, Feng J, Zhang L, Brookes PC, Xu J, He Y (2019) Synchronous response in methanogenesis and anaerobic degradation of pentachlorophenol in flooded soil. J Hazard Mater 374:258–266

Sustainable Solid Waste Management in India: Practices, Challenges and the Way Forward

17

Hardeep Rai Sharma, Balram Bhardwaj, Bindu Sharma, and C. P. Kaushik

Abstract

Increasing population, volume, and complex nature of generated solid waste, improper implementation of existing rules, failure of waste disposal techniques, limitation of funds and infrastructure are the common causes of unsustainable solid waste management in many countries of the world including India. Further, traditional beliefs and approaches such as "out of sight, out of mind," "not in my backyard (NIMBY)" and "flame, flush or fling" even towards the generated solid waste results in an unsustainable society which hinders sustainability. According to the World Bank forecast, annual global waste generation is expected to increase to 3.40 billion tonnes in 2050 (considering 2016 as a reference year). As per Waste Atlas 2017, India is the third-largest generator (volume basis) of municipal solid waste and will play an important role in addressing this global issue. A management system from generation up to final disposal of wastes in an environment friendly, economically affordable, and socially acceptable way is termed "Sustainable." According to the Indian Planning Commission report of 2014, if effectively managed, the unused MSW will generate about 439 MW of

H. R. Sharma
Institute of Environmental Studies, Kurukshetra University, Kurukshetra, Haryana, India

B. Bhardwaj
University School of Environment Management, GGS Indraprastha University, Dwarka, Delhi, India

B. Sharma
Department of Management Studies, Graphic Era (deemed to be university), Dehradun, Uttarakhand, India

C. P. Kaushik (✉)
Environment Committee, PHD Chamber of Commerce and Industry, Delhi, India
e-mail: cpkaushik@rediffmail.com

A. Kaushik et al. (eds.), *Climate Resilience and Environmental Sustainability Approaches*, https://doi.org/10.1007/978-981-16-0902-2_17

power, 1.3 million m^3 of biogas/day, or 72 MW of electricity from biogas and 5.4 million metric tonnes of compost annually. Many times improper use and ineffectiveness of different available techniques of harnessing energy and wealth from the wastes cause more environmental costs than economic gains. Due attention must be paid to waste composition, geographical conditions, people's attitude, adoption of economically feasible and environment friendly disposal and treatment options to bring sustainability in solid waste management. This chapter addresses the present scenario of solid waste generation, management practices, challenges, and sustainable management systems in the Indian context.

Keywords

Compost · Disposal · Energy · Policies · Solid waste · Sustainable management

17.1 Introduction

Basically, any substance or material mainly in solid form which is no longer required by the user or generator is considered as solid waste, e.g. materials such as food wastes, glass and plastic bottles, rags or used cloths, paper, containers, packages, electronic items, chemicals, expired medicines, equipment, and devices produced by day to day humans activities. The solid waste can be categorized in the following ways, based on its:

1. Origin (domestic, industrial, commercial, institutional, or construction)
2. Contents (organic material, glass, paper, plastic, metal, etc.)
3. Hazard potential (toxic, non-toxic, radioactive, flammable, and infectious, etc.)

In India, legally solid waste consists of solid or semi-solid domestic, commercial, institutional, sanitary, catering, market, and other non-residential wastes, silt removed or collected from the surface drains, street sweepings, horticulture waste, agriculture, and dairy waste, treated bio-medical waste generated in the area under the local authorities and other entities mentioned in the rule-2 of the Solid Waste Management Rules (2016). It excludes industrial waste, bio-medical wastes, hazardous chemicals, hazardous waste, batteries, radioactive waste, and e-waste for which separate provisions are made under the Environment (Protection) Act (1986).

As per records, the city of Mohenjo-daro, an ancient Indus Valley civilization city, had practiced solid waste management by 2000 BC (Vesilund et al. 2002). Generally, solid waste is viewed as an urban problem. Improvement of living standard, economic wealth, use of various goods and services add to the type and amount of the generated waste viz., domestic sewage, municipal solid waste, agriculture waste, industrial waste, plastic waste, e-waste, etc. and makes the collection, waste treatment and management, and disposal very difficult (Khan et al. 2016). It is estimated that globally, still three billion people do not have access to controlled waste disposal facilities (Wilson et al. 2015).

The unmanaged waste blocks sewers, transmits diseases, pollutes our oceans, causes respiratory problems, affects animals and the economic development of the nations (Kumar et al. 2017). Uncontrolled organic wastes become habitat and breeding places for vectors of diseases by creating favourable growth conditions for insects, rodents, and other animals. If not managed properly solid waste may cause serious problems such as unpleasant odour, risk of explosion in landfill areas, as well as groundwater contamination due to leachate percolation (Mor et al. 2006), and economic losses due to failures in recycling and composting. Besides causing different types of pollution and low aesthetic values, the waste piles also pose risk to humans. Two people died, five injured and many people on road swept into the nearby Kondli canal when about 50 tonnes of waste garbage collapsed down at the Ghazipur landfill site in East Delhi, India in early 2007. This site is spread over 70 acres and with an approximate height of about 50 m. Waste dumping sites emit greenhouse gases and large amount of leachate (contaminated liquid). Agriculture waste burning during the winter seasons, especially crop residues in the Punjab and Haryana states has been identified as the major reason for formation of smog in the National Capital Delhi and the surrounding areas. Burning, open dumping, or unmanaged landfilling cause air pollution by the emission of harmful gases, particulate matter, toxins, dioxins, furans, polycyclic aromatic hydrocarbons, other volatile organic compounds, etc. Exposure to such emissions causes toxic effects on human health. It may also cause neurological and respiratory diseases (Kaza et al. 2018). Currently, non-existent or inefficient recycling systems, not only results in resource exhaustion but also environmental damage due to emissions from landfills responsible for climate change, ecosystem damages by the release of gases and heavy metals leading to pollution of air, water, and soil, and also health problems to the workers during waste collection, transportation, and treatment. Therefore, proper waste management needs urgent attention. This chapter examines the present solid waste generation scenario, management practices and challenges, and sustainable waste management system in the Indian context.

17.2 Solid Waste Generation Scenario

According to former US President Jimmy Carter, "This (solid waste) growing mountain of garbage and trash represents not only an attitude of indifference toward valuable natural resources, but also a serious economic and public health problem" (Carter 1977). These words reflect the environmental, economical, and social impact of improper waste disposal. The waste problem is one of the major worldwide concerns. As per the latest World Bank estimation, the annual rate of solid waste generation by the world cities has reached around 2.01 billion tonnes amounting to a footprint of 0.74 kg/person/day and is expected to increase to about 3.4 billion tonnes by 2050 (The World Bank 2019). In three large Indian cities of Chennai, Kolkata, and Mumbai the waste generation is 1.01, 0.84, and 0.82 kg/capita/day, respectively which is almost reaching the level of Organization for Economic Co-Operation and Development (OECD) member countries (Dutta and Jinsart

2020). The solid waste generation rate and its composition corresponds to the socio-economic status of a country. High-income countries comparatively produce higher quantities of plastic, paper, and other non-organic materials, while, lower-and middle-income nations add higher organic matter in the waste stream (Slorach et al. 2019). Stock-building, import, or production of materials generate additional waste, especially in megacities with population of more than 10 million, which are projected to increase in number from 33 in 2018 to 43 by 2030 (United Nations 2014).

India is a fast-developing country and has faced many issues with waste management such as the indifferent attitude of people towards waste management, informal sector, and poor infrastructure (Kumar et al. 2017). During mid-1970s, the per capita solid waste generation for various Indian cities varied from 150–350 g/day (Bhide et al. 1975); and in late 1980s, it ranged between 320 and 530 g/day (Jain 1994). As per NEERI's study, in 2004–2005, this figure from 59 Indian Metro Cities and State Capitals varied between 120 and 600 g/day (Ministry of Urban Development 2016). According to Central Pollution Control Board (CPCB 2012), it varied from 200 to 800 g/day. Solid waste generation depends upon population density, cultural habits, socio-economic conditions, location, climate, and urban structure. India's population will grow continuously at the rate of 3–3.5% annually. In India waste generation, per capita will increase by 1.3% per annum, with an annual increase of 5% (Ghatak 2016). According to a report of Waste Atlas (2017), India, in terms of volume of waste generation, is in third position. India's challenges in waste management have increased due to rapid urbanization. The reasons for this are alarming population growth, scarce job opportunities in villages, and better prospects of growth with better-paying jobs compared to stagnation with low-profit agriculture, etc. Currently, people living in rural India account for 65% of the population, and remaining 35% live in urban areas. The present urban population has doubled from 17% in 1950 and is expected to increase to 50% by 2050 (Ministry of Statistics and Programme Implementation 2020). More than 377 million people reside in urban settlements of India producing 62 million tonnes (MT) of municipal solid waste annually (Down To Earth, India 2020), out of which, the collected waste is only 43 MT, treated is 11.9 MT and about 31 MT of waste is dumped in the landfill sites. Out of 62 MT of the generated waste, 5.6 MT is plastic, 0.17 and 7.90 MT are bio-medical and hazardous waste, respectively and 15 lakh tonnes is e-waste. The sources of waste and various modes of waste collection in Indian cities are illustrated in Fig. 17.1.

It is predicted that the amount of annual waste generation in India will increase from the present 62 MT to about 165 MT in the year 2030 (Press Information Bureau 2016). According to Planning Commission (2014) data, if MSW continues to be dumped without treatment, it will need 340,000 m^3 of landfill space everyday (1240 ha per year) for the current annual generation of 62 MT, and 66,000 ha of land for setting up 10 m high landfill for the proposed 165 million tonnes by 2030. India cannot afford to waste this huge area for solid waste dumping.

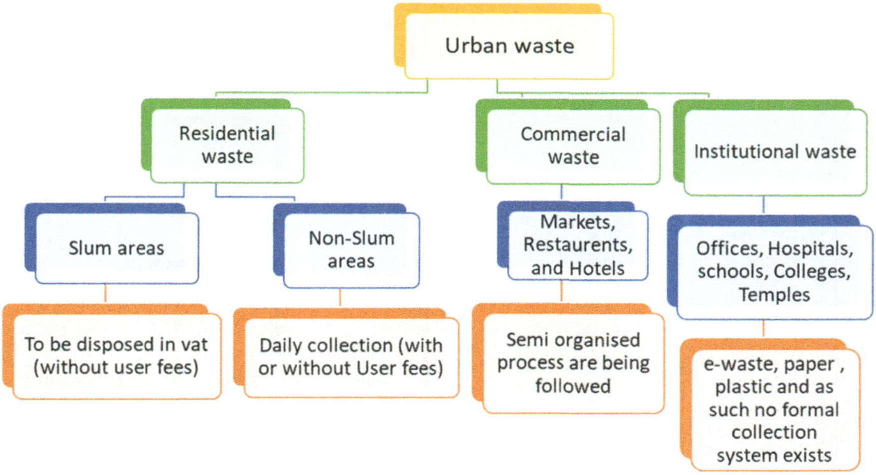

Fig. 17.1 Sources of waste and modes of waste collection in Indian cities [modified from Ghatak (2016)]

17.3 Solid Waste Management Practices and Challenges in India

The common practices in solid waste management generally start from waste generation, waste handling and processing at source, waste collection, sorting and processing of waste (away from the source), transfer, and final disposal as mentioned in Fig. 17.2. Usually, the first two steps of waste generation, handling, and processing at source depend on the individual i.e. the producer of the waste, while the rest are taken care by the concerned waste management authority of the area. Further, there are different waste disposal practices like open dumping, controlled tipping, sanitary landfilling, incineration, gasification pyrolysis, plasma pyrolysis, composting, vermi-composting, anaerobic digestion/biomethanation, production of refuse derived fuel (RDF), based upon availability of funds, technology, and rules and regulations.

17.3.1 Urban Areas

Some decades back the Indian cities have never given priority to waste management because they termed it as "Kachra" or "Koora" in native language Hindi which refers to dirty things and the people involved in its management are named as "Kachra Wala." The citizens were least bothered about the fate of the waste they generated and assigned a low social status to the function of waste management which made the situation worse. However, the onset of Municipal Waste Management

Fig. 17.2 Activities involved in solid waste management

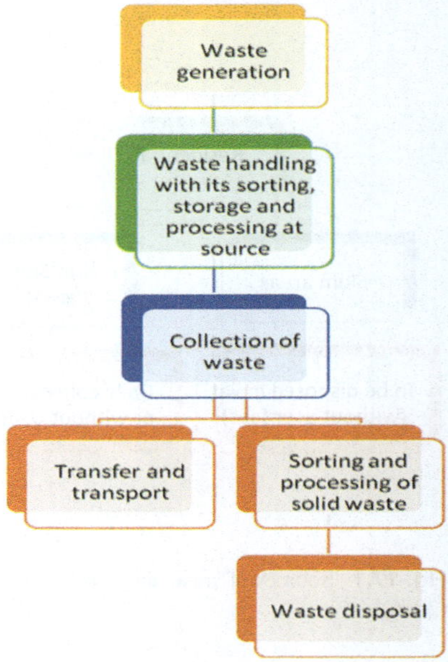

(Management and Handling) Rules (2000) changed the situation and a good amount of funds was allotted for waste management. According to the World Bank (2015), "the overall goal of urban solid waste management is to collect, sort, treat and dispose of solid wastes generated by all urban population groups in an environmentally and socially satisfactory manner using the most economical means available." The main goal of urban solid waste management (USWM) is to protect public health and the environment through natural resource conservation (Allesch and Brunner 2014).

The mode of waste collection in Indian cities is mainly the community bin collection system. In this collection method, the residents dispose their waste at the nearest located bins frequently. That waste should be collected daily from those bins, because the degradable or organic waste starts decomposing. In many cities, the waste generated from households is collected from door-to-door for which users pay the charges. However, no single collection mode can be effective in India because of narrow and congested roads. Moreover, at source waste segregation is negligible. The predominantly followed methods for urban solid waste disposal are (a) open dumping, (b) sanitary land filling, (c) incineration or thermal processing (d) composting and (e) Waste-to-Energy (WTE) technology.

(a) *Open Dumping*: A report of Planning Commission (2014) stated that annual disposal of MSW at open dump sites, without any treatment, is more than 81%. Currently, according to the Housing and Urban Affairs Ministry (2020), out of about 147,613 metric tonnes daily generated solid waste across the States, 58,000 metric tonnes (MT) i.e. 39% solid waste is dumped daily without

Fig. 17.3 A typical solid waste dumpsite

processing and 61% is processed. Chhattisgarh is the only State which processes 90% of the daily solid waste generated followed by Gujarat and Madhya Pradesh (87%), Himachal Pradesh and Telangana (78%) Rajasthan, Kerala, and Goa (70–72%), Tamil Nadu processes (68%), Andhra Pradesh (63%), Punjab (61%) and Jharkhand and Nagaland (60%).

This kind of open dumping, mostly on land as shown in Fig.17.3, in an uncontrolled manner is most practiced because it is easy and cheap with the requirement of less expertise. Many studies have reported that, blowing away of light weight paper and plastic wastes through wind, scattered wastes with uncontrolled burning, foul odour from the piles of waste, presence of cattle, vultures, pigs, and dogs, etc., and absence of fencing on dumping sites without accessible roads make an unaesthetic scene of the dumping sites (Sharma et al. 2011, 2013; Tyagi et al. 2014).

(b) *Sanitary Landfill*: Due to its lower initial cost, the sanitary landfill is the most dominant method for waste disposal. The sites for the modern landfill as shown in Fig.17.4, are selected carefully. Several impermeable layers of thick plastic, clay, and sand line the bottom of the landfill to prevent contamination of groundwater with leachate. Thin layers of solid waste are spread out, compacted, and covered with daily fresh layer of plastic foam or clay. It helps to keep the waste material dry, reduce the volume of waste, decrease odour, reduce accessibility to worms, and reduce the risk of fire. For monitoring of contaminants of groundwater due to any leakage, some wells are drilled near the landfill site. Pipes in the landfills are laid to collect inflammable methane gas which is used to generate electricity. However, transportation of waste to the landfill sites, which is clumsy, tedious, and costly process, is the main limitation of this method.

In India, most of the sanitary landfills are not set up as delineated above. The waste on the site is simply compacted and covered without proper technical care and treatment technology. No provision is made for leachate and landfill gas control. Kumar et al. (2009) reported that waste dumping is uncontrolled in 47 main cities of India. Out of these, in 26 cities waste is compacted with

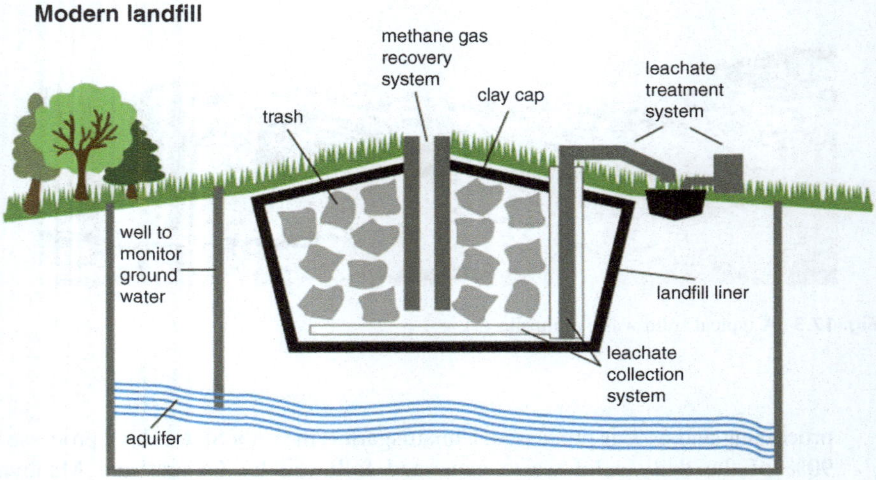

Fig. 17.4 Schematic of a modern landfill (Adapted from National Energy Education-Project; public domain)

bulldozers and only 16 used the earth covering method to avoid contamination. However, out of these only nine cities viz. Thiruvananthapuram, Jamshedpur, Amritsar, Jaipur, Shimla, Ahmedabad, Allahabad, Greater Mumbai, and Nashik, the collected waste was compacted and had earth covering.

(c) *Incineration*: In this process, the garbage is burnt to ashes until nothing is left. The material of the incinerator is quite heavy and insulated to withstand high heat. It is also referred to as a "thermal treatment." In this method, the waste gets converted into ash, heat, steam, and gas. This method is usually adopted for the disposal of hazardous solid, gaseous, and liquid waste. The greenhouse gases are the by-products when the waste burnt is generated from plastics and synthetic fabrics based on the fossil fuels. This method is commonly adopted in developed countries that are highly industrialized and where space, high cost of land, and political oppositions act as a limiting factor. Whereas in developing countries, the loophole is less availability of material with high heat value i.e. plastic and paper in the waste and high capital price of incinerator which discourages this option for waste disposal. In many cases, this practice is mainly limited up to biomedical wastes and other kinds of hospital waste. The reason behind this may be the high moisture content (40–60%), high organic content (40–60%), and low calorific value 800–1100 kcal/kg (Rajput et al. 2009) as against the required value of 1500–2400 kcal/kg which can be attained by adding the low-cost additive materials such as wood coal, bagasse, rice husk, coconut husk, engine oil, etc. in different proportions (Dhanbhar et al. 2018). Incineration produces energy and reduces the volume of solid waste by 90% so that the minimum amount goes to landfills. It, however, produces particulates,

Table 17.1 Waste-to-energy plants (operational)

Location	Capacity (TPD)	Electricity generation (MW)
Delhi–Bawana	2000	24.0
Delhi–Okhla	1950	16.0
Delhi–Ghazipur	1300	14.0
Hyderabad	2400	20.0
Hyderabad	1000	11.0
Jabalpur (MP)	600	11.5
Chennai	300	2.9

Adopted from: Patel (2019) Solid Waste Management in India, ICRIER, New Delhi

carbon monoxide, toxic metals, dioxins, and furans for which air pollution control devices need to be employed.

(d) *Composting*: The municipal solid waste of different Indian cities comprises about 37–60% of compostable matter (Jha et al. 2003; Sharholy et al. 2008), hence composting can be a good option for manure production and income generation. It is a method where waste is degraded biologically into humus-like substances with the help of microorganisms like bacteria and fungi. The decomposed residue is rich in carbon and nitrogen and if contaminant free than it can be applied as a manure for agriculture. The earthworms can also be used in the composting, referred as "vermi-composting" and adds more fertility values (Sangwan et al. 2010). For the prevention of nitrite pollution in water and soil, the compost pit is lined with a granite layer.

(e) *Waste to Energy Plants*: Implementation of WTE technologies is one of the effective ways by which waste volume will be reduced to significant levels, treatment and processing of wastes need to be carried out before their disposal and energy generation. Various technologies that are in use in India to produce energy from waste are biomethanation (anaerobic digestion), mass burn/RDF incineration, gasification, pyrolysis, and plasma arc gasification. According to Central Pollution Control Board (CPCB 2019) report, there are about 2426 compost plants, 534 vermi-compost plants, 426 biogas plants, and 40 Refuse Derived Fuel/Palletization plants in different states of India. As per the Minister of State for Power, New and Renewable Energy in a reply to the Lok Sabha in November 2019, 56.34 MW W-to-E power generation plants from waste have been set up during the last 3 years and the current year in India (The Economics Times 2019). The waste-to-energy plants operational in India are given in Table 17.1.

17.3.1.1 Challenges in Urban India

In India, SWM services are provided by Municipal Corporations or Municipal Panchayats. However, at some places, the implementation is extremely poor and requires some changes. In most of the municipalities, no separate department for waste management is there. The activity of waste collection is labour-intensive and demands two to three workers per 1000 residents served. The expenditure on waste management is 5–25% of the allocated budget on SWM, which is 75–250 INR per capita per annum (Kumar and Gaikwad 2004). However, despite spending such a huge amount, the desired services are not up to the level of satisfaction. Based on research papers and reports of different Indian cities Sharma (2017), identifies issues and challenges related to MSW management. The issues reported were related to policy, finance, technical, management whereas, the high proportion of organic constituents, high moisture content, low calorific value, and presence of heterogeneous materials were the main challenges observed. The challenges regarding SWM in India are as follows:

1. Unsegregated/Mixed Waste: Wastes of all kinds, such as recyclable, biodegradable, hazardous or soiled and construction waste are mixed in urban areas. In most cases, the waste is not segregated at source. Though in few cases and especially in rural areas, the kitchen waste is used to feed cattle and other animals but with advancement and changing lifestyles, the use of more plastic and packaging material results in mixed waste which is put into one dustbin that is difficult to handle. The waste contains a high percentage of organic content (about 60–70%), high moisture level, and low calorific value. The combination of different materials such as discarded pharmaceuticals, and sometimes hazardous substances, reduce the economic value and resource recovery.
2. Increasing Population and Waste: Extrapolating from census 2011, the current population in India is 1.35 billion and is increasing rapidly. The population burden is more in newly urbanized and semi-urban areas. The waste management has become a challenge in densely populated areas. The quantity of per capita waste is increasing constantly due to increased income and consumerism and the absence of any system of Integrated Waste Management (IWM). Large amount and volume of waste generated make it difficult to manage or dispose of.
3. Lack of Proper Management: In residential areas such as apartments or flats, at the time of passing the plan, provision is not made for the collection and disposal of waste. There is no site designed for disposal, separation, and treatment of the waste. It results in the disposal of waste anywhere in the open and the private developers are benefitted from it. Due to lack of management, solid waste can be seen littered or is disposed of in low-lying areas, outer areas of cities, nearby roads, or any vacant place from where the waste cannot be seen easily or cannot be objected to by any person. Even this problem is also faced by the newly developed cities where dumping of city waste in landfills becomes difficult.
4. Lack of Resources: The SWM is assigned very less budget by the civic bodies and most of the budget amount (approx. 90%) gets used up for collection and transportation of waste and meagre funds are left for the process of treatment and recycling (Viz 2012).
5. Old Technology and Equipment: No specific landfill sites are designed for dumping of the waste in the metropolitan cities of India. Old equipment are used for the collection and transportation of the waste. The disposal of waste is done

by incineration which can cause serious environmental and health hazards if mixed waste is burnt. Also, the workers do not take any protective measures or precautions while handling waste which may cause different diseases.

6. Social Apathy: The task of SWM is considered degraded and is assigned low social status. People try to leave waste at any vacant area or secluded place, alongside roads and outside bins etc. It increases the work of waste collectors and results in scattered waste. This process is repeated even if the waste is picked up from the site.
7. Unplanned Collection System: In India, the community bins are used for collecting waste but these bins are not designed properly for the required quantity. Moreover, they are not placed at proper sites. Some managers provide waste bins or waste bags for waste collection. However, in developing countries, nothing is provided. For example, in high-income countries, waste bins in different colours are used to collect different wastes such as green bins for papers, blue bins for glasses and other sharp objects, and red bins for other hazardous and harmful wastes (Viz 2012).
8. Large Travel Time to Landfill Sites: The one-way travel time to landfill sites in large municipalities varies from 30 to 75 min to pass through a crowded city including the loading, unloading, and the waiting time at the site. It results in huge waste accumulation in the collection as well as transfer site. The design of vehicles for transporting waste is not proper. They are not duly covered. Non-availability of proper instruments required for waste collection leads to scattering of waste during collection and transportation.
9. Narrow Roads: The Indian cities have narrow roads with less than 4 m width in about 25–30% of areas. Hence, it becomes difficult for any mechanized vehicle to collect waste from all the sites. Moreover, these areas are mostly densely populated with authorized and unauthorized residential settlements that produce a huge amount of daily waste. It poses a new challenge in the waste management system of urban areas (Ghatak 2016). For example, it is exceedingly difficult to collect waste from slum areas.
10. Pricing: The pricing method for waste collection service is not uniform as there is no standard. As per the World Bank report of 2012, the user charge for waste collection varied from 41% in developing countries to 98% in developed countries (Hoornweg and Bhada-Tata 2012). Only the collection of solid waste accounts for around 80% of all associated costs for waste disposal (Belien et al. 2012). However, the environmental costs are also important that are associated with the collection of waste, for example, carbon emissions from heavily polluting vehicles need seriously consideration (Ramos et al. 2014; Sandhu et al. 2015). Further, it is quite difficult to collect fees from waste generators as they refuse to pay charges. To enforce payment, some enabling laws must be there which will urge the waste collectors to demand charges.
11. Management of Landfills: In many cases, the landfill sites are also not designed scientifically and managed properly. Landfill gas and leachate generated from these mismanaged sites cause air and water pollution. Sometimes uncontrolled

burning and odour at landfill sites make them a nuisance for the nearby communities. The problem is more in developing nations as they lack institutional or legal framework for managing their urban wastes.

17.3.2 Rural Areas

In rural areas, the solid waste is generally composed of kitchen and garden waste, house sweeping, cattle dung and cattle sheds, agricultural waste, broken glass and metal, waste papers, plastics, cloths, rubber, including the waste from shopping areas and markets, etc. Although, the solid waste produced from the rural settings is mostly organic and biodegradable in nature, still it has increasingly become a major problem. The main reason behind it is that the garbage generated is not segregated at source. In rural areas, the goal of SWM is collection of the waste at source, recovery, and recycling of recyclable materials, making compost from organic waste and proper disposal of left-over waste (Sutar and Gawande 2013). For management of solid waste in Haryana state, a project was launched in 2012, named "Wealth from Waste" under the "Nirmal Bharat Abhiyan (NBA)" of the Ministry of Drinking Water and Sanitation (MDWS). A total number of 1336 solid waste management projects have been sanctioned in 22 districts of Haryana under this project. However, out of the total 662 completed projects, only 39.42% of projects were functional up to 2019.

17.3.2.1 Challenges in Rural India

The main challenges in the management of solid waste faced by village councils are:

1. Unsegregated Solid Waste: In rural areas where there is a waste collection system, unsegregated wastes are handed over to the sanitary worker most of the times and if some separation is done then only animal dung is separated. Majority of the households preserve the metal containers, plastic bottles, and other saleable items for the vendors. That mixed waste requires additional manpower and lots of time to segregate.
2. Lack of Funds and Common Land: The villagers do not have any village fund except from the state government grants. For example the project on solid-liquid waste management (SLWM) started by the central government in Indian villages was funded under Swachh Bharat Mission-Grameen (SBM-G). However, there are also some other expenditures such as payment of remuneration to sanitary workers, maintenance of equipment, etc. Further, some village councils (Gram Panchayats in native language) do not have land availability to construct the sheds for waste storage and segregation (Kumar and Sharma 2017).
3. Lack of Proper Planning: The rural waste management projects started by the village councils, in many cases, lack proper planning and management. The solid waste management projects require preparation, detailed survey, group discussions, etc. A project can never become successful if it gets started without estimates from the generation of solid waste, requirement of manpower, and

transportation (Kumar et al. 2018). The site selection should also be appropriate. For example the waste dumping site should not be close to residential areas and village ponds otherwise they will produce foul odour and can get inundated in the rainy season.

4. Low Market Value of Compost and Recyclable Material: Other important issue observed in the rural areas is the low economic value of non-biodegradable dry waste, i.e. rubber, plastics, polythene, glass, paper, etc. Hence, these kinds of wastes remain stored in the sheds for longer periods resulting in shortage of space. Further, in many village councils, the compost made from waste does not get a local buyer leading to no income generation. All these factors affect the sustainability of a waste management project.
5. Lack of Priority in Village Councils: Many village councils are still not interested in implementing waste management practices in their villages. They believe in traditional practices such as the construction of drains and streets as their first priority. Hence, there is a need to change the mindset of rural people and village council members to focus on this issue.
6. Reluctant Behaviour of Households: A large fraction of the population in villages is reluctant to pay some fees for door-to-door collection and transportation of waste, if Gram Panchayat makes any arrangement for waste collection. Insanitary and unhygienic behaviour of some villagers also poses a big problem because they dispose non-separated waste in the bins and sometimes, they put all their waste into polythene bags and disposes of it.
7. Lack of Technical Skills in Sanitary Workers: In the rural areas, the engaged sanitary workers for waste management do not have the proper technical skills to handle the different kind of waste. The dry waste with various electronic and hazardous waste material must be handled separately. However, it is observed that it is kept in mixed form in dry pits. Similarly, other kinds of wet waste (organic and degradable waste) also require different kinds of management (Kumar and Sharma 2019).
8. Increased Quantity of Waste: The increasing population and commercial activities result in the production of more quantity of waste in rural areas. The organic solid waste generated from rural areas in India is estimated at 0.3–0.4 million metric tonne per day (Ministry of Drinking Water and Sanitation (MDWS) (2014). Although the waste generation has increased in rural settings still it is lower than that in urban areas.

17.3.3 Solid Waste Management in Hilly Regions

An important source of waste generation in hills is tourism. There are 300 hill stations in India. Many hill stations are famous for temples that generate lot of organic waste including flowers offered to deities. At Chamundeswari Temple situated at Chamundi hill top, near Mysore (Karnataka) collection chain and a biogas plant has been developed for this purpose (Ghosh et al. 2018). Uttarakhand has many tourist spots such as Bhimtal, Ranikhet, Mussoorie, Nainital, Yamunotri, Gangotri.

The state has many rivers and a large area of forests, rich in flora and fauna. Out of 13 districts, 09 are hill districts that attract tourists. It is the worst-performing state in respect of solid waste management. Leachate contaminates the precious water of rivers in which flows water from glaciers. Large heaps of unsegregated solid waste can be seen at most of the places, especially near Bhimtal and Ranikhet. Animals including migratory birds get attracted to the left-over food and ingest polythene and plastic (especially multi-layered plastic which is not recyclable and is of no use) which becomes the cause of death of these animals. Garbage poses threat to the ecosystem of the hills.

17.3.3.1 Challenges in Hilly Regions

Solid waste management in the hilly regions is a real challenge because of the rough terrain, scarcity of land, insufficiency of technical manpower and funds, small-town size, etc. Therefore, the 3R strategy is the best approach for small hill towns as has been successfully implemented in other towns, as mentioned earlier in this chapter, where the Zero landfill approach has been demonstrated. Collection of solid waste from houses is difficult because these are situated on hills which can be approached by climbing on the steps or footpaths. Here the use of carts or tricycles is not easy. Therefore, it is advisable to segregate waste into wet and dry waste so that the waste from door-to door can be collected in separate bags or baskets which can be carried on the backs of animals or human waste collectors or through the lightweight containerized handcarts. The collection should start from the upper reaches and brought to the lower levels from where it can be transferred to the vehicle on the motorable road. Direct loading on the vehicle would not cause littering on the ground and will save time during the transfer. Vehicle movement down the slope will ensure that the vehicles do not waste extra fuel and cause less exhaust gases.

Landfill is not advised on the hill. It is constructed on plain area down the hill within 25 km area. A transfer station is, however, constructed at a flat area somewhere near the processing facility to collect residual waste and inert material. At the processing facility, which is set up by the local body, the biodegradable waste is converted into value-added product and the non-biodegradable material is used in road building or filling up purposes. The undesired material not considered useful for either function is sent to the landfill for final disposal.

17.4 Sustainable Waste Management System

In United Nation World Commission on Environment and Development report "Our Common Future," sustainability was defined as development that meets the needs of the present without compromising the ability of future generations to meet their own needs (Brundtland 1987). As per Satyro et al. (2017), sustainability indicates the capability of a dynamic system to continue into the future by integrating environmental, economic, and social principles in waste management practices. Prior to year 2000, traditional MSW management systems were more focused on economic efficiency than environmental sustainability and social well-being (Morrissey and

Browne 2004; Marshall and Farahbakhsh 2013). The main objectives of sustainable waste management are to reduce the consumption of natural resources, to confirm that any materials that are taken from nature must be reused as many times as possible, and to keep waste generation at minimum (SRWS 2018). One of the priority global issues in the sustainable development goals is effective sustainable waste management practices in place (Hosono and Aoyagi 2017). Further, it is evident that at least 12 sustainable development goals and their relevant targets have a direct link to solid waste management (Rodic and Wilson 2017).

The European Directive (2008) confirmed the following two approaches for sustainable waste management.

1. The concept of self-sufficiency and proximity where each member state is liable to define an integrated and adequate network for waste disposal and recovery. This concept led to the planning of the Integrated Municipal Solid Waste Management (IMSWM) systems
2. The concept of Extended Producer Responsibility (EPR) which involves both the reverse logistics and the design phase of a product. EPR forces producers to supply information about the reuse and recycling of their products. This concept will promote eco-design procedures for several products.

17.5 Sustainable Waste Management Principles and Policies

Several principles and policies that may be closely related to SWM sustainability (Chang and Pires 2015) are:

1. Waste Hierarchy and Sustainability Hierarchy principle: Waste Hierarchy Principle (WHP) was first introduced by the European Union Waste Framework Directive in 1975 as a guiding principle for managing MSW worldwide. The aim of the WHP is to achieve maximum benefits with the generation of minimum amount of waste for a product (Singh 2010; McElhatton and Pizzuto 2012). For waste minimization, it relies on 3Rs—reduce, reuse, and recycle (Singh 2010) where top priority is given to avoid waste generation in the very beginning. Therefore, to achieve this objective, the priority is on reuse, recycling, recovery, and disposal at the end. With due course the 3Rs were raised to 7Rs-rethink, redesign, reduce, reuse, renew, refurbish, and recycle. Though Sustainability Hierarchy (SH) and the Waste Management Hierarchy, both are compatible principles, the SH concentrates more particularly on waste prevention options. SH is established on some of the most promising waste prevention options of natural resource conservation and protection, manufacturing efficiency, product quality, judicious product selection, product repair, and restoration.
2. Polluter-Pays Principle: To eradicate pollution, this principle fixes financial responsibility on the producer. The polluter-pays principle was defined by Organization for Economic Co-Operation and Development (OECD) as the "principle to be exercised for allocating charges of pollution prevention and control

measures to encourage reasonable use of scarce environmental resources and to avoid distortions in international trade and investment" (OECD 1972). Regarding SWM, this principle entails that all types of waste generators, i.e. citizens, industries, companies, are to pay the price for the waste generated by them (Nahman and Godfrey 2010).

3. Extended Producer Responsibility (EPR): EPR is the perception of manufacturers and brand owners taking environmental responsibility for their manufactured goods and the associated packaging when they become waste. It involves activities like arranging waste collections, creating take-back programmes, recycling, or suitable disposal, and designing products for reuse or recyclability. According to OECD (2001) "EPR is an environmental policy in which a producer's responsibility, physically and/or financially, for a product is extended to the post-consumer stage of a product's life cycle." EPR shifts the responsibility to the producer instead of users, and the environmental impacts related to the end-of-life, in such a way that could influence the design of the products (e.g. by promoting eco-design) (OECD 2001, 2006). In India, the EPR programme and the EPR policy find place in Plastic Waste Management Rules (2016) and E-Waste (Management) Rules (2016).
4. Precautionary Principle-Protection of Human Health and Environment: This principle can be described as environmental safeguard based on precaution, even where there is no clear indication of risk or harm from an activity (Bell and McGillivray 2006). The precautionary principle in SWM context is considered as its policy when the information is not adequate on waste management operations and their casual effects on human and environment. Application of this principle can lessen the burden of supports and evidences for those who wish to regulate an action, even when the casual scientific link with of harm and the activity has not been established (Kuhlau et al. 2011).
5. Principles of Self-sufficiency and Proximity: The EU has considered this principle to find a solution to their solid waste within the community. According to the proximity principle most waste should be treated and managed within the region of its generation, provided there are no unacceptable adverse effects—on public, environment, or transportation systems. The transportation of waste can cause significant environmental and nuisance impacts in addition to unwanted costs. This principle will favour and encourages processing, reuse, recycling, and waste disposal in the vicinity of its production point as possible. Many cities in India have shown success stories in solid waste management. Alappuzha city in Kerala is managing its own waste. It has no landfills and no processing units. The municipality here does not collect waste. Waste management is done by the local people. They manage the waste they produce, effectively for the last many years. The waste is segregated (started in 2002), and compost is produced from the biodegradable waste while the non-biodegradable waste consisting of paper, plastic, metal, etc. is recycled by the waste pickers. About 80% of the waste is used for composting or sent to decentralized biogas plants. As per UNEP, 2017, the city is included in top five cities of the world with sustainable solid waste management. Panaji in Goa also started segregation of waste in 2003. It is a

bin free, zero landfill city where waste management was decentralized. Similarly, Mysuru in Karnataka started waste segregation in 2014, on-site composting of waste is done by hotels and markets. The non-biodegradable waste is separated at the household and garbage is collected from door-to-door.

6. Zero Waste Principle: Zero Waste is a holistic approach to address the problem of unsustainable resource use. According to Glavic and Lukman (2007), zero waste has not been defined by the UNEP, United States Environmental Protection Agency (USEPA), or European Environment Agency. However, it is a state when all materials in various stages of existence are converted into products so that nothing remains useless. The peer-reviewed internationally accepted definition of zero waste is by the Zero Waste International Alliance (ZWIA 2018). According to ZWIA (2018), zero waste means the conservation of all resources by means of responsible production, consumption, reuse, and recovery of products, packaging, and materials without burning and with no discharges to land, water, or air that threaten the environment or human health. It involves moving from the back end of the waste disposal to the front end of resource management. The lyrics of the song by Pete Seeger (American Singer 1919–2014) are worth mentioning that "if a product cannot be reused, repaired, rebuilt, refurbished, refinished, resold, recycled or composted, then it should be restricted, redesigned, or removed from production." Furthermore, the zero waste principle does not believe to dispose of waste material but considers it a resource to be reused to take full advantage of its potential (Glavic and Lukman 2007). This principle has its application in some SWM case studies, based on industrial symbiosis with industrial ecology and in the perspective of circular economy in zero waste cities. Civic EXNORAs is one such successful example that supports composting at home and "zero waste" for the municipality, where residents of 30 localities in Chennai, India, segregate at source, their waste into recyclable and organic fractions.
7. Integrated Product Policy (IPP): Environmental degradation is caused by all products from their manufacturing, use, and disposal. IPP aims to reduce this by considering all aspects of a product's life cycle for taking the most effective actions. There are both voluntary and mandatory tools used to achieve such as substance bans, economic instruments, voluntary agreements, environmental labelling, and product design guidelines (European Commission 2020). According to European Commission (1998) the Five building blocks of IPP are:
   (a) Ways to plan innovation of more environment friendly products.
   (b) Measures to generate demand and offer environmentally sound products.
   (c) Ways designed to reduce and manage waste generated from products' consumption.
   (d) Measures to circulate information up and down the product chain.
   (e) Measures that assign responsibility for controlling the environmental burden of product systems.

17.6 Elements of a Sustainable Waste Management System

As stated above, the sustainability of a waste management system requires to be economically affordable, environmentally effective, and socially acceptable (Seadon 2010). For integrating sustainability concepts with SWM, it is essential to firmly stick to the economic, environmental, and social performance and compliances for all the processes it adopts (Nilsson-Djerf 2000; Petts 2000). Environmental protection, waste as a valued resource, public health, closing the loop, institutional and responsibility issues, and public awareness are the six factors that influence sustainable waste management (Wilson 2007). The sustainable waste management elements are related to sustainable living systems developed by Vester (2007) and are explained by Seadon (2010) as:

1. Negative Feedback Loops Dominate Positive Feedback Loops: Negative feedback provides a component of self-regulation and self-monitoring. Fiscal disincentives (e.g. environmental fees, user charges, levies, and liability costs for waste generation) act as negative feedback loops (Vester 2007). For example in Auckland City, New Zealand, negative feedback of reducing mobile garbage bin size to its half along with the additional cost for disposing of excess weight had an enhanced effect on waste recycling than that of praising messages of recycling virtues for years as positive feedback (Seadon 2006).
2. System Vitality Is Independent of Quantitative Growth: Waste Management Systems that depend on growth are unsustainable. The Waste Recycling Centre in Portland, Oregon, the USA, for example successfully utilized the collected instead of requiring waste in larger quantities (Suzuki and Dressel 2002).
3. Function-Oriented System Rather than Product-Oriented: A basic SWM operation comprises waste collection, transportation, and storage. A product-oriented approach means assessment of impacts on environmental and scope for improvement in the production and life cycle of the product—and incorporating the learnings in the system (Kirsten et al. 2002). A product's total environmental impacts can be reduced by designing a new product or a new concept with the similar function as an existing product. In a functional economy, consumers are buying mobility as an alternative to cars, cleaning services in place of washing powders, and movies instead of videocassettes (Friend 1994; Pantzar et al. 1994; Popov and DeSimone 1997). Functional economy generates positive environmental effects as it shares product use, remanufacturing processes, and slower product cycles, which help to reduce the use of virgin raw materials in making the products. Car sharing service according to a research, for example resulted in a reduction of 44% in the number of cars, requiring less space for parking (Meijkamp 2000).
4. Multiple Uses of Products, Functions, and Organizational Structures: The assimilation of products, functions, and structures permit the best use of available resources in a waste management system. As a concept of industrial ecology, the waste heat, wastewater, or waste material of one industry can be used by another industry thereby reducing the demand for virgin material and resources.

5. Diversion Processes for Waste Utilization: Waste is a resource that can be reused, recovered, or recycled. As per USEPA (2004) food waste recovery hierarchy, the food can firstly be feed to people, to animals, to industries for animal feed production or biofuel generation, in compost generation. There are many common examples where glass and plastic bottles are reused. Many discarded materials are reprocessed into new useful products having some market value (Kaushik and Kaushik 2021).
6. Symbiotic Usage by Employing Coupling and Exchange: Coupling relationship is the relationship between two things that interact and influence each other. In Eco-industrial parks, industrial coupling symbiosis is the relationship between different eco-industrial enterprises. A symbiotic relationship in waste management is an industrial ecology concept which attempts to utilize the waste of an ecosystem in an industrial context (Suzuki and Dressel 2002). This concept of symbiotic exchange is very well utilized in case of using treated wastewater for irrigation purpose but need to be practised effectively in sustainable solid waste management.
7. Biological Design of Products, Procedures, and Organizations: The basic biological design processes are illustrated by the application of The Natural Step programme (Korhonen 2004). The Natural Step is a theoretical model of sustainability developed in 1988 by cancer physician Dr. Karl-Henrik Robèrt along with one hundred eminent Swedish scientists (Kranz and Burns 1997; Tynberg 2001). It refines complex scientific principles into four easily understandable concepts that can guide society toward sustainability.

   The first three concepts or conditions presented by Robert et al. (2002) state that the functions and biodiversity of the ecosphere must not be systematically undermined by:

   (a) Increasing concentrations of substances extracted from the lithosphere.
   (b) Increasing concentrations of substances generated in the technosphere.
   (c) Physical impoverishment from manipulation or overharvesting, and the fourth one states:
   (d) Use of resources is reasonable and efficient enough to meet essential human needs everywhere.

17.7 Legal and Policy Framework for SWM in India

Legislative provisions governing waste management and disposal are very important in sustainable waste management. It also guides and unites various stakeholders in waste management. Sustainable waste management must be incorporated in local legislation that can be geographically and culturally practicable, reasonable, and far-sighted. Number of Acts and Regulations have been made from time to time by the Govt of India for proper management of municipal solid waste. The laws concerning waste management in India were initiated only in the third 5-year plan during 1961–1966 which encouraged composting of MSW in urban areas, mainly in the cities having population of 0.3 million. However, due to poor planning, MSW composting did not meet the envisaged level of success (Priti and Mandal 2019). In the mid-1980s the Bhopal Gas tragedy on 3 December 1984 first time drew the government's attention towards environmental damage and fatalities due to toxic wastes and hazardous chemical substances. After this disaster, the Ministry of Environment and Forests [now Ministry of Environment, Forest & Climate Change (MoEF&CC)] ratified an umbrella Act is known as the Environment (Protection) Act 1986. Under this umbrella Act, the Hazardous Waste (Management and Handling) Rules 1989 and Bio-medical Waste (Management and Handling) Rules 1998 were enacted to reduce environmental and public health hazards. The ineffective management of MSW was considered as a main reason for the plague incident of Surat, Gujarat, India in 1994 which again drew government attention towards the MSW management. Bajaj Committee, a high-powered committee was constituted in 1995 after this incident which gave several recommendations for proper management of solid waste. These recommendations emphasize waste segregation; formation of cooperatives for door-to-door collection of waste; integration of waste pickers; ward level recovery centres; incentives for recycling units; composting; disposal tax on waste-creating products, and user fees for waste collection. As an outcome, the Municipal Solid Waste (Management and Handling) Rules were notified in the year 2000 by the Ministry of Environment and Forests. The MSW Rules mandated all municipal authorities in the country to manage MSW in a time-bound manner and entrusted the State government to ensure implementation of the (management and handling) Rules 2000. It prescribed the way by which authorities should commence collection, segregation, storage, transportation, processing, and disposal of the MSW generated in their respective areas following the respective governing legislation. However, it was observed that the authorities are unable to implement and sustain projects related to MSW disposal due to lack of financial and institutional means. Solid waste management was also a sub-mission under Jawaharlal Nehru National Urban Renewal Mission which was launched in 2005 by the Ministry of Housing and Urban Affairs, Govt. of India.

On 2 October 2014, the Govt. of India launched The Swachh Bharat Mission (SBM 2014) with an aim to make India clean. The main objectives under this mission were to eliminate open defecation, eradicate manual scavenging, and effect behavioural change regarding healthy sanitation practices. A national campaign covering 4041 statutory towns was launched with an aim to ensure door-to-door collection and proper disposal of municipal solid waste in urban areas by 2019 (Ghosh 2016). On 2 October 2019, the country has achieved the goal of making India Open Defecation Free under this mission. Keeping the objective of SBM, and in supersession of the Municipal Solid Waste (Management and Handling) Rules (2000), the Central Government has issued and implemented the new Solid Waste Management Rules (2016). For the first time different duties were fixed under this rule for waste generators and various urban local bodies to manage the solid waste properly. Different criteria for waste-to-energy process, special provisions for hilly areas, sanitary landfills including standards of processing and treatment of solid waste are made available in SWM Rules (2016). Unlike the Municipal Solid Waste

(Management and Handling) Rules (2000), the new Solid Waste Management Rules (2016) are applicable beyond municipal areas including urban agglomerations, census towns, notified industrial areas, areas under Indian Railways, airports, special economic zones, places of pilgrimages, religious, and historical importance, state and central government organizations. Under these rules, segregated solid waste, i.e. biodegradable or wet waste, non-biodegradable or dry waste, and domestic hazardous waste should be stored in different coloured bins. The non-biodegradable waste is to be handed over to authorized waste pickers or waste collectors. The generator of solid waste must store the waste in own premises and dispose it off. In Special Economic Zone (SEZ), industrial estate, industrial parks at least 5% of the total area is to be earmarked for recovery and recycling facility. Likewise, street vendors have to store the waste in containers and deposit it at the notified waste storage depot. No waste generator can throw, burn, or burry the generated waste on streets, open public spaces, in drain or any water body. Biodegradable waste is to be processed, treated, and disposed of through composting or bio-methanation within the premises as far as possible. The residual waste is to be given to notified waste collector. With sanitary products, the manufacturer should provide a pouch for proper disposal. The non-biodegradable packaging is to be collected by the brand owners. The local bodies should decide the user fees from the bulk generators for collection, processing, and disposal. There is a provision for the local bodies for "spot fine" for littering and non-segregation of the solid waste. Timeline for setting up of waste processing facility, depending on population strength, and for bioremediation has been spelt out. The compost thus produced will be co-marketed with chemical fertilizers. The refuse derived fuel (RDF) will be made from non-recyclable waste having calorific value of 1500 Kcal/Kg or more, useful for co-processing in cement or thermal power plants. Industries near the RDF plant should replace at least 5% of their fuel requirement by RDF. Conditions for setting up landfill sites near water bodies, highways, airports, hill stations etc. have been provided. Standards for emission of dioxins, furans, particulate matter have been amended. Provision of adequate funds, central monitoring committee, and frequency of their meetings have been specified.

In addition to the above-mentioned SWM rules, there are many other Indian rules and regulations related to solid waste management. The Water (Prevention and Control of Pollution) Act 1974 (amended in 1978 and 1988) specify standards for the discharge of sewage and industrial effluents into water bodies. There is a provision for levying and collection of cess on water consumed for waste processing and disposal facility mentioned under the Water (Prevention and Control of Pollution) Cess Act (1977, and amended in 1991). The Batteries (Management and Handling) Rules (2001) fix the responsibilities of manufacturer, importer, and assembler to collect used batteries and sent them back only to registered recyclers. The Environmental Impact Assessment (EIA) notification (2006) provides the requirement for environmental clearance from the statutory authorities for setting a municipal solid waste processing and disposal facility. After Australia and New Zealand, India became the third country in the world to set up a specialized environmental tribunal, i.e. the National Green Tribunal Trust under the National

Green Tribunal Act (2010). It is meant for effective and expeditious clearance of cases relating to environmental protection and conservation of forests and other natural resources.

The Plastic Waste Management (Handling & Handling) Rules (2016) provide specific rules for the manufacture and reuse of plastic bags and containers. The Bio-Medical Waste Management and Handling Rules (2016), describes various management aspects related to segregation, collection, treatment, processing, and disposal of bio-medical wastes. The Hazardous and Other Wastes (Management and Trans-boundary Movement) Rules (2016) ensure resource recovery and disposal of hazardous waste in environmentally sound manner. The Construction and Demolition Waste Management Rules (2016) specify that a waste generator should properly collect and store the waste within his/her premises and ensure no spill over or mixing with MSW. The E-Waste (Management) Rules (2016) which replaces E-waste (Management and Handling) Rules (2011) have comprehensive provisions on the management of e-waste in India. The E-waste (management) Rules (2016) assign responsibilities of the manufacturer to collect e-waste produced during the manufacture of any electrical and electronic equipment and channelize it for recycling or disposal. The Department of Labour of the concerned state government will ensure the health and safety of workers engaged in dismantling and recycling of E-wastes. Many workers stress that urban local bodies fail to implement these laws adequately. Further, the infrastructure created is not properly utilized and the return on investment is often neglected resulting in failure of the whole efforts (Shenoy 2018). Issues like separate authorizations, and not a single licence for disposal of different kinds of waste in place, long time taken by officials to issue or renew authorizations (Arene and Mbata 2008) needed to be rectified for effective implementation of rules towards sustainable solid waste management.

17.8 Integrated Solid Waste Management System

Integrated Solid Waste Management (ISWM) has emerged as a possible solution to deal with the increasing problems MSW disposal in the world's mega cities. In general, it involves waste reduction, recycling, recovery, treatment, and disposal of waste to attain operational efficiency, enhance society's well-being and minimize environmental risks (USEPA 2002). ISWM is a comprehensive approach to prevent, manage, and recycle solid waste in a manner that most effectively protects the environment and human health (Van de Klundert and Anschütz 2001). ISWM lays more importance on social acceptance in the design and development of a sustainable MSW management system. The social resistance and environmental movements like *Not In My Back Yard (NIMBY)* and *Build Absolutely Nothing Anywhere Near Anything (BANANA) syndromes* may be the driving factors behind social resistance and environmental movements resulted in ISWM. The Integrated Waste Management suggests incorporating different management approaches and technologies and deciding on the most appropriate one (Ikhlayel and Nguyen 2017). In its simplest sense, IWM incorporates the waste management hierarchy (Turner and Powell 1991) by considering direct impacts (collection, transportation, treatment, and disposal of waste) and indirect impacts (use of waste materials and energy outside the waste management system) (Korhonen et al. 2004). According to the UNEP (1996), IWM is a "framework of reference for designing and implementing new waste management systems and for analysing and optimising existing systems." Integration requires the organization of the community, governmental bodies, and commerce each of which is an agent for change. Chua et al. (1992) suggested that in addition to community awareness and participation, and a determined political motivation, the key drivers for an IWM approach are (1) co-operation to develop multi-national waste management strategies, which comprise suitable technology, manpower availability and exchange of information, and (2) a multi-sectoral approach to environmental projects. In 2002, UNEP referred to adopt Regional IWM Strategy to promote the application of environmentally sound technologies for sustainable IWM which recommends regional collaboration among governments and partner organizations to tackle the growing problem of waste. This approach was meant to help policy makers and urban managers to identify the appropriate measures, techniques, and technologies that should be used and applied to deal with every kind of waste and to manage waste in a holistic manner (UNEP 2002).

The IWM highlights that waste management systems consist of interrelated functions. There is no individual waste management method that is appropriate to process all kinds of waste, therefore, there is no perfect IWM system. Individual IWM systems may differ across regions and organizations. According to McDougall et al. (2001), the establishment of IWM systems in developing economics will require data collection on waste composition, development from uncontrolled dumping to simple sanitary landfills, organic waste separation from MSW, and formal involvement of scavengers in recyclable materials collection. The various key features of IWM are:

- To assess the overall environmental burdens and economic costs of the system.
- To use a range of collection and treatment methods which focus on generating less waste and effectively managing already generated waste.
- To handle all the materials, present in the solid waste instead of focusing only on specific materials or sources of materials. The hazardous materials should be dealt separately but within the system.
- To reduce the environmental impacts in water, air, and soil by controlling emissions.
- Must be economically affordable by driving costs down and applying a market-oriented approach with the waste products having end uses and can thus create income.
- Enhance social acceptability by integrating community participation techniques and ensuring that individuals must understand their responsibility in the waste management system (McDougall et al. 2001).

The Bali Declaration of 2008 also discussed integrated waste management. The Ninth Meeting of the Conference of Parties (CoP) to the Basel Convention at Bali,

Indonesia, in June 2008 adopted the Bali Declaration on waste management for human health and livelihood (UNEP 2008). The declaration recognized the danger to the environment, human health, and sustainable livelihoods from mismanaged wastes. It discussed waste prevention and minimization, environmentally sound management of waste and management of the trans-boundary movement of waste as contribution to sustainable development. Further, the IWM gets support as per the 25/8 decision on waste management adopted in the 25th meeting of the UNEP Governing Council in 2010, which supported the implementation of the Bali Declaration (Hyman et al. 2013).

Based on a systematic literature review by Gahana et al. (2018), enormous population, poor social values and ethics, illiteracy, poor environmental commitment, unscientific planning, lack of technical expertise, and ineffective processes and administration are the different barriers of Integrated Sustainable Solid Waste Management. Waste management commercialization is one of the most advantageous methods for ensuring an integrated and sustainable waste management system (Qotole et al. 2001; Suocheng et al. 2001). This could enhance the involvement of the government, community, and private organizations in solving the waste problem (Wang and Nie 2001). Implementation of ISWM can strategically reduce the government expenditure on waste disposal practices and will further raise the options of development of more commercial models. This will motivate the private organizations and companies to come up with various start-ups in the area of waste management especially in developing nations like India.

17.9 Conclusions and the Way Forward

There must be sincere efforts for solid waste processing, recycling, and recovery. Above all, sincere and collective efforts of the community, authorities, NGO's working in waste management, and regional and national governments can bring a big change in efficient waste management. There is a need for innovative approaches in all the areas of the waste hierarchy of reduction, reuse, recycling, recovery, and disposal to achieve sustainable waste management. The sustainability in SWM cannot be attained till we consider wastes as a liability; it must be considered a resource for which a more effective massive public outreach programme is required.

For a sustainable solid waste management, the authorities need to strictly follow SWM Rules, 2016, and regularly redesign their strategies according to the changing lifestyles and solid waste generated. Waste management should be reinvented in rural and urban areas so that waste can be segregated at source, processed instead of open dumping, burning, or landfilling. Certain municipalities have started levying fine on societies for not segregating domestic waste at source. This shows that despite the rules being in place, people are not sensitive to the issue of waste management. The disposal is centralized where the major portion of funds is spent on collection and transportation and a small amount is left for final treatment and proper disposal of the waste. This increases the size of waste dumps. The process should be decentralized so that funds are utilized for treatment, recovery, and recycling of the waste.

Effective management is possible if the public and administration work in tandem. The success stories of decentralized management of solid waste in Alappuzha, Panaji, and Mysuru have been discussed earlier in this chapter. Indore and Bhopal are segregating 100% solid waste at source. There is decentralized management of solid waste from households, shopping complexes, and temples. In Indore, less than 10% of waste goes to landfills. Many cities are integrating ragpickers, waste pickers, and kabadiwalas from the informal sector into the formal sector and renaming them as safaimitra or swachhdoots. This helps in improving their social and financial status. In Mysuru, ragpickers are involved in waste management. Hyderabad Municipal Corporation has also proved this. The collected "user fees" goes to the ragpickers. Kanpur Nagar Nigam has adopted public private partnership (PPP) model.

In some cities, a lot has been done in the field of waste management over time. New concepts, principles, strategies, techniques, and policies have been developed. Surely, there has been an increase in public awareness and involvement in achieving sustainable waste management in such cities. However, more efficient and SWM practices need to be developed in other cities to recover valuable materials from the waste stream and simultaneously solving the problem of waste disposal. For example recycled plastics can be used in road making along with bitumen. Many municipalities such as Chennai (among the first cities globally), Pune, Mumbai, Surat, Jamshedpur, Indore, Lucknow, Delhi, etc. are using this process. Interlocking tiles for footpath paving may be prepared from construction waste. More innovative uses of different types of solid waste can be invented.

By increasing and improving solid waste collection services, especially in hilly, remote, and rural areas, waste segregation at household level and adopting kitchen composting will surely help in achieving sustainability in waste management. Zero waste principle, ISWM, Waste-to-Energy technologies, circular economy concept, and effective implementation of existing rules and regulations will bring a remarkable change in the solid waste management status. The circular economy concept is a new operating method that helps in processing large volumes of waste, thereby creating more free space, protecting the environment from pollution, saving resources, reducing residues, and reducing health risks (European Environment Agency 2017). Circular economy is based on the principles of designing out waste and pollution, keeping materials and products at their highest use, and regenerating natural systems (Ellen MacArthur Foundation 2017). Its aim is to design entire production systems in the closed-loop type cycles in which material, energy, emissions, and waste losses are minimised (Bruggemann 2019). The role model examples of solid waste management in the Indore city and Civic EXNORA organization in Chennai, Kancheepuram, and Tiruvallur can be adopted and applied in other Indian cities.

References

Allesch A, Brunner PH (2014) Assessment methods for solid waste management: a literature review. Waste Manag Res 32(6):461–473

Arene C, Mbata G (2008) Determinants of profitability and willingness to pay for metropolitan waste-use in urban agriculture of the Federal Capital Territory, Abuja, Nigeria. Agro-Science 7:41–46

Batteries (Management and Handling) Rules (2001) Ministry of Environment and Forests, New Delhi, India. The Gazette of India, Notification 16 May 2001

Belien J, De Boeck L, Van Ackere J (2012) Municipal solid waste collection and management problems: a literature review. Transp Sci 48(1):78–102

Bell S, McGillivray D (2006) Environmental law. Oxford University Press, Oxford

Bhide AD, Titus SK, Alone BZ, Dixit RC, Bhoyar RV, Mothghare LM, Gautam SS, Patil AD (1975) Studies on refuse in Indian cities part II—variation in quality and quality. Indian J Environ Health 17(3):215–222

Bio-Medical Waste (Management and Handling) Rules (1998) Ministry of Environment and Forests, New Delhi, India. The Gazette of India, No 460, Notification 27 July 1998

Bio-Medical Waste (Management and Handling) Rules (2016) Ministry of Environment, Forest and Climate Change, New Delhi, India. The Gazette of India, Notification 28 March 2016

Bruggemann A (2019) The circular economy—pivotal to sustainability and resource security. KfW Research—Focus on Economics. https://wwwkfwde/PDF/Download-Center/Konzernthemen/Research/PDF-Dokumente-Fokus-Volkswirtschaft/Fokus-englische-Dateien/Fokus-2019-EN/Fokus-No-258-July-2019-Circular-economypdf. Accessed 14 July 2020

Brundtland GH (1987) Report of the World Commission of Environment and Development: our common future. https://sustainabledevelopmentunorg/content/documents/5987our-common-futurepdf. Accessed 24 June 2020

Carter J (1977) Public papers of the Presidents of the United States: Jimmy Carter. Office of the Federal Register, National Archives and Records Service, General Service Administration, United States Government Printing Office, Washington

Chang N-B, Pires A (2015) Legal and institutional concerns. In: Sustainable solid waste management: a systems engineering approach. John Wiley & Sons, Hoboken, pp 141–170

Chua TE, Garces LR, Edmonds CJ (1992) Waste management in the coastal areas of the ASEAN region: a need for action. In: Chua TE, Garces LR (eds) Proceedings of the conference on waste management in the coastal areas of the ASEAN region. Singapore ICLARM Manila, pp 197–205

Construction and Demolition Waste Management Rules (2016) Ministry of Environment, Forest and Climate Change, New Delhi, India. The Gazette of India, Notification 29 March 2016

CPCB (Central Pollution Control Board) (2012) Consolidated annual review report on implementation of municipal solid wastes (management and handling) rules—2000, Ministry of Environment and Forest, Govt of India, Delhi

CPCB (Central Pollution Control Board) (2019) Annual report for the year 2018–19 on implementation of solid waste management rules. Central Pollution Control Board, Ministry of Environment, Forest and Climate Change, Govt of India, Delhi, Annexure-III, pp 30–36

Dhanbhar A, Ghatge S, Jadhav K, Pawar S, Zarad P, Khambekar TS (2018) Improve the calorific value of municipal solid waste by adding additive material for incineration process. IOSR J Mech Civil Eng (IOSR-JMCE) 15(2):77–82

Down To Earth (2020) India's challenges in waste management. https://wwwdowntoearthorgin/blog/waste/india-s-challenges-in-waste-management-56753. Accessed 5 June 2020

Dutta A, Jinsart W (2020) Waste generation and management status in the fast-expanding Indian cities: a review. J Air Waste Manag Assoc. https://doi.org/10.1080/10962247.2020.1738285

Ellen MacArthur Foundation (2017) What is the circular economy. https://wwwellenmacarthurfoundationorg/circular-economy. Accessed 14 July 2020

Environment (Protection) Act (1986) Ministry of Environment and Forests, New Delhi, India. Act No 29 of 1986, The Gazette of India No 525 Notification 12 November 1986
Environmental Impact Assessment Notification (2006) Ministry of Environment and Forests, New Delhi, India. The Gazette of India, Notification 14 September, 2006
European Commission (1998) Integrated product policy: a study on analysing national and international developments with regard to integrated product policy in the environmental field and providing elements for an EC policy in this area. Final Report, Brussels
European Commission (2020) Integrated Product Policy (IPP). https://ec.europa.eu/environment/ipp/. Accessed 30 June 2020
European Directive (2008) Directive 2008/98/EC of the European Parliament and of the Council of 19 November 2008 on waste and repealing certain directives (Text with EEA relevance). http://data.europa.eu/eli/dir/2008/98/oj. Accessed 20 July 2020
European Environment Agency (2017) Circular by design-products in the circular economy. EEA Report No 6/2017, Publications Office of the European Union, Luxembourg
E-Waste (Management and Handling) Rules (2011) Ministry of Environment and Forests, New Delhi, India. The Gazette of India, Notification 12 May 2011
E-Waste (Management) Rules (2016) Ministry of Environment, Forest and Climate Change, New Delhi, India. The Gazette of India, Notification 23 March 2016
Friend G (1994) The end of ownership? Leasing, licensing, and environmental quality. The New Bottom Line, vol 3.11. https://natlogic.com/resources/publications/new-bottom-line/vol3/
Gahana GC, Patil YB, Shibin KT, Prakash A (2018) Conceptual frameworks for the drivers and barriers of integrated sustainable solid waste management-a TISM approach. Manag Environ Qual 29(3):516–546
Ghatak TK (2016) Municipal solid waste management in India: a few unaddressed issues. International conference on solid waste management 5IconSWM 2015. Procedia Environ Sci 35:169–175
Ghosh SK (2016) Swachhaa Bharat Mission (SBM)—a paradigm shift in waste management and cleanliness in India. Procedia Environ Sci 35:15–27
Ghosh A, Bebnath B, Ghosh SK (2018) 3R opportunities in hilly regions of India: a case study of Chamundi Hills, Mysore. February 2018, conference: recycle 2018 project: waste management in hilly regions in India
Glavic P, Lukman R (2007) Review of sustainability terms and their definitions. J Clean Prod 15 (18):1875–1885
Hazardous and Other Wastes (Management and Transboundary Movement) Rules (2016) Ministry of Environment, Forests and Climate Change, New Delhi, India. The Gazette of India, Notification 04 April 2016
Hazardous Waste (Management and Handling) Rules (1989) Ministry of Environment and Forests, New Delhi, India. The Gazette of India, No 694, Notification 28 July 1989
Hoornweg D, Bhada-Tata P (2012) What a waste—a global review of solid waste management. Urban development series knowledge papers no 15, the World Bank, Washington, DC
Hosono T, Aoyagi K (2017) Effectiveness of interventions to induce waste segregation by households: evidence from a randomized controlled trial in Mozambique. J Mater Cycles Waste Manag 20(2):1143–1153
Housing and Urban Affairs Ministry (2020) 58,000 MT solid waste dumped daily without processing: The Hindu businessline Published on July 15, 2020
Hyman M, Turner B, Carpintero A (2013) Guidelines for national waste management strategies: moving from challenges to opportunities. UNITAR, UNEP
Ikhlayel M, Nguyen LH (2017) Integrated approaches to water resource and solid waste waste management for sustainable development. Sus Dev 25:467–481
Jain AP (1994) Solid waste management in India. 20th WEDC conference on affordable water supply and sanitation, Colombo, Sri Lanka, pp 177–182
Jha MK, Sondhi OAK, Pansare M (2003) Solid waste management—a case study. Indian J Environ Prot 23(10):1153–1160

Kaushik A, Kaushik CP (2021) Perspectives in Environmental Studies, 7th Edn. New Age International Publishers, London, New Delhi, Nairobi
Kaza S, Yao LC, Bhada-Tata P, Van Woerden F (2018) What a waste 2.0: a global snapshot of solid waste management to 2050. World Bank, Washington, DC. https://openknowledgeworldbankorg/handle/10986/30317. Accessed 19 June 2020
Khan D, Kumar A, Samadder SR (2016) Impact of socioeconomic status on municipal solid waste generation rate. Waste Manag 49:15–25
Kirsten S, Frans MC, Lotte J, Henriette O, Charlotte BN (2002) Manual on product-oriented environmental work. Environment News 64, Ministry of Environment, Danish Environmental Protection Agency, Copenhagen
Korhonen J (2004) Industrial ecology in the strategic development model: strategic applications of industrial ecology. J Clean Prod 12(8–10):809–823
Korhonen J, Okkonen L, Niutanen V (2004) Industrial ecosystems indicators—direct and indirect effects of integrated waste- and by product management and energy production. Clean Techn Environ Policy 6(3):162–173
Kranz D, Burns S (1997) Combining the natural step and ISO 14001. Persps Bus Global Change. Dec 1997
Kuhlau F, Hoglund AT, Evers K, Eriksson S (2011) A precautionary principle for dual use research in the life sciences. Bioethics 25(1):1–8
Kumar S, Gaikwad SA (2004) Municipal solid waste management in Indian urban centres: an approach for betterment. In: Gupta KR (ed) Urban development debates in the new millennium. Atlantic Publishers & Distributors, New Delhi, pp 100–111
Kumar S, Sharma HR (2017) Issues and challenges in rural sanitation in Haryana and role of gram panchayat in its management. In: Kaushik et al (eds) Climate change, resource conservation and sustainability strategies. DBH Publishers and Distributors, New Delhi
Kumar S, Sharma HR (2019) Wealth from waste an initiative for solid waste management in rural areas of Haryana: issues, challenges and way forward. Indian J Waste Manag 3(2):61–67
Kumar S, Bhattacharyya JK, Vaidya AN (2009) Assessment of the status of municipal solid waste management in metro cities, state capitals, class I cities, and class II towns in India: an insight. Waste Manag 29:883–895
Kumar SJ, Smith SR, Fowler G, Velis C, Jyoti Kumar S, Arya A, Rena, Kumar R, Cheeseman C (2017) Challenges and opportunities associated with waste management in India. R Soc Open Sci 4:160764. https://doi.org/10.1098/rsos.160764
Kumar S, Mal G, Sharma HR (2018) Waste stabilization ponds: a technical option for liquid waste management in rural areas in Haryana under Swachh Bharat Mission-Gramin. Environ We An Int J Sci Tech 13:177–187
Marshall RE, Farahbakhsh K (2013) Systems approaches to integrated solid waste management in developing countries. Waste Manag 33(4):988–1003
McDougall FR, White PR, Franke M, Hindle P (2001) Integrated waste management. In: Integrated solid waste management: a life cycle inventory, 2nd edn. Blackwell Science Ltd, pp 15–31
McElhatton A, Pizzuto A (2012) Waste and its rational management. In: Mcelhatton A, Sobral PJA (eds) Novel technologies in food science-their impact on products, consumer trends and the environment. Springer, New York, pp 3–20
Meijkamp R (2000) Changing consumer behaviour through eco-efficient services: an empirical study of car sharing in the Netherlands. Design for Sustainability Research Programme. Delft University of Technology, Delft
Ministry of Drinking Water and Sanitation (2014) Guidelines on solid and liquid waste management (SLWM) in rural areas. Govt of India. https://jalshakti-ddwsgovin/sites/default/files/Guldelines_on_Solid_and_liquid_1_0pdf. Accessed 20 July 2020
Ministry of Statistics and Programme Implementation and UN (2020) (World Population Prospects 2019) 17 May 2020
Ministry of Urban Development (2016) Municipal solid waste management manual part II. Central Public Health and Environmental Engineering Organisation (CPHEEO), Ministry of Urban

Development, Govt of India. http://mohuagovin/upload/uploadfiles/files/Part2pdf. Accessed 15 July 2020
Mor S, Ravindra K, Visscher A, Dahiya RP, Chandra A (2006) Municipal solid waste characterization and its assessment for potential methane generation: a case study. Sci Total Environ 371:425–432
Morrissey AJ, Browne J (2004) Waste management models and their application to sustainable waste management. Waste Manag 24(3):297–308
Municipal Solid Wastes (Management and Handling) Rules (2000) Ministry of Environment and Forests New Delhi, Notification 25th September 2000, pp 1–28
Nahman A, Godfrey L (2010) Economic instruments for solid waste management in South Africa: opportunities and constraints. Resour Conserv Recycl 54(8):521–531
National Green Tribunal Act (2010) Ministry of Law and Justice New Delhi, The Gazette of India, No 25 Notification 02 June 2010
Nilsson-Djerf J (2000) Social factors in sustainable waste management. Warmer Bulletin 73:18–20
OECD (Organisation for Economic Co-Operation and Development) (1972) Recommendations of the council on guiding principles concerning international aspects of environmental policies, C (72)128. https://legalinstruments.oecd.org/en/instruments/OECD-LEGAL-0102. Accessed 19 July 2020
OECD (Organisation for Economic Co-Operation and Development) (2001) Extended producer responsibility: a guidance manual for governments. OECD Publications, Paris
OECD (Organisation for Economic Co-Operation and Development) (2006) Working group on waste prevention and recycling—EPR policies and product design: economic theory and selected case studies. Environment Directorate, Environment Policy Committee, ENV/EPOC/WGWPR(2005)9/FINAL, OECD Publications, Paris
Pantzar M, Raijas A, Heiskanen E (1994) Green consumers? Greening consumption? The instruments to promote sustainable patterns of consumption and production. Seminar, Stockholm
Patel U (2019) Solid waste management in India. ICRIER, New Delhi. http://icrierorg/pdf/4-Jan-2019/Utkarsh_Patel-SWM_%20in_Indiapdf. Accessed 30 Sep 2020
Petts J (2000) Municipal waste management: inequities and the role of deliberation. Risk Anal 20 (6):821–832
Planning Commission (2014) Report of the task force on waste to energy, vol I. Govt. of India, New Delhi
Plastic Waste Management Rules (2016) Ministry of Environment, Forest and Climate Change, New Delhi, India. The Gazette of India, Notification 18 March 2016
Popov F, DeSimone D (1997) Eco-efficiency—the business link to sustainable development. The MIT Press, Cambridge, MA
Press Information Bureau (2016) Solid waste management rules revised after 16 years. Rules now extend to urban and industrial areas. Ministry of Environment and Forests, Government of India. http://home.iitk.ac.in/~anubha/H15.pdf. Accessed 19 July 2020
Priti MK (2019) Review on evolution of municipal solid waste management in India: practices, challenges and policy implications. J Mater Cycles Waste Manag 21:1263–1279
Qotole M, Xali M, Barchiesi F (2001) Commercialisation of waste management in South Africa. Occasional papers series no. 3, Municipal Services Project, Cape Town
Rajput R, Prasad G, Chopra AK (2009) Scenario of solid waste management in present Indian context. Casp J Environ Sci 7(1):45–53
Ramos TRP, Gomes MI, Barbosa-Povoa AP (2014) Planning a sustainable reverse logistics system: balancing costs with environmental and social concerns. Omega 48:60–74
Robert K-H, Schmidt-Bleek B, Aloise de Larderel J, Basile G, Jansen JL, Kuehr R, Price Thomas P, Suzuki M, Hawken P, Wackernagel M (2002) Strategic sustainable development—selection, design and synergies of applied tools. J Clean Prod 10:197–214
Rodic L, Wilson D (2017) Resolving governance issues to achieve priority sustainable development goals related to solid waste management in developing countries. Sustainability 9(3):1–18

Sandhu GS, Frey HC, Bartelt-Hunt S, Jones E (2015) In-use activity, fuel use, and emissions of heavy-duty diesel roll-off refuse trucks. J Air Waste Manag Assoc 65(3):306–323

Sangwan P, Kaushik CP, Garg VK (2010) Growth and yield response of marigold to potting media containing vermicompost produced from different wastes. Environmentalist 30(2):123–130

Satyro WC, Sacomano JB, Contador JC, Almeida CMVB, Giannetti BF (2017) Process of strategy formulation for sustainable environmental development: basic model. J Clean Prod 166:1295–1304

SBM (Swachh Bharat Mission) (2014) Department of drinking water & sanitation. Ministry of Jal Shakti, Govt of India

Seadon JK (2006) Integrated waste management-looking beyond the solid waste horizon. Waste Manag 26(12):1327–1336

Seadon JK (2010) Sustainable waste management systems. J Clean Prod 18(16–17):1639–1651

Sharholy M, Ahmad K, Mahmood G, Trivedi RC (2008) Municipal solid waste management in Indian cities—a review. Waste Manag 28:459–467

Sharma HR (2017) Issues and challenges of municipal solid waste management in Indian cities. In: Kaushik A, Garg JK, Bhattacharya P, Gupta NC, Singh R, Joshi V (eds) Climate change, resource conservation and sustainability strategies. DBH Publishers and Distributors, New Delhi, pp 143–147

Sharma HR, Abebe T, Admassu M, Teshaye T, Aseffa T, Emana M (2011) Municipal solid waste management and community awareness and involvement in management and practice: an overview and a case study from Gondar town of Ethiopia. Int J Environ Waste Manag 7 (3/4):294–304

Sharma HR, Destaw B, Negash T, Negussie L, Endris Y, Meserte G, Fentaw B, Ibrahime A (2013) Municipal solid waste management in Dessie City, Ethiopia. Manag Environ Qual 24 (2):154–164

Shenoy J (2018) Lack of access to finance key hurdle in municipal solid waste management: study. The Times of India, Mangalore City edition, 5 June 2018

Singh J (2010) Different methods in solid waste management. In: Singh J, Ramanathan AL (eds) Solid waste management: present and future challenges. IK International Publishing House Pvt Ltd, New Delhi, pp 84–89

Slorach PC, Jeswani HK, Cuellar-Franca R, Azapagic A (2019) Environmental and economic implications of recovering resources from food waste in a circular economy. Sci Total Environ 693:133516. doi.org/10.1016/j.scitotenv.2019.07.322

Solid Waste Management Rules (2016) Ministry of Environment, Forest and Climate Change, New Delhi, India. The Gazette of India, No 861, Notification 8 April 2016

SRWS (Select Recycling Waste Services) (2018). https://www.srwscorp.com/what-is-sustainable-waste-management. Accessed 24 June 2020

Suocheng D, Tong KW, Yuping W (2001) Municipal solid waste management in China: using commercial management to solve a growing problem. Util Policy 10(1):7–11

Sutar SA, Gawande SM (2013) Solid waste management in rural areas emerging towards growth centre through GIS system-Mahalung, Solapur. Int J Sci Res 4(7):475–478

Suzuki D, Dressel H (2002) Good news for a change: hope for a troubled planet. Stoddart Publishing, Toronto

The Economic Times (2019) Government sets up over 56 MW waste-to-energy capacity in last 3 years. The Economic Times-E-paper, Nov 22, 2019

The Water (Prevention and Control of Pollution) Act (1974) Ministry of Law, Justice and Company Affairs, New Delhi, India. Act No 6 of 1974, Notification 23 March 1974

The Water (Prevention and Control of Pollution) Cess Act (1977) Ministry of Law, Justice and Company Affairs, New Delhi, India. Act No 36 of 1977, Notification 7 December 1977

The World Bank (2019) Brief-solid waste management. https://www.worldbank.org/en/topic/urbandevelopment/brief/solid-waste-management. Accessed 19 June 06 2020

Turner RK, Powell J (1991) Towards an integrated waste management strategy. Environ Manag Health 2(1):6–12

Tyagi V, Fantaw S, Sharma HR (2014) Municipal solid waste management in Debre Berhan city of Ethiopia. J Environ Earth Sci 4(5):98–103
Tynberg A (2001) The natural step and its implication for a sustainable future. Hastings West Northwest J Environ Law Policy 73. https://repositoryuchastingsedu/hastings_environmental_law_journal/vol7/iss1/8. Accessed 11 July 2020
UNEP (United Nations Environmental Programme) (1996) International source book on environmentally sound technologies for municipal solid waste management. International Environmental Technology Centre, Technical Publication Series-6
UNEP (United Nations Environmental Programme) (2002) Annual Report 2002. Division of Technology, Industry and Economics: International Environmental Technology Centre, Osaka
UNEP (United Nations Environmental Programme) (2008) Basel Convention. Conference of the parties to the Basel convention on the control of transboundary movements of hazardous wastes and their disposal, Ninth meeting, Bali, UNEP/CHW.9/39
UNEP (United Nations Environmental Programme) (2017) Solid approach to waste: how 5 cities are beating pollution. https://wwwunenvironmentorg/news-and-stories/story/solid-approach-waste-how-5-cities-are-beating-pollution. Accessed 30 Sep 2020
United Nations (2014) United Nations, Department of Economic and Social Affairs, Population Division. World Urbanization Prospects: The 2014 Revision, Highlights (ST/ESA/SER.A/352)
USEPA (2002) What is integrated solid waste management? Solid waste and emergency response. Environmental Protection Agency (EPA), United States
USEPA (2004) Putting surplus food to good use. U.S. Environmental Protection Agency, 530-F-006-04: Washington, DC
Van de Klundert A, Anschutz J (2001) Integrated sustainable waste management—the concept; tools for decision-makers—experiences from the urban waste expertise programme (1995–2001). Waste, Gouda
Vesilund PA, Worrell WA, Reinhart DR (2002) Solid waste engineering. Brookes/Cole, California
Vester F (2007) The art of interconnected thinking. Ideas and tools for tackling with complexity. MCB-Verlag, Munich
Viz D (2012) Urbanization and solid waste management in India: present practices and future challenges. Procedia Soc Behav Sci 37:437–447
Wang HT, Nie YF (2001) Remedial strategies for municipal solid waste management in China. J Air Waste Manag Assoc 51(2):264–272
Waste Atlas (2017) Municipal solid waste generation: country charts. wwwatlasd-wastecom/. Accessed 19 July 2019
Wilson DC (2007) Development drivers for waste management. Waste Manag Res 25:198–207
Wilson DC, Rodic L, Modak P, Soos R, Rogero AC, Velis C, Iyer M, Simonett O (2015) Global waste management outlook. United Nations Environment Programme, Nairobi
ZWIA (Zero Waste International Alliance) (2018) Zero waste definition. http://zwiaorg/zero-waste-definition/. Accessed 30 June 2020

[illegible] (2001) [illegible] Western Cape
[illegible] (2012) [illegible] solid waste [illegible]
[illegible] (2012) [illegible] AICE [illegible]

Social Enterprises as an Emerging Platform in Waste Management 18

Poulomy Banerjee and Kisslay Anand

Abstract

Globally, societies have been riddled with multiple problems and long-standing issues that remain unresolved by the Government and its institutions. In the Indian context, waste management is one such area, where the journey towards building clean, safe and sustainable communities seems to be a race with no finish line. With the Swachh Bharat Mission (SBM) creating a nation-wide surge and mass awareness on cleanliness and healthy sanitation practices in India, substantial focus has been laid to this formerly overlooked sector. Various solutions and best practices on sustainable solid waste management have emerged. However, often such solutions tend to be infrastructure-based requiring large investments, which become infeasible for deployment in smaller Urban Local Bodies (ULBs). Hence, context-specific, innovative solutions for waste management is the need of the hour. This is where social entrepreneurs have stepped in, challenging and rethinking concepts and assumptions. The significance of social entrepreneurs lies in the fact that they have familiarity with unmet local needs and the entrepreneurial zeal to drive and facilitate extensive social impact. SBM has also given the much-needed impetus to such entrepreneurial initiatives, as a result of which a number of ventures focusing on innovative waste management solutions have grown rapidly in the past few years. This chapter dwells upon certain selected social enterprises that have revolutionized the waste management practices by diffusing innovations using both emerging and disruptive technologies.

Keywords

Waste management · Social enterprise · Innovation · Urban local bodies · Social impact · Informal sector · Employment generation · Sustainable solution

P. Banerjee (✉) · K. Anand
Swaniti Initiative, Delhi, India
e-mail: banerjeepoulomy@gmail.com

A. Kaushik et al. (eds.), *Climate Resilience and Environmental Sustainability Approaches*, https://doi.org/10.1007/978-981-16-0902-2_18

18.1 Introduction

The population of India had crossed the 1.2 billion mark in 2013, and now stands at 1.33 billion. With a population density of 325/km^2 (GOI 2011), India's 2.9 million km^2 of land area holds close to 18% of the world's population. Although the Government administration has been constantly trying to keep up with the expanding cities through various reforms, it is either the poor policy framework or their improper implementation that has resulted in a prolonged delay in addressing the basic public services. The current waste generation in the country is 1.33 T per day (MoHUA 2020), and has an average annual growth rate of 4%.

Whilst managing this huge quantum of waste becomes a problem, it has also opened large gates of opportunities in the country. According to a report published by NOVONOUS, India's waste management market will be worth US$13.62 billion by 2025 (Cision 2020). The novel Smart Cities Mission as well as the Swachh Bharat Mission over the years have been aiming at more public–private partnerships, digital governance, social innovation, and people-centric development. This has created enabling environment for engagement of social enterprises, start-ups, and NGOs in taking a lead in bringing radical outlook and prompt action for the provision of basic services like waste management to the citizens.

18.2 Social Enterprises: Its Characteristic Features and Objectives

According to Asian Development Bank, Social Enterprises (SEs) are defined as organizations that have a triple bottom line approach to address social and environmental needs and have financially sustainable revenues (ADB 2020). Social Enterprises are the clear leaders in the category and should serve the bottom of the pyramid. They do not operate only on grants as they cover their costs and ensure a small surplus. One of the most important aspects of social enterprise is employment generation. Social Enterprises might be registered as private limited companies, cooperatives, not-for-profit organizations, trusts or other sorts of legal entities (Social Enterprise, An Overview of the Policy Framework in India 2015). A social enterprise, though it may respond to social needs, is a privately owned, profit-oriented venture which markets its own products and services, blending business interests with social ends. The Canadian Centre for Social Entrepreneurship considers social enterprises as fitting the notion of "hybrid" organizational models which "fuse innovative, entrepreneurial practices with a commitment to both social and economic return on investment" (Westley 2017).

Whereas firms in the private sector mostly focus on financial returns, social enterprises are also concerned with social and/or environmental impacts. In addition, they are autonomous organizations with governance and ownership structures often based on participation by stakeholder groups (users or clients, local community groups etc.) or by trustees. As a result, there is a perception that social enterprises are more innovative and dynamic than their counterparts in the public sector as they tend to offer more flexibility and tailored services to maximize client satisfaction. The focus of such enterprises on social and environmental values and their understanding of community/client group issues mean that social enterprises are perceived as potential partners by the state to help deliver public services. This also fits in with the government's agenda to externalize, where appropriate, public service provision.

Social enterprises also exhibit some characteristics similar to organizations in the private sector. In particular, they operate in specific markets, providing goods and services that enable them to be commercially sustainable organizations to a greater or lesser extent. They also contribute to the development of new markets (e.g. ethical markets), responding to changing consumer trends. Social enterprise, therefore, bridges the gap between public and private sectors being to some extent driven by market forces but also having regard for social and environmental issues.

The interest of the public sector in social enterprises may be attributed to the fact that they have the capability to contribute to socially inclusive wealth creation, regenerating local communities by bringing them together to work for local causes and leading the way in delivering and reforming public services (British Council 2015).

At present, India's rapidly growing economy has been unable to alleviate the extreme poverty and deal with other serious social issues (Hashimoto 2016). Some of the major explanations for this inability can be non-uniform infrastructures, low-quality public services such as health, education and environment and resource limitations. One such concern that has a substantial negative impact on the environment is improper solid waste management, especially in large cities where its ill effects are found to be more pronounced.

By employing innovative business models, social entrepreneurs are able to address the country's massive development needs, whilst maintaining sustainability through viable revenue models. A study by the British Council published in 2016 (British Council 2016) has an interesting finding. It states that this growing and dynamic sector has successfully contributed to social empowerment of semi-skilled workforces, creating jobs for the disadvantaged groups and addressing social exclusion across the country. According to the study, the average number (median) of employees in a social enterprise in India is 19 (17 full-time employees and four part-time employees).[1] On an average, 25% of the employees in a social enterprise are women. The total number of new employees added by the social enterprises in 2015 was 13,077, whereas the total number of employments generated by these is 61,548, comprising both full-time and part-time personnel. There was a 27% increase in the total number of employees in social enterprises from 2014 to 2015, whilst 89% of social enterprises expected their staff numbers to increase in the next financial year, as per this study.

[1]Two part-time employees are treated as equivalent to one full-time employee. The mean average number of employees in social enterprises in India: full-time = 177; part-time = 70.

18.3 Assessment of the Role of Social Enterprises in Solid Waste Management

The research has been based on a mixed-method approach comprising of quantitative and qualitative research methods. The research gives the insight generated through Focus Group Discussion (FGD), informed interviews and structured observations. Additionally, the research referred to secondary data sources to supplement the data available.

Through the broad scan of social enterprises working in the Solid Waste Management (SWM) Sector, case studies were selected based on the author's acquaintance with the organizations and first-hand interactions. In addition, an attempt was made to ensure a distribution across the value chain of SWM, segregation, recycling, treatment and value creation. In addition, a fair geographical spread within the country was maintained during the selection of the organizations.

For the assessment, the following information was collected for each enterprise: name and age of the enterprise, mission, size, legal structure, whether or not the enterprise has raised capital in the past and target segment in terms of geography, demography and socio-economic profile. It should be noted that the sample of 10 social enterprises is not representative of the entire Indian SE landscape in SWM sector, but rather a combined sampling of several new and mature enterprises in the sector. The objective of the study is to provide an understanding of how the intervention by SEs has innovated ways of managing and handling waste and turn it into valuable resources. It also tries how it has positively impacted the lives of informal workers.

18.4 Solid Waste Management Scenario in India and Role of Social Enterprises

As per the Central Pollution Control Board (CPCB), 143,449 TPD of Municipal Solid Waste (MSW) was generated for 34 states and union territories. The average rate of waste generation in India, based on this data, is 0.11 kg/capita/day. Out of the total waste generated, approximately 117,644 TPD (82%) of MSW was collected and 32,871 TPD (22.9%) was processed or treated. Waste generation rate is between 200 and 300 gm/capita/day in small towns and cities with a population below 200,000. It is usually 300–350 gm/capita/day in cities with a population between 200,000 and 500,000; 350–400 gm/capita/day in cities with a population between 500,000 and 1,000,000; and 400–600 gm/capita/day in cities with a population above 1,000,000. However, as per estimates available in recent years from Swachh Survekshan data (MoHUA 2020), the waste management scenario in the country's situation has made a significant improvement. The data claims 33,420 urban wards in the country to have achieved 100% door-to-door collection and more than 50,600 wards to have 100% source segregation in place. It also states that around 84,000 informal Waste Pickers have been integrated into the formal system within ULBs.

Social enterprises operational in Solid Waste Management (SWM) in India have adopted innovative methods to not only promote basic sanitary behaviour amongst common masses but also have successfully worked towards creating effective waste collection, recycling, segregation and value creation by integrating the waste pickers in their business model. The enterprises have also tried to leverage digital technology through the use of Information and Communications Technology (ICT) ICT and Internet of Things (IoT) as a more integrated approach towards waste management. The case studies provide an insight into the operation of social enterprises in the SWM sector and their engagement across the value chain with the ULBs and end customers. The enterprises have been selected based on their operations along the SWM value chain. This has been tabulated below:

Waste collection	Sanshodhan, Hyderabad, Telangana
Waste handling and management	Brook and Bloom, Ahmedabad, Gujarat
	Conserve India, Delhi
Waste to value	Dye manufacturing from floral waste, Bodh Gaya, Bihar
	Floral waste to Incense Sticks, Mango Foundation, Lucknow, Uttar Pradesh
	Eco-friendly stationery, Chanu Associates, Imphal, Manipur
	Brick Manufacturing from Plastic waste by Zerund Bricks, Guwahati, Assam
	Thunk in India, Bengaluru, Karnataka

18.4.1 Case Study 1: E-Waste Management by Sanshodhan, Hyderabad, Telangana

"Sanshodhan", a not-for-profit organization, that is formed of a group of young professionals from Hyderabad, have developed a simple and efficient Internet of Things (IoT)-based platform, called Sanshodhan E-waste Exchange. The platform that works by creating a sustainable ecosystem around e-waste. The organization aims at supporting the government in developing robust data and monitoring mechanisms. Keeping this in mind, the digital platform they designed currently enables corporates and societies across Hyderabad in identifying and directly transferring their e-waste to government-authorized, technically competent recyclers. The E-waste Exchange operates on three models:

- *A citizen-centric model*: This offers a platform for residents of the city to dispose their e-waste conveniently and sustainably from home. This method is not only transparent but also provides legitimate support to the legitimate recycling sector.
- *A bulk consumer-based model*: This supports industrial units such as the FMCG and the IT FMCG companies. It also includes large-scale electrical and electronics users and helps them to reduce their carbon footprint and contribute to the nation's Sustainable Development Goals.

- *Circular Economy-based standards for e-waste*: This helps in transforming and uplifting the informal sector.

Sanshodhan has been working in cities like Mumbai, Chennai, Delhi, amongst others. Smart Cities in India, that pivots around ICT application in the field of waste management, are in need of similar platforms too. The enterprise has built a network of around 180 refurbishers, dismantlers and recyclers to work with. This online facility for e-waste disposal has been quite beneficial to the Residents' Welfare Societies (RWAs) in Hyderabad. The organization was awarded the Swachh Bharat Grand Challenge Award 2018 by the Ministry of Industry and Commerce and has also been recognized by the World Economic Forum as a "Highly Commended" entrant in the People's Choice category of the Circulars 2019.

18.4.2 Case Study 2: Composting of Flower Waste by "Brook and Bloom", Ahmedabad, Gujarat

"Brook and Bloom", a social enterprise based in Ahmedabad in the state of Gujarat, has been carrying out research in the sustainable management of waste generated from religious premises since 2015 (Raval 2018). The flower waste generated from religious premises is manually segregated and shredded into particles of 2 cm size. Following this, the shredded matter is channelized into the mixing unit through a conveyor belt, where it is mixed with water and culture for around 10 min. The mixture is turned every 8 days for 20 days followed by intermittent drying. The dried mixture is then screened mechanically through 4 mm screens and packed in packs of 750 g each. Currently, the facility has the capacity to process 100 kg of waste in an hour. The waste is collected from the "*Kalash*" (pot) placed by Ahmedabad Municipal Corporation (AMC) on the 8 bridges of Sabarmati river and also from "*Pipal Kalash*" which is placed in around 50 temples of Ahmedabad. The total quantity of waste collected amounts to 18 TPD on an average. This waste is brought to the facility at Victoria Garden, Ellis Bridge. It is further segregated and only 0.9 T is found suitable for making compost. The compost thus generated is partly sold to AMC (50%) and partly to the temples. AMC has provided 100 yards land to Brook and Bloom on operational basis. However, no rent or lease needs to be paid to the corporation. The infrastructure with a few initial equipment was set up by AMC. Other costs have been taken over by Brook and Blooms.

18.4.3 Case Study 3: Conserve India, Delhi

"Conserve India", a Delhi-based organization founded by Anita and Shalabh Ahuja, is based on the idea of upcycling plastics and reinventing them into fashion accessories. The team at Conserve India experimented with plastic bags by cleaning, drying and pressing them into sheets and converting them into useful products. They named this material as Handmade Recycled Plastic (HRP).

This unique venture by Conserve India has not only been instrumental in cleaning the streets of Delhi but has also provided hundreds of jobs (British Council 2016) for some of the poorest people living in their city. The income generated by the sale of the upcycled products made from HRP is spent on social welfare projects. HRP is made from discarded polythene bags picked from the streets, rubber from tyres of old trucks and also uses old denims and saris. It is ensured that the process employed in making bags from reusing these materials is energy efficient and harmful chemicals and dyes are avoided as much as possible. This not only makes the products eco-friendly but also cuts costs, giving the organization a higher capital to invest in other projects.

This initiative has led to better wages for the employees—a waste picker collecting bags for Conserve India earns on an average three times more money than they would earn elsewhere. Secondly, proper training opportunities are provided to the staff in order to improve their skills so that they can get more skilled jobs either within the organization or elsewhere. Arrangements have been made to provide loans to the waste pickers in order to enable them to start their own businesses. Furthermore, a school and a health clinic have been set up for these waste pickers within their slum using the profits earned by the organization.

18.4.4 Case Study 4: Dye Manufacturing from Floral Waste, Bodh Gaya, Bihar

The Mahabodhi Temple at Bodh Gaya generates around 50–60 kg (Chakraborty 2018) of floral waste on a daily basis, which rises to 300–400 kg during peak season, particularly between December and February. These were earlier disposed of without any treatment. Taking a cue from the problem, Praveen Chauhan, a NIFT alumnus started a social enterprise "MATR" to pursue two goals through the organization; recycle the floral waste as well as promote the use of khadi in the international market. MATR manages the temple's waste by converting floral waste into natural dyes and increase Khadi sales by banking on peoples' sentimental value towards the flowers and leaves of the temple. MATR initiated a joint venture with an Australia-based sustainable clothing brand "Because of Nature". The "Happy Hands Project" launched by the brand aimed at repurposing the discarded waste from the temple to make natural dyes for garments made from khadi. To operationalize the project around 50 women from the local population were engaged in the project. Initially, 50 women were trained on the manufacture of floral dyes from different shades of marigold flowers, which formed a majority of the offerings. Five women from the organization visit the temple every morning, collect the floral waste from the temple and bring it to a small cluster, 6 km outside the town. The marigold flowers are segregated colour-wise and kept for drying. Once dried, the flowers are ground manually and then kept for boiling, which is then followed by mordanting. Before drying the fabric, it is covered with amla juice and washing soda. Depending on the skill of the dyer, hundreds of shades are being produced on Khadi from marigolds.

The Bodhgaya Temple Management Committee (BTMC) signed a Memorandum of Understanding (MoU) with "MATR" and "Because of Nature", to implement the project "Happy Hands Project" that was launched on 15 September 2018. MATR manufactures several Khadi garments ranging from women's clothes to robes used by the monks, and most of them are sold internationally at Australia and Japan where there is a huge demand for authentic khadi clothes. The prices range between 100 and 400 Australian dollars. From the profit earned 90% is divided amongst "MATR", "Because of Nature" and the women. The remaining 10% of the money earned on khadi sales go to the Bodhgaya Temple.

The future plan of the project is to expand the initiative to cater to the 60–70 (Karelia 2019) small and big temples in and around Bodh Gaya. The "Happy Hand Project" hopes to reuse each of the temple's floral waste and hire 200 more women from the villages and expand the enterprise. Currently, the fabric is purchased from retailers. From 2019 to 2020 onwards, the team has the plan to spin and weave the fabric on their own. The enterprise is helping in solving the floral waste problem in the temple and providing sustainable livelihood opportunities to the local artisans and women through a social revenue-generating model without increasing the burden on the temple trust. The idea of the project is simple and could be easily scaled up in other pilgrimage sites of the country.

18.4.5 Case Study 5: Floral Waste to Incense Sticks, Mango Foundation, Lucknow, Uttar Pradesh

The Lucknow Municipal Corporation (LMC) in Uttar Pradesh has moved into collaboration with a city-based social enterprise "The Mango Foundation" for converting the huge amounts of floral waste generated from the temples and shrines into organic incense sticks and compost. The Mango Foundation has been involved in improving the living standards of people in the slums of Lucknow for the past 3 years. The Central Institute for Medicinal and Aromatic Plants (CSIR-CIMAP) has also been roped in the project by LMC for scientific expertise. LMC has also approached the management of the major temples and shrines to help in the project. The space for processing of waste has been allotted by LMC and a sum of Rs. 10,000 was invested by the enterprise to initiate the project.

Besides handling the waste generated from the temples, this initiative has also provided employment opportunities to the local slum dwellers. Scientists from the CIMAP provided training to the women from the Nirala Nagar slum to prepare incense sticks from the floral waste. Currently, a team of eight women handle the processing works whilst four other members are involved in the marketing of the product. The earnings from the sale of the incense sticks are utilized in providing the wages to the workers. The compost produced is sold to the farmers at a nominal price.

Floral waste collected from specially designated dustbins from the temples and shrines are transported to the Kanha Upvan area by small carrier vehicles, where fresh flowers are segregated from the wilted ones. These are used for making the incense sticks and the remaining are composted.. The segregated flowers are then sun dried and crushed to a paste, to which a small amount of saw dust and fragrant oil is added. Coal dust is used as coating for the incense sticks. Around 30–35 kg of incense can be obtained from 100 kg of fresh flowers. So far, the project has converted over 150 kg of floral waste into 1.2 lakh incense sticks.

The enterprise plans to expand the project in various phases. It aims to eliminate all the floral waste and convert them into other useful products and provide employment to around 100 people, majorly women, and create market of the products in the cities surrounding Lucknow.

18.4.6 Case Study 6: Eco-Friendly Stationery by Chanu Associates, Imphal, Manipur

Chanu Associates is a Manipur-based social enterprise that deals with a plethora of products ranging from eco-friendly pens with seeds, pencils, direct filling pens, paper bags, non-woven bags and so on under the brand name "Chanu". The enterprise started its operation from June, 2017 with a nominal investment of Rs. 20,000.

The innovative eco-friendly pen made from compressed paper, is a great alternative to plastic ballpoint pens that not only addresses the plastic menace but also contain seeds that, when disposed of, can grow into saplings. Once the ink of the pen gets over or the pencil is used up, they can be planted at a 30° angle manually without any tools.

Presently the enterprise produces 200–300 units and earns a profit of Rs. 50,000 per month. The State Government of Manipur has provided loan to the enterprise under the Startup Conclave 2018 to upgrade machinery in order to achieve the target of producing 3000–5000 units per day. This enterprise has also been instrumental in job creation for the locals.

18.4.7 Case Study 7: Brick Manufacturing from Plastic Waste by Zerund Bricks, Guwahati, Assam

Guwahati generates 37,000 kg of plastic waste every day, which is about 12.37% of the state's plastic production. Taking the seriousness of the situation, three engineers Rupam Choudhury, David Pratim Gogoi and Mousum Talukdar from Assam Engineering College established Zerund Bricks Manufacturing Pvt. Ltd. in order to utilize multilayered non-recyclable plastic waste in manufacturing concrete bricks. The enterprise is a centralized unit setup at an initial capital expenditure of Rs. 50 lakhs (Anand and Banerjee 2019). The unit also acts as an ideal centralized dry waste recycling centre. Unlike the usual clay bricks, Zerund bricks are made of cement, fly ash, organic chemicals and plastic waste and are available at prices 15–20% lower than that of the regular bricks available in the market. The plastic is supplied to the organization by Urban Local Bodies (ULBs). The fly ash is supplied free of cost by National Thermal Power Corporation (NTPC) as a gesture to support the innovation. The enterprise helps in managing around 600 kg of plastic waste by utilizing them in manufacturing of bricks.

These bricks are larger than conventional bricks and also have several advantages over the latter. The plastic component of the bricks reduces water absorption and also increases the drying shrinkage of the brick reducing the occurrence of hairline cracks. Being lightweight, total dead load of the structure reduces up to 40% in comparison to clay bricks thereby reducing structural cost. The size being larger reduces the total number of mortar joints in the walls which results in less cement consumption.

18.4.8 Case Study 8: Thunk in India, Bengaluru, Karnataka

Thunk in India Pvt. Ltd. was founded by Suren Vikhash in order to make eco-friendly shoulder bags, pouches, laptop cases, table mats and so on by upcycling waste materials collected from hotels, residential properties, IT parks and coffee shops in the city of Bengaluru. The waste materials used for the purpose are mostly plastic and Tetra Pak that is segregated, cleaned, shredded and woven to make the upcycled products. Most of these products are made of 90% (Pulakkat 2018) waste and 10% new material. Thunk has been in negotiations with two large IT services firms to buy waste from them and sell them corporate gifts made of the same material.

Thunk, which engages four designers and nine weavers in Pollachi, Tamil Nadu, is exploring varied weaves fusing shredded plastic with rubber from automobile tyres, plastic with banana fibres, corn leaves and coconut fibres to see what works best. Vikhash, who started the firm borrowing Rs.10 lakh from his parents, is scouting for more funds to scale up operations. The organization, serves underprivileged communities, thereby empowering their livelihoods including slum families, HIV patients and tribal communities by engaging them in various aspects of the upcycling project.

18.5 Conclusion and Way Forward

Resources in our modern world are rapidly depleting, without substantial recovery. Governments and businesses are currently concerned about the increasing pressures on our global resources due to anthropogenic activities. Our current economies are linear, that has excessively high material and energy consumption, and eject large proportions of material as waste. To build a sustainable future, we require strategies and instruments that seek to keep resources and products in use, at as high a value as we can, for as long as possible, instead of merely disposing them as waste after a single use. This calls for creation of new forms of business, new technologies and processes, and new forms of governance that aim to reduce resource consumption and utilize waste as a resource and input in production processes. Such changes offer the potential to generate value to society via stimulation of employment, and an increased demand for educated and skilled workers. This evolution also needs new thinking, new social systems, new forms of engagement and new institutions.

Although, at present individual social enterprises may seem to have a smaller and localized impact, however, scaling up their initiatives may contribute to add a significant social and economic value in the future. This can be made possible through exchange of knowledge and meaningful partnerships with key players in the upstream and downstream of the value chain. Beyond value chain collaborations within industry, collaboration with government authorities such as urban local bodies, NGOs and even, consumers will enable such enterprises to have a measurable impact through their service provisions. At a larger scale, such partnerships shall pave the way for transitioning to a circular economy in the future that governments of both developed and developing nations are aiming at. A favourable policy landscape is another driver that would foster business innovations and guide sustainable consumption of resources.

Social enterprises are also successful in developing mechanisms, which leverage the capabilities of unorganized and informal sector whilst providing them an occupational identity.

Scalability of such initiative is need of the hour in the country for the government and other private players through Impact Investment or other alternative source of financing. The government should identify such initiatives and facilitate the entrepreneurs in receiving government supports to scaleup the business as it would help in generating a twofold effect as they proactively intend to create positive impact as well as generate profits.

The Central Government along with the Urban Local Bodies should focus towards Credit Guarantee Scheme (CGS) to increase the flow of funds towards social enterprises in the sanitation and solid waste management sector. The CGS would help in making the lending for social enterprises by absorbing and sharing the risks associated with lending. The CGS would also help in increasing the amount of funds lent to the social enterprises beyond their collateral limits as the risk associated with the project would be shared by Central Government and Urban Local Bodies. In this way, the Central Government along with Urban Local Body could play the additional role of monitoring body for the social enterprises so that the growth of enterprises is on track and the government could gain from the spillover effects of the social enterprises in terms of cleaner society and livelihood options to become at the bottom of the pyramid.

Similarly, the Urban Local Bodies along with Donors (Corporate Sector, CSR Funds, Central and State Government, NGOs) and Investors can come forward in establishing Waste Management Trust Funds for financing the working capital for social enterprises in Solid Waste Management sector at seed stage. The working methods of the funds can be established in the lines of Hometown Trust Funds (Sachs 2019). The Hometown Trust Funds are widely used by the Japanese government to support risky sectors such as solar and wind power projects (Naoyuki Yoshino 2019) by providing working capital to the enterprises. The working capital of the enterprise at the seed stage would help in initial operations. If these enterprises are successful in bringing about desired results then they would be well received by individual investors, venture capitalists and banks to grant loan for the scalability. The share of profit earned by the social enterprises can be taken in the Waste Management Trust Funds at later stage.

First, the financial return potential increases the attractiveness of opportunities that produce a positive impact, drawing more private sector capital to areas that promote development. Secondly, private sector participation, and the opportunity to generate returns, spurs innovation and growth. The commercial capital pushes enterprises to experiment with new business models, capture new opportunities and drive for greater impact. Both private and public entities could benefit from viewing social enterprise development not only as a responsibility but as a financially or strategically valuable investment.

References

ADB (2020, May 12). https://www.adb.org/themes/social-development/inclusive-business/social-enterprises. Accessed from www.adb.org

Anand K, Banerjee P (2019, Sept). https://drive.google.com/file/d/1bC0FgKlnOyDzBHtiBwtmo4cyWML-NQUs/view. Accessed from http://swachhbharaturban.gov.in/

British Council (2015, Oct). https://www.britishcouncil.in/sites/default/files/social_enterprise_policy_landscape_in_india_0.pdf. Accessed from https://www.britishcouncil.in/

British Council (2016) A survey of the social enterprise landscape in India. Accessed from https://www.britishcouncil.in/sites/default/files/british_council_se_landscape_in_india_-_report.pdf

Chakraborty S (2018, Sept 4) Mahabodhi comes up with flower-to-fabric idea. The Telegraph Online Version

Cision (2020, May 10). https://www.prweb.com/releases/2014/waste-management-market/prweb12206976.htm. Accessed from www.prweb.com

GOI (Government of India), Ministry of Home Affairs (2011). https://censusindia.gov.in/2011census/population_enumeration.html?q=density. Accessed from https://censusindia.gov.in/

Hashimoto Y (2016, Apr 1). https://eurasianpublications.com/Eurasian-Journal-of-Business-and-Management/Vol.4-No.1-2016-5.pdf. Accessed from https://eurasianpublications.com/

Karelia G (2019, Nov 28). https://swachhindia.ndtv.com/waste-management-bihar-natural-dyes-colour-for-khadi-from-floral-waste-28379/. Accessed from https://swachhindia.ndtv.com/

MoHUA (Ministry of Housing and Urban Affairs) (2020, Month 12) http://swachhbharaturban.gov.in/dashboard/. Accessed from http://swachhbharaturban.gov.in/. Accessed 04 May 2020

Naoyuki Yoshino FT-H (2019) Modelling the social funding and spill-over tax for addressing the green energy financing gap. Econ Model:34–41

Pulakkat H (2018) Global Innovation Index ranks India the 57th most innovative nation. [online] The Economic Times. Available at: https://economictimes.indiatimes.com/news/economy/indicators/india-maintains-position-as-top-it-exporter-lags-in-ease-of-doingusiness/articleshow/64935501.cms?from=mdr. Accessed 15 July 2019

Raval G (2018, Nov 27). https://www.dnaindia.com/ahmedabad/report-ahmedabad-start-up-makes-incense-compost-from-puja-flowers-2689720. Accessed from https://www.dnaindia.com/

Sachs JW-H (2015) Social enterprise, an overview of the policy framework in India. British Council

Sachs JW-H (2019) Handbook of green finance, energy security and sustainable development. Springer, New York

Westley NA (2017, Jul). https://mcconnellfoundation.ca/wp-content/uploads/2017/07/Funding-Social-Innovation.pdf. Accessed from https://mcconnellfoundation.ca/

Understanding Linkages Between Sustainability and Traditional Ethnoecological Knowledge (TEK): A Case Study of Paudi Bhuyans in Northern Odisha, India

19

Livleen K. Kahlon and Rita Singh

Abstract

Studies based on Traditional Ethnoecological Knowledge (TEK) systems help in development of critical, evaluative, and creative thinking skills, so as to aid policy decisions around sustainable solutions. Sustainable development framework outlined under Sustainable Development Goals (SDGs) presents an opportunity to experiment and evolve pathways promoting resource efficiency measures that hold a hope for a better tomorrow. A systematic study of ancient wisdom embedded in indigenous communities is a step in this direction, and this chapter explores lifestyles of populations of Paudi Bhuyans, a particularly vulnerable tribal group (PVTG) living in northern parts of Odisha. Variations in the resource use practices are also described in detail. Socio-ecological uncertainties due to changes in land use patterns, habitat fragmentation, and other forces have influenced members of Paudi Bhuyan tribe to move from hills to plains triggering a slow erosion of their ethnicity. The chapter also advocates a need to study ethnoecological knowledge of Paudi Bhuyan tribe as a step towards achievement of sustainable development goals (SDGs), and also highlights unique relationships between Paudi Bhuyan populations and their natural settings. Both tribal and non-tribal communities in the study location were included for

L. K. Kahlon (✉)
The Energy and Resources Institute (TERI), Delhi, India

University School of Environment Management (USEM), Guru Gobind Singh Indraprastha University, Dwarka, Delhi, India
e-mail: livleen.kahlon72@gmail.com

R. Singh
University School of Environment Management (USEM), Guru Gobind Singh Indraprastha University, Dwarka, Delhi, India
e-mail: rsinghipu@gmail.com

A. Kaushik et al. (eds.), *Climate Resilience and Environmental Sustainability Approaches*, https://doi.org/10.1007/978-981-16-0902-2_19

collection of information through participatory rural appraisals (PRA), interview schedules and semi-structured and open-ended questionnaires. The chapter concludes with a need to consider that Indigenous Development Goals (IDG) may be adopted as a new mantra for sustainable development. Both tribal and non-tribal communities in the study location were included for collection of information through participatory rural appraisals (PRA), interview schedules, and semi-structured and open-ended questionnaires.

Keywords

Ethnicity · Ethnoecological · Indigenous development goals (IDG) · Particularly vulnerable tribal group (PVTG) · Resource efficiency

19.1 Introduction

Nature and natural resources are a source of several direct and indirect benefits. Apart from a source of nutritiously varied and abundant food, medicines, and water, there are also other valuable services such as crop pollination, disease-control, food-chain stability, supply of oxygen and cultural diversity. Understanding the impacts of rapid urbanization on environment and experiencing growing losses of natural resources, human race is now exploring adoption of resource efficiency models of development. Overexploitation of natural resources due to anthropogenic causes is moving our planet Earth towards the sixth mass extinction (Kolbert 2014). It is in this context that sustainable development framework presents an opportunity to experiment and evolve pathways towards the achievement of sustainable development goals (SDGs) to have a meaningful and transformative impact on society. Understanding the impact of anthropogenic activities on biodiversity, and comprehending dangers linked with sixth mass extinction, this chapter explores the lifestyle of a particular population of Paudi Bhuyans, an indigenous particularly vulnerable tribal group (PVTG) found in northern parts of Odisha. These indigenous communities are reservoirs of ecological knowledge and wisdom and have been following sustainable practices that are closely linked to their beliefs and traditions evolved in a particular ecosystem of a landscape. Several factors have forced Paudi Bhuyan tribe to move from hills to plains and thereby triggered a slow erosion of their traditional knowledge. In order to achieve SDGs in indigenous settings, it is important to first align the knowledge base so that ideas and practices to face global sustainability challenges can be prioritized. Documentation helps to record and understand practices that were in the past following the mode of oral transmission. In order to prioritize the overall development of indigenous communities, there is also a need to identify indigenous development goals (IDGs) and place more focus on these.

19.1.1 Understanding Traditional Ethnoecological Knowledge

Indigenous communities have safeguarded a reservoir of knowledge evolved and inherited over generations. This knowledge is closely linked with cultural values and is also the social capital of the communities which have their special set of experiences in relation with their immediate environment (Roue and Nakashima 2002). Berkes et al. (2000) postulated that indigenous populations have a range of knowledge about resources that are locale-specific and experience about ecosystem management. They have evolved a unique perspective about ecosystem processes and functions. Traditional ethnoecology is thus proposed by Bettencourt and Kaur (2011) as a system balanced on twin pillars of ecology and anthropology since it combines both scientific and traditional methods for the study of biodiversity and community interactions thereby allowing new opportunities for adaptive management practices. Yet another concept which gained momentum in the last quarter of twentieth century and first decade of the twenty-first century is biocultural diversity (BCD). It is a composite representation of diversity of languages, culture, and life forms. It recognizes the inclusion of human cultures and languages within the diversity of flora, fauna, habitats, and ecosystems. With this understanding of biocultural diversity, it is also inferred that its various elements interact and affect each other in complex ways, which are proposed by Maffi (2005) to have seen evolutionary trends with active interactions between humans and the environment. For the formulation of indigenous development goals within the aegis of the sustainable developmental goals, and for mapping the needs of traditional communities, understanding of biocultural diversity is essential.

19.1.2 Nature of Traditional Ethnoecological Knowledge

Perusal of literature on traditional ethnoecological knowledge (TEK) systems reveals a growing interest in research community on reporting on this subject. Description of existing biological resources and vegetation patterns form the core of ecological activities. Description of landform, physical features, and documentation of ethnic culture and practices are also included within the ambit of the TEK. For a holistic handling of this subject, it is important to understand the various segments covered in this concept. In her paper, Houde (2007) presents six faces of TEK as "factual observations, management systems, past and current land uses, ethics and values, culture and identity, and cosmology". She has further emphasized a strong need for research on nature of TEK and recording of field observations systematically and accurately. TEK is all about the indigenous people, the people of the land. Indigenous populations are well integrated with the land they are borne on. They are accurately known as people of the soil. They do not look at land as a provider and source to draw all their needs, but regard it as their parent and in turn respect it accordingly. Land is a living commodity and all its rituals and activities are incomplete without a direct intervention from Mother Earth. They are well bound to their land of origin and revere it as their ancestral root. Traditional communities maintain a direct dialogue with nature. A lot of what they do is in silent communication with what they perceived from their environment. Socio-cultural practices with links to natural environment are also evident. Seed sowing pattern, time of harvest, shift in crop cultivation depending on water availability, preservation of seeds for next season, are a few examples related to this. Drew and Henne (2006) have proposed that traditional knowledge is not a static component but dynamic since it includes interactions between living and nonliving components and with the environment around them. There is a very close relationship between traditional ethnoecological knowledge (TEK) and the traditional knowledge and sometimes these terms are used synonymously. Both encompassed cultural values and ethics in relation to the environment. There are two important dimensions in TEK. These according to Berkes (1999) are inherited cultural wisdom and a pool of knowledge spread over several generations. Hence intergenerational connections are the building blocks in any form of TEK and these need to be adequately recognized and preserved.

TEK encapsulates collective autochthonous knowledge of a community in a particular geographic environment, which is majorly unwritten cumulative account of culturally transmitted, multigenerational knowledge and which is a dynamic and evolving pool of knowledge (Berkes 1999; Mazzochi 2006). Advocating a concept of isolated TEK pockets globally, Baggethun et al. (2013) have proposed a hybridization model of TEK wherein a mixing of old and new schools of thought happens simultaneously. It is through TEK that ancient wisdom is inherited and encapsulated mostly in form of nonformal media such as folklores, songs, and beliefs. Drew and Henne (2006) advocate a need for multidisciplinary research based on evidences from ecological processes, pure sciences, and sociological structures. Since TEK is unique for a geographical location, hence it is regarded as a nationss' composite knowledge system. TEK is a form of practical knowledge that helps communities to live in harmony with nature adapting and improving their living conditions. Accurate documentation of TEK should automatically pave the way to define the route for achievement of IDG.

19.1.3 Understanding SDGs

There are many challenges that the world is facing today, like climate change, lack of access to clean water and sanitation, energy crisis, food insecurity, and other environmental and societal challenges. Since the Sustainable Development Goals (SDGs) have been formally adopted by the world in 2015, the governments have been calling out to all sections of the society for meaningful participation in decision-making and implementation of post 2015 agenda. Indigenous communities play an important role in these new settings since principles of sustainability are ingrained in them since time immemorial. Gadgil, Berkes and Folke (1993) have provided information about these relationships. It is important that we collate such cases so that the data helps in proper analysis and interpretation of sustainability principles with a historian perspective. This places faith on knowledge created by indigenous communities staying in close proximity to least disturbed ecosystems. This knowledge is a means for their survival and holds a key for wellness of these communities. This enhances an understanding at field level and assists modern society to adapt and adopt sustainability pathways. The international political frameworks supporting principles of sustainability are aptly captured by the 17 SDGs. Sustainability education plays an important role in this global agenda (Rieckmann 2018). It is important to recognize indigenous perspectives identified through traditional knowledge systems (Shava 2005). A challenge is in linking these with SDGs (Rieckmann 2018). But at the same time, this is a unique opportunity to connect Western and traditional methods of research pathways. By establishing linkages between traditional wisdom and global sustainability agenda, while we are facilitating transformative impacts and moving toward a sustainable future.

19.1.4 A Case Study of Paudi Bhuyans

Paudi Bhuyan is a hilltop tribe and an important PVTG found in the state of Odisha. Like all other tribes, the livelihood of Paudi Bhuyans is dependent on agriculture, harvest of non-timber forest products (NTFP), hunting, engagement in arts and craft, and involvement in other nonagricultural labor.

Paudi Bhuyan tribe is included in both tribal sub-plan (TSP) and non-tribal sub-plan (non-TSP) and are included in the Techno-economic stage of Shifting Cultivators. Paudi Bhuyan falls under the Indo-Aryan linguistic group. Perusal of literature on PVTGs of Northern Odisha revealed almost negligible account exists on their ethnoecological knowledge systems and how it will impact the knowledge systems of Paudi Bhuyans in coming future. Limited studies are available on social and cultural aspects of the Paudi Bhuyan tribe. It is also important to mention that Paudi Bhuyan is a unique and sensitive tribe that shares genetic similarity with three other important tribes, namely the Jarawas, the Great Andamanese, and the Onge endemic to Andaman Islands (Acharya et al. 2015). Despite their genetic similarities with such far-off endemic island tribes in the Bay of Bengal, there is absolutely no information on their migration patterns or routes. In the period between 1980s and early 1990s, the Indian government unified tribal communities of Northern Odisha belonging to Bhuyans, Juangs, Panos, Hadis, Doms, Patras, Chasas, Mundas, and Kolhs. This intermixing of tribes and subsequent assimilation initiatives by the Indian government led to changes in levels of socio-cultural practices observed in their natural surroundings. Paudi Bhuyans are slowly changing their hilltop dwelling habitat and have started moving downward and have started mixing with other native tribal populations and rural residents. How this conglomeration is impacting the TEK and their biocultural identity lead us to investigate on some of the villages at different altitudes inhabited in the recent past by Paudi Bhuyans in Northern Odisha.

19.2 Approaches

Based on an inductive research approach, information related to traditional knowledge was collected and analyzed subsequently. As also supported by Hasan and Khazei (2016), data capture was followed by laying out ideas to reach conclusions.

19.2.1 Selection of Informants

TEK is highly situated and hence it is a complex and specialized information available with a particular target group. An example to support this is information about the type of plants used for medicinal purposes. Only practitioners have knowledge about the correct plants that can be used for treatment purposes and hence selection of the right informant is extremely essential.

19.2.2 Sampling Techniques

Tongco (2007) has put forward the technique of purposive sampling for both qualitative and quantitative research. Purposive sampling technique also known as judgmental sampling or nonrandom technique is most effective when a researcher has complete familiarization of the study area. Number of field trips, ease of communication with respondents, and a check on final results help in the collection of right data. The non-random purposive sampling technique is a recognized and an accepted research protocol. Bendasolli (2013) has also supported the right of a researcher to be able to exercise discretion in collection of data that is linked through knowledge or experience. Tongco (2007) has repeatedly supported purposive sampling over random sampling as a more accepted methodology, especially in studies related to social sciences, natural sciences, and ecology. In several instances, results derived from purposive and random sampling are similar and, in fact, case studies demonstrate that purposive sampling is superior to random sampling. The findings discussed in this chapter deal with TEK collected through purposive sampling, and selection of the defined population, since these support cost-effectiveness in data collection and sudden loss of respondents (due to physical movement or deaths). In order to support the need for formulation of IDGs, it is important to map the situated resources and also ascertain various factors that contribute towards the changes observed in study areas. Earnest efforts are also required to publish socioeconomic data about tribal communities because this facilitates creation of interest toward improving and mainstreaming development of these "vulnerable" tribes. There is also a need to document the impact of migration on traditional family setups with a focus on its impact on local biodiversity.

Table 19.1 List of villages selected for the study

S. No.	Name of the village	Altitude (msl) (m)	Distance from motorable roads (km)	Percentage Paudi Bhuyan households(~)	Quantifiable knowledge of edible plant species	Quantifiable knowledge of medicinal plant species	Quantifiable knowledge of biocultural practices
1.	Balibahal	613	18.71	81	24	10	5
2.	Sibida	605	20.66	97	17	22	13
3.	Namagaon	583	16.41	85	28	20	9
4.	Nagira	542	14.45	94	25	15	7
5.	Bandabhuyan	305	10.64	100	15	13	4
6.	Rugudhi	254	10.09	97	18	13	9
7.	Bhimkhand	239	7.94	93	12	13	5
8.	Dudipani	225	7.47	97	13	12	8
9.	Jamardihi	211	5.77	64	8	10	10
10.	Itee	198	8.08	17	22	5	15
11.	Siaria	180	7.65	52	9	13	8

19.2.3 Field Assessment

This chapter highlights our findings of a research carried out in 11 villages in district Angul situated in the state of Odisha, India. Names of these villages, their altitudes, distances from motorable roads, number of households, and percentage of Paudi Bhuyan households are provided in Table 19.1. Focus group discussions and interviews with respondents from these villages were conducted over a period of 3 years (2015–2018). Based on responses received from villages, lists of plant based knowledge and their uses were prepared. These were further verified by going back to the village and doing a second-level validation with respondents. A list of plant species and a detailed account of edible and ethnomedicinal values and biocultural practices observed by these villagers were subsequently compiled. After verification with experts, a quantification of this knowledge was done and is also presented in Table 19.1.

19.3 Categorization of Plant-Based Knowledge

Detailed plant lists (species with edible, medicinal, and biocultural value) were compiled at the village level. A subsequent quantification was done and the knowledge about number of species under each category—edible, medicinal, and biocultural value was done.

19.3.1 Knowledge About Edible Plants

A gradation in the knowledge levels was observed from the study area. The highest knowledge about edible plants was documented from villages Nagira and Namagaon. Altitudes (msl) of these were 542 m and 583 m, respectively. Mixed populations of Paudi Bhuyan with members of Munda tribe in Namagaon and members of Kolha tribe in Nagira were recorded. This intermixing is one of the possible reasons behind hybridization of knowledge, further leading to a greater information about new consumption practices. There were other villages also depicting a similar trend of collective wisdom, particularly Itee. Villages closer to roads displayed a lower level of knowledge about edible plants than those located farthest. A possible reason is an easy access to improved means of transportation, which pushes villagers to cities and towns for better livelihood opportunities. In contrast, a higher knowledge of edible plants was noted from villages in higher altitude category (Namagaon, Nagira, Balibahal, Sibida). These villages were also located farthest from motorable roads.

19.3.2 Ethnomedicinal Knowledge About Plant Species

Guniyas (medicine men) were present in all Paudi Bhuyan villages. Knowledge about plant species with medicinal value still exists in villages covered in this research. However, the presence of mixed population of Paudi Bhuyan tribe along with members of Munda, Kolha, Gandia, and Ho tribes have led to cultural assimilations. Losses and gains in knowledge pools were reported by Guniyas in the study area. Based on variable patterns of usage, a list of 50 plant species with medicinal value was prepared. Villagers used various plant parts such as bark, root, leaf, seed, flower, fruit, or entire plant to cure different ailments. Some ailments cured through plant-based medications were respiratory problems, digestive issues, and skin ailments. On further investigation, a village-level knowledge about medicinal plants was deciphered. Only 10–12 species were being used by inhabitants of Dudipani and Sibida. Namagaon reported around 20 species of medicinal importance. Village Itee had the lowest level of usage of medicinal plants. Three factors emerged as probable contributors for higher level of knowledge in Namagaon and Sibida-domination by members of Paudi Bhuyan tribe members (Sibida), intermixing between Paudi Bhuyan and Munda tribe (Namagaon), and elevation levels (both Namagaon and Sibida are located at higher altitudes).

19.3.3 Biocultural Knowledge About Plants

Traditionally, tribal societies rejoice in festivities and rituals that are closely linked with natural resources and have evolved over a period of time in a particular landscape. Biocultural knowledge recognizes interlinkages between culture and biodiversity and is an integral component of traditional ethnoecological knowledge (TEK). Paudi Bhuyan tribe members were well versed in dance forms imitating different animals. These dance performances were done to welcome guests in their villages. The tribe was found to be God fearing and worshipped supreme power in form of Earth, Sun, hills, forests, fields, and water bodies. Along with continuation of their age-old practices, members of the tribe have adopted new ways of worship, wherever there is an evidence of intermixing of tribes and with other members of the society. Hence two types of cultural festivities were practiced in the study area. These were festivals endemic to tribal population and those that were celebrated by other rural populations and members of general caste. In the former category Gramshree, Badam, Bisiri, Mahula, Karam, Nuakhai, Dihabandha, Basana, Rajo parab, Nadi puja, Jatal Puja, Bhimbudha, Dhana Nua, and Sarapta puja were included. Akshay Tritiya, Pushpurnima, Kartik Purnima, Jagannath puja, and Makar Sankranti were included in the latter category. They have now also started worshipping various Hindu deities like Laxmi, Durga, Saraswati, and Ganesha. Observance of new rituals and festivals were more evident in villages located in lower altitudes and closer to comparatively dense human habitations. Inhabitants of villages Itee and Jamardihi, which were closer to motorable roads and more exposed to urbanization followed more cultural festivals in both the categories mentioned above. However, in Bandabhuyan (100% PB presence) only endemic festivals were practiced. The villagers also regularly worship trees such as mahul (*Madhuca longifolia* A.Chev), kendu (*Diospyros melanoxylon*), and sal (*Shorea robusta*). Religious festivals that are commonly celebrated include Karam, Bisiri, Gramshree, and Banadurga puja. All these festivals were linked to forests, because the inhabitants directly depend on them for meeting their daily requirements. The villagers have a strong belief that observance of these festivals will also protect them from any harm. They have inherited these traditions from their forefathers and are continuing to do so. Some of the villages have formed an active Van Sangrakshak Samiti (VSS) and they guard and protect the forests adjoining their villages. They do not allow any trespassers to move into the forest. A survey and discussion with all the villages revealed active use of certain plant species for customary cultural practices. However, there are many other species of plants that are also used regularly in rituals and cultural festivities observed by members of Paudi Bhuyan tribes in the study villages. These are mango (*Mangifera indica*), rajkusumo (*Schleichera oleosa*), genda (*Tagetes erecta*), muchkundu (*Pterospermum acerifolium*), barh (*Ficus benghalensis*), hambu (*Cryptolepis brazzaei*), limbo (*Azadirachta indica*), and palasho (*Butea monosperma*). The Paudi Bhuyan tribal society is patriarchal. Women are forbidden to attend many festivals and customs like Jatal puja, Nadi puja, and Dihibandha. Results about observance of cultural practices affirm earlier conclusions related to knowledge about edible plant species recorded in the study area. Collective wisdom of mixed population groups has led to an increase in observance of cultural practices.

19.4 Understanding SDGs Against a Background of TEK

While this chapter brings forth the TEK encapsulated in the Paudi Bhuyan tribal populations, it also encourages the readers to balance this knowledge system against the SDG framework so that a stronger case for IKS could be built. TEK plays an important role in the global stride for achievement of several SDGs, since it considers sustainability to be a paradigm for thinking about a future in which environmental, societal, and economic considerations are balanced in the pursuit of development, and an improved quality of life. Knowledge about TEK helps in development of critical, evaluation, and creative thinking skills, so as to aid policy decisions around development of sustainable solutions. SDGs provide a holistic view of the most pressing global issues ranging from poverty, gender inequalities, and to socio-environment mismanagement. Investment in TEK research is a step in this direction. Interactions with tribal populations to study situated knowledge and their adaptive responses are important because they practice several resource efficiency measures that are replicable with positive impact. "Leave no one behind" is the global sustainability slogan of agenda 2030 and extra efforts to include "indigenous people" are being taken across the world. However, global development agenda is not entirely a boon for indigenous community (www.un.org/indigenous). There are certain limitations that emerge out of the 2030 agenda which offer major

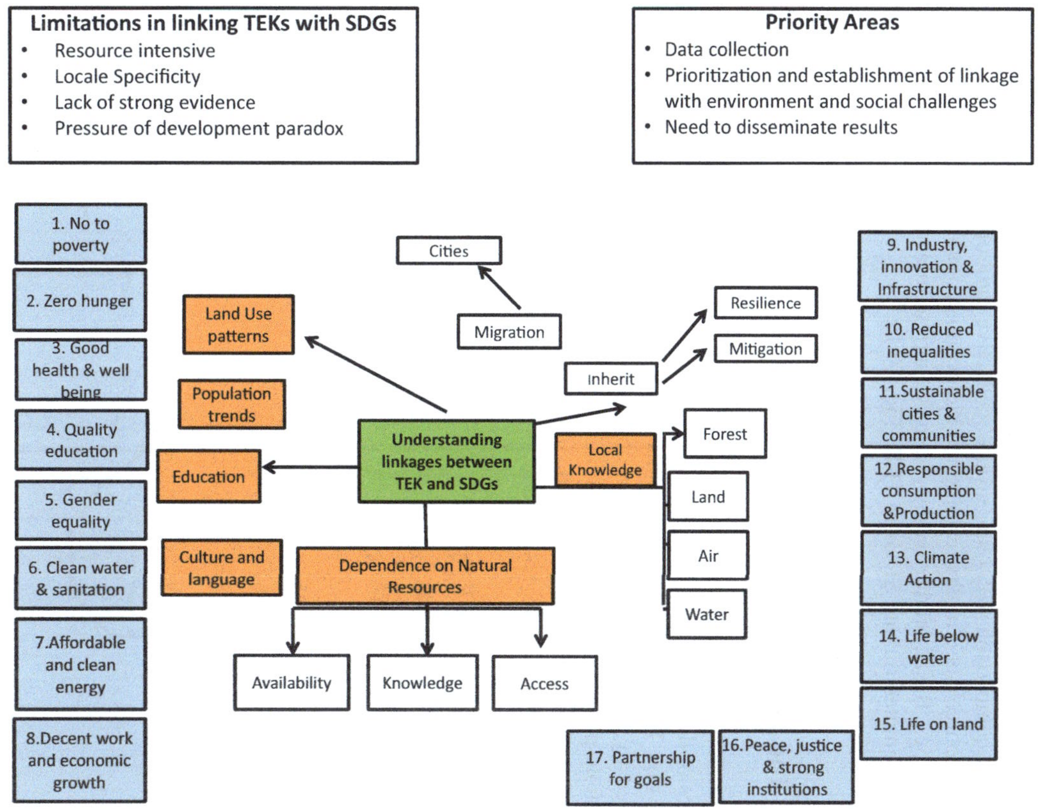

Fig. 19.1 Understanding connections between SDGs and TEK: The 17 SDGs are listed in the blue boxes, boxes in orange shade represent components of TEK under stress, and the white boxes represent the sectors that are linked with SDGs

bottlenecks for indigenous communities and their mainstreaming into society. These limitations need to be addressed case by case, to propose a new framework of development based on a repository of indigenous knowledge systems.

Cases of isolation, lack of education in mother tongue, inability to determine their own priorities for development, respect for ethnic culture, are some of the issues that are majorly disregarded in the global 2030 agenda. Specific indicators are laid out in the SDGs that track income of marginalized farmers, access of indigenous people to education in comparison with other groups, collective right of indigenous people to land and resources, and mainstreaming the indigenous communities thereby reducing inequalities. Based on the findings from the current research a basic framework has been attempted linking SDGs to TEK (Fig. 19.1).

19.5 The Future of TEK

In the current Anthropocene era where change is ceaseless, TEK is subjected to dynamics of development. It was evident through current research presented in this chapter that Paudi Bhuyan populations residing in selected 11 villages were influenced by several exogenic factors that are impacting their TEK (knowledge about edible plant species, plants with medicinal value, and plants with biocultural importance). It is a complex process, therefore it was difficult to accurately predict all issues and factors that are responsible for percolation of the biological knowledge

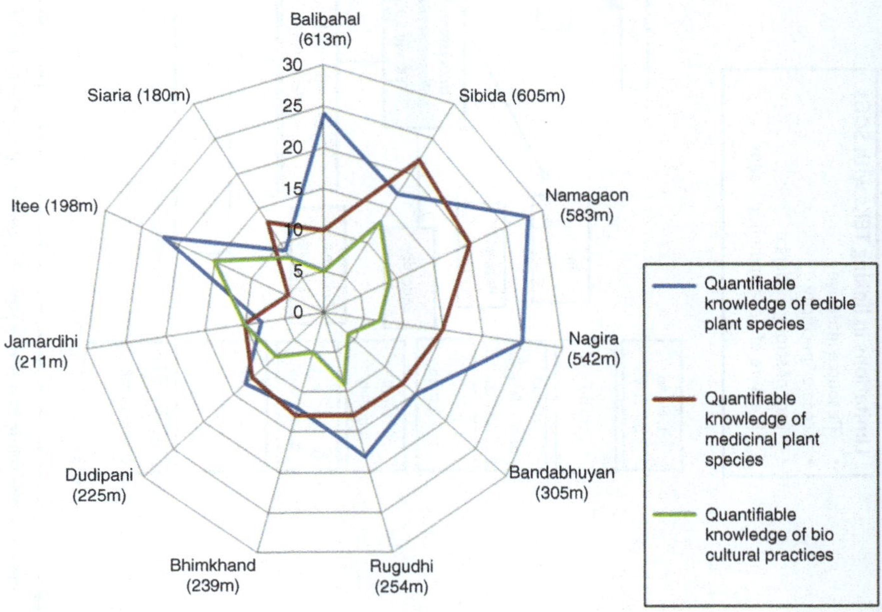

Fig. 19.2 Intricate and dynamic web of TEK levels in villages covered in the study (2015–2018)

systems among and between indigenous populations of Paudi Bhuyans. However, an attempt has been made to provide an understanding about TEK prevalent in Paudi Bhuyan tribe. Observations have led to an understanding that each Paudi Bhuyan tribal population possesses a unique ethnoecological knowledge base and the knowledge is less vulnerable to be lost in near future particularly in those settlements which are, firstly, are far away from the road; and secondly, have a mixed population of different age groups (Fig. 19.2). In this category of villages, interactions with forests and other natural resources are spontaneous and continuous and thus transmission is occurring on daily basis both vertically between generations and horizontally among different age groups within the villagers. Each influencing factor has a unique footprint, and can also be linked to the SDGs.

In villages of Itee and Siaria where populations have come in contact with other tribes or rural or non-tribes and had /have exposure to factors like alternate agriculture practices, education, health facilities and, telecommunications, have caused tremendous changes not only in their traditional way of living but also in their relationship with the forest produce and other natural resources. In a process of adaptation to new conditions, these tribes have slowly detached from many of their traditional practices and have evolved a new/hybrid version of their socio-cultural factors. According to Baggethun et al. (2013), these forms of partial adaptations to fit into new lifestyles in new ecological environments are a means of sustenance of indigenous communities. A tribe's ability to adapt and adopt will determine the degree to which it is linked to the development goals and hence will help in charting out a unique framework of Indigenous Development Goals (IDGs).

19.6 Conclusion

The aim of this chapter has been to present the integrated nature of SDGs by a comprehensive assessment of TEK against three dimensions—economic, environmental, and social using Paudi Bhuyans as a case study. This reveals that while TEK can be mapped against several SDGs, there are certain gap areas that can be filled up only if we layout a framework of indigenous knowledge system.

Acknowledgments The authors would like to thank the local communities for sharing knowledge. We would also like to thank officials at Paudi Bhuyan Development Agency, Pallahara for their valuable guidance in the selection of villages and for sharing information generously. We are grateful to staff at Odisha State Tribal Museum, Bhubaneshwar for their guidance in accessing library resources and USEM (University School of Environment Management), GGSIPU for providing the required facilities for undertaking this study.

References

Acharya SK, Das PK, Kshatriya GK (2015) Love and marriage practices among the Pauri Bhuyan tribe in Keonjhar district of Odisha, India. East Anthropol 68(1):26–43

Baggethun EG, Corbera E, García VR (2013) Traditional ecological knowledge and global environmental change: research findings and policy implications. Ecol Soc 18(4):Art.72

Bendasolli PF (2013) Theory building in qualitative research: reconsidering the problem of induction. Forum Qual Soc Res 14(1):Art. 25

Berkes F (1999) Sacred ecology: traditional ecological knowledge and resource management. Taylor & Francis, Philadelphia

Berkes F, Colding J, Folke C (2000) Rediscovery of traditional ecological knowledge as adaptive management. Ecol Appl 10(5):1251–1262

Bettencourt LMA, Kaur J (2011) Evolution and structure of sustainability science. Proc Natl Acad Sci 108(49):19540–19545

Drew JA, Henne AP (2006) Conservation biology and traditional ecological knowledge: integrating academic disciplines for better conservation practice. Ecol Soc 11(2):Art.34

Gadgil M, Berkes F, Folke C (1993) Indigenous knowledge for biodiversity conservation. Ambio 22(2/3):151–156

Hasan Z, Khazei M (2016) The role of deductive and inductive reasoning in accounting research and standard setting. Asian J Finan Account 8(1):227–241

Houde N (2007) The six faces of traditional ecological knowledge: challenges and opportunities for Canadian co-management arrangements. Ecol Soc 12(2):Art.34

Kolbert E (2014) The sixth extinction: an unnatural history. Bloomsbury Publishing, New York

Maffi L (2005) Linguistic, cultural, and biological diversity. Annu Rev Anthropol 34(1):599–617

Mazzochi F (2006) Western science and traditional knowledge: despite their variations, different forms of knowledge can learn from each other. EMBO Rep 7(5):463–466

Rieckmann M (2018) Learning to transform the world: key competencies in education for sustainable development. Available at https://unesdoc.unesco.org/ark:/48223/pf0000261802. Last accessed Dec 2019

Roue M, Nakashima D (2002) Knowledge and foresight: the predictive capacity of traditional knowledge applied to environmental assessment. Int Soc Sci J 54(173):337–347

Shava R (2005) Research on indigenous knowledge and its application: a case of wild food plants of Zimbabwe. S Afr J Environ Educ 22:73–86

Tongco MDC (2007) Purposive sampling as a tool for informant selection. Ethnobot Res Appl 5:147–158

20 Achieving Sustainable Development Goals (SDGs): Challenges and Preparation in Bangladesh

Amir Mohammad Nasrullah

Abstract

Sustainable Development Goals (SDGs) are a universal call for appropriate actions by the nations in the world under a global partnership to end poverty and to protect the planet, adopted by the United Nations in 2015. These were approved to the descendant context of the Millennium Development Goals (MDGs). However, SDGs include some new areas such as climate change, economic inequality, innovation, sustainable consumption, peace, and justice among others which are interconnected and commonly associated with one another. They provide opportunities, as well as, challenges for the nations based on their own distinctiveness and perceptions. It is a matter of hope that Bangladesh has been moving forward toward balanced development supported by these universal goals. As of the United Nations Committee for Development Policy (UNCDP), Bangladesh has recently achieved an implausible "development" to get the developing country status from the Least Developed Countries (LDCs). It is a graduation process that the country met all the requirements for upgrading the status. However, which is very important here is that how sustainable could be this "development," while Bangladesh is waiting to get the official status of a "Developing Country" by 2024. Nonetheless, the progress of development in Bangladesh brings confidence toward accomplishing the SDGs. Bangladesh has already taken good preparation and remarkable policy initiatives to achieve the goals, but in reaching the goals, like others, Bangladesh has been facing different significant challenges. Consequently, the aim of this chapter is to critically figure out the key challenges as well as to evaluate the preparation in achieving the SDGs in Bangladesh. More specifically, this study intends to

A. M. Nasrullah (✉)
Department of Public Administration, University of Chittagong, Chittagong, Bangladesh
e-mail: amir.nasrullah@cu.ac.bd

A. Kaushik et al. (eds.), *Climate Resilience and Environmental Sustainability Approaches*, https://doi.org/10.1007/978-981-16-0902-2_20

identify the policy instruments and institutional mechanisms to face the challenges and suggest a way out. For the purpose of the study, a list of existing literature on the issue, such as government and international documents, and research articles were collected and extensively reviewed. The study hopes to contribute to the policy as well as others concerned by the findings and their analyses.

Keywords

Sustainable Development Goals · Challenges · Bangladesh

20.1 Introduction

Bangladesh is moving forward toward balanced development underpinned by the Sustainable Development Goals (SDGs). The SDGs are the blueprint to attain a better and more sustainable future for all the nations in the world. The goals address the global challenges that we have been facing, including those related to poverty, inequality, climate change, environmental degradation, peace, and social justice. The UN Committee for Development Policy (UNCDP) has recently declared that Bangladesh has recently achieved an incredible "development" to get the developing country status from the list of least developed countries (LDCs) in the world. It is a graduation process for Bangladesh that the country met the required eligibilities for upgrading the status. However, the important questions are: "how sustainable could be this development" and what should Bangladesh do in the next phase to be sustained in all about her development, while she is waiting to get an official status for a "Developing Country" by 2024. Nevertheless, this success indeed brings confidence for Bangladesh toward achieving the SDGs. Some shrewd examples exemplify the country's development model that has been witnessed globally after 1990, such as, the fastest reduction in poverty (currently 22.4%), infant mortality (46 now from 92 per 1000 live births in 1990), underweight of children (32.6% now from 66% in 1990 for less than 5 years of age), escalating student enrollment (almost 100%) in primary schools, gender balance and women empowerment in every sector, public health through improved water sources and sanitation, and many more. Huge infrastructural changes of the country have been accomplished in the last 10 years and also going up with many mega projects (Saha 2018).

But in achieving the SDGs, Bangladesh has been facing significant challenges. A global report on the SDGs published recently by The UN Sustainable Development Solution Network (UNSDSN) shows an SDG index of 149 countries based on their past status in 2015, on each of the goals where Bangladesh ranked last (118) among the BIMSTEC countries (India, Bhutan, Nepal, Bangladesh, Myanmar, Thailand, Sri Lanka), and second last among the SAARC countries (Afghanistan 139, Myanmar 117, Pakistan 115, India 110, Nepal 103, Bhutan 82). However, these ranking scenarios evidently remind Bangladesh that she has to strive a lot in achieving the SDGs in the coming future (Saha 2018). However, the government of Bangladesh has taken enough preparation to face these challenges in achieving these SDGs.

The aim of this chapter is to critically figure out the key challenges in achieving the SDGs in Bangladesh, as well as, evaluate the preparation of Bangladesh in facing those challenges. More specifically, this chapter tries to identify the policy instruments and institutional mechanisms to face the challenges in achieving SDGs. The nature of the study is analytical and descriptive; hence reviewed available secondary sources of data such as various government documents, research articles, reports of different national and international organizations, etc.

20.2 Why Does the World Need the SDGs?

Since the last two decades the world has been changing dramatically. Sustainable Development Solution Network (SDSN) identified four shifts in these changes that are particularly important for the nations around the world. First, the world is closer than ever to eradicate extreme poverty; secondly, human societies are causing more damage than ever to the planet and the environment; thirdly, inequality among and between the countries is increasing; and finally, governance is becoming more complex, especially with the growing influence of technology (SDSN 2014).

Today, more than 1 billion people live in extreme poverty (less than $1.25 a day), while 8 men have as much wealth as the bottom 3.5 billion poorest people (half the world population); in 2017, more than 750 million people went to bed hungry every night and at the same time, there are approximately 2 billion people that are overweight or obese, while more than 6 million children died within 5 years because of preventable diseases. That is more than 15,000 deaths each day; about 263 million children and youth are out of school, including 61 million children of primary school age; 49 countries have no laws specifically protecting women from domestic violence; in 2012 at least 1.8 billion people were exposed to drinking water that was contaminated; today, more than a billion people do not have access to electricity; the number of people living in slums and shantytowns is now estimated at 863 million, in contrast to 760 million in 2000 and 650 million in 1990; in 2017, youth were almost three times more likely than adults to be unemployed (12.8%); the number of deaths from natural disasters continues to rise (SDSN 2014).

From 1990 to 2015, more than 1.6 million people died in internationally reported natural disasters; each year, the world generates roughly 1.3 billion tons of waste, but that is expected to soar to 4 billion tons by 2100. In the United State alone, about $200 billion a year is spent on solid waste management; and tens of thousands of species, including 25% of all mammals and 13% of birds are now threatened with extinction. This is because of hunting, poaching, and pollution, loss of habitat, the arrival of invasive species, and other human-caused problems (SDSN 2014).

However, today's problems will deteriorate quickly and perilously if we do not urgently and radically change the advancement. To do this we need the world to be oriented in one direction and socially mobilized, but it is not easy. It is very hard in this noisy, disparate, divided, crowded, congested, distracted, and often overwhelmed world to mount a consistent effort to achieve any of our common purposes. In this context, SDGs help individuals, organizations, and governments worldwide to agree on the direction, essentially, to focus on what really matters for our future (www.weforum.org). Consequently, the Sustainable Development Goals (SDGs) have been taken by the United Nations (UN) to fight these problems (SDSN 2014).

However, as of UN the SDGs are a call for action by all countries, poor, rich, and middle-income to promote prosperity while protecting the planet. They recognize that ending poverty must go hand-in-hand with strategies that build economic growth and address a range of social needs including education, health, social protection, and job opportunities, while tackling climate change and environmental protection (www.un.org). There are 17 global goals set by the United Nations in 2015. The goals are broad and somewhat interdependent, yet each has a separate list of 169 targets to achieve, and 231 indicators for monitoring the implementation of the 2030 Agenda. Achieving all 169 targets would signal accomplishing all 17 goals that are concerned with pressing development issues (Khaled 2018).

20.3 SDGs and Bangladesh

Bangladesh has been widely commended as one of the forerunners of MDGs implementation. It made outstanding progress in achieving the goals and targets. Many of the targets were achieved ahead of time and some others were achieved within the 2015 deadline. Kept by its success Bangladesh became an active participant in the global process underlying the preparation of post-MDG agenda with its domestic and global actions. Internally, the General Economics Division (GED) of the Planning Commission, with the support of UNDP, steered the preparation of a draft of the post-2015 Development Agenda along with Goals, Targets, and several indicators.

Several rounds of consultations with multiple stakeholders including government officials, public representatives, civil society organizations, and media representatives took place in 2013 at both the national and the sub-national levels. Bangladesh's proposals to the UN were strikingly similar to the proposals made by the Open Working Group (OWG) of the United Nations where 9 out of 11 goals were similar and other goals proposed by OWG were also these in Bangladesh proposals, but as targets of different goals. Bangladesh could thus articulate the sustainable development problems that the world would have to grapple within the coming one and a half decade (GED 2018).

20.4 Key Challenges in Achieving SDGs in Bangladesh

The SDGs provide opportunities as well as challenges for developing countries, because the cost of its implementation is very high. But it can bring long-term benefits for the nations around the globe. The nations around the globe are different and have their distinctiveness and different perspectives. Therefore, it is important that each nation should interpret each relevant SDG into their own National Plan of

Action (Jaiyesim 2016). However, Bangladesh like other developing nations has been facing many challenges in achieving the SDGs. Some of which are discussed below goal by goal.

20.4.1 Goal 1: No Poverty

This goal focuses on eradicating poverty, currently which is measured as people living on less than $1.25 a day, in all its forms everywhere around the globe by 2030. Globally, the number of people living in extreme poverty declined from 36% in 1990 to 10% in 2015. Currently, more than 700 million people, or 10% of the world population, still live in extreme poverty, struggling to fulfil the most basic needs like health, education, and access to water and sanitation, to name a few. Worldwide, the poverty rate in rural areas is 17.2%, which is more than three times higher than in urban areas. Moreover, for working people, having a job does not guarantee a decent living. In 2018, in fact, 8% of employed workers worldwide and their families lived in extreme poverty. It indicates that one out of five children lives in extreme poverty around the globe. Therefore, ensuring social protection for all children and other vulnerable groups is very important. However, to achieve this goal it is necessary to focus on economic growth which must be inclusive to provide sustainable jobs and promote equality for all (www.un.org). Nonetheless, Bangladesh has been striving to achieve this goal by emphasizing on job creation, social protection, human capital development, improving private investment, and mitigating the impact of various shocks. But, constraints on mobilizing external resources, implementation of National Social Security Schemes (NSSS), enhancing the professional capacity of Bangladesh Bureau of Statistics (BBS), and preventing slippage into poverty are considered the main challenges in achieving this goal in Bangladesh.

20.4.2 Goal 2: Zero Hunger

This goal focuses on ending hunger by achieving food security and improved nutrition, and promoting sustainable agriculture by 2030. After decades of steady decline, the number of people who suffer from hunger, as measured by the prevalence of undernourishment, began to slowly increase again in 2015. Today, more than 820 million people regularly go to bed hungry, of whom about 135 million suffer from acute hunger largely due to man-made conflicts, climate change, and economic downturns. At the same time, a profound change of the global food and agriculture system is needed if we are to nourish the more than 820 million people who are hungry and the additional 2 billion people the world will have by 2050. Increasing agricultural productivity and sustainable food production are crucial to help alleviate the perils of hunger (www.un.org). As the food and agriculture sector offers key solutions for development, and is central for hunger and poverty eradication, Bangladesh has been trying to focus on that. However, Bangladesh has been facing many uncertainties due to climate change, population growth, addressing hunger in lagging regions and of disadvantaged groups, building resilience of poor people, and problems emerging from urbanization. Consequently, these all create severe challenges to achieve SDG 2 in Bangladesh.

20.4.3 Goal 3: Good Health and Well-Being

Goal 3 focuses on ensuring healthy lives and promote well-being for all at all ages. Ensuring healthy lives and promoting the well-being for all at all ages is essential to sustainable development. Currently, the world is facing a global health crisis that is spreading human suffering, destabilizing the global economy, and upending the lives of billions of people around the world. Earlier, major progress was made in improving the health of millions of people. Significant strides were made in increasing life expectancy and reducing some of the common killers associated with child and maternal mortality. But more efforts are needed to fully eradicate a wide range of diseases and address many different persistent and emerging health issues. By focusing on providing more efficient funding of health systems, improved sanitation and hygiene, and increased access to physicians, significant progress can be made in helping to save the lives of millions (www.un.org). Health sector in Bangladesh has been facing considerable challenges concerning increasing access to, improving quality of and achieving equity in health care services for all. There is also increasing incidence of various injuries, drowning, aging and geriatric diseases, spread of infectious diseases, health effects of geo-climatic disasters and arsenicosis, and re-emergence of traditional diseases such as Malaria and TB. These all are big threats to achieving SDG 3 in Bangladesh.

20.4.4 Goal 4: Quality Education

This goal focuses on ensuring inclusive and equitable quality education and promote lifelong learning opportunities for all. Education enables upward socioeconomic mobility and is a key to escaping poverty. Obtaining a quality education is the foundation to improving people's lives and sustainable development. Over the past decade, major progress was made toward increasing access to education and school enrollment rates at all levels, particularly for girls. Nevertheless, about 260 million children were still out of school in 2018, nearly one-fifth of the global population in that age group. And more than half of all children and adolescents worldwide are not meeting minimum proficiency standards in reading and mathematics (www.un.org). In this world context of education, the governments of Bangladesh have been trying to ensure quality education. Despite various government efforts and progress achieved in the education sector in different dimensions in Bangladesh, significant challenges still remain in achieving SDG 4. These include inclusive and equitable education, quality of education at all levels, quality of teaching, adult literacy, and lifelong learning.

20.4.5 Goal 5: Gender Equality

This goal focuses on achieving gender equality and empower all women and girls. Gender equality is not only a fundamental human right, but a necessary foundation for a peaceful, prosperous, and sustainable world. There has been progress over the last decades, such as more girls are going to school, fewer girls are forced into early marriage, more women are serving in parliament and positions of leadership, and laws are being reformed to advance gender equality. Despite these gains, many challenges remain, such as discriminatory laws and social norms remain pervasive, women continue to be underrepresented at all levels of political leadership, and 1 in 5 women and girls between the ages of 15 and 49 years report experiencing physical or sexual violence by an intimate partner within a 12-month period (www.un.org). Gender issue is very importantly focused in Bangladesh. However, like others, still there are few challenges present in achieving SDG 5. These include eradication of violence against women, prevention of child marriage, and addressing gender digital divide.

20.4.6 Goal 6: Clean Water and Sanitation

Goal 6 focuses on ensuring the availability and sustainable management of water and sanitation for all. Clean, accessible water for all is an essential part of the world we want to live in. While substantial progress has been made in increasing access to clean drinking water and sanitation, billions of people, mostly in rural areas, still lack these basic services. Worldwide, one in three people do not have access to safe drinking water, two out of five people do not have a basic handwashing facility with soap and water, and more than 673 million people still practice open defecation (www.un.org). However, Bangladesh has got a good progress in this issue of water resources management. But in achieving goal 6, ensuring availability of water and sanitation facilities for all, still she has been facing some challenges, such as improvement of water quality and protection of water ecosystem.

20.4.7 Goal 7: Affordable and Clean Energy

This goal focuses on ensuring access to affordable, reliable, sustainable, and modern energy for all. It is an encouraging sign that energy is becoming more sustainable and widely available around the globe. Access to electricity in the poorer countries has begun to accelerate, energy efficiency continues to improve, and renewable energy is making impressive gains in the electricity sector. Nevertheless, more focused attention is needed to improve access to clean and safe cooking fuels and technologies for three billion people, to expand the use of renewable energy beyond the electricity sector, and to increase electrification in sub-Saharan Africa (www.un.org). Bangladesh has taken its own action plan for modern, affordable, and clean energy. However, finance, gas exploration, and pricing and subsidy are the main challenges in implementing the action plan in Bangladesh. Nonetheless, energy efficiency remains a big challenge as it is central to nearly every major challenge and opportunity all over the world.

20.4.8 Goal 8: Decent Work and Economic Growth

This goal focuses on promoting inclusive and sustainable economic growth, full and productive employment, and decent work for all. Sustainable and inclusive economic growth will require the nations to create the conditions that allow people to have quality jobs, hence improving living standards. However, in achieving this goal Bangladesh has been trying her best, but still faces some key challenges, such as, increasing informality in job market, high youth unemployment rate, skills demand and supply mismatch, low female labor force participation, low inflow of foreign resources particularly FDI, workplace injuries, and child labor and problems involving migration of workers.

20.4.9 Goal 9: Industry, Innovation, and Infrastructure

This goal focuses on building resilient infrastructure, promoting inclusive and sustainable industrialization, and fostering innovation. Inclusive and sustainable industrialization, together with innovation and infrastructure, can unleash dynamic and competitive economic forces that generate employment and income. They play a key role in introducing and promoting new technologies, facilitating international trade, and enabling the efficient use of resources. However, the world still has a long way to go to fully tap this potential. Least developed countries, in particular, need to accelerate the development of their manufacturing sector if they are to meet the 2030 target, and scale up investment in scientific research and innovation (www.un.org). In this regard Bangladesh has been facing financing and capacity constraints of implementation agencies and complexity of land acquisition. However, in the face of vulnerability to natural disasters that are worsening due to climate change impact Bangladesh has been building resilient infrastructure.

20.4.10 Goal 10: Reduced Inequalities

It focuses on reducing income inequality within and among countries. Reducing inequalities and ensuring no one is left behind are integral to achieving the Sustainable Development Goals. Inequality within and among countries is a persistent cause for concern. Despite some positive signs toward reducing inequality in some dimensions, such as reducing relative income inequality in some countries and preferential trade status benefiting lower-income countries, inequality still persists. To reduce inequalities, policies should be universal in principle, paying attention to the needs of disadvantaged and marginalized populations. However, the increasing inequality poses a big challenge to Bangladesh. Partly it is a problem of the inability to bring all types of income under progressive taxation and partly a problem of not being able to appreciably increase the share of government expenditure on education, health, rural development, and social protection in total government expenditure.

20.4.11 Goal 11: Sustainable Cities and Communities

This goal focuses on making cities and human settlements inclusive, safe, resilient, and sustainable. The world is becoming increasingly urbanized. Since 2007, more than half of the world's population has been living in cities, and that share is projected to rise to 60% by 2030. Cities and metropolitan areas are powerhouses of economic growth, contributing about 60% of global GDP. However, they also account for about 70% of global carbon emissions and over 60% of resource use. Rapid urbanization is resulting in a growing number of slum dwellers, inadequate and overburdened infrastructure and services, such as waste collection and water and sanitation systems, roads and transport, worsening air pollution, and unplanned urban extension (www.un.org). In order for the cities and towns to play the role of a driving force behind economic and social development, Bangladesh has been facing a number of challenges, including ensuring adequate, safe and affordable housing, ensuring affordable, accessible and sustainable urban transport, ensuring urban resilience, and resource constraints.

20.4.12 Goal 12: Responsible Consumption and Production

This goal focuses on ensuring sustainable consumption and production patterns. Worldwide consumption and production is a driving force of the global economy that rests on the use of the natural environment and resources in a way that continues to have destructive impacts on the planet. Economic and social progress over the last century has been accompanied by environmental degradation that is endangering the very systems on which our future development, indeed, our very survival, depends. It is estimated that each year, one-third of all food produced around the globe, equivalent to 1.3 billion tons worth around $1 trillion, ends up rotting in the bins of consumers and retailers, or spoiling due to poor transportation and harvesting practices; if people worldwide switched to energy-efficient light bulbs the world would save US$120 billion annually; should the global population reach 9.6 billion by 2050, the equivalent of almost three planets could be required to provide the natural resources needed to sustain current lifestyles (www.un.org). In achieving this goal in Bangladesh the main challenges are related to harvesting, storing, transporting, and also to market mechanisms.

20.4.13 Goal 13: Climate Action

This goal focuses on taking urgent action to combat climate change and its impacts by regulating emissions and promoting developments in renewable energy. 2019 was the second warmest year on record and the end of the warmest decade (2010–2019) ever recorded. Carbon dioxide (CO_2) levels and other greenhouse gases in the atmosphere rose to new records in 2019 (www.un.org). Climate change is affecting every country on every continent. It is disrupting national economies and affecting lives. Weather patterns are changing, sea levels are rising, and weather events are becoming more extreme. In Bangladesh, disaster management infrastructures are suffering from a lack of proper operation and maintenance. As a result, when disaster strikes the structures fail to provide adequate protection. Rehabilitation of these structures after a disaster is also becoming a major problem due to requirement of large funds.

20.4.14 Goal 14: Life Below Water

This goal focuses on conservation and sustainable use of the oceans, seas, and marine resources for sustainable development. The ocean drives global systems that make the Earth habitable for humankind. Our rainwater, drinking water, weather, climate, coastlines, much of our food, and even the oxygen in the air we breathe, are all ultimately provided and regulated by the sea. Careful management of this essential global resource is a key feature of a sustainable future. However, at the current time, there is a continuous deterioration of coastal waters owing to pollution, and ocean acidification is having an adversarial effect on the functioning of ecosystems and biodiversity. This is also negatively impacting small-scale fisheries. Saving our ocean must remain a priority. Marine biodiversity is critical to the health of people and our planet. Marine protected areas need to be effectively managed and well-resourced and regulations need to be put in place to reduce overfishing, marine pollution, and ocean acidification. In Bangladesh, exploitation of gas resources may pose a grave danger to the biological resources. As a result, sustainable management of these resources is now a big challenge for the country.

20.4.15 Goal 15: Life on Land

Goal 15 focuses on protecting, restoring, and promoting sustainable use of terrestrial ecosystems. It also focuses on the sustainable management of forests, combating desertification, reversing and halting land degradation and biodiversity loss. Nature is very critical to our survival as it provides us with oxygen, regulates our weather patterns, pollinates our crops, and produces our food, feed, and fiber. But it is under increasing stress. Human activity has altered almost 75% of the earth's surface, squeezing wildlife, and nature into an ever-smaller corner of the planet (www.un.org). According to IPBES (2019), around one million animal and plant species are threatened with extinction and many within decades. They called for transformative changes to restore and protect nature. It found that the health of ecosystems on which we and all other species depend is deteriorating more rapidly than ever, affecting the very foundations of our economies, livelihoods, food security, health, and quality of life worldwide. Deforestation and desertification, caused by human activities and climate change, pose major challenges to sustainable development and have affected the lives and livelihoods of millions of people. Forests are vitally important for sustaining life on Earth, and play a major role in the fight against climate change. And investing in land restoration is critical for improving livelihoods, reducing vulnerabilities, and reducing risks for the economy (IPBES 2019). The health of our planet also plays an important role in the emergence of zoonotic diseases, i.e., diseases that are transmissible between animals and humans. As we continue to encroach on fragile ecosystems, we bring humans into ever-greater contact with wildlife, enabling pathogens in wildlife to spill over to livestock and humans, increasing the risk of disease emergence and amplification. However, in this global context, in achieving goal 15, the major challenges in Bangladesh are absence of proper documentation and conspicuous demarcation of forest lands and lengthy procedure of disposing up land related cases, land tenure conflict in Chittagong Hill Tracts, inadequate human resources and logistics, demand for timber, fuel wood (mainly for cooking), poles and saplings, brick burning and tobacco curing, making huts, thatches and sheds of betel leaf plantations. Moreover, the recent influx of Rohingya refugees from Myanmar is putting tremendous pressure on remaining forests in Teknaf-Cox's Bazar range in Bangladesh (GED 2015).

20.4.16 Goal 16: Peace, Justice, and Strong Institutions

This goal focuses on promoting peaceful and inclusive societies for sustainable development, providing access to justice for all and build effective, accountable, and inclusive institutions at all levels. Conflict, insecurity, weak institutions, and limited access to justice remain a great threat to sustainable development. The number of people fleeing war, persecution, and conflict exceeded 70 million in 2018, the highest level recorded by the UN refugee agency (UNHCR) in almost 70 years. In 2019, the United Nations tracked 357 killings and 30 enforced disappearances of human rights defenders, journalists, and trade unionists in 47 countries. Moreover, the births of around one in four children aged below 5 years worldwide are never officially recorded. This has been depriving them to get proof of legal identity in their country which is crucial for the protection of their rights and for access to justice and social services (www.un.org). In Bangladesh, some important challenges in ensuring peace, justice, and strong institutions include lack of comprehensive and updated data on various types of offences, proper enforcement of laws, capacity of the judicial system to handle heavy volume of cases, and lack of reporting/timely reporting of incidence of violence/crime.

20.4.17 Goal 17: Partnerships for the Goals

Goal 17 focuses on strengthening the means of implementation and revitalize the global partnership for sustainable development. A successful development agenda requires inclusive partnerships, at the global, regional, national and local levels, built upon principles and values, and upon a shared vision and shared goals placing people and the planet at the center. Without strong global partnerships and cooperation it is impossible to realize the SDGs. However, many countries require Official Development Assistance to encourage growth and trade. Yet, aid levels are falling and donor countries have not lived up to their pledge to ramp up development finance. However, despite impressive country performance, there are challenges, especially in the areas of resource mobilization in order to implement the required interventions toward achieving the SDG targets in Bangladesh. Available estimates indicate that domestic resource mobilization potential is yet to be utilized. Mobilizing external resources is also posing a big challenge. Remittance sources are undiversified. FDI is not registering high growth due to lack of appropriate investment climate in the country. These pose a big challenge in achieving this goal in Bangladesh.

20.5 Commitment toward Achieving SDGs in Bangladesh

Bangladesh has been facing numerous challenges in achieving SDGs. Moreover, there are also challenges of coordination between various ministries and agencies responsible for implementing projects and different programs contributing toward the attainment of SDGs in Bangladesh. However, considering all the challenges, Bangladesh has demonstrated her commitment toward achieving SDGs and has taken different steps for monitoring and evaluating the progress on the achievement of SDGs at the national level.

20.5.1 Formation of SDGs Implementation Committees

An inter-Ministerial Committee on SDGs implementation and review has been established. The Committee comprising Secretaries from 20 Ministries/Divisions coordinates SDGs monitoring and implementation. The Principal Coordinator (SDGs Affairs), a newly created high-level position in the Prime Minister's office is the head of the Committee. The General Economic Division (GED) of the government is the secretariat for the committee to coordinate implementation at the policy level along with monitoring and reporting SDGs status. The Committee has been working on prioritizing and contextualizing the SDGs with the national ambitions requesting all ministries to identify relevant goals and targets and reflect these in their respective sector plans as well as in the newly introduced Annual Performance Agreement (APA). The Committee is supposed to report the SDGs implementation status to the Cabinet every 6 months GED (2018).

Moreover, there are some other committees formed for the attainment of the SDGs. They are: "SDGs Working Team" hosted by the Governance Innovation Unit, Prime Minister's Office; "SDGs Implementation Sub-committee" by the NGO Affairs Bureau; "National Data Coordination Committee" to deal with *data gap* under Statistics and Informatics Division; and Voluntary National Review (VNR) Committee.

20.5.2 Approach to Achieving SDGs in Bangladesh

Bangladesh has consistently applied "whole of society" approach to the preparation of national development plans and policy documents of national importance. The Government has been applying this approach throughout the processes of SDGs preparation. For example, the Post-2015 Development Agenda: Bangladesh Proposal to UN (GED 2013) was prepared with inputs from multiple stakeholders including national experts, private sector and CSO representatives, and development partners. The Government has extended this strong tradition to the implementation of ambitious SDGs. Several consultations on "Stakeholders" engagement on the SDGs implementation in Bangladesh' have been held with representatives from all the relevant stakeholders.

20.5.3 Policies and Strategies in Achieving SDGs in Bangladesh

Commitment and enthusiasm of Bangladesh in achieving SDGs have been also reflected at the policy level. One of the major important development policies of Bangladesh is its 7th Five Year Plan (FYP) 2016–2020. All the 17 goals of SDGs are already integrated into the running 7th FYP. Among the 17 goals, 14 goals (82%) are thematically fully aligned, 3 goals (Goal 14, Goal 16, and Goal 17) of the SDGs (18%) are partially aligned with the 7th FYP. It is rationally expected that achievement of the FYP objectives and targets will contribute toward achievement of SDGs. The 7th FYP has made a reflection on the core goal of SDGs, especially in ending poverty in all its forms. The Government of Bangladesh had earlier prepared a National Sustainable Development Strategy (2010–2021) which is also helpful in the successful attainment of the SDGs (GED 2018).

However, the Planning Commission of Bangladesh has conducted a mapping of all the concerned existing policies and actions for the attainment of SDGs. The existing policies and strategies are: (a) Bangladesh Climate Change Strategy and Action Plan (BCCSAP) 2009; (b) National Education Policy 2010; (c) National Health Policy 2011; (d) National Women's Development Policy (NWDP) 2011; (e) National Skills Development Policy (NSDP) 2011; (f) National Labor Policy 2012; (g) National Social Security Strategy (NSSS) 2015; (h) National Food Policy Plan of Action (2008–2015); (i) Health, Population and Nutrition Sector Development Program (HPNSDP) 2011–2016; (j) Primary Education Development Program-Revised (PEDP III) 2011–2017; (k) Power System Master Plan (PSMP)

2016; (l) National Industrial Policy 2016; (m) Revised Strategic Transport Plan 2016; (n) Expatriates' Welfare and Overseas Employment Policy 2016; (o) National Plan for Disaster Management (NPDM) 2016–2020; (p) Health, Population and Nutrition Sector Program (HPNSP) (2017–2022); and (q) Bangladesh Delta Plan 2100.

20.5.4 Lead Ministries and Divisions Involved in Achieving SDGs in Bangladesh

The Planning Commission of Bangladesh has conducted a mapping of all the concerned ministries/divisions by SDGs areas. Economic Relations Division (ERD) is the lead Division of the Government for mobilizing external resources in the form of development cooperation from different sources. According to "Handbook on Mapping of Ministries by Targets in the implementation of SDGs aligning with 7th Five Year Plan (2016–2020)," ERD is the lead organization for 15 different targets, similarly Co-lead for 2 targets and Associate for another 29 targets. ERD has undertaken several strategic institutional and policy measures to harness resources to support Bangladesh's SDGs implementation. It has strengthened its collaboration with different line Ministries/Divisions to speed up the process of resource mobilization and project implementation (GED 2016).

Table 20.1 shows some of the related existing policies/acts/strategies and lead ministries/divisions working for the achievement of SDGs in Bangladesh.

20.5.5 National Action Plan for the Achievement of SDGs in Bangladesh

As a continuation to the mapping exercise, the government emphasized on preparing respective action plans by the ministries/divisions/organizations, which would have specific actions/activities and interventions to achieve their respective goals/targets in line with the SDGs and the 7th FYP. They have been formulating short, medium, and long-term sector-specific plans for the 7th FYP period and beyond.

The National Action Plan for the implementation of the SDGs has been prepared by GED, which has coordinated the Action Plans of the 43 lead Ministries/Divisions through a rigorous process of consultations, review, and feedback. The Plan lists the ongoing projects/programs that contribute to the achievement of a particular goal and its targets, identifies new projects/programs that will need to be undertaken during the remaining period of the 7th Plan as well as beyond the current Plan period with indicative costs.

New policies/strategies that might be needed in the process are also stipulated. This Plan will guide the Ministries/Divisions/Agencies to determine their respective investment portfolio that will be realized to attain SDGs as well as the related objectives of Five Year Plans and help assess the performance of the ministries in achieving goals/targets (GED 2018).

Table 20.1 SDGs related national policy/act/strategy and lead ministries/divisions

SDGs	Related national policy/act/strategy	Responsible lead ministries/ divisions
Goal 1: No poverty	The Government has adopted different policies and programs to address multidimensional nature of poverty in the country	Cabinet Division General Economic Division Ministry of Environment and Forest Ministry of Disaster Management and Relief Finance Division
Goal 2: Zero hunger	Delta Plan 2100	Ministry of Agriculture Ministry of Food Ministry of Health and Family Welfare Ministry of Industries Economic Relations Division Ministry of Commerce Finance Division
Goal 3: Good health and well-being	National Health Policy 2011 Health, Population and Nutrition Sector Program (HPNSP) (2017–2022)	Ministry of Health and Family Welfare Ministry of Home Affairs Road, Transport and Highway Division Ministry of Labour and Employment Ministry of Environment and Forest Economic Relations Division
Goal 4: Quality education	National Education Policy 2010 Primary Education Development Program-Revised (PEDP III) 2011–2017	Ministry of Primary and Mass Education Ministry of Education Economic Relations Division
Goal 5: Gender equality	National Women's Development Policy (NWDP) 2011 National Social Security Strategy (NSSS) 2015	Ministry of Women and Children Affairs Ministry of Social Welfare
Goal 6: Clean water and sanitation	Some of the key challenges of achieving gender equality in the country are: eradication of violence against women, prevention of child marriage, and addressing gender digital divide	Local Government Division Ministry of Agriculture Ministry of Water Resources Economic Relations Division
Goal 7: Affordable and clean energy	Power System Master Plan (PSMP) 2016	Power Division Energy and Mineral Resource Division Economic Relations Division
Goal 8: Decent work and economic growth	National Skills Development Policy (NSDP) 2011 National Labor Policy 2012	Finance Division Ministry of Commerce Ministry of Industries Ministry of Agriculture General Economic Division

(continued)

Table 20.1 (continued)

SDGs	Related national policy/act/strategy	Responsible lead ministries/ divisions
		Ministry of Labor and Employment Ministry of Youth and Sports Ministry of Civil Aviation and Tourism; Ministry of Cultural Affairs
Goal 9: Industry, innovation and infrastructure	National Industrial Policy 2016	Road, Transport and Highway Division Local Government Division Ministry of Industries Ministry of Environment and Forest Ministry of Science and Technology Ministry of Agriculture Economic Relations Division Information and Communication Technology Division
Goal 10: Reduced inequalities	Expatriates' Welfare and Overseas Employment Policy 2016	General Economic Division Law and Justice Division Finance Division Economic Relations Division Ministry of Overseas Employment Ministry of Commerce
Goal 11: Sustainable cities and communities	Revised Strategic Transport Plan 2016	Ministry of Housing and Public Works Road, Transport and Highway Division Ministry of Railways Ministry of Cultural Affairs Ministry of Disaster Management and Relief Local Government Division Ministry of Women and Children Affairs Economic Relations Division
Goal 12: Responsible consumption and production	National Food Policy Plan of Action (2008–2015) National Social Security Strategy (NSSS) 2015	Ministry of Foreign Affairs Ministry of Environment and Forests; Ministry of Finance Ministry of Agriculture Local Government Division Ministry of Industries Ministry of Energy Ministry of Primary and Mass Education Economic Relations Division Ministry of Civil Aviation and Tourism Finance Division

(continued)

Table 20.1 (continued)

SDGs	Related national policy/act/strategy	Responsible lead ministries/ divisions
Goal 13: Climate action	Bangladesh Climate Change Strategy and Action Plan (BCCSAP) 2009 Disaster Risk Reduction Strategies of Bangladesh (2016–2020) Bangladesh Delta Plan 2100 National Adaptation Plan (NAP) 2015 National Plan for Disaster Management (NPDM) 2016–2020	Ministry of Disaster Management and Relief General Economic Division Ministry of Environment and Forests
Goal 14: Life below water	Bangladesh Delta Plan 2100	Ministry of Water Resources Ministry of Shipping Ministry of Science and Technology Ministry of Fisheries and Livestock Ministry of Environment and Forests Ministry of Defense
Goal 15: Life on land	Bangladesh Delta Plan 2100	Ministry of Environment and Forests; Economic Relations Division Finance Division Ministry of Foreign Affairs
Goal 16: Peace, Justice and strong institutions	National Social Security Strategy (NSSS) 2015 Annual Performance Agreement (APA), Citizen Charter, National Integrity Strategy (NIS) Grievance Redress System (GRS)	Ministry of Home Affairs Ministry of Women and Children Affairs Ministry of Foreign Affairs Law and Justice Division Law and Parliamentary Affairs Division Bangladesh Bank Cabinet Division Ministry of Public Administration; General Economic Division Economic Relations Division Local Government Division Ministry of Information
Goal 17: Partnerships for the goals	National Sustainable Development Strategy 2010–2021	Internal Resource Division Finance Division Economic Relations Division Prime Minister's Office (Board of Investment) Bangladesh Bank Ministry of Science and Technology Ministry of Environment and Forests Information and Communication Technology Division

(continued)

Table 20.1 (continued)

SDGs	Related national policy/act/strategy	Responsible lead ministries/ divisions
		General Economic Division; Ministry of Commerce Prime Minister's Office (PPPA) Statistical Informatics Division (BSS)

Source: Compiled by the author from different government documents and GED 2018

20.5.6 Monitoring and Evaluation Framework for the Achievement of SDGs in Bangladesh

The Monitoring and Evaluation Framework of SDGs has been developed to track progress on implementation and achievement of SDGs in Bangladesh in the next 13 years. It provides baseline data for each indicator for which data are available and the target for the terminal year of SDGs, i.e., 2030 with two milestones: 2020 and 2025, in the intervening period (GED 2018). Information on relevant agency currently responsible for generating data along with the ministry/division to which the agency belongs as well as the title of the publication where the data appear are provided for quick and easy identification of data sources. Lacking availability of data uniform baseline could not be set for all the indicators. The indicators, for which annual data are available 2014–2015, the final year of MDGs, have been taken as the baseline. The baseline data is set for 127 indicators and M&E framework is designed for 108 indicators. Lacking data from national sources, international sources like WB, FAO, WHO, and ILO have been utilized to set the baseline for 22 indicators (GED 2018).

20.6 Potentials in Achieving SDGs in Bangladesh

Bangladesh is one of the 47 countries that has presented its Voluntary National Review (VNR) on SDGs to the High-Level Political Forum (HLPF). Such reviews, however, do not contain any future projections of SDG implementation in the country. Nonetheless, based on the availability of data and essence of the goals, some targets and indicators were chosen to reflect the status of each SDG by different research organizations. However, the Centre for Policy Dialogue (CPD) initiated a research to identify the future potentials of Bangladesh in achieving SDGs based on some past progress of each and every SDGs related indicators. Khatun and Saadat (2018) used Auto Regressive Integrated Moving Average (ARIMA) model to forecast these potentials under that CPD initiative. The past progresses and future potentials of Bangladesh in achieving SDGs as identified by Khatun and Saadat (2018) are briefly shown in Table 20.2.

However, Khatun and Saadat (2018) summarized that among the 17 targets chosen from the 17 SDGs some are on-track, some are partially on-track and some

Table 20.2 Past performances and future potentials in Achieving SDGs in Bangladesh

Past progress of SDG related indicators in Bangladesh	Future potentials in achieving SDGs in Bangladesh
Goal 1: No poverty	Goal 1: No poverty
Target and indicator 1.1: Poverty headcount ratio at national poverty line In 2000, almost half of the population (48.9%) was below the national poverty line. But in 2015, the share of population below the national poverty line fell to 24.3%. However, national trends fail to account for regional differences	*Target and indicator 1.1: Poverty headcount ratio at national poverty line* The target for 2030 is to have 0% of the population living below the national poverty line. However, if current trends continue, then 1.7% of the population will be below the national poverty line in 2030
Goal 2: Zero hunger	Goal 2: Zero hunger
Target and indicator 2.1: Prevalence of undernourishment Incidence of hunger fell from 20.8% in 2000 to 16.3% in 2006, and between 2007 and 2012 incidence of hunger increased. However, 15.2% of the population were undernourished in 2016	*Target and indicator 2.1: Prevalence of undernourishment* The target for 2030 is to have 0% prevalence of undernourishment. The forecast shows that 11.7% of the population will still remain undernourished in 2030, if current trends continue
Goal 3: Good health and well-being	Goal 3: Good health and well-being
Target and indicator 3.1: Maternal deaths Maternal deaths have been falling continuously since 1986 in Bangladesh and only 2001 and 2006 are exceptional years when maternal deaths rose. However, the rate of decline has been more gradual in recent years	*Target and indicator 3.1: Maternal deaths* The target for 2030 is to have less than 70 maternal deaths per 100,000 births. If the current rate continues the target will be achieved within 2026
Goal 4: Quality education	Goal 4: Quality education
Target and indicator 4.1: Lower secondary completion rate Lower secondary completion rates rose from 48.8% in 1998 to 58.9% in 2003 and 76.5% of the target population had completed lower secondary education in 2016. However, completion rates do not account for the quality of education	*Target and indicator 4.1: Lower secondary completion rate* The target for 2030 is to have a 100% lower secondary completion rate. However, the forecast shows that lower secondary completion rate will be 97.6% in 2030, if current trends continue
Goal 5: Gender equality	Goal 5: Gender equality
Target and indicator 5.3: Women Who Were First Married By Age of 18 years 73.3% of women aged 20–24 years were married by age 18 in 1994 and overall, early marriage shows a declining trend but the incidence is still alarmingly high. However, 58.6% of women aged 20–24 years were married by age of 18 years in 2014	*Target and indicator 5.3: Women Who Were First Married By Age of 18 years* The target for 2030 is to have 0% of women between the ages 20–24 years getting married by age of 18 years. However, 48.9% of women aged between 20 and 24 years will get married by age of 18 years in 2030, if current trends continue
Goal 6: Clean water and sanitation	Goal 6: Clean water and sanitation
Target and indicator 6.2: People Using At Least Basic Sanitation Services Access to basic sanitation has nearly doubled	*Target and indicator 6.2: People Using At Least Basic Sanitation Services* The target for 2030 is to ensure that 100% of

(continued)

Table 20.2 (continued)

Past progress of SDG related indicators in Bangladesh	Future potentials in achieving SDGs in Bangladesh
in the last 15 years and 46.9% of the population had access to basic sanitation in 2015, compared to 25.3% in 2000. However, more than half of the population still do not have access to basic sanitation	the population has access to basic sanitation services. However, 68.7% of the population will have access to basic sanitation services in 2030, if current trends continue
Goal 7: Affordable and clean energy	Goal 7: Affordable and clean energy
Target and indicator 7.1: Access to electricity Access to electricity has been increasing almost every year since 1990 and more than average increases were observed in 2006 and 2015. However, affordability of electricity has been compromised with higher tariffs	*Target and indicator 7.1: Access to electricity* The target for 2030 is to provide electricity access to 100% of the population. However, if current trends continue, 95.6% of the population will have access to electricity in 2030
Goal 8: Decent work and economic growth	Goal 8: Decent work and economic growth
Target and indicator 8.1: GDP Growth Rate Economic growth in Bangladesh has been rapid and resilient, and GDP has been growing at more than 6% since 2011. However, economic growth has failed to create decent jobs, especially for the large population of youth	*Target and indicator 8.1: GDP Growth Rate* The target for 2030 is to achieve and sustain 7% annual GDP growth and annual GDP growth has been over 7% since 2016 and is projected to reach 9.1% in 2030. Similar forecasts were obtained by the General Economics Division (GED)
Goal 9: Industry, innovation, and infrastructure	Goal 9: Industry, innovation, and infrastructure
Target and indicator 9.2: Share of Manufacturing in GDP Rate of industrialization picked up pace from 2005 onward and share of manufacturing in GDP increased from 14.4% in 1996 to 21% in 2016. However, such structural change of the economy presents a challenge for the labor market	*Target and indicator 9.2: Share of Manufacturing in GDP* The target for 2030 is to have a share of manufacturing in GDP which is double that of the levels in 2015 and for Bangladesh, this would mean a manufacturing share of GDP of 40%. However, by 2030, manufacturing will be 31% of GDP, if current trends continue
Goal 10: Reduced inequalities	Goal 10: Reduced inequalities
Target and indicator 10.1: Income Share of Bottom 40% Income share of the poorest 40% of the population has been declining and declining income shares for the poor imply rising income shares for the rich. However, if poor get poorer, then overall inequality increases	*Target and indicator 10.1: Income Share of Bottom 40%* The target for 2030 is to increase the share of income held by the poorest 40% of the population. This means that the growth in the share of income held by the poorest 40% of the population should be greater than zero. However, if current trends continue, income share of the poorest 40% will decrease by 1% annually in 2030
Goal 11: Sustainable cities and communities	Goal 11: Sustainable cities and communities
Target and indicator 11.1: Population Living in Slums Urban slum population has declined significantly and 55.1% of the urban population lived in slums in 2014, compared to 87.3% in 1990. However, the share of urban	*Target and indicator 11.1: Population Living in Slums* The target for 2030 is to ensure that 0% of the urban population are living in slums. However, 34% of the urban population will be living in slums in 2030, if current trends continue

(continued)

Table 20.2 (continued)

Past progress of SDG related indicators in Bangladesh	Future potentials in achieving SDGs in Bangladesh
population living in slums is still unacceptably high	
Goal 12: Responsible consumption and production	Goal 12: Responsible consumption and production
Target and indicator 12.2: Total Natural Resources Rents Economic rents from natural resource extraction soared between 2003 and 2006. However, there has been a sharp decline from 2011 onward. As natural resources become scarcer, extraction costs increase since the low-hanging fruits have already been depleted	*Target and indicator 12.2: Total Natural Resources Rents* Natural resource rents as a percentage of GDP will fall to 0.9% in 2030, if current trends continue. However, extraction of natural resources is forecasted to be less profitable in 2030 compared to 2015
Goal 13: Climate action	Goal 13: Climate action
Target and indicator 13.2: Net Carbon Emissions/Removals Net carbon emissions/removals were largely stable from 1990 to 2005. However, there was a big drop in 2006 followed by a jump in 2011. Bangladesh is one of the least carbon emitting yet most climate vulnerable countries in the world	*Target and indicator 13.2: Net Carbon Emissions/Removals* The target for 2030 is to reduce carbon emissions. By 2030, net carbon emissions/removals will rise back to levels similar to 2015
Goal 14: Life below water	Goal 14: Life below water
Target and indicator 14.4: Total Fisheries Production Fisheries production shows increasing trend and between 1990 and 2016, fisheries production has increased by a factor of 4. However, trends indicate increasing tendency of overfishing which threatens fish stocks	*Target and indicator 14.4: Total Fisheries Production* The target for 2030 is to pursue fishing in a sustainable way. However, total fisheries production is forecasted to keep rising till 2030, although the rate of increase will slow down gradually
Goal 15: Life on land	Goal 15: Life on land
Target and indicator 15.2: Forest Area Forest's share of land area has been declining and in 1990, forests covered only 11.4% of land, which was already quite low. Destruction of primary forests can cause irreversible damage to biodiversity	*Target and indicator 15.2: Forest Area* The target for 2030 is to stop deforestation. This means that percentage change in forest area as a share of total land area should be greater than or equal to zero. However, if current trends continue, forest area is forecasted to decline continuously up to 2030
Goal 16: Peace, Justice, and strong institutions	Goal 16: Peace, Justice, and strong institutions
Target and indicator 16.1: Intentional Homicides Murder rates have been around 2.5 to 2.8 per 100,000 people during the period 2000 and 2015 and rates are low compared to more unstable or violent countries. However, data are estimated, since crime is generally underreported	*Target and indicator 16.1: Intentional Homicides* The target for 2030 is to reduce the murder rate. Projecting the current trends shows an increasing trend in the murder rate. However, generally, criminal activity is unpredictable by nature

(continued)

Table 20.2 (continued)

Past progress of SDG related indicators in Bangladesh	Future potentials in achieving SDGs in Bangladesh
Goal 17: Partnerships for the goals	Goal 17: Partnerships for the goals
Target and indicator 17.1: Tax Revenue Tax revenue as a share of GDP was stagnant at around 7% between 2001 and 2007 and in 2016, tax revenue as a share of GDP could not reach even 9%. However, domestic revenue mobilization in Bangladesh is poor compared to regional standards	*Target and indicator 17.1: Tax Revenue* The target for 2030 is to have tax revenue that is double the level of tax revenue in 2015 and for Bangladesh, this would be mean that tax revenue as a percentage of GDP would have to be 17%. However, tax revenue as a percentage of GDP will be 10.7% in 2030 if current trends continue

Source: Compiled by the author from Khatun and Saadat (2018)

are off-track. The targets of on-track SDGs that will be achieved or nearly achieved in 2030 are SDG target 1.1: Poverty headcount ratio at national poverty line; SDG target 3.1: Maternal deaths; SDG target 4.1: Lower secondary completion rate; SDG target 7.1: Access to electricity; and SDG target 8.1: GDP growth rate.

The targets of partially on-track SDGs that will require intervention to be achieved or nearly achieved by 2030 are SDG target 2.1: Prevalence of undernourishment; SDG target 5.3: Women who were first married by age of 18 years; SDG target 6.2: People using at least basic sanitation services; SDG target 9.2: Share of manufacturing in GDP; SDG target 11.1: Population living in slums; SDG target 12.2: Total natural resources rents; SDG target 14.4: Total fisheries production; and SDG target 17.1: Tax revenue.

The targets of off-track SDGs that will not be achieved by 2030, and require immediate attention are SDG target 10.1: Income Share of Bottom 40%; SDG target 13.2: Net Carbon Emissions/Removals; SDG target 15.2: Forest area; and SDG target 16.1: Intentional Homicides.

20.7 Concluding Remarks

One of the major concerns in achieving the SDGs is the lack of integration of the goals into the country's strategic planning process. A recent study shows that about 20% of the SDGs targets are missing in the national priorities and only 8 of 17 goals are comparatively well integrated into the existing development process (Bhattacharya et al. 2018). Moreover, in most of the cases, the development will not be sustainable until Bangladesh considers a proper distribution of resources and economic growth as a shared uniformity and fairness. But in reality, the development trend does not correspond to income equality of a person, group, or company, which is pretty much manifesting a malady competition in Bangladesh, even in government, private, or at a personal level, which has turned in spin-off another challenging goal in the SDGs. Nakicenovic et al. (2015) laid emphasis on five priorities to achieve the SDGs, such as devise metrics, establish monitoring mechanisms, evaluate progress, enhance infrastructure, standerdize and verify data. Bangladesh as well as other developing nations should focus on these priorities to face the challenges in achieving the SDGs.

However, Bangladesh should obviously take over a balanced income generating sources in both urban and rural areas. Lack of integration process with coastal people in the priority of the government's national plan is also present. About 20% of the country's geographical area is covering the coastal belt where 50 million people are living and most of them are below the poverty line. Migration of the environmentally affected people is gradually increasing, while thousands of people from these areas are migrating to urban areas and creating the ancillary challenges for SDG implementations in the cities mostly in Dhaka and Chittagong (Saha 2018). It is also very important to build democratic institutions for governing rules everywhere where downward accountability is necessary to achieve the SDGs. Nonetheless, Bangladesh has recently planned a Voluntary National Review (VNR) 2017 on SDGs, but attention to the findings of the review is not mentionable.

For the successful achievement of SDGs in Bangladesh, there should be a revolution in the current education system in line with the needs of the modern labor market which is the root of achieving all the SDGs (Equity 2017). The position of Bangladesh in the ease of doing business index is almost at the bottom. Therefore, the process of starting a business should be simplified to encourage entrepreneurship and create job opportunities for skilled youth, as well as encouraging foreign investments. Moreover, to reduce early marriage and involve women in income-earning activities awareness should be created. Food security is of very important for all the nations. So, investment in agricultural research and technology should also be increased to upsurge food yields.

Local government institutions and nonprofit organizations along with the government should be involved with their full capacity to improve water and sanitation facilities all over the country. Decentralized governance system is very supportive to ensure development in the local areas. Therefore, public institutions should be truly decentralized to reduce pressure on cities and improve management of natural resources.

There should be a revolution of change in the current tax system of the country. However, Bangladesh should reduce tax evasion, and increase the number of direct tax payers. Taxes should be shifted away from consumption and income and toward wealth and capital gain. Natural disaster is very common to Bangladesh. To fight natural disasters Bangladesh should increase the number of protected areas and nature reserves.

Overall, Bangladesh needs an increased amount of investment for the development of health and education sector, agriculture and rural development sector as well as basic infrastructure development like roads and highways, railway and ports, and power sector. Moreover, it is necessary to invest more in the other important sectors like water and sanitation, and climate change mitigation and adaptation (Greenwatch 2019). Recent trends in sector-wise budgetary allocation in these sectors are found not up to the desired level. In the national budget of Fiscal Year 2019–2020, Bangladesh government allocated 11.7% for education sector, 3.7% for agriculture sector, 4.9% for health sector, 7.2% for local government and rural development, 5.4% for fuel and electricity, 12.2% for transport and communication, and 4.91% for social welfare and safety among others (www.nbr.gov.bd).

In analyzing the sector-wise budgetary allocation, very interestingly, it was found that about 52% of the total allocation in education sector was for nondevelopment spending. On the other hand, nearly 60 lac people in Bangladesh have been suffering from malnutrition among other health-related problems. Under these circumstances, budget allocation for health sector was not adequate at all. Moreover, 40.6% of the total workforce in Bangladesh depends on agricultural sector, the contribution of which to GDP is 13.61% (www.icab.org.bd). In this context, allocation of national budget in agriculture was also far behind than their contribution to the GDP. However, specifically the allocation of national budget in agriculture, health, and education sector are found very poor, though these sectors are very important for the overall growth and development of the country. If this trend continues it will be impossible for Bangladesh to achieve the SDGs within the timeframe. Therefore, the government of Bangladesh needs an increase in specific sector-wise budgetary allocation related to SDGs. Side by side it is also important for appropriate implementation of all the policies and strategies taken earlier for the achievement of SDGs in Bangladesh.

References

Bhattacharya D, Rezbana US, Khatun A (2018) Interpreting SDGs for South Asia: in search of a regional framework country study: Bangladesh. CPD, Dhaka

Equity BD (2017) SDGs and country process in Bangladesh: the missing links and where assistance is an obligation of international community especially in climate compensation and stopping illicit flow. EquityBD Campaign, Dhaka

GED (General Economics Division) (2013) Post 2015 development agenda: Bangladesh proposal to UN. Planning Commission, Dhaka

GED (General Economics Division) (2016) A Handbook on mapping of ministries by targets in the implementation of SDGs aligning with 7th five year plan (2016-20). Planning Commission, Dhaka

GED (General Economics Division) (2015) Millennium development goals - Bangladesh progress report 2015. Planning Commission, Dhaka

GED (General Economics Division) (2018) Sustainable development goals: Bangladesh progress report 2018. MoP, Dhaka

Greenwatch (2019) SDGs and Bangladesh: progress, challenges, missing links. In Bangladesh development and sustainability section. www.greenwatchbd.com. Accessed 1 Apr 2019

Intergovernmental Science-Policy Platform on Biodiversity and Ecosystem Services (IPBES) (2019) Global assessment report 2019 on biodiversity and ecosystem service. IPBES, Germany

Jaiyesim R (2016) The challenge of implementing the sustainable development goals in Africa: the way forward. Afr J Reprod Health 2016, 20(3). https://doi.org/10.29063/ajrh2016/v20i3.1

Khaled SMS (2018) SDGs: challenges Bangladesh faces in achieving the goals. Published in February 16, 2018, in an Online Newspaper: Dhaka (Found in the internet)

Khatun F, Saadat SY (2018) What are the prospects of achieving SDGs in Bangladesh? CPD, Dhaka

Nakicenovic NL, Visbeck M, Stevance A (2015) Policy: five priorities for the UN sustainable development goals. Nature 520(7548):432–433

Saha PK (2018) Implementation of SDGs in Bangladesh. Published in September 26, 2018, in a National Online Newspaper, Dhaka (found in the internet)
Sustainable Development Solution Network (SDSN) (2014) An action agenda for sustainable development, 2014. UNSDSN, New York

Web Sources

http://www.bdplatform4sdgs.net
http://www.cpd.org.bd
http://www.gov.bd
http://www.greenwatchbd.com
http://www.icab.org.bd
http://www.ipbes.net
http://www.rsources.unsdsn.org
http://www.southernvoice.org
http://www.un.org
http://www.weforum.org

Sustainable Health Care Under Ayushman Bharat Initiative in India: Role of Institutional CSR

21

Viraja Prasanna Bhat, Jeevan Nagarkar, and Prakash Rao

Abstract

Access and availability of basic education and universal health care to all the citizens of the nation have been considered as significant parameters for the nation to be considered as developed. With technological and economic development, India has been aspiring to be a developed nation however fragmented healthcare schemes and poorly supported basic education dominated by the private sector pose as a major challenge for the aspiration. With due cognizance to the shortcomings, the Government of India has introduced the Right to Education Act to provide education to all and Ayushman Bharat, for universal health care as an affordable scheme. Education gets financial support from the government and has been showing sign of improvement. Significant requirement of funds and facilities for Ayushman Bharat require support and participation from other stakeholders such as corporates through their corporate responsibility funds and schemes. With the implementation of the CSR Act in 2013, companies have been spending CSR funds in areas of their own interest which also includes health care services. In order to strengthen the Ayushman Bharat scheme the contribution through CSR funds can show a direction and hence authors in this chapter have analyzed the use of CSR budget for the past last 4 years. As a conclusion, a thematic PPP model has been proposed for utilization of the CSR budget and various stakeholder participation.

Keywords

Ayushman Bharat · CSR · Universal Healthcare

V. P. Bhat (✉) · J. Nagarkar · P. Rao
Symbiosis Institute of International Business, Symbiosis International (Deemed University), Pune, Maharashtra, India
e-mail: viraja@siib.ac.in

A. Kaushik et al. (eds.), *Climate Resilience and Environmental Sustainability Approaches*, https://doi.org/10.1007/978-981-16-0902-2_21

21.1 Introduction

In the Indian context rise of the literacy rate to 65.4% from 18.31 in the post-Independence period during 1951–2001 has been an indication of improvement in education as a result of successful implementation of education schemes. The health care scenario has not been encouraging as the scope of the health care facilities was mainly narrowed down to insuring the human life mainly in the workable population. Initiatives both at the central and state level in the form of State Insurance Scheme, Central Government Health Scheme, Janashree Bima Yojana, and Rashtriya Swasthya Bima Yojan have been in place with an attempt to improve the health care scenario. Focus of many centralized schemes to provide improved healthcare services to safeguard the deprived sections of the society have been more on insurance-based health care. Effectiveness of these schemes has not been very encouraging due to the poor implementation and out of reach to the actual beneficiaries. Ineffective public health care infrastructure and efficient expensive private health care services have led to a situation of out of reach accessibility and affordability of universal health care for the majority of the population.

In order to help the needy in healthcare schemes such as Rashtriya Swastha Bima Yojana targeted towards below poverty families at the central level has been one of the initiatives in the right direction. With every state having their own schemes targeted towards weaker sections of the society, multiplicity, and complexity of schemes lead to a situation where every state attempts to implement schemes as per their own suitability. Magnitude of Indian population and the choice of private sector reach in every small village complicates the matter even further. Although, many private sector health care facilities have been catering to the underprivileged at a lower cost by way of philanthropy (Mohanan et al. 2016). It can be found that many reputed doctors provide free consultation to the weaker section of the society in designated places or days. These services have not been adequate and hence the poor usually get unacceptably low quality healthcare services leading to loss of life (Das and Mohpal 2016). Gap in public health care infrastructure and expertise of private health care expertise needs to be bridged with a common agenda and policy at the national level. At the global level, a similar experience suggests that many countries are taking steps in providing universal health care to the underprivileged section of the society (Nafsiah Mboi 2015). The universal medical care schemes work better than targeted reformed in the health care sector (McIntyre et al. 2013). The political will and commitment in the coverage of such schemes and changing the rules and scope of the scheme plays an important role in universal healthcare as seen in a study of nations like Chile, Thailand, Vietnam, China, and Indonesia (Aregbeshola 2017).

Historically, India has over the last few decades, shown a significant interest in corporate philanthropy and giving as part of its cultural traditions and culture. Scriptures cutting across religions have cited the role of an individual in giving back to society as an important value system in the cultural fabric of the country. This tent of Indian society has been reflected in various activities giving rise to social responsibility as a meaningful thought process across, individuals, organizations, and society. Post-independence, corporate India initiated the idea of philanthropic giving through a structured process called Corporate Social responsibility. However, for several decades, these activities remained unregulated and disorganized. This often led to serious accountability issues in the way CSR was being implemented.

The Companies Act, 2013 created a mandate for companies to spend 2% of their preceding 3 years' average profit to be spent on Corporate Social Responsibility (CSR) initiatives. The typical CSR initiative includes education and skill development, women empowerment, environment, health and sanitation, sports, rural areas, etc. The importance of health and sanitation has been paramount for a nation like India with a population equivalent to nearly 1/6 of the humanity on earth. The CSR mandate for the corporate sector should not be viewed as one more government dictate or as simply a grant-making initiative for a sector that qualifies the criteria under the mandate. Community development in a sustainable manner should be the idea (Ghosh 2018). The collaboration between government and corporate on health and sanitation through CSR funding can bring about positive changes and help society (Lawania and Kapoor 2018). The companies can tie up with different stakeholders in bringing about change in the society i.e. Self-Help Groups are connected usually to the last person in the economic strata. If these groups can be connected to the healthcare mission in India, it will definitely help in reaching the right beneficiary (Lakshmi 2018).

Government of India's National Health mission advocates affordable health care services to all and the Ayushman Bharat scheme has been a significant step in providing Universal Healthcare to 10 crore families with the healthcare cover of up to 5 lakh rupees per year per family basis. The scheme will be funded in 60:40 ratio between Center and state governments. The current central schemes namely Rashtriya Swastha Bima Yojana and senior citizen health insurance scheme will get subsumed in Ayushman Bharat. The key to the success of this scheme depends on the political will of individual states in implementing this scheme. There are many novel ideas introduced in the scheme like the portability of the scheme at the national level, cashless treatment, the beneficiaries will be decided on socio-economic caste censes basis, the fund transfer would happen through an escrow account directly. Since the scheme talks about the vast majority of people, government alone may not be able to fund this scheme entirely. There are many ways in which various sections of society can help this mission. One such way could be re-proportioning the CSR budget of healthcare and pharmaceutical companies, which otherwise is required by law to spend the amount on CSR initiatives.

This chapter analyzes the data of company expenditure on healthcare and sanitation in last 4 years as also discussing key thematic areas of health care of these companies. The present chapter seeks to study and provide a model approach to rethinking the Ayushman Bharat model through reallocation of CSR budget funds of companies towards key thematic health care services and suggests how companies could tie up with Ayushman Bharat initiative of GoI to further the cause of universal healthcare.

The study has considered the CSR data from the top 300 market listed companies. The objectives of the study comprised analysis of the CSR budget spend in the health care areas to explore areas of spending in health care and sanitation with linkages to

Ayushman Bharat and proposing a thematic PPP Model for the strengthening of Ayushman Bharat Scheme.

Top 300 market listed companies were chosen as the sample for the current research in order to find out their CSR spend focus on health care and allied areas. For the data collection sources considered have been organizations' official annual reports from company websites and Bloomberg- online financial platforms.

21.2 India's Universal Health Care Program: Ayushman Bharat

Ayushman Bharat scheme is a step in providing Universal Healthcare to 10 crore families with the healthcare cover of up to 5 lakh rupees per year per family basis. The current central schemes namely Rashtriya Swastha Bima Yojana and senior citizen health insurance scheme will get subsumed in Ayushman Bharat. There are many novel ideas introduced in the scheme like the portability of the scheme at the national level, cashless treatment, the beneficiaries will be decided on socio-economic caste censes basis, the fund transfer would happen through an escrow account directly. Since the scheme talks about millions and millions of people, government alone may not be able to fund this scheme entirely. There are many ways in which affluent sections of society can help this mission. One of such ways could be the CSR budget of healthcare and pharmaceutical companies, which in any case need to spend the amount on CSR initiatives. This paper analyzes the data of such companies' expenditure on healthcare and sanitation in the last 4 years since the act mandates companies to spend 2% on CSR and will suggest how these companies tie up the Ayushman Bharat initiative of GOI to further the cause of universal healthcare.

21.2.1 Ayushman Bharat Budget Allocation

Focus on health care and allied services by the Government of India through Ayushman Bharat—Swasth for everyone needs a huge investment and active participation from many stakeholders. The budget allocated for the scheme in 2018–19 was Rs. 2400 crore and has seen increase of 166% for the year 2019–20, the allocation is Rs. 6400 crore for 5 months. The allocation for 2019–20 in the national budget for the year is expected to be more than Rs. 1000 crores. There have been funds crunch and issues in the implementation and hence the scheme has still not reached the stakeholders. The state government participation would aid in the coverage of beneficiaries but funding the scheme would still be a challenge and the business organizations through their CSR funds can contribute positively.

Since 2014, with the shift in mandatory spend by the business organizations towards community development through CSR activities attributing to the CSR Act has also been in the area of health care and allied activities. Organizing health camps, helping in the promotion of preventive health care, spending in improving nutrition, vaccination, and contributing towards health care infrastructure such as constructing hospitals, mobile clinics, ambulance, toilets have been some of the major activities undertaken by the business organizations through their CSR funds contributing to the improving health care scenario. Instead of carrying out these activities on their own, business organizations can join hands with the government and spend or fund the Ayushman Bharat scheme directly or indirectly. In this context, the behavior and trend of CSR funds deployment are vital to understand the budget under CSR spending by organizations pre and post CSR Act implementation year.

21.3 Perspectives of Business Organizations: CSR Spent on Health Care

Community development through CSR activities has been made mandatory for all organizations falling under the scope of the CSR Act with flexibility in the area of spend. The CSR policy guidelines have outlined broad areas where the budget can be used. These include education, environment, health, rural development, and women empowerment-focused activities. These have been carried out by organizations as they help in brand building, increase visibility and help the development of the community.

Data has been collected from the official and authentic sources which include the company websites, annual reports, and Bloomberg platform to get the perspective of spending of the CSR allocated budget. Among the organizations considered topmost 300 market cap listed companies data have been used for analysis. Gathering of data for CSR areas spend, the spend trend, CSR budget uses, and unused funds for the financial years 2013–14 till 2017–18 has been conducted. Various types of graphs, table, and models have been used for representing the discussion and result analysis.

The study has representation of non-pharma (92%) and pharma (8%) companies (Fig. 21.1). The CSR funds spent in the health care and allied areas were widespread and hence in the study, all these activities have been categorized in four broader

Fig. 21.1 Number of companies studied ($n = 300$)

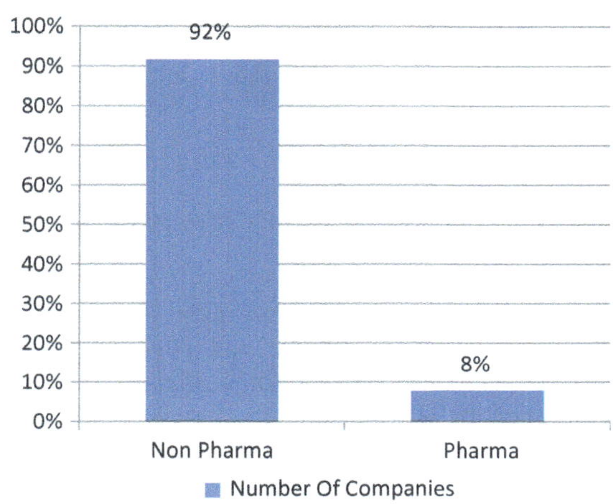

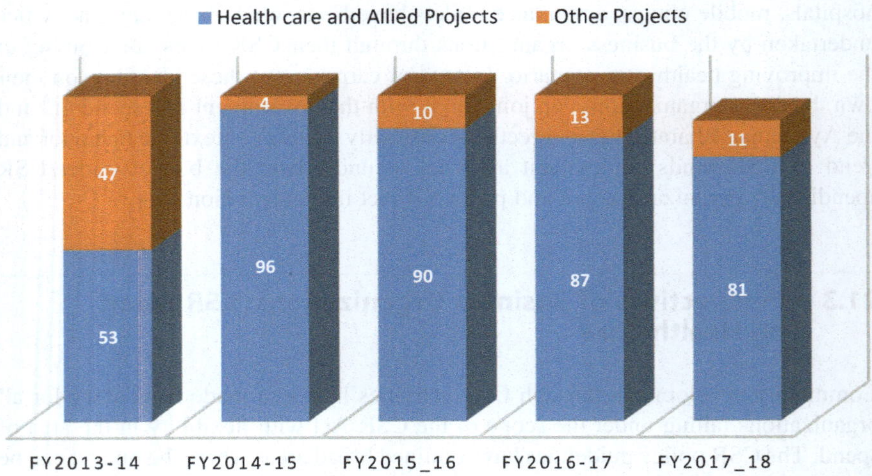

Fig. 21.2 CSR spend in FY 13–FY 18 by Non-Pharma companies

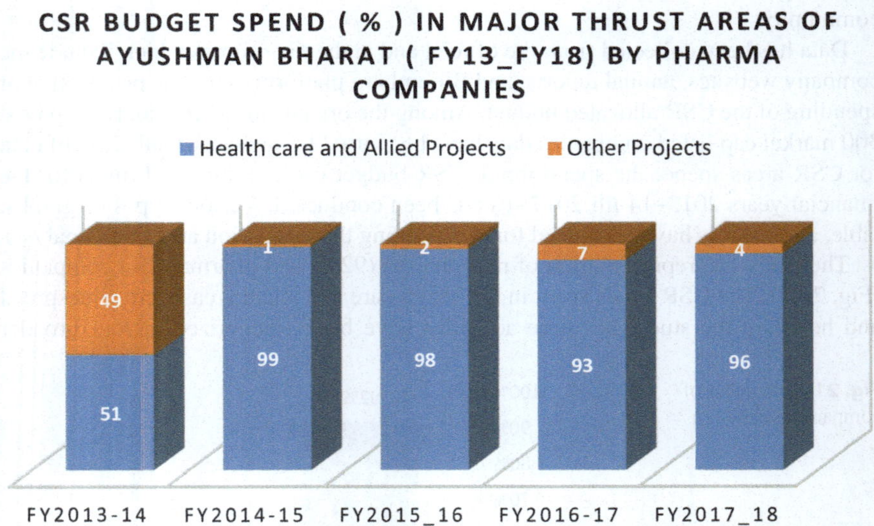

Fig. 21.3 CSR spend in FY 13–FY 18 by Pharma companies

areas in alignment to the areas under Ayushman Bharat as Health Awareness and Preventive Care, Health care, and allied services, Health care and sanitation Infrastructure and other CSR activities. Figures 21.2, 21.3, 21.4, 21.5, and 21.6 show the CSR spent by pharma and non-pharma companies in these areas.

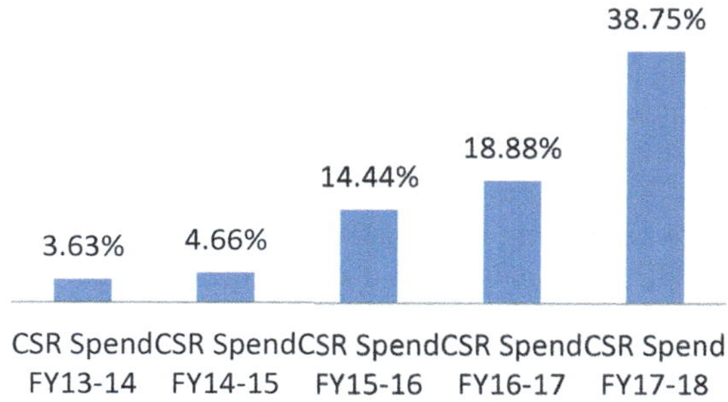

Fig. 21.4 CSR spend (%) in health awareness and preventive care activities

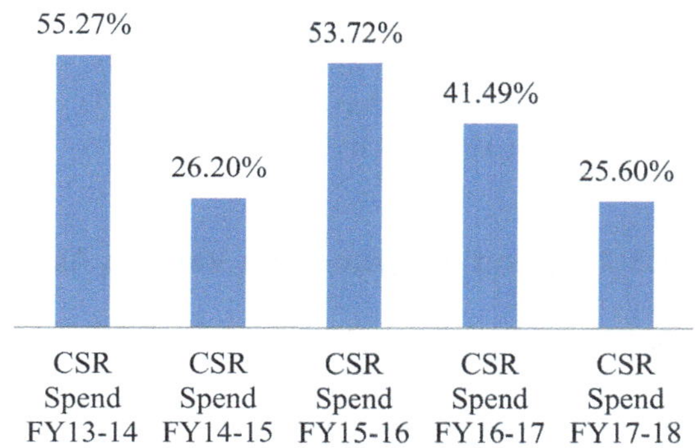

Fig. 21.5 CSR spend (%) in health care and allied services

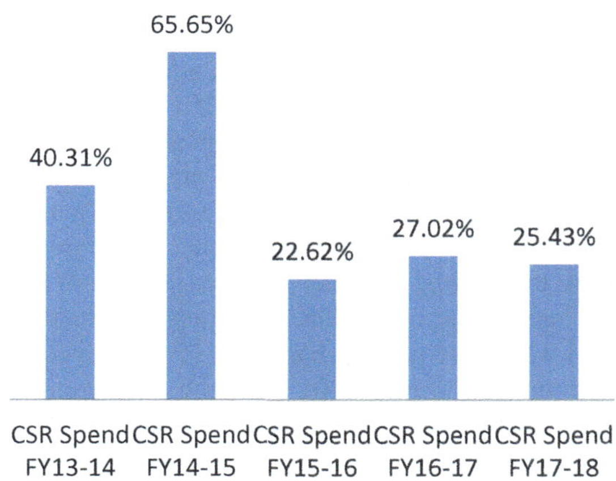

Fig. 21.6 CSR spend in health care and sanitation infrastructure

21.3.1 Trends in Pharma and Non-Pharma Companies

Figure 21.2 indicates a trend of focus spend by the Non-Pharma companies for the financial year 2013 to the financial year 2018. It can be observed that there has been a positive sign of increase in spend in health care-related projects from 53%

(FY 2013–14) to 81% (FY 2017–18). In this duration, a health care and allied areas have been given focus and more funds have been allocated.

Figure 21.3 indicates a trend of focus spends by the Pharma companies for the financial year 2013 to the financial year 2018. It can be observed that there has been a positive sign of increase in spend in health care-related projects from 51% (FY 2013–14) to 98% (FY 2017–18). In this duration health care and allied areas have been given focus and more funds have been allocated.

The data analyzed and represented in Figs. 21.2 and 21.3 show the CSR funds utilization in the health care and allied areas, projects by both pharma and non-pharma sector businesses. The focus of the pharma companies can be seen shifting more towards the health care projects and hence suggestive of an alignment towards the implementation of the CSR Act.

It has been a very positive step that both pharma and non-pharma sector contributes and work in the varied activities in the health care area. Hence it has been a challenge to understand and conclude whether these activities in real sense impact society by the activities carried out by these companies.

21.3.2 Spending Trends on Various Aspects of Healthcare

The CSR projects in health care were further categorized into four categories namely Health Awareness and Preventive care projects, Health care and Allied services projects, and Health Care and sanitation Infrastructure Projects. Figures 21.4, 21.5, and 21.6 depict the data of CSR spend in % over the years 2013–18 in these areas.

On further analyzing the data in order to understand the spending in various areas related to Health care and allied areas, it was found that the spend in health care and preventive services (Awareness Campaigns, Blood donations, etc.) has gradually increased year on year. From FY 2013–14 to FY 17–18, there has been an increase from 3.63 to 38%. Whereas the spend in health care and allied areas (vaccination, support for medicines or disease prevention, etc.) has not been consistent with a variation between 55% in FY 13–14 to 25.6% in FY 2017–18. This area has seen a range of 225–56% in the last 5 years. The non-consistent and diverse spend do not serve the focused approach. Whereas in case of health care and sanitation infrastructure area spend (Ambulance, Facility construction, Mobile vans, Dedicated wards, etc.) was at the peak in FY 2014–15 starting from 40.31% in FY 13–14 and again reaching 25.3% in FY 2017–18 with fluctuation every year. It can be observed that both pharma and non-pharma companies spend on other CSR activities such as education, environment, and many other non-health care-related activities.

21.3.3 Challenges for Effective Implementation of Ayushman Bharat Scheme

From the data analysis, it can be observed that, the companies do have inclination towards investing their CSR funds and efforts in health care and related activities, but they have not been focused and consistent. The reason for this behavior can be attributed to factors such as lack of capacity and interest as these activities need a dedicated and committed effort. In such scenario, instead of only the business houses either themselves or with the help of NGO's trying to bring in change in health care services of India, can be just one of the major stakeholders in the health care service scenario. It is necessary that all stakeholders join hands to realize the Ayushman Bharat by a model and project approach.

21.4 Public Private Partnership (PPP) Model for Strengthening Ayushman Bharat Scheme

Corporate community in the initial phase of CSR Act mandate had raised concerns, in recent times there has been a renewed focus by business entities and organizations to focus on projects in areas of, health care and allied areas. From the current study, CSR spends analysis of 300 top market cap indicated that health care and allied areas have been receiving a positive response in terms of CSR expenditure from both pharma and non-pharma sector. Even though the approach of many companies has been merely to ensure compliance towards the statutory obligation instead of a sustainable and proactive efforts towards social welfare, crores of rupees have been allocated and spent through the CSR budget have materialized into high budgeted expenditure and lower actual utilization displaying no impact of these funds on the societal development. In this context, it has been proposed that a cooperative strategy through a public private partnership would help in strengthening the Ayushman Bharat scheme and help the beneficiaries.

The model proposes four major players as key stakeholders for the Ayushman Bharat scheme effective implementation in the Indian context (Fig. 21.7). In order to bring in the holistic health perspective, the scheme requires the infrastructure and hence funds for the same. In the current situation, most of the raw materials for drug manufacturing gets imported and hence the drugs get highly priced. Affordable drugs to cater to the larger population can be supported by the pharma companies by extensive research, capacity building, and build a sustainable ecosystem. Government has to be the pivotal player by bringing together the pharma, non-pharma companies, the academic institutions through collaborations, and the large private hospitals through attractive incentives. The gap in building the necessary physical infrastructure and supply chain strengthening can be taken by non-pharma companies. All the players should be part of projects to spread awareness and promotion campaigns The expertise of pharma companies would be put into use and would help strengthen the Ayushman Bharat domestic drug manufacturing objective. Non-pharma CSR funds can be channelized to fulfill the other major requirement of Ayushman Bharat by funding the infrastructure such as hospitals, medical devices, ambulances, and wellness center. Funds can also be utilized in building the capacity required at many levels not just medical staff but also health workers, social workers, and distributors. The awareness among masses related to Ayushman Bharat also needs to be strengthened and hence this task can also be

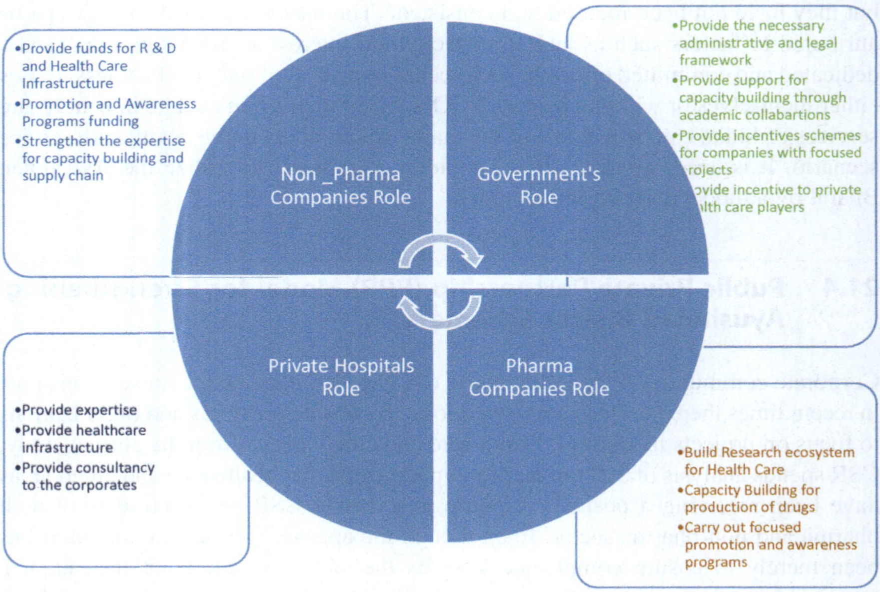

Fig. 21.7 PPP model for Strengthening Ayushman Bharat

carried out using various advertising on several media which can be funded by so non-pharma companies. The model proposes a roadmap for pharma and non-pharma sector CSR funds utilization which authors believe would strengthen the Ayushman Bharat scheme and provide universal health care to the Indian Population.

21.5 Conclusion

Indeed, business organizations have engaged in CSR as largely a compliance mechanism and require a focused approach with the involvement of all the stakeholders for mutual benefit. Although there have been efforts to engage in health care and allied activities with varied spending and diverse activities, the sustainability of these efforts can be questionable and of concern and hence CSR funds can be channelized for better use in the areas aligned for health care for the realization of "Swasth Bharat" through "Ayushman Bharat scheme." A tangible model with multiple stakeholder participation would strengthen the Ayushman Bharat scheme.

References

Aregbeshola BS (2017) Enhancing political will for universal health coverage in Nigeria. MEDICC Rev 19:42–46

Das J, Mohpal A (2016) Socioeconomic status and quality of care in rural India: new evidence from provider and household surveys. Health Aff 35(10):1764–1773

Ghosh A (2018) Corporate social responsibility: going beyond cheque book philanthropy. Glob Med J 9(1):1–11

Lakshmi CR (2018) Role of women self-help groups in the holistic health care in India. Editorial Board 47(2):91

Lawania BD, Kapoor S (2018) Leveraging corporate social responsibility for the advancement of development goals in India: sanitation and cleanliness movement in India. Aust Account Bus Financ J 12(2):46–70

Mboi N (2015) Indonesia: on the way to universal health care. Health Syst Reform 1(2):91–97. https://doi.org/10.1080/23288604.2015.1020642

McIntyre D, Ranson MK, Aulakh BK, Honda A (2013) Promoting universal financial protection: evidence from seven low-and middle-income countries on factors facilitating or hindering progress. Health Res Policy Syst 11(1):36

Mohanan M, Hay K, Mor N (2016) Quality of health care in India: challenges, priorities, and the road ahead. Health Aff 35(10):1753–1758

The manufacturer's authorised representative in the EU is Springer Nature Customer Service Centre GmbH, Europaplatz 3, 69115 Heidelberg, Germany. If you have any concerns regarding our products, please contact ProductSafety@springernature.com

Printed and bound by CPI Group (UK) Ltd, Croydon, CR0 4YY
30/09/2026
02225293-0015